Air Pollution and Global Warming
History, Science, and Solutions
Second Edition

This new edition of Mark Z. Jacobson's textbook provides a comprehensive introduction to the history and science of the major air pollution and climate problems that face the world today, as well as the energy and policy solutions to those problems. Every chapter has been updated with new data, figures, and text. There is also a new chapter on large-scale solutions to global warming and air pollution. Color photographs, diagrams, examples, and homework problems have also been added. This is an ideal introductory textbook on air pollution and global warming for students taking courses in atmospheric chemistry and physics, meteorology, environmental science, earth science, civil and environmental engineering, chemistry, environmental law and politics, and city planning and regulation. It also forms a valuable reference text for researchers and an introduction to the subject for general audiences.

Mark Z. Jacobson is Director of the Atmosphere/Energy Program and Professor of Civil and Environmental Engineering at Stanford University. He is also a Senior Fellow of the Woods Institute for the Environment and of the Precourt Institute for Energy. Professor Jacobson has been on the faculty at Stanford since 1994. His research involves the development and application of numerical models to understand the effects of energy and transportation systems on climate and air pollution and the analysis of renewable energy resources. Professor Jacobson received the 2005 American Meteorological Society Henry G. Houghton Award, based in part on his discovery that black carbon may be the second leading cause of global warming after carbon dioxide. He coauthored a 2009 cover article in *Scientific American* (with Dr. Mark Delucchi) on how to power the world with renewable energy. In 2010, Professor Jacobson was appointed to the Energy Efficiency and Renewables Advisory Committee by the U.S. Secretary of Energy. He has taught courses on atmospheric pollution and climate, weather and storms, air pollution modeling, and numerical weather prediction. Professor Jacobson has also published more than 120 peer-reviewed journal articles and the textbook *Fundamentals of Atmospheric Modeling* (2nd edition, 2005; Cambridge University Press).

Air Pollution and Global Warming
History, Science, and Solutions

Second Edition

Mark Z. Jacobson

Stanford University

CAMBRIDGE
UNIVERSITY PRESS

CAMBRIDGE UNIVERSITY PRESS
Cambridge, New York, Melbourne, Madrid, Cape Town,
Singapore, São Paulo, Delhi, Mexico City

Cambridge University Press
32 Avenue of the Americas, New York, NY 10013-2473, USA

www.cambridge.org
Information on this title: www.cambridge.org/9781107691155

First edition published 2002, under the title *Atmospheric Pollution: History, Science, and Regulation*
Second edition first published 2012

Printed in the United States of America

A catalog record for this publication is available from the British Library.

Library of Congress Cataloging in Publication data

Jacobson, Mark Z. (Mark Zachary)
Air pollution and global warming : history, science, and solutions / Mark Z. Jacobson. – 2nd ed.
    p.   cm.
Rev. ed. of: Atmospheric pollution. 2002.
Includes bibliographical references and index.
ISBN 978-1-107-02161-7 (hardback) – ISBN 978-1-107-69115-5 (paperback)
1. Air pollution.   2. Atmospheric chemistry.   3. Air – Pollution – Law and legislation.
I. Jacobson, Mark Z. (Mark Zachary). Atmospheric pollution.   II. Title.
TD883.J37   2012
363.739′2–dc23   2011044340

ISBN 978-1-107-02161-7 Hardback
ISBN 978-1-107-69115-5 Paperback

Cover photograph: Late afternoon sunlight peering through smoke billowing from a power plant against the backdrop of a sky enhanced in
its redness by air pollution. (Photo taken on December 10, 2009, by Mikhail Didarev/Dreamstime.com)

Brief Contents

Contents

Foreword

Atmospheric chemistry, as a modern discipline, can be considered to have originated in 1931, when Sydney Chapman, distinguished British physicist, formulated a chemical mechanism for the formation of stratospheric ozone. The foundations of understanding tropospheric chemistry were laid in the early 1950s by Arie Haagen-Smit, a bioorganic chemist at the California Institute of Technology, who described ozone formation in the Los Angeles Basin as resulting from reactions involving volatile organic compounds and oxides of nitrogen. The essential reactive species in tropospheric chemistry remained unknown until the early 1970s, when the central role of the hydroxyl radical as the troposphere's "detergent" was revealed. The existence of particles in the air (aerosols) had long been recognized, but it was not until the past 50 years that instrumentation was developed that is capable of determining the size distribution and composition of atmospheric aerosols.

Threats to stratospheric ozone made headlines in the early 1970s, when Harold Johnston at the University of California, Berkeley, published calculations of the effect on stratospheric ozone of a proposed fleet of supersonic aircraft. Johnston's work was followed shortly thereafter by the revelation of the stratospheric chemical impact of chlorofluorocarbons, widely used as refrigerants and in consumer products, by F. Sherwood Rowland and Mario Molina of the University of California, Irvine. For their penetrating insights into atmospheric chemistry, Rowland, Molina, and Paul Crutzen of the Max Planck Institute for Chemistry in Mainz, Germany, received the 1995 Nobel Prize in Chemistry.

Mathematical models that describe the three-dimensional transport and chemistry in the atmosphere were first developed in the early 1970s. Such computer models have played a key role in representing our understanding of atmospheric processes and in planning emission controls to achieve desired levels of air quality. Three-dimensional numerical atmospheric chemical transport models are mandated in the U.S. Clean Air Act as the tool that must be used to design emission control strategies to attain national ambient air quality standards in urban and regional areas. Virtually all aspects of atmospheric chemistry and physics are now embodied in these models.

Humans have been concerned with weather and climate for millennia. Over geologic time, the Earth's climate variations have been a result of changes in the Earth's orbit, in the sun's output, or of volcanic eruptions that inject large amounts of material into the atmosphere. Although warming of the Earth by an increase of atmospheric CO_2 was calculated in 1896 by Svante Arrhenius, it was not until 1958 that David Keeling of the Scripps Institute of Oceanography began making precise measurements of atmospheric CO_2 at the top of Hawaii's Mauna Loa Volcano. The continuous record of CO_2 measurements on Mauna Loa, now overseen by the U.S. National Oceanographic and Atmospheric Administration (NOAA), constitutes the most profound environmental dataset in existence. That CO_2 record, when reconciled with estimates of CO_2 emissions from fossil fuel burning, establishes the unequivocal effect of humans on Earth's climate over the past 60 years.

Atmospheric science now involves thousands of scientists, in academic disciplines ranging over physics, chemistry, engineering, and health sciences. Writing an introductory text that spans the important components of atmospheric science, air pollution, climate, and effects is a daunting task indeed. Professor Mark Z. Jacobson has produced a text of remarkable breadth, one that can be appreciated by first-year college students and professionals alike. The book is alive with historical vignettes, photos, and figures. The reader gains an appreciation of the elegance of the science of the atmosphere, as well as of the role of humans in perturbing the atmosphere's composition and the effects of those perturbations. It is a pleasure to recommend this book to those with an interest in understanding Earth's precious atmosphere.

John H. Seinfeld
Louis E. Nohl Professor
California Institute of Technology

Preface

Natural air pollution problems on the Earth are as old as the planet itself. Volcanos, fumaroles, natural fires, and desert dust have all contributed to natural air pollution. Humans first emitted air pollutants when they burned wood and cleared land (increasing wind-blown dust). More recently, the burning of coal, chemicals, oil, gasoline, kerosene, diesel, jet and alcohol fuel, natural gas, biomass, and waste, as well as the release of chemicals into the environment, have contributed to several major air pollution problems on a range of spatial scales. These problems include outdoor urban smog, indoor air pollution, acid deposition, the Antarctic ozone hole, global stratospheric ozone reduction, and global warming.

Urban smog is characterized by the outdoor buildup of gases and particles that are either emitted from vehicles, homes, industrial facilities, power plants, incinerators, or land-clearing and natural fires or formed chemically in the air from emitted pollutants. Smog affects human and animal health, structures, and vegetation. Urban smog occurs over scales of meters to hundreds of kilometers.

Indoor air pollution results from the emission of pollutant gases and particles in enclosed buildings and the transport of pollutants from outdoors to indoors. Worldwide, indoor air pollution is responsible for about 1.6 million premature deaths per year, mostly from the burning of wood, animal and agricultural waste, and coal for home heating and cooking in developing countries. Indoor air pollution occurs over scales of meters to tens of meters.

Acid deposition occurs when sulfuric, nitric, or hydrochloric acids in the air deposit to the ground as a gas or dissolve in rainwater, fog water, or particles.

Acids harm soils, lakes, forests, and structures. In high concentrations, they can also harm humans. Acid deposition occurs over scales of tens to thousands of kilometers.

The **Antarctic ozone hole** and **global stratospheric ozone reduction** are caused, to a large extent, by human-produced chlorine and bromine compounds that are emitted into the air and break down only after they have traveled to the upper atmosphere (the stratosphere). Ozone reduction increases the intensity of ultraviolet (UV) radiation from the sun reaching the ground. Intense UV radiation destroys microorganisms on the surface of the Earth and causes skin cancer in humans and animals. The Antarctic ozone hole occurs over a region the size of North America. Global stratospheric ozone reduction occurs globally.

Global warming is the increase in lower atmospheric (tropospheric) global temperatures and the resulting increase in ice melt, sea level, coastal flooding, heat stress, air pollution, malaria, influenza, severe storminess, and starvation due to shifts in agriculture caused by human emission of both greenhouse gases and particles. Greenhouse gases include carbon dioxide, methane, nitrous oxide, and chlorofluorocarbons. Major particle constituents contributing to global warming include black and brown carbon. Global warming is a global problem with regional impact.

Air is not owned privately; instead, it is common property (accessible to all individuals). As a result, air has historically been polluted without limit. This is the classic **tragedy of the commons**. The only known mechanism of limiting air pollution, aside from volunteerism and the fortuitous development of inexpensive, clean technologies, is government intervention. Intervention can take the form of setting up economic markets for the rights to emit pollution, providing subsidies for the development and implementation of clean technologies, limiting emissions from specific sources, requiring certain emission control technologies, or setting limits on pollutant concentrations and allowing the use of any emission reduction method to meet those limits.

Because government action usually requires consensus that a problem exists, the problem is severe enough to warrant action, and action taken will not have its own set of adverse consequences (usually economic), national governments did not act aggressively to control global air pollution problems until the 1970s and 1980s. For the most part, action was not taken earlier because lawmakers were not always convinced of the severity of air pollution problems. Even when problems were recognized, action was often delayed because industries used their political strength to oppose government intervention. Even today, government intervention is opposed by many industries and politicians out of often misplaced concern that intervention will cause adverse economic consequences. In many developing countries, intervention is sometimes opposed because of the concern that developed countries are trying to inhibit economic expansion of the less developed countries. In other cases, pollution is not regulated strictly due to the perceived cost of emission control technologies and enforcement.

Despite the opposition to government intervention, such intervention has proven effective in mitigating various major air pollution problems facing humanity in some countries or on a global scale. For example, outdoor and indoor air pollution and acid deposition in many industrialized countries, including the United States, Japan, and most European countries, have decreased since the 1970s due to the development and use of emission control technologies and more efficient devices. However, such problems have increased in most of the rest of the world due to rapidly rising populations, higher energy demand, and low transfer rates of emission control technologies. Indoor air pollution, in particular, has become more severe in developing countries as populations have expanded and indoor burning of fuel for heating and cooking has continued.

The main cause of the Antarctic ozone hole and stratospheric ozone reduction, the emissions of classes of chemical compounds called chlorofluorocarbons and bromocarbons, has been substantially addressed. However, stratospheric ozone levels continue to stay low, in part due to the long lifetime of existing chlorofluorocarbons and partly due to global warming, which warms the lower atmosphere (the troposphere) but cools the stratosphere. Stratospheric cooling exacerbates damage to the ozone layer caused by chlorofluorocarbons and bromocarbons.

Addressing the problem of global warming is a process in its infancy. Despite modest efforts, emissions and global temperatures continue to rise rapidly. The consequences of higher temperatures are readily visible. The solution to global warming, although clear-cut in concept, is the challenge for the current and future generations.

This book discusses the history and science of major air pollution problems, the consequences of these problems, and efforts to control the problems through government intervention and existing clean technologies. The book then presents a proposed solution to global warming and air pollution, namely, the conversion of the world's energy infrastructure to a large-scale, clean, renewable one. Because air pollution and global warming, in particular, are so severe, a rapid and large-scale conversion is needed. *The main barriers to conversion are not technical, resource based, or even economic. Instead, they are social and political.*

The book synthesizes knowledge in the fields of chemistry, meteorology, radiation science, aerosol sciences, cloud physics, soil science, microbiology, epidemiology, energy, materials science, economics, policy, and law. The study of air pollution and climate is truly interdisciplinary.

This book is directed at students in environmental, Earth, atmospheric, and energy sciences; engineering; and policy. It was designed to be general enough for the interested layperson, yet detailed enough to be used as a reference text. The text uses chemical symbols and chemical equations, but all chemistry required is introduced in Chapter 1. No previous knowledge of chemistry is needed. The text also describes a handful of physical laws. No calculus, geometry, or higher math is needed.

Acknowledgments

I want to thank several colleagues who reviewed different sections of this text. In particular, I am indebted to (in alphabetical order) Cristina Archer, Mary Cameron, Joe Cassmassi, Andrew Chang, Mark Delucchi, Frank Freedman, Ann Fridlind, Elaine Hart, Lynn Hildemann, Gerard Ketefian, Jinyou Liang, Nesrin Ozalp, Ana Sandoval, Roberto San Jose, Alfred Spormann, Amy Stuart, Azadeh Tabazadeh, John ten Hoeve, Daniel Whitt, and Yang Zhang, all of whom provided comments, suggestions, and/or corrections relating to the text. I also want to show my gratitude to Jill Nomura, Daniel and Dionna Jacobson, William Jacobson, and Yvonne Jacobson for helping with graphics and/or editing, as well as to the students who have used this text and then provided suggestions and corrections. Finally, I want to thank several anonymous reviewers, in particular one who provided many inspirational comments.

Air Pollution and Global Warming

Chapter 1

Basics and Discovery of Atmospheric Chemicals

The study of air pollution begins with the study of chemicals that comprise the air. These chemicals include molecules in the gas, liquid, or solid phases. Because the air contains so many different types of molecules, it is helpful to become familiar with the more important ones through the history of their discovery. Such a history also gives insight into characteristics of atmospheric chemicals and an understanding of how much our knowledge of air pollution today relies on the scientific achievements of alchemists, chemists, natural scientists, and physicists of the past. This chapter begins with some basic definitions, and then examines historical discoveries of chemicals of atmospheric importance. Finally, types of chemical reactions that occur in the atmosphere are identified, and chemical lifetimes are defined.

1.1. Basic Definitions

Air is a mixture of gases and particles, both of which are made of atoms. In this section, atoms, elements, molecules, compounds, gases, and particles are defined.

1.1.1. Atoms, Elements, Molecules, and Compounds

In 1913, **Niels Bohr** (1885–1962), a Danish physicist, proposed that an **atom** consists of one or more negatively charged electrons in discrete circular orbits around a positively charged nucleus. Each **electron** carries a charge of −1 and a tiny mass.[1] The **nucleus** of an atom consists of 1 to 118 protons and 0 to 165 neutrons. **Protons** have a net charge of +1 and a mass 1,836 times that of an electron. **Neutrons** have zero net charge and a mass 1,839 times that of an electron. For the net charge of an atom to be zero, the number of electrons must equal the number of protons. Positively charged atoms have fewer electrons than protons. Negatively charged atoms have more electrons than protons. Positively or negatively charged atoms are called **ions**.

The average mass of protons plus neutrons in a nucleus is called the **atomic mass**. Electrons are not included in the atomic mass calculation because the summed mass of electrons in an atom is small in comparison with the summed masses of protons and neutrons. The number of protons in an atomic nucleus is the **atomic number**.

An **element** is a single atom or a substance composed of several atoms, each with the same atomic number (the same number of protons in its nucleus). Whereas all atoms of an element have a fixed number of protons, not all atoms of the element have the same number of neutrons. Atoms of an element with the same number of protons but a different number of neutrons are **isotopes** of the element. Isotopes of an element have different atomic masses but similar chemical characteristics.

---

[1] Mass is an absolute property of a material. Mass, multiplied by gravity, equals weight, which is a force. Because gravity varies with location and altitude, weight is a relative property of a material. A person who is nearly "weightless" in space, where gravity is small, has the same mass, whether in space or on the surface of the Earth.

Table 1.1. Characteristics of the first ten elements in the periodic table

Element	Symbol	Number of protons (atomic number)	Number of neutrons in main isotope	Atomic mass (g mol^{-1})	Number of electrons
Hydrogen	H	1	0	1.00794	1
Helium	He	2	2	4.00206	2
Lithium	Li	3	4	6.941	3
Beryllium	Be	4	5	9.01218	4
Boron	B	5	6	10.811	5
Carbon	C	6	6	12.011	6
Nitrogen	N	7	7	14.0067	7
Oxygen	O	8	8	15.9994	8
Fluorine	F	9	10	18.9984	9
Neon	Ne	10	10	20.1797	10

The **periodic table of the elements**, developed in 1869 by Russian chemist **Dmitri Mendeleev** (1834–1907), lists elements in order of increasing atomic number. Table 1.1 identifies the first ten elements of the periodic table and some of their characteristics. The atomic mass of an element in the periodic table is the sum, over all isotopes of the element, of the percentage occurrence in nature of the isotope multiplied by the atomic mass of the isotope.

The simplest element in the periodic table is **hydrogen** (H), which contains one proton, no neutrons, and one electron. Hydrogen occurs in three natural isotopic forms. The most common is **protium** (one proton and one electron), shown in Figure 1.1. The other two are **deuterium**, which contains one proton, one neutron, and one electron, and **tritium**, which contains one proton, two neutrons, and one electron. **Helium** (He), also shown in Figure 1.1, is the second simplest element and contains two protons, two neutrons, and two electrons.

When one atom bonds to another atom of either the same or different atomic number, it forms a molecule. A **molecule** is a group of atoms of like or different elements held together by chemical forces. When a molecule consists of different elements, it is a compound. A **compound** is a substance consisting of atoms of two or more elements in definite proportions that cannot be separated by physical means.

1.1.2. Gases and Particles

Gases are distinguished from particles in two ways. First, a **gas** consists of individual atoms or molecules that are separated, whereas a **particle** consists of an aggregate of atoms or molecules bonded together. Thus, a particle is larger than a single gas atom or molecule. Second, whereas particles contain liquids or solids, gases are in their own phase state. Particles may be further segregated into aerosol particles and hydrometeor particles.

An **aerosol** is an ensemble of solid, liquid, or mixed-phase particles suspended in air. An **aerosol particle** is a single liquid, solid, or mixed-phase particle among an ensemble of suspended particles. The term *aerosol* was coined by British physicochemist **Frederick George Donnan** (1870–1956) near the end of World War I (Green and Lane, 1969).

A **hydrometeor** is an ensemble of liquid, solid, or mixed-phase water particles suspended in or falling through the air. A **hydrometeor particle** is a single such particle. Examples of hydrometeor particles are

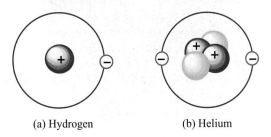

(a) Hydrogen   (b) Helium

Figure 1.1. Simplified configuration of protons, neutrons, and electrons in (a) protium, an isotope of the hydrogen atom, and (b) a helium atom.

cloud drops, ice crystals, raindrops, snowflakes, and hailstones. The main difference between an aerosol particle and a hydrometeor particle is that the latter contains much more water than the former.

Liquids in aerosol particles and hydrometeor particles may be pure or may consist of a solution. A **solution** is a homogeneous mixture of substances that can be separated into individual components on a change of state (e.g., freezing). A solution consists of a **solvent**, such as water, and one or more **solutes** dissolved in the solvent. Solids may be mixed throughout a solution but are not part of the solution. In this text, pure water and solutes dissolved in water are denoted with "(aq)" for **aqueous** (dissolved in water). Gases are denoted with "(g)", and solids are denoted with "(s)".

Gases and aerosol particles may be emitted into the air naturally or anthropogenically or formed chemically in the air. **Anthropogenic emissions** are human-produced emissions, such as from fossil fuel combustion or industrial burning. Hydrometeor particles generally form from physical processes in the air. **Air pollution** occurs when gases or aerosol particles, *emitted anthropogenically, build up in concentration sufficiently high to cause direct or indirect damage to humans, plants, animals, other life forms, ecosystems, structures,* or works of art.

1.2. History of Discovery of Elements and Compounds of Atmospheric Importance

Reactive elements that comprise most gases in the air are hydrogen (H), carbon (C), nitrogen (N), oxygen (O), fluorine (F), sulfur (S), chlorine (Cl), and bromine (Br). **Unreactive elements** in the air include helium (He), argon (Ar), krypton (Kr), neon (Ne), and xenon (Xe). Two radioactive elements of importance are polonium (Po) and radon (Rn). Aerosol particles contain the elements present in gases and often sodium (Na), magnesium (Mg), aluminum (Al), silicon (Si), potassium (K), calcium (Ca), iron (Fe), lead (Pb), and/or phosphorus (P). Tables 1.2 and 1.3 summarize the dates of discovery of elements and compounds, respectively, of atmospheric importance.

1.2.1. Solids and Liquids, Ancient World–1690

The first elements in the periodic table to be identified were the metals gold (Au), silver (Ag), lead (Pb), mercury (Hg), iron (Fe), copper (Cu), and tin (Sn). Many cultures, including the Egyptians and the Chaldeans,

were aware of these metals. Of note were the Chaldeans (612–539 BC), who connected them with planets, identifying gold as the sun, silver as the moon, lead as Saturn, mercury as Mercury, iron as Mars, copper as Venus, and tin as Jupiter. Of these seven metals, lead, mercury, and iron are the most relevant to air pollution today.

1.2.1.1. Lead

Lead (*plumbum* in Latin) is a dense bluish-white metal element. It was discovered before 6400 BC in modern-day Turkey, probably during the heating of lead ore (lead bound with sulfur, copper, zinc, or silver) in a campfire. Lead melts at a temperature of 327°C. In a fire, lead liquefies, separating from its ore. Because it is so dense (11.3 times the density of water), pure lead would flow to the bottom of a campfire. Because resolidified lead is malleable (Figure 1.2), early users of lead molded it in into cookware and jewelry. The Romans molded it into pipes.

Lead was referred to in the Books of Job and Numbers as *biblicalx*. The Roman Pliny the Elder (23–79 AD) called it *plumbum nigrum*, and the English word *plumber* describes a person who installs or fixes lead

Figure 1.2. Melted and shaped lead. © Shootzpics/ Dreamstime.com.

Table 1.2. Dates of discovery of elements of atmospheric importance

Element	Origin of name or previous name	Year discovered	Discoverer
Lead (Pb)	Previously *biblicalx, plumbum nigrum*	c. 6400 BC	?
Iron (Fe)	Named after *Iarn*	c. 2700 BC	?
Mercury (Hg)	Means "runny silver" (*hydrargyrum*)	c. 2500 BC	?
Carbon (C)	Named from *carbo,* "charcoal"	BC	?
Sulfur (S)	Named from *sulvere, sulphurium,* previously *brimstone*	BC	?
Phosphorus (P)	Means "light bearer"	1669	Brand (Sweden)
Hydrogen (H)	Means "water producer"	<1520, 1766	Paracelsus (Switzerland), Cavendish (England)
Fluorine (F)	Named from *fluere,* "flow" or "flux"	1771	Scheele (Sweden)
Nitrogen (N)	Means "nitre maker"	1772	Rutherford (England)
Oxygen (O)	Means "acid maker"	1774, 1772–1775	Priestley (England), Scheele (Sweden)
Chlorine (Cl)	Means "green gas"	1774	Scheele (Sweden)
Sodium (Na)	Named from *soda*	1807	Davy (England)
Potassium (K)	Named from *potash*	1807	Davy (England)
Calcium (Ca)	Named from *calx*	1808	Davy (England)
Silicon (Si)	Named from *silex,* "flint"	1823	Berzelius (Sweden)
Bromine (Br)	Means *stench*	1826	Balard (France)
Aluminum (Al)	Found in *alum*	1827	Wöhler (Germany)
Magnesium (Mg)	Named after Magnesia, Greece	1830	Bussy (France)
Helium (He)	Named from *Helios,* Greek sun god	1868	Janssen (France), Lockyer (England)
Argon (Ar)	Named from *argos,* "lazy"	1894	Rayleigh (England), Ramsay (Scotland)
Krypton (Kr)	Named from *kryptos,* "concealed"	1898	Ramsey, Travers (Scotland)
Neon (Ne)	Named from *neos,* "new"	1898	Ramsey, Travers (Scotland)
Xenon (Xe)	Named from *xenos,* "guest"	1898	Ramsey, Travers (Scotland)
Polonium (Po)	Named after the country of Poland	1898	Curie, Curie (France)
Radon (Rn)	Originally named *radium emanation*	1900	Dorn (Germany)

pipes. The pervasive use of lead by the Romans in cookware, pipes, face powders, rouges, and paints is believed to have contributed to the slow poisoning of the aristocracy in the Roman Empire. In particular, the wine sweetener and preservative *defrutum,* produced in part by the boiling must in lead pots, was consumed in large quantities by the wealthy. The Romans knew of the dangers of lead. Vitruvius, in *De Architectura,* for example, professed,

> Water conducted through earthen pipes is more wholesome than that through lead; indeed that conveyed in lead must be injurious, because from it white lead (lead carbonate) is obtained, and this is said to be injurious to the human system.... This may be verified by observing the workers in lead, who are of a pallid colour; for in casting lead, the fumes from it fixing on the different members, and daily burning them, destroy the vigour of the blood; water should therefore on no account be conducted in leaden pipes if we are desirous that it should be wholesome. (VIII.6-10-11)

Nevertheless, the Romans continued with their addiction to the metal. The consequence was epidemics of

Table 1.3. Dates of discovery of compounds of atmospheric importance

Molecule	Chemical formula	Mineral name	Former name, alternate name, or meaning	Year discovered	Discoverer
Calcium carbonate	$CaCO_3(s)$	Calcite, aragonite	Calcspar	BC	?
Sodium chloride	$NaCl(s)$	Halite	Rock salt, common salt	BC	?
Potassium nitrate	$KNO_3(s)$	Nitre	Saltpeter, nitrum	BC	?
Sulfurous acid	$H_2SO_3(aq)$		Oil of sulfur, acidum volatile	BC	?
Sodium carbonate	$Na_2CO_3(s)$	Natrite	Nitrum, nator, nitron, natrum, soda ash, washing soda, salt-cake, calcined soda	BC	?
Calcium sulfate dihydrate	$CaSO_4-2H_2O(s)$	Gypsum	"plaster"	315 BC	Theophrastus (Greece)
Sulfuric acid	$H_2SO_4(aq)$		Oil of vitriol, acidum fixum, vitriolic acid, spirit of alum, spirit of vitriol	<1264	de Beauvais (France)
Ammonium chloride	$NH_4Cl(s)$	Sal ammoniac		<1400	Geber or later author
Molecular hydrogen	$H_2(g)$		Inflammable air	<1520, 1766	Paracelsus (Switzerland), Cavendish (England)
Nitric acid	$HNO_3(aq)$		Spirit of nitre	1585	Libavius (Germany)
Hydrochloric acid	$HCl(aq)$		Spirit of salt	1611	Libavius (Germany)
Sodium sulfate	$Na_2SO_4(s)$	Thenardite	Sal mirabile, Glauber's salt	1625	Glauber (Germany)
Carbon dioxide	$CO_2(g)$		Gas silvestre, Fixed air	<1648, 1756	Van Helmont (Belgium), Black (Scotland)
Ammonia	$NH_3(g)$		Gas pingue, alkaline acid air	<1648, 1756	Van Helmont (Belgium), Black (Scotland)
Ammonium nitrate	$NH_4NO_3(s)$	Nitrammite	Nitrum flammans, ammonia-nitre, ammonia-saltpeter	1648	Glauber (Germany)
Ammonium sulfate	$(NH_4)_2SO_4(s)$	Mascagnite	Secret sal ammoniac	<1648	Glauber (Germany)
Calcium nitrate	$Ca(NO_3)_2-4H_2O(s)$	Nitrocalcite	Baldwin's phosphorus	<1669	Baldwin (Germany)

(continued)

Table 1.3 (*continued*)

Molecule	Chemical formula	Mineral name	Former name, alternate name, or meaning	Year discovered	Discoverer
Magnesium carbonate	MgCO$_3$(s)	Magnesite	Magnesia alba	c. 1695	Grew (UK)
Nitrogen dioxide	NO$_2$(g)		Nitrous gas, red nitrous vapor	<1714, 1774	Ramazzini (Italy), Priestley (England)
Molecular nitrogen	N$_2$(g)		Mephitic air	1772	Rutherford (England)
Nitric oxide	NO(g)		Nitrous air	1772	Priestley (England)
Nitrous oxide	N$_2$O(g)		Diminished nitrous air, laughing gas	1772	Priestley (England)
Hydrochloric acid	HCl(g)		Marine acid air, Muriatic gas	1772	Priestley (England)
Hydrofluoric acid	HF(g)		Fluor acid	1773	Scheele (Sweden)
Molecular oxygen	O$_2$(g)		Dephlogisticated air	1774, 1772–1775	Priestley (England), Scheele (Sweden)
Chlorine gas	Cl$_2$(g)		Dephlogisticated marine (muriatic) acid gas, "green gas"	1774	Scheele (Sweden)
Acetaldehyde	CH$_3$CHO(g)			1774	Scheele (Sweden)
Carbon monoxide	CO(g)			1772–1779	Priestley (England)
Sulfur dioxide	SO$_2$(g)		Vitriolic acid air	1774–1779	Priestley (England)
Nitric acid	HNO$_3$(g)			1784	Priestley (England), Cavendish (England)
Hypochlorous acid	HOCl(g)			1830	Balard (France)
Ozone	O$_3$(g)		Ozein, "to smell"	1840	Schonbein (Germany)

kidney damage (saturnine gout), sterility, infertility, stillbirths, and mental disorders among aristocrats in particular (Lewis, 1985).

During the Renaissance, lead was used in pistols, rifles, cannons, and ammunition. It was mined and forged in Virginia beginning in 1621. In the 1920s, lead was first added to gasoline as an antiknock, octane-boosting agent, resulting in it being emitted to the air from vehicle exhaust as an aerosol particle. By the 1980s, the U.S. production of lead was equivalent to 5.2 kilograms per person per year. Due to its serious health effects and the fact that lead deactivates the catalyst in an automobile catalytic converter, most countries have since banned leaded gasoline. However, lead particulate matter is still emitted worldwide during lead acid battery manufacturing and lead ore crushing and smelting. **Smelting** is a method of extracting a metal from its ore.

1.2.1.2. Mercury

Mercury is a heavy silvery-white metal that is liquid at room temperature, often aggregating into beads (Figure 1.3). Its elemental symbol (Hg) derives from the Latinized Greek word, *hydrargyrum*, which means watery or runny (*hydr*) silver (*argyros*). Mercury was originally discovered upon the heating of the ore cinnabar, made of mercury (II) sulfide. Cinnabar is bright scarlet to brick red in color and is often located near recent volcanic activity and alkaline hot springs. A cinnabar mine at least 2,500 years old can be found

Figure 1.3. Shiny mercury beads. © Ventin/Dreamstime.com.

Figure 1.4. Pigments due to iron and manganese oxides in fine-grained rock layers in the Painted Desert, Arizona. © Maniscalco/Dreamstime.com.

in Almadén, Spain. The heating of cinnabar produces gas-phase mercury by the reaction $HgS(s) + O_2(g) \rightarrow Hg(g) + SO_2(g)$. Upon cooling, gas-phase mercury condenses to a liquid. The ancient Greeks used mercury in ointments, and the ancient Egyptians and Romans used it in cosmetics. Mercury has been found in Egyptian tombs dating back to 1500 BC. Mercury enters the atmosphere primarily as particulate matter. Today, most mercury emissions occur during stationary power plant combustion (primarily from coal-fired power plants), gold production, metal production in smelters, cement production, waste disposal, sodium hydroxide production, and iron and steel production.

1.2.1.3. Iron

Iron (*ferrum* in Latin; *iarn* in Scandinavian) is a dense metal element that is the primary component of the Earth's core and the fourth most abundant element in the Earth's crust. In pure form, it appears lustrous silvery gray. However, in the form of iron ore [iron oxide, $Fe_2O_3(s)$], it is reddish (e.g., Figure 1.4).

Meteorites consisting primarily of iron-nickel alloys, called **iron meteorites**, were the only source of iron before the discovery of iron ore. However, iron meteorites were rare, so only specialized meteoritic iron swords were forged from them, usually for ritual. Meteoric iron beads were also crafted, as evidenced by an archeological find dating to 3500 BC, in Gerzah, Egypt.

Between 3000 and 2700 BC, iron was purified from iron ore in Asmar, Mesopotamia, and Tall Chagar Bazaar, northern Syria. Pure iron was obtained by first heating iron ore to high temperature for several hours, then cooling, pounding, and reheating the compound to release oxygen and brittle impurities to produce shiny, pure iron. This technology transferred slowly, making its way to Scandinavia only by 50 BC. Today, iron is the element emitted in the greatest abundance in aerosol particles from industrial processes worldwide. Iron is also emitted naturally when the wind lifts iron-containing desert soil dust particles into the air.

1.2.1.4. Sulfur

Elemental **sulfur** (*sulvere* in Sanskrit; *sulphurium* in Latin) is a nonmetallic, pale yellow, crystalline mineral found in volcanic and hot spring deposits (e.g., Figure 1.5), sedimentary beds, and salt domes. Sulfur burns with a weak blue flame, and burning continues for a long time before the sulfur is consumed. Ancient Egyptian alchemists were well aware of sulfur (Brown, 1913). The word **brimstone** (or "burn-stone," referring to its combustibility) is an Old English word for sulfur. In the Book of Genesis, "brimstone and fire" were said to have rained down on the cities of Sodom and Gomorrah, located near the southern end of the Dead Sea, destroying them. The destruction of the cities, estimated to have occurred c. 1900 BC, may have been due to a volcanic eruption in which various forms of sulfur emanated. Sulfur in the air is primarily in the form of sulfur dioxide gas [$SO_2(g)$] and aqueous sulfuric acid [$H_2SO_4(aq)$].

Figure 1.5. Worker breaking sulfur inside volcanic crater of Kawa Ijen, Indonesia. © Henri Faure/ Dreamstime.com.

1.2.1.5. Carbon

Elemental **carbon** (*carbo* in Latin, meaning "charcoal") was well known in the ancient world, although it is unlikely that alchemists at the time knew that diamonds, graphite (plumbago), and charcoal all contained carbon. Carbon in diamonds and graphite is in pure crystalline form. In charcoal, coal, and coke, it takes on a variety of shapes and structures. In the ancient world, diamonds were valued only for their rarity, not for their beauty, because diamonds were not cut (and thus did not shine) until the fifteenth century (Figure 1.6). Graphite was used to make black marks on paper, and charcoal was used as a fuel. Today, the emission of elemental carbon (also called "black carbon") in the form of soot particles exacerbates global warming and visibility and causes health problems.

1.2.1.6. Sodium Carbonate (Solid)

Sodium carbonate [$Na_2CO_3(s)$] is a crystal mineral first found by the Egyptians in the Lakes of Natron, a group of six lakes to the west of the Nile Delta. The Egyptians called it "nitrum." Its name was modified to "nator" by the Hebrews, "nitron" by the Greeks, and "natrum" in the fifteenth century. Today, its mineral name is **natrite**. For centuries, it has been used as an ingredient in soaps. Some chemical industry names for it have been **washing soda, soda ash,** and **salt cake**. The manufacture of sodium carbonate for use in soaps resulted in acid deposition problems in England and France in the nineteenth century (Chapter 10). In the air, sodium carbonate is present in soil dust particles.

1.2.1.7. Calcium Carbonate (Solid)

Calcium carbonate [$CaCO_3(s)$] is a crystal present in pure form in the minerals **calcite** and **aragonite** and in mixed form in limestone, marble, chalk, and shells and skeletons of invertebrates. **Limestone** is sedimentary rock containing calcite or **dolomite** [$CaMg(CO_3)_2(s)$], **marble** is recrystallized limestone, and **chalk** (Figure 1.7) is fine-grained rock made of skeletons of microorganisms. In the ancient world, chalk was used for writing. In the air, calcium carbonate is present in soil dust particles. The name calcite originates from the word "calcspar," itself derived from the Greek word for limestone, *khálix*.

Figure 1.7. Calcite rock formations in form of a mushroom and a chicken in White Desert of Egypt. © Time De Boeck/Dreamstime.com.

Figure 1.6. Uncut diamond stones embedded in soil within a mine. © Kheng Ho Toh/Dreamstime.com.

Figure 1.8. Salt piles in Salt Desert of Chile. © Ferrucio Zuccolotto/Dreamstime.com.

Figure 1.9. Gypsum rock. © Viktor Onyshchenko/ Dreamstime.com.

1.2.1.8. Sodium Chloride (Solid)

Sodium chloride [$NaCl(s)$] is a crystal mineral (Figure 1.8) formed from the evaporation of ocean water. It was well known in the ancient world, found mixed with earthy material and mentioned in the Old Testament to "lose its savor" on exposure. It was often referred to as "rock salt." Today, it is the main ingredient in table (common) salt. Its mineral name is **halite**, from the Greek word *hals* ("salt"). In the air, sodium chloride is emitted in sea spray particles.

1.2.1.9. Potassium Nitrate (Solid)

Potassium nitrate [$KNO_3(s)$] is a crystal mineral also called **saltpeter** ("salt of rock") because it was often found as a saltlike crust on rocks. Because it is combustible, saltpeter was an ingredient of Greek fires (an incendiary weapon used by the Byzantine Empire). In the fifteenth century, it was called "nitrum" (the same early name as sodium carbonate). Today, its mineral name is **nitre** and it is used in rocket propellant, fireworks, gunpowder, and fertilizers. Plants benefit from both the potassium and nitrate in fertilizer provided by saltpeter. Potassium nitrate forms chemically in sea spray and soil dust particles by the reaction of the potassium ion, which is a natural constituent of both, with nitric acid gas. Potassium nitrate is one of the most abundant nitrogen-containing solids in the air.

1.2.1.10. Sulfurous Acid (Aqueous)

Ancient Egyptian alchemists obtained **sulfurous acid** [$H_2SO_3(aq)$] ("oil of sulfur") by combusting elemental sulfur in the presence of water. Such burning was also carried out in Homer's time for the purpose of fumigation. Sulfurous acid's use in bleaching wool is mentioned by Pliny the Elder. In the air, sulfurous acid, a precursor to acid deposition, forms when sulfur dioxide gas dissolves in water-containing aerosol particles or cloud drops.

1.2.1.11. Calcium Sulfate Dihydrate (Solid)

Calcium sulfate dihydrate [$CaSO_4–2H_2O(s)$] is a crystal mineral, more commonly known as **gypsum** (*gypsos* ("plaster") in Greek) (Figure 1.9). Gypsum was first referred to in 315 BC by Greek botanist and alchemist **Theophrastus** (371–286 BC), born in Lesbos, who wrote ten books on botany, stones, metals, and minerals. Gypsum is a naturally occurring mineral that appears worldwide in soils and aerosol particles. It forms chemically when aqueous sulfuric acid reacts with the mineral calcite.

When aerosol particles containing sulfuric acid deposit onto marble statues (which contain calcite), a gypsum crust also forms. Gypsum soil beds are mined to produce plaster of Paris, obtained by heating pure gypsum and adding water. **Plaster of Paris** was named such because early Parisians found gypsum in the clays and muds of the Paris Basin and used the gypsum to make plaster and cement. Gypsum is possibly the most common sulfur-containing solid in the atmosphere.

1.2.1.12. Ammonium Chloride (Solid)

Geber (or Abu Abdallah Jaber ben-Hayyam al-Kufi) was an Arabian alchemist who lived about 750–800 AD (Figure 1.10). Although the writings attributed to him may have been forged in the thirteenth century, it is clear that Geber or the writer was aware of

Figure 1.10. Geber (c. 750–800). Edgar Fahs Smith Collection, University of Pennsylvania Library.

Figure 1.11. Andreas Libavius (1540–1616). Edgar Fahs Smith Collection, University of Pennsylvania Library.

sal ammoniac [$NH_4Cl(s)$, **ammonium chloride**], a mineral crystal obtained from the Libyan Desert near the temple of Jupiter Ammon (the ultimate source of the name for the gas ammonia). Ammonium chloride can form when ammonia gas enters sea spray particles, which contain chlorine. It may be the most abundant ammonium-containing solid in the air.

1.2.1.13. Sulfuric Acid (Aqueous)

Vincent de Beauvais (1190–1264), a French philosopher, mentions the solvent power of the liquid acid distilled (separated by heating) from the natural crystal, potassium alum [$KAl(SO_4)_2-12H_2O(s)$]. The acid was probably dissolved **sulfuric acid** [$H_2SO_4(aq)$], and de Beauvais may have been the first to record its observation.

In 1585, **Andreas Libavius** (1540–1616; Figure 1.11), a German chemist who in 1597 wrote one of the first organized textbooks on chemistry, *Alchemia*, found that sulfuric acid, referred to at the time as "oil of vitriol," could also be extracted from "green vitriol" (ferrous sulfate, $FeSO_4-7H_2O(s)$, a blue-green natural crystal) and obtained by burning elemental sulfur with saltpeter [$KNO_3(s)$] in the presence of liquid water.

Sulfuric acid is present in aerosol particles and responsible for most acid deposition problems today.

1.2.1.14. Nitric Acid (Aqueous)

Libavius also reacted elemental sulfur with dissolved **nitric acid** [$HNO_3(aq)$, **spirit of nitre**], indicating that he was aware of this compound. Nitric acid was most likely formed from the reaction of $H_2SO_4(aq)$ with $KNO_3(s)$. Nitric acid is a strong acid that is corrosive. It enters aerosol particles by the dissolution of nitric acid gas into liquid water–containing bases (e.g., Na^+, Ca^{2+}, Mg^{2+}, K^+, NH_4^+) in the particles. Gas-phase nitric acid forms chemically from nitrogen oxide gases, emitted during combustion of fossil fuels and wood and produced by lightning.

1.2.1.15. Hydrochloric Acid (Aqueous)

In his book, *Syntagma*, published between 1611 and 1613, Libavius recorded the first description of the formation of **hydrochloric acid** [$HCl(aq)$], called *spiritus salis*, or "spirit of salt." He prepared it by heating rock salt [$NaCl(s)$] in water in the presence of clay (in clay crucibles). $HCl(aq)$ can also be obtained by reacting rock salt with green vitriol [$FeSO_4-7H_2O(s)$].

Figure 1.12. Johann Rudolf Glauber (1604–1670). Edgar Fahs Smith Collection, University of Pennsylvania Library.

Hydrochloric acid is an abundant component of sea spray particles.

1.2.1.16. Sodium Sulfate, Ammonium Nitrate, and Ammonium Sulfate (Solids)

Sodium sulfate, ammonium nitrate, and ammonium sulfate are all solid compounds found in atmospheric aerosol particles. **Johann Rudolf Glauber** (1604–1670; Figure 1.12), a German chemist, played a role in understanding all three of these compounds. At age twenty-one, Glauber was traveling to Vienna when he came down with a severe fever that reduced his ability to digest food. Inhabitants of the town he was passing through suggested he drink from a well an hour's walk away. Upon drinking from the well, Glauber regained

his appetite and ability to digest food. He became curious about the content of the water, so he set out to experiment with it. While evaporating the water, he discovered that it contained a solid material that had a salty taste, melted on his tongue, and did not burn quickly or crackle when placed on a fire (Hill, 1979). He also found that it lost weight when heated. He called the salt *sal mirabile* and referred to it as a universal medicine due to its healing ability. What he had discovered was **sodium sulfate decahydrate** [$Na_2SO_4-10H_2O(s)$], later referred to as **Glauber's salt**. Although the salt he found was natural, he was able to manufacture it in a laboratory by combining common salt (sodium chloride) with oil of vitriol (sulfuric acid). Without the hydrated water, sodium sulfate is a mineral now called **thenardite**, named after Baron Louis Jacques Thenard (1777–1857), who found it in Espartinas salt lake, near Madrid, Spain.

Glauber was also aware of the mineral **ammonium nitrate** [$NH_4NO_3(s)$], which he called *nitrum flammans*. Ammonium nitrate is not a common naturally occurring mineral in soil, although it has been found in Nicojack Cavern, Tennessee. Its mineral name is **nitrammite**, after its composition.

In his book, *Miraculum Mundi*, Glauber provided a recipe for producing the mineral **ammonium sulfate** [$(NH_4)_2SO_4(s)$], which he called *secret sal ammoniac*, and stated that two alchemists, Paracelsus and Van Helmont, may have used it. Ammonium sulfate is also a natural sublimation product of the fumaroles of Mount Vesuvius and Mount Etna. Paolo Mascagni (1752–1815) first described the natural occurrence of this salt; therefore, its mineral name today is **mascagnite**.

1.2.1.17. Calcium Nitrate (Solid)

In 1675, **Christopher Baldwin** (1600–1682) wrote a book in which he discussed the reaction of chalk [calcite, $CaCO_3(s)$] with nitric acid [$HNO_3(aq)$] to produce the crystal **calcium nitrate** [$Ca(NO_3)_2(s)$], which is phosphorescent (glows in the dark). Because of its appearance, he named the substance phosphorus, meaning "light bearer." It is now known as **Baldwin's phosphorus** because it differs from elemental phosphorus. Calcium nitrate forms chemically in aerosol particles from the reaction of the calcium ion and the nitrate ion.

Elemental **phosphorus** (P), a nonmetallic substance that also releases light, was discovered in Germany in 1669 by an alchemist, **Hennig Brand** (c. 1630–c. 1710) of Sweden, who distilled a mixture of sand and evaporated urine. The extraction of phosphorus was replicated by Johann Kunckel (1630–1750) of Germany, who knew both Baldwin and Brand, and

Figure 1.13. Sketch of hygroscope (circular), together with *Study for the Adoration of the Shepherds*, 1478–1480, by Leonardo Da Vinci. Encore Editions, www.encore-editions.com.

called phosphorus the "**phosphorus of Brand**." Kunckel published a treatise on phosphorus in 1678. The light due to phosphorus results from the chemical reaction of oxygen from the air with the surface of the phosphorus to create a product that emits visible radiation. Elemental phosphorus is a component of the Earth's crust and of soil dust particles.

1.2.2. Studies of Gases in the Air, 1450–1790

Gases were more difficult to observe and isolate than were liquids or solids, so the study of gases occurred only after many liquids and solids had been investigated. In this subsection, the history of discovery of gases from the fifteenth through eighteenth centuries is discussed.

1.2.2.1. Water Vapor

Although water vapor was known in the ancient world, changes in its abundance were not detected until the fifteenth century. In 1450, **Nicolas Cryfts** suggested that

changes in atmospheric water vapor could be measured with a **hygroscope** constructed of dried wool placed on a scale. A change in weight of the wool over time would represent a change in the water vapor content of the air. Capitalizing on Cryfts' notes, **Leonardo da Vinci** (1452–1519) drew such a hygroscope in a sketch that also contained drawings for his *Study for the Adoration of the Shepherds* (Figure 1.13) and in his *Codex Atlanticus* (1481), but using a sponge instead of wool. Wood and seaweed were later used in place of wool. In 1614, the Italian physician **Santorio Santorre** invented the hygrometer, which provided the relative water vapor content of the air by the contraction or elongation of a cord or lyre string, which absorbs water vapor. Gut and hair were used in later hygrometers.

1.2.2.2. Molecular Hydrogen (Gas)

Paracelsus (1493–1541; Figure 1.14), an alchemist born near Zurich, may have been the first to observe what is now known as **hydrogen gas** or **molecular hydrogen** [$H_2(g)$]. He discovered that when sulfuric

Figure 1.14. Paracelsus (1493–1541). Edgar Fahs Smith Collection, University of Pennsylvania Library.

acid was poured over the metals iron (Fe), zinc (Zn), or tin (Sn), it gave off a highly flammable vapor. In 1766, Henry Cavendish, who is more well known for the discovery of molecular hydrogen, also made this determination and isolated the vapor's properties. Molecular hydrogen is a well-mixed gas in today's lower atmosphere. It is produced naturally by ocean and soil bacteria and anthropogenically by fossil fuel combustion. It is removed from the air primarily by soil bacteria.

1.2.2.3. Ammonia and Carbon Dioxide (Gases)

John Baptist Van Helmont (1577–1644), born in Belgium, introduced the term **gas** into the chemical vocabulary. He produced what he called *gas silvestre* ("gas that is wild and dwells in out-of-the-way places") by fermenting alcoholic liquor, burning charcoal, and acidifying marble and chalk. The gas he discovered in all three cases, but did not know at the time, was **carbon dioxide** [$CO_2(g)$]. Another gas he produced was an inflammable vapor evolved from dung. He called this *gas pingue*, which was probably impure **ammonia** [$NH_3(g)$]. Today, carbon dioxide is known to be the main cause of global warming. Ammonia, produced naturally and anthropogenically, dissolves and reacts in aerosol particles.

1.2.2.4. Fire-Air

In 1676, **John Mayow** (1643–1679), an English physician, advanced the study of the composition of the air slightly by suggesting that air appeared to contain two components, one that allowed fire to burn and animals to breathe (which Mayow called *nitro-aereo* or "**fire-air**"), and another that did not. When he placed a lighted candle and a small animal in a closed vessel, the lighted candle went out before the animal died. When he placed only the animal in the vessel, the animal took twice as long to die. Thus, Mayow hypothesized that air was diminished by combustion and breathing. Fire-air later turned out to be **molecular oxygen** [$O_2(g)$].

1.2.2.5. Phlogisticated Air

In 1669, **Johann Joachim Becher** (1635–1682), a German physician, took what proved to be a step backward in the understanding of the composition of air. That year, he completed a book, *Physica Subterranea*, in which he introduced an incorrect concept that dominated the chemical world for the next 100 years. He suggested that every combustible material contains different amounts of three components, *terra mercurialis* ("fluid" or "mercurial earth," believed to be mercury),

terra lapidia ("strong or vitrifiable earth," believed to be common salt), and *terra pinguis* ("fatty earth," believed to be sulfur). He claimed that, during combustion, *terra pinguis* was expelled to the air. The principle that every combustible material releases its "source" of combustion, albeit incorrect, was not new, but it was more specific than were previous theories.

One of Becher's followers was **Georg Ernst Stahl** (1660–1734). In 1702, Stahl published *Specimen Becherianum*, in which he restated that every material contains a special combustible substance that escapes to the air when the material is burned. Stahl called the combustible substance, previously named *terra pinguis* by Becher, *phlogiston* after the Greek word *phlogizein*, "to set on fire." Stahl believed that *phlogiston* escaped primarily as either fire or **soot**, which he believed was the purest form of phlogiston. According to phlogiston theory, the following processes produced phlogiston:

$$\text{Metal} + \text{fire} \rightarrow \text{phlogiston} + \text{"calx" (residue)}$$
$$\text{Sulfur} + \text{fire} \rightarrow \text{pure phlogiston}$$
$$\text{Phosphorus} + \text{fire} \rightarrow \text{phlogiston} + \text{powder}$$
$$\text{Animal respiration} \rightarrow \text{pure phlogiston}$$

As demonstrated by the French chemist **Antoine Laurent de Lavoisier** (1743–1794), more than 100 years after Becher's first suggestion of the phlogiston theory, the real processes are as follows:

$$\text{Metal} + \text{fire} + O_2(g) \rightarrow \text{metal oxide (residue)}$$
$$\text{Sulfur} + \text{fire} + O_2(g) \rightarrow \text{sulfur dioxide gas } [SO_2(g)]$$
$$4P(s) + \text{fire} + 5O_2(g) \rightarrow P_4O_{10}(s)$$

Animal respiration:
$$C_6H_{12}O_6(s) + 6O_2(g) \rightarrow 6CO_2(g) + 6H_2O(g)$$

In other words, when a substance is burned, it combines with oxygen to form an oxide of the substance that either remains as a solid or is released to the air as a gas. In both cases, the oxide of the substance weighs more than the original substance. In the case of respiration, the reaction is effectively the slow combustion of organic material by inhaled oxygen. Thus, Becher and Stahl's theories of *terra pinguis* and phlogiston turned out to be incorrect.

To his credit, Stahl, in *Specimen Becherianum*, was the first to point out that sulfurous acid is more volatile (evaporates more readily) than is sulfuric acid.

Figure 1.15. Joseph Black (1728–1799). Edgar Fahs Smith Collection, University of Pennsylvania Library.

He called the former *acidum volatile* and the latter *acidum fixum*. He also noted that sulfuric acid is the stronger acid.

1.2.2.6. Carbon Dioxide Again – Fixed Air

In 1756, **Joseph Black** (1728–1799; Figure 1.15), a Scottish physician and chemist, performed an experiment in which he heated **magnesium carbonate** [$MgCO_3(s)$], an odorless white powder called magnesia alba ("white magnesia") at the time. On heating, $MgCO_3(s)$ lost weight, producing a heavy gas that neither sustained a flame nor supported life. When the gas was exposed to quicklime [$CaO(s)$, calcium oxide], a white-gray crystal, the weight was reabsorbed. Black called the gas "**fixed air**" because of its ability to attach or "fix" to compounds exposed to it. The fixed air turned out to be **carbon dioxide** [$CO_2(g)$], and when it was reabsorbed on exposure to $CaO(s)$, it was really forming calcium carbonate [$CaCO_3(s)$]. Fixed air was renamed to carbon dioxide in 1781 by Lavoisier. What Black did not recognize was that fixed air, or $CO_2(g)$, had been discovered by Van Helmont more than a century before. In 1756, Black also isolated ammonia gas [$NH_3(g)$], previously observed by Van Helmont and later called "**alkaline acid air**" by Joseph Priestley. Black is separately known for making the first systematic study of a chemical reaction and developing the concepts of latent heat and specific heat.

1.2.2.7. Molecular Hydrogen Again – Inflammable Air

In 1766, Henry Cavendish (1731–1810), an English chemist and physicist, followed up Black's work by producing a gas he called "**inflammable air**" (highly flammable gas). This gas was obtained by diluting either sulfuric acid [$H_2SO_4(aq)$] or hydrochloric acid [$HCl(aq)$] with water and pouring the resulting solution on a metal, such as iron, zinc, or tin. This experiment was similar to that of Paracelsus, who also observed an inflammable vapor. Cavendish believed that "inflammable air" was phlogiston, but this turned out to be incorrect. Nevertheless, Cavendish isolated the properties of the gas. In 1783, he found that exploding a mixture of the gas with air produced water. Subsequently, Lavoisier called the gas **hydrogen**, the "water producer." More specifically, the gas was **molecular hydrogen** [$H_2(g)$].

In other experiments, Cavendish exposed marble, which contains $CaCO_3(s)$, to hydrochloric acid [$HCl(aq)$] to produce $CO_2(g)$, as Van Helmont had done previously. Cavendish took the further step of measuring the properties of $CO_2(g)$. He is also known for studying the weights of gases and the density of the Earth. In 1783, after oxygen had been discovered, Cavendish calculated that air contained 20.83 percent oxygen by volume, close to the more accurate measurement today of 20.95 percent.

1.2.2.8. Molecular Nitrogen (Gas) – Mephitic Air

In 1772, **Daniel Rutherford** (1749–1819; Figure 1.16), a student of Joseph Black, performed an experiment by which he allowed an animal to breathe the air in an enclosed space until the animal died, removing the molecular oxygen, $O_2(g)$, which had not yet been discovered. He then exposed the remaining air to the crystal **caustic potash** [$KOH(s)$, potassium hydroxide or pot ashes], obtained by burning wood in a large iron pot. $CO_2(g)$ in the remaining air reacted with caustic potash, forming **pearl ash** or **potash** [$K_2CO_3(s)$, potassium carbonate]. The residue after $CO_2(g)$ was removed could not sustain life; thus, Rutherford called it "**mephitic (noxious or poisonous) air**." Mephitic air is now known as **molecular nitrogen** gas [$N_2(g)$], which makes up nearly 80 percent of air by volume. Nitrogen is tasteless, colorless, and odorless. The name **nitrogen**, the "nitre maker" was given in 1790 by **Jean-Antoine**

Figure 1.16. Daniel Rutherford (1749–1819). Edgar Fahs Smith Collection, University of Pennsylvania Library.

Chaptal (1756–1832), a French industrial chemist, because nitrogen was found to be a constituent of the crystal nitre [$KNO_3(s)$].

1.2.2.9. Molecular Oxygen (Gas) – Dephlogisticated Air

Two chemists discovered **molecular oxygen** gas [$O_2(g)$] independently: **Joseph Priestley** (1733–1804; Figure 1.17) on August 1, 1774, and **Karl Wilhelm Scheele** (1742–1786; Figure 1.18) from Sweden, sometime between 1772 and 1775. Although both chemists discovered oxygen near the same time, Priestley announced his discovery in 1774, and Scheele published his discovery in 1777.

Priestley was born in 1733 in Birstall, England, the eldest of six children in a dissenting family (one that did not conform to the Church of England). At age nine, he was sent to live with his aunt and uncle, who provided him with an education in preparation for the ministry. In 1749, he became seriously ill and, thereafter, had a permanent stutter. Soon, he began to study higher math, natural philosophy, logic, and metaphysics. He then returned to theological studies as a Rational Dissenter, one who believed in rational analysis of the natural world and the Bible. In 1758, Priestley moved to Nantwich, Chesire, and started a theological school in which he taught natural philosophy. After writing *The Rudiments of English Grammar*, he was offered a teaching position at Warrington Academy in 1761. While there, he wrote several books on history. Of note was *The History and Present State of Electricity*, published in 1767. This book was inspired by a meeting between him and Benjamin Franklin, who had recently invented the lightning rod. Franklin suggested that Priestley perform several experiments, which Priestley did along with designing experiments of his own. In one of these, Priestley found that **graphite conducted electricity**, overturning a theory that only water and metals could conduct electricity. He further proposed that electrical forces followed an **inverse square law**, although Coulomb later proved this in the 1780s.

In 1767, Priestley moved to Leeds and lived in a home next to a brewery. Curious as to the bubbly gas emanating from the brew, he found a method to produce the gas at home. Upon experimenting, Priestley learned that it extinguished lighted wood chips. When he dissolved the gas in water, he noted it had a tangy taste and caused the water to bubble. The gas was Van Helmont's and Black's carbon dioxide, and Priestley had inadvertently invented a drinkable form of **carbonated soda water**. In 1772, he published a pamphlet called *Directions for Impregnating Water with Fixed Air*, but he did not earn any royalties in his lifetime for this invention. Instead, **Johann Jacob Schweppe** (1740–1821) capitalized on this invention to build a soda water empire that started in Geneva in 1783 and move to London in 1792.

On April 15, 1770, Priestley discovered that Indian gum could be used to rub out lead pencil marks. This fortuitous finding was the **discovery of the eraser**, or "rubber." In 1772, he performed a completely different experiment in which he placed a small green plant in a container in which he also lit a candle. Several days after the candle had burned out, he found that he could light the candle again because the plant had produced oxygen (which he had not yet discovered). Hence, he had discovered **photosynthesis**.

In 1773, Priestley moved to Calne and began to focus more on laboratory experimentation of gases. In his most important experiment, he burned the element mercury (Hg) in air to form bright red mercuric oxide [$HgO(s)$], a powder. He then heated the mercuric oxide in a container from which all air had been

(a)

(b)

Figure 1.17. (a) Joseph Priestley (1733–1804). (b) Reconstruction of Priestley's oxygen apparatus. Edgar Fahs Smith Collection, University of Pennsylvania Library.

removed. Burning mercuric oxide in a vacuum released oxygen so that it was the only gas in the container. Due to the container's high oxygen content, flammable material burned more readily in the container than in regular air. He also found that the gas was insoluble in water, supported combustion, and "invigorated a mouse."

Priestley called the new gas, discovered August 1, 1774, "**dephlogisticated air**" because he incorrectly believed that the gas burned so brightly because it contained no phlogiston. He believed that the burning of a substance emitted phlogiston, causing the flame to die out eventually.

Scheele independently isolated molecular oxygen in at least three ways: heating manganic oxide [$Mn_2O_3(s)$] (a black powder), heating red mercuric oxide [$HgO(s)$], and heating a mixture of nitric acid [$HNO_3(aq)$] and potassium nitrate [$KNO_3(s)$].

Shortly after his discovery of oxygen in August 1774, Priestley went on a tour of Europe. One of his stops was in Paris, where he met Lavoisier. **Antoine-Laurent de Lavoisier** was born into a wealthy French family on August 26, 1743, in Paris. He studied chemistry, botany,

astronomy, and mathematics at the Collège Mazarin from 1754 to 1761. His early scientific work was on street lighting, for which he was admitted to the French Academy of Sciences at age twenty-five, and on geological surveying. In 1771, at twenty-eight, he married the thirteen-year-old daughter of one of the coowners of the Ferme Générale, an agency that set and collected taxes for the king of France from 1726 to 1790. Lavoisier became a member of the group, setting himself up for his ultimate demise.

Lavoisier experimented with combustion and, on November 1, 1772, found that phosphorus and sulfur weighed more after than before they were burned. The reason, although he did not know it at the time, was that both reacted with oxygen during combustion by $4P(s) + fire + 5O_2(g) \rightarrow P_4O_{10}(s)$ and $S(s) + fire + O_2(g) \rightarrow SO_2(g)$, respectively. In early 1774, he also discovered that lead and tin weighed more and consumed one-sixth to one-fifth of the volume of air when they were burned. Again, he did not yet realize that oxygen comprised about 21 percent of air and was consumed during combustion.

Figure 1.18. (a) Karl Wilhelm Scheele (1742–1786). (b) Scheele's laboratory, with oven in center. Edgar Fahs Smith Collection, University of Pennsylvania Library.

With these results in hand, Lavoisier's meeting with Priestley in October 1774 was a perfect alignment of the stars. During their meeting, Priestley replicated his own experiment, producing oxygen from mercuric oxide. Lavoisier was surprised that the gas was insoluble and was not fixed air (Rodwell, 1882). He immediately realized, though, that the gas Priestley found completed the explanation of why substances gained weight and reduced the volume of air on combustion.

Lavoisier subsequently experimented with the gas for twelve years and revised its name to **oxygen**, the "acid maker," because he believed (incorrectly) that all acids contained oxygen. Almost all oxygen in the air is in the form of **molecular oxygen** [$O_2(g)$]. During this period, Lavoisier formalized the **oxygen theory of combustion** and proved the **law of conservation of mass**, which states that, in a chemical reaction, mass is conserved. For example, when sulfur, phosphorus, carbon, or another solid is burned, its gas- plus solid-phase mass increases by an amount equal to the loss in mass of oxygen from the air. Lavoisier used the fact that oxygen combines with a solid to form an oxide of the solid that is released to the air as a gas during combustion to disprove the theory of phlogiston, which was premised on the belief that only material in the original solid was released on combustion. He similarly showed that rusting is a mass-conserving process by which oxygen from the air combines with a solid to form an oxide of the solid.

In 1775/1776, Lavoisier found that diamonds contain pure carbon and produce carbon dioxide when heated. In 1781, he renamed Black's fixed air **carbon dioxide** and determined its elemental composition. Lavoisier also devised the first chemical system of nomenclature and specified that matter exists in at least three states: gas, liquid, and solid. He found that gases could be reduced to liquids or solids by cooling the air. Lavoisier is said to be the founder of modern chemistry.

However, because of his membership in the Ferme Générale, Lavoisier was arrested during the French Revolution (Figure 1.19). On May 8, 1794, after a trial of less than a day, he and twenty-seven other members of the group were guillotined, and their bodies thrown into a common grave.

Ironically, Priestley was attacked for his staunch defense of the principles of the French Revolution and

Figure 1.19. "The Arrest of Lavoisier." Photograph of painting (1876) by L. Langenmantel. Edgar Fahs Smith Collection, University of Pennsylvania Library.

his dissenting religious views. On July 14, 1791, he lost his house, library, and laboratory in Birmingham, England, to a fire set by a mob angry at his public support of the revolution (Figure 1.20). Priestley moved to London, but further persecution led him to flee to Pennsylvania in 1794, where he lived until his death in 1804.

1.2.2.10. Additional Discoveries by Priestley

During his career, Priestley discovered several additional gases significantly relevant to air pollution. Between 1767 and 1773, while working at Mill Hill Chapel, in Leeds, Yorkshire, he isolated **nitric oxide** [NO(g), "nitrous air"], **nitrogen dioxide** [NO$_2$(g), "red nitrous vapour"], **nitrous oxide** [N$_2$O(g), "diminished nitrous air"], **hydrochloric acid** gas [HCl(g), "marine acid air"], and **ammonia** [NH$_3$(g), "alkaline acid air"]. Nitrogen dioxide may have been previously observed by **Bernardo Ramazzini** (1633–1714), an Italian medical doctor and early pioneer in industrial medicine, and ammonia was previously observed by Van Helmont and Black. Priestley also discovered **carbon**

monoxide [CO(g)], and, in 1775, **sulfur dioxide** [SO$_2$(g), "vitriolic acid air"]. Additionally, he formed gas-phase **nitric acid** [HNO$_3$(g)], although Cavendish uncovered its composition.

Today, NO(g), NO$_2$(g), CO(g), N$_2$O(g), and SO$_2$(g) are emitted significantly during fossil fuel, biofuel, and biomass burning. **Fossil fuel combustion** is the burning of coal, oil, natural gas, gasoline, kerosene, and diesel for transportation, heating/cooling, industrial processes, and energy production. **Biofuel burning** is the burning of liquid biofuels, such as ethanol and biodiesel, for transportation, or solid biofuels, such as wood, grass, agricultural waste, and dung, for heating and cooking in homes or producing energy in power plants. **Biomass burning** is the outdoor burning of vegetation, either naturally, such as with a forest fire, or anthropogenically in order to clear land, stimulate grass growth, manage forest growth, or satisfy a ritual.

N$_2$O(g) is also produced from microbial metabolism, and SO$_2$(g) is emitted by volcanos. HCl(g) is emitted by volcanos, evaporates from sea spray particles, and is

Figure 1.20. Drawing of destruction of Priestley's house, library, and laboratory, Fair Hill, Birmingham, 1791. Edgar Fahs Smith Collection, University of Pennsylvania Library.

a product of chlorine reactions in the upper atmosphere. $NO_2(g)$ and $SO_2(g)$ are both precursors of acid deposition. $NH_3(g)$ is a major aerosol particle precursor in photochemical smog.

1.2.2.11. Hydrofluoric Acid (Gas)

A meticulous artist at his craft, Scheele also discovered **hydrofluoric acid** gas [HF(g)] in 1773. Scheele named HF(g) "fluor acid" after the crystal mineral **fluorspar** [$CaF_2(s)$, fluorite], which contains it. The name fluorspar was coined in 1529 by Georgius Agricola from the Latin and French word *fluere*, which means "flow" or "flux," because fluorspar appeared to flow. Elemental **fluorine** (F) was isolated from HF(g) only in 1886 by French chemist **Henri Moissan** (1852–1907). Prior to that time, at least two chemists died from toxic exposure trying to isolate F from HF(g). Moissan

won a Nobel Prize for isolating fluorine and inventing the electric arc furnace. Today, HF(g) is a product of chemical reactions in the upper atmosphere involving anthropogenically emitted fluorine compounds.

1.2.2.12. Chlorine (Gas)

In 1774, Scheele discovered **chlorine gas** [$Cl_2(g)$], and thus the element chlorine (Cl), by reacting dissolved hydrochloric acid [HCl(aq)] with pyrolusite [$MnO_2(s)$]. Chlorine gas is a dense, odorous, greenish-yellow, corrosive, toxic gas that chokes its victims' lungs. It also bleaches green grass to white. He called it "dephlogisticated marine acid gas." Lavoisier changed the name to "oxymuriatic acid" because he incorrectly believed that it contained oxygen and chlorine. In 1810, the name was changed to chlorine, the "green gas," by Sir Humphry Davy, who showed that chlorine was an element and did not contain oxygen. Today, $Cl_2(g)$ is a product of chemical reactions, primarily in the upper atmosphere.

1.2.3. Discoveries after 1790

After 1790, the pace at which gas, liquid, and solid chemicals were discovered increased. In the following subsections, a few more chemicals of atmospheric importance are discussed.

1.2.3.1. Elemental Potassium, Sodium, Calcium, and Chlorine

In 1807/1808, **Sir Humphry Davy** (1778–1829; Figure 1.21), who along with Priestley is the most well-known British chemist, developed electrolysis, which led to the discovery of the elements **potassium** (K), **sodium** (Na), **calcium** (Ca), and **barium** (Ba). **Electrolysis** is the passage of an electric current through a solution to break down a compound or cause a reaction. Potassium was isolated from **caustic potash** [potassium hydroxide, KOH(s)] by electrolysis. Potassium is the seventh most abundant element in the Earth's crust and is emitted

Figure 1.21. Sir Humphry Davy (1778–1829). Edgar Fahs Smith Collection, University of Pennsylvania Library.

into the air in soil dust and sea spray particles. Sodium was isolated from **caustic soda** [sodium hydroxide, NaOH(s)] by electrolysis. The name "sodium" derives from the Italian word *soda*, a term applied to all alkalis in the Middle Ages. Sodium is the sixth most abundant element in the Earth's crust and is emitted in soil dust and sea spray particles. Calcium was isolated from quicklime [CaO(s)] by electrolysis. The name "calcium" was derived from the word *calx*, the name the Romans used for lime. Calcium is the fifth most abundant element in the Earth's crust and is emitted in soil dust and sea spray particles.

In 1810, Davy also named the element **chlorine**, previously called oxymuriatic acid. He proved that chlorine was an element and that muriatic gas [HCl(g), hydrochloric acid gas] contains chlorine and hydrogen, but no oxygen. He similarly proved that hydrofluoric acid gas [HF(g)] contains no oxygen. Both proofs contradicted Lavoisier's theory that all acids contained oxygen.

1.2.3.2. Elemental Silicon and Chemical Symbols

A contemporary of Davy, **Jöns Jakob Berzelius** (1779–1848) of Sweden discovered the elements **silicon** (Si) (1823), selenium (Se) (1817), and thorium (Th) (1828). He also spent ten years determining the atomic or molecular weights of more than 2,000 elements and compounds, publishing the results in 1818 and 1826. Berzelius isolated silicon, a name derived from the Latin word *silex*, meaning "flint," by fusing iron, carbon, and the crystal **quartz** [SiO$_2$(s)]. Silicon is the second most abundant element in the Earth's crust, after oxygen, and is present in soil dust particles.

Berzelius's most well-known achievement was to invent a system of chemical symbols and notation. For elements, he used the first one or two letters of the element's Latin or Greek name. For example, oxygen was denoted with an O, hydrogen with an H, mercury with Hg (hydrargyrum), and lead with Pb (plumbum). For compounds with more than one atom of an element, he identified the number of atoms of the element with a subscript. For example, he identified water with H_2O.

1.2.3.3. Elemental Bromine and Hypochlorous Acid (Gas)

In 1826, **Antoine-Jérôme Balard** (1802–1876), a French apothecary, accidentally discovered the element **bromine** (Br) after analyzing the "bittern" (saline liquor) that remained after common salt had crystallized out of concentrated water in a salt marsh near the Mediterranean Sea. Bromine means "stench" in Greek. It is a heavy, reddish-brown liquid that evaporates at room temperature to a red gas that irritates the throat and eyes and has a strong smell. It is the only nonmetallic element that can be in the liquid phase at room temperature. Balard also discovered **hypochlorous acid** HOCl(g)]. Bromine and hypochlorous acid contribute to ozone destruction in the upper atmosphere today.

1.2.3.4. Ozone (Gas)

On March 13, 1839, the German chemist **Christian Friederich Schönbein** (1799–1868; Figure 1.22) discovered one of the most important atmospheric trace gases, **ozone** [O_3(g)]. He discovered ozone by passing an electric current through water (electrolysis). The resulting gas had a pungent, sweet smell, and he named it ozone after the Greek word, "ozein," which means "to smell." Although the Dutch chemist Martinus Van Marum (1750–1837) had previously detected "the odor of electrical matter" associated with electrical sparks in

Figure 1.22. Christian Friederich Schönbein (1799–1868). Edgar Fahs Smith Collection, University of Pennsylvania Library.

1785, he had not associated the odor with a gas in the air. In 1840, Schönbein proposed that the odor he found was due to a distinct chemical. In an 1841 report to the British Association for the Advancement of Science, he further wrote:

> The peculiar smell makes its appearance as soon as the electrolysis of water begins and continues to be perceived for some time after stopping the flow of electricity. . . . The odor must be due to some gaseous substance disengaged (conjointly with oxygen) from the fluid due to the decomposing power of the current.

In 1845, the Swiss scientists Auguste de la Rive (1801–1873) and Jean-Charles de Marignac (1817–1894) suggested that ozone is a form of oxygen. However, only in 1865 did Swiss chemist Jacques-Louis Soret (1827–1890) identify the structure and molecular formula of ozone as O_3. In 1879, the French physicist Alfred Cornu (1841–1902) found a sharp cutoff in the ultraviolet light spectrum reaching the Earth's surface. In 1881, Walter N. Hartley explained the cutoff as due to absorption of short ultraviolet wavelengths by ozone high in the atmosphere. In 1913, French physicists Charles Fabry (1867–1945) and Henri Buisson (1873–1944) quantified the thickness of this layer of ozone in the upper atmosphere, which is now referred to as the **stratospheric ozone layer**.

In 1846, Schönbein also discovered **guncotton**, the first nitro-based explosive and precursor to trinitrotoluene (TNT). Nitro-based explosives were smokeless. Previously, gunpowder, a black powder mixture of sulfur, charcoal, and saltpeter (potassium nitrate) was the explosive and gun propellant of choice, but it resulted in poor battlefield visibility. Schönbein discovered guncotton after spilling nitric acid and sulfuric acid on his kitchen counter, and then wiping up the mess with his wife's cotton apron. After hanging the apron to dry over the stove, it readily caught fire. The nitric acid had reacted with the cellulose in the cotton to form explosive nitrocellulose and water. The sulfuric acid absorbed the water, preventing the water from diluting the nitric acid. Because guncotton was so dangerous to handle, factories producing it tended to explode, so it was not widely used on its own. In 1889, it was used more safely in a mixture called cordite, which contained guncotton, nitroglycerine, and Vaseline in long cords.

1.2.3.5. Noble Gases

The air contains several inert noble gases in trace quantities, including helium (He), argon (Ar), neon (Ne), krypton (Kr), and xenon (Xe), all of which were discovered between 1868 and 1898. In 1868, **Pierre Janssen** (1824–1907), a French astronomer, observed a yellow line in the spectrum of the sun's chromosphere. Because no known element on Earth could account for this line, he believed that it was due to an element unique to the sun. **Joseph Norman Lockyer** (1836–1920), an English astronomer, confirmed Janssen's findings and named the new element **helium** (He), after *Helios*, the Greek god of the sun. The element was not discovered on Earth until 1895, when **Sir William Ramsay** (1852–1916), a Scottish chemist, found it in the mineral **clevite**. Swedish chemists Per Theodor Cleve (after whom clevite is named) and Nils Abraham Langlet found helium in the mineral at about the same time. Helium is the most abundant element in the universe after hydrogen. On Earth, helium is emitted to the air following the decay of radioactive minerals.

In 1894, **Lord Baron Rayleigh**, an English physicist born John William Strutt (1842–1919), discovered that nitrogen gas from the air was 0.5 percent heavier than was that prepared chemically. He and Sir Ramsay found that the difference was due to a previously undiscovered gas that they called **argon** (Ar), after the Greek word *argos*, meaning "lazy" in reference to the inert qualities of the gas. The two shared a Nobel Prize for their discovery. Argon forms from the radioactive decay of potassium (K). In his 1898 book *War of the Worlds*, H. G. Wells wrote that Martians used "toxic brown argon gas" to attack London, but were subdued by the common cold. However, argon is neither brown nor poisonous at typical atmospheric concentrations. It is colorless and odorless as a gas and liquid. Sir Ramsay, together with M. W. Travers, went on to discover the elements **neon** (Ne), **krypton** (Kr), and **xenon** (Xe), all in 1898. The three were named after Greek words: *neos* ("new"), *kryptos* ("concealed"), and *xenos* ("guest"), respectively. The source of krypton and xenon is the radioactive decay of elements in the Earth's crust, and the source of neon is volcanic outgassing.

1.2.3.6. Radioactive Gases

Two radioactive elements of atmospheric importance, **polonium** (Po) and **radon** (Rn), were discovered in the late twentieth century. These carcinogenic elements are found in the air of many homes overlying uranium-rich soils. In 1898, French chemists **Pierre Curie** (1859–1906) and **Marie Curie** (1867–1934; Figure 1.23) discovered polonium, which was named after Marie Curie's native country, Poland. In 1903, Pierre and Marie Curie, along with French physicist **Antoine Henri Becquerel** (1852–1908), won a Nobel

Figure 1.23. Marie Curie (1867–1934). Edgar Fahs Smith Collection, University of Pennsylvania Library.

Prize for their fundamental research on radioactivity. In 1911, Marie Curie won a second prize for her discoveries of polonium and **radium** (Ra), a radon precursor. Radon, itself, was discovered in 1900 by German physicist **Friedrich Ernst Dorn** (1848–1916), who called it **radium emanation** because it is a product of radioactive decay of radium. The name radium is from the Latin word *radius*, meaning "ray." Ramsay and Gray, who isolated radon and determined its density, changed its name to **niton** in 1908. In 1923, niton was renamed radon.

1.3. Chemical Structure and Reactivity

In this section, the structure and reactivity of some of the compounds identified previously are discussed. Table 1.4 shows the chemical structure of selected compounds. Single, double, and triple lines between atoms denote single, double, and triple bonds, respectively. For some compounds [OH(g), NO(g), NO$_2$(g)], a single dot is shown adjacent to an atom. A single dot indicates that the atom has a free electron. Compounds with

Table 1.4. Structures of some common compounds

Compound name	Structure showing bonds and free electrons	Formula with free electrons	Formula without free electrons
Molecular oxygen	$O=O$	$O_2(g)$	$O_2(g)$
Molecular nitrogen	$N\equiv N$	$N_2(g)$	$N_2(g)$
Ozone	$-O \diagup \overset{+}{O} \diagdown\!\!\diagdown O$	$O_3(g)$	$O_3(g)$
Hydroxyl radical	$\dot{O}-H$	$\dot{O}H(g)$	$OH(g)$
Water vapor	$O \diagdown^{H}_{\,H}$	$H_2O(g)$	$H_2O(g)$
Nitric oxide	$\dot{N}=O$	$\dot{N}O(g)$	$NO(g)$
Nitrogen dioxide	$-O \diagup \overset{\cdot}{\overset{+}{N}} \diagdown\!\!\diagdown O$	$\dot{N}O_2(g)$	$NO_2(g)$
Sulfur dioxide	$O \diagup\!\!\diagup S \diagdown\!\!\diagdown O$	$SO_2(g)$	$SO_2(g)$
Carbon monoxide	$^-C\equiv O^+$	$CO(g)$	$CO(g)$
Carbon dioxide	$O=C=O$	$CO_2(g)$	$CO_2(g)$
Methane	$H-\overset{\displaystyle H}{\underset{\displaystyle H}{C}}-H$	$CH_4(g)$	$CH_4(g)$
Sulfate ion	$O=\overset{\displaystyle O}{\underset{\displaystyle O-}{S}}-O-$	$SO_4{}^{2-}$	$SO_4{}^{2-}$

a free electron are called **free radicals** and are highly reactive. Some non–free radicals that have a single bond [e.g., $O_3(g)$] are also reactive because single bonds are readily broken. Compounds with triple bonds [$N_2(g)$ and $CO(g)$] are not so reactive because triple bonds are difficult to break. **Noble elements** (He, Ar, Ne, Kr, Xe) have no free electrons and no potential to form bonds with other elements; thus, they are chemically unreactive (inert).

For some compounds in Table 1.4 [$NO_2(g)$, $O_3(g)$, $CO(g)$], positive and negative charges are shown. Such a charge distribution arises when one atom transfers charge to another atom during molecular formation. During $NO_2(g)$ formation, for example, a net negative charge is transferred to an oxygen atom from the nitrogen atom, resulting in the charge distribution shown. Compounds with both positive and negative charges have zero net charge and are not ions, but the positive (negative) end of the compound is likely to attract negative (positive) charges from other compounds, enhancing the reactivity of the compound. For $SO_4{}^{2-}$, a net negative charge is shown, indicating that it is an ion.

When oxygen combines with an element or compound during a chemical reaction, the process is called **oxidation**, and the resulting substance is **oxidized**. The gases $O_2(g)$, $O_3(g)$, $OH(g)$, $H_2O(g)$, $NO(g)$, $NO_2(g)$, $SO_2(g)$, $CO(g)$, and $CO_2(g)$ are all oxidized gases. When oxygen is removed from a substance during a reaction, the process is called **reduction**, and the resulting element or compound is **reduced**. The gases $H_2(g)$, $N_2(g)$, $NH_3(g)$, and $CH_4(g)$ are all reduced gases.

Table 1.4 shows structures of inorganic compounds and methane, an organic compound, whereas Table 1.5 shows structures of additional organic compounds. **Inorganic compounds** are compounds that contain any element, including hydrogen (H) or carbon (C), but not both. **Organic compounds** are compounds that contain

Table 1.5. Structures of some common organic compounds found in air

Alkane	Alkene	Cycloalkene	Hemiterpene
Ethane $C_2H_6(g)$	Ethene $C_2H_4(g)$	Cyclopentene $C_5H_8(g)$	Isoprene $C_5H_8(g)$
Aromatic	Alcohol	Aldehyde	Ketone
Toluene $C_6H_5CH_3(g)$	Methanol $CH_3OH(g)$	Formaldehyde $HCHO(g)$	Acetone $CH_3COCH_3(g)$

both C and H, but may also contain other elements. Methane is the simplest organic compound.

Organic compounds that contain only H and C are **hydrocarbons**. Hydrocarbons include alkanes, cycloalkanes, alkenes, cycloalkenes, alkynes, aromatics, and terpenes. Examples of some of these groups are given in Table 1.5. **Alkanes** (paraffins) are open-chain (noncyclical) hydrocarbons with a single bond between each pair of carbon atoms and the molecular formula C_nH_{2n+2}. **Cycloalkanes** (not shown) are similar to alkanes, but with a cyclical structure. **Alkenes** (olefins) are open-chain hydrocarbons with a double bond between one pair of carbon atoms and the molecular formula C_nH_{2n}. **Cycloalkenes** are similar to alkenes, but with a cyclical structure. **Alkynes** (acetylenes, not shown) are open-chain hydrocarbons with a triple bond between at least one pair of carbon atoms. **Terpenes** are a class of naturally occurring hydrocarbons that include hemiterpenes (C_5H_8), **monoterpenes** ($C_{10}H_{16}$), **sesquiterpenes** ($C_{15}H_{24}$), **diterpenes** ($C_{20}H_{32}$), and so on. **Aromatic hydrocarbons** are hydrocarbons with a benzene ring and possibly other H and C atoms attached to the ring. Two representations of a benzene ring are shown in Figure 1.24.

Aromatics are so named because the first aromatics isolated were obtained from substances that had a pleasant fragrance, or aroma. Around 1868, Austrian chemist **Joseph Loschmidt** (1821–1895) found that such aromatic compounds could be obtained by replacing one or more hydrogen atoms on a benzene ring with another atom or group. The name aromatic was subsequently applied to any compound that had a

benzene ring in its structure. Loschmidt was the first to explain the structure of benzene, toluene, and ozone. He is also the first to quantify accurately Avogadro's number (Section 3.4).

When methane, a slowly reacting hydrocarbon, is excluded from the list of hydrocarbons, the remaining hydrocarbons are called **nonmethane hydrocarbons (NMHCs)**. When oxygenated functional groups, such as aldehydes, ketones, alcohols, acids, and nitrates, are added to hydrocarbons, the resulting compounds are **oxygenated hydrocarbons**. In Table 1.5, the alcohol, aldehyde, and ketone are oxygenated hydrocarbons. Nonmethane hydrocarbons and oxygenated hydrocarbons are **reactive organic gases (ROGs)**. **Total organic gas (TOG)** is the sum of ROGs and methane. **Volatile organic compounds (VOCs)** are organic compounds with relatively low boiling points that, therefore, readily evaporate. Although all VOCs are not necessarily ROGs, these terms are often interchanged. Finally, aldehydes and ketones are called **carbonyls**. The sum of nonmethane hydrocarbons and carbonyls is **nonmethane organic carbon (NMOC)**.

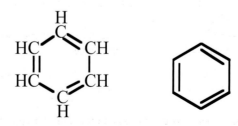

Figure 1.24. Two representations of benzene ring.

1.4. Chemical Reactions and Photoprocesses

Many of the pollution problems today are exacerbated by atmospheric chemical reactions. Reactions are initiated by sunlight, lightning, changes in temperature, or molecular collisions. In this section, chemical reactions are briefly discussed.

Gas-phase chemical reactions are conveniently divided into photolysis reactions (also called photoprocesses, photodissociation reactions, or photolytic reactions) and chemical kinetic reactions. Photolysis reactions, which are **unimolecular** (involving one reactant), are initiated when solar radiation strikes a molecule and breaks it into two or more products. An example of a **photolysis reaction** is

$$\dot{N}O_2(g) + h\nu \rightarrow \dot{N}O(g) + \cdot\dot{O}(g) \quad \lambda < 420 \text{ nm} \quad (1.1)$$

Nitrogen dioxide, Nitric oxide, Atomic oxygen

where the $h\nu$ implies a photon of solar radiation and λ is the wavelength of the radiation (defined in Chapter 2).

Chemical kinetic reactions are usually **bimolecular** (involving two reactants) and include thermal decomposition, isomerization, and standard collision reactions. Thermal decomposition and isomerization reactions occur when a reactant molecule collides with an air molecule. The kinetic energy of the collision elevates the reactant to an energy state high enough that it can thermally decompose or isomerize. **Thermal decomposition** occurs when the excited reactant dissociates into two or more products. **Isomerization** occurs when the excited reactant changes chemical structure, but not composition or molecular weight.

An example of a bimolecular **thermal decomposition reaction** is

$$N_2O_5(g) + M \rightarrow \dot{N}O_2(g) + N\dot{O}_3(g) + M \quad (1.2)$$

Dinitrogen pentoxide, Nitrogen dioxide, Nitrate radical

where M, which can be any molecule, provides the collisional energy. Because molecular oxygen $[O_2(g)]$ and nitrogen $[N_2(g)]$ together comprise more than 99 percent of the gas molecules in the air today, M is most likely to be $O_2(g)$ or $N_2(g)$.

Because M in Reaction 1.2 does not change concentration, the reaction can also be written as

$$N_2O_5(g) + \xrightarrow{M} \dot{N}O_2(g) + N\dot{O}_3(g) \quad (1.3)$$

Dinitrogen pentoxide, Nitrogen dioxide, Nitrate radical

Thermal decomposition reactions are temperature dependent. At high temperatures, they proceed faster than at low temperatures. Isomerization reactions are similar to Reaction 1.3, except that an isomerization reaction has one product, which is another form of the reactant. An example of an **isomerization reaction** is

Excited Criegee biradical   Excited formic acid   (1.4)

The bimolecular **collision reaction**, the most common type of kinetic reaction, may occur between any two chemically active reactants that collide. A prototypical collision reaction is

$$CH_4(g) + \dot{O}H(g) \rightarrow \dot{C}H_3(g) + H_2O(g) \quad (1.5)$$

Methane, Hydroxyl radical, Methyl radical, Water vapor

In some cases, bimolecular reactions result in **collision complexes** that ultimately break into products. Such reactions have the form $A + B \rightleftharpoons AB^* \rightarrow D + F$, where AB^* is a molecule that has weak bonds and is relatively unstable, and the double arrow indicates that the reaction is **reversible**.

Termolecular (involving three reactants) collision reactions are rare because the probability that three trace gases collide simultaneously and change form is not large. For descriptive purposes, however, pairs of reactions can be written as termolecular **combination reactions**. For example, the combination of the bimolecular kinetic reaction $NO_2(g) + NO_3(g) \rightleftharpoons N_2O_5(g)^*$ with the isomerization reaction $N_2O_5(g)^* + M \rightleftharpoons N_2O_5(g) + M$ gives

$$\dot{N}O_2(g) + N\dot{O}_3(g) + M \rightleftharpoons N_2O_5(g) + M \quad (1.6)$$

Nitrogen dioxide, Nitrate radical, Dinitrogen pentoxide

In this case, M is any molecule, whose purpose is to carry away energy released during the reaction. The purpose of M in Reaction 1.6 differs from its purpose in Reaction 1.2, where it provided collisional energy for the reaction. In both cases, M is usually either $N_2(g)$ or $O_2(g)$. Reactions 1.2 and 1.6 are pressure dependent because the concentration of M is proportional to the air pressure. Because M in Reaction 1.6 does not change concentration, Reaction 1.6 can also be written as

$$\dot{N}O_2(g) + N\dot{O}_3(g) \xrightarrow{M} N_2O_5(g) \quad (1.7)$$

Nitrogen dioxide, Nitrate radical, Dinitrogen pentoxide

1.5. Chemical Lifetimes

Some gases are important in the atmosphere because they persist in high concentrations since they have either a high production rate or a low removal rate. Gases are produced by either emissions or chemical reaction. Gases are removed by chemical reaction, scavenging by rain or aerosol particles, or deposition to a surface. Some gases that persist in the air for a long time due to a slow removal rate include $N_2(g)$, $CO_2(g)$, $CH_4(g)$, $N_2O(g)$, and $H_2(g)$.

Some gases are important in the atmosphere because they chemically react quickly with other gases to form one or more harmful products. For example, $OH(g)$ reacts quickly with many organic gases to produce organic radicals that ultimately contribute to ozone formation in polluted urban air.

The rate at which a chemical species is lost from the air is characterized by its *e-folding lifetime*. This is the time required for the concentration of a gas to decrease to $1/e$ its original concentration as a result of chemical reaction. It is similar to the **half-lifetime** of a gas, which is the time required for the gas's concentration to decrease to one-half its original value. The *e-folding* lifetime is a lifetime against loss only and is independent of a gas's emission rate or rates of production from other sources. It is possible to estimate the *e-folding* lifetimes of a gas against loss by an individual reaction or against all reactions.

The *e-folding* lifetime of a gas against loss by a single chemical reaction is calculated as follows. Suppose species A is lost by the photolysis reaction A + $h\nu \rightarrow$ products (e.g., Reaction 1.1). This reaction, which has one reactant, has rate coefficient J, in units of s^{-1}. The rate of change of the concentration of A, denoted by [A], which has units of molecules per cubic centimeter of air (molec. cm^{-3}), is

$$\frac{d[A]}{dt} = -J[A] \tag{1.8}$$

Integrating Equation 1.8 from concentration $[A]_0$ at time $t = 0$ to concentration [A] at time $t = h$ gives $[A] = [A]_0 e^{-Jh}$. The *e-folding* lifetime of A (denoted by τ_A) is the time h at which

$$\frac{[A]}{[A]_0} = \frac{1}{e} = e^{-Jh} \tag{1.9}$$

This occurs when the lifetime of A against loss by the reaction is

$$\tau_A = h = \frac{1}{J} \tag{1.10}$$

In the case of a collision reaction, B + C $\rightarrow$ products, which has a rate coefficient of k in units of cm^3 molec.$^{-1}$ s^{-1}, the rate of change of concentration of [B] is first estimated by writing

$$\frac{d[B]}{dt} = -k[B][C]_0 \tag{1.11}$$

where [C] is held constant (as denoted by the subscript 0). Integrating Equation 1.11 gives $[B] = [B]_0 e^{-k[C]_0 h}$. The resulting *e-folding* lifetime of B occurs when $[B]/[B]_0 = 1/e$, which occurs when

$$\tau_B = h = \frac{1}{k[C]_0} \tag{1.12}$$

The **overall chemical lifetime** of a species is determined by calculating the lifetime of the species against loss due to each individual chemical reaction or other loss process that it is involved in, and then applying

$$\tau_A = \frac{1}{\frac{1}{\tau_{A1}} + \frac{1}{\tau_{A2}} + \cdots \frac{1}{\tau_{An}}} \tag{1.13}$$

In this equation, τ_A is the overall chemical lifetime of species A and $\tau_{A1}, \ldots, \tau_{An}$ are the lifetimes of A due to loss from chemical reactions $1, \ldots, n$, respectively. Equation 1.13 can account for losses due not only to chemistry, but also to rainfall and removal by biological consumption of the species at the ground.

Example 1.1

The main loss of methane in the atmosphere is due to the chemical reaction $CH_4(g) + OH(g) \rightarrow CH_3(g) + H_2O(g)$. Calculate the e-folding lifetime of methane against loss by this reaction when the rate coefficient for this reaction is $k = 6.2 \times 10^{-15}$ cm^3 molec.$^{-1}$ s^{-1} at 298 K and the concentration of OH in the air is $[OH] = 5.0 \times 10^5$ molec. cm^{-3}. If the e-folding lifetime of methane against loss by soil bacteria metabolism at the surface is 160 years, calculate the overall lifetime of $CH_4(g)$ against loss by both processes combined.

Solution

From Equation 1.12, $\tau_{CH4} = 1/(6.2 \times 10^{-15} \times 5.0 \times 10^5) = 3.23 \times 10^8$ s = 10.2 years due to chemical reaction alone. From Equation 1.13, the overall lifetime of methane is thus $\tau_{CH4} = 1/(1/10.2 + 1/160) = 9.6$ years. The long lifetime of methane combined with its ability to warm the atmosphere per molecule make it an important contributor to global warming (Chapter 12).

1.6. Chemical Units

Throughout this text, the atmospheric abundance of a chemical will generally be quantified in terms of its mixing ratio or concentration. A **mixing ratio** is the number of molecules of a chemical per molecule of **dry air** (total air minus water vapor). Mixing ratios can be in units of **parts per trillion volume** (pptv), **parts per billion volume** (ppbv), **parts per million volume** (ppmv), parts per thousand volume, or **fraction of air**. For example, 40 ppbv of ozone gas indicates that, of every billion molecules of dry air in a given volume, 40 are ozone molecules. The mixing ratio of oxygen gas is so large that it is expressed as a fraction (0.2095) or percent (20.95 percent) of dry air.

Concentrations of a substance can be expressed in terms of either number or mass concentration. **Number concentration** is the number of molecules of a gas per unit volume of air (e.g., molec. cm^{-3}) or the number of aerosol particles per unit volume of air (e.g., particles cm^{-3}). An example of the conversion between mixing ratio and number concentration is given in Example 3.6. **Mass concentration** of a gas or particle is usually expressed in units of micrograms per cubic meter of air ($\mu g\ m^{-3}$). The conversion from mixing ratio (χ, expressed as a fraction) to mass concentration (m, expressed in $\mu g\ m^{-3}$) is

$$m = \frac{\chi m N_d z}{A} \quad (1.14)$$

where m is molecular weight of the substance of interest (g mol^{-1}), A is Avogadro's number (6.02252×10^{23} molec. mol^{-1}), N_d is the number concentration of dry air molecules (molec. cm^{-3}), and $z = 10^{12}\ \mu g\ cm^3\ g^{-1}\ m^{-3}$ is a conversion constant. Some of these parameters are defined more precisely in Section 3.4. Example 1.2 illustrates the conversion from mixing ratio to mass concentration.

Example 1.2
Convert 40 ppbv O_3(g) to μg-O_3(g) m^{-3} under typical near-surface conditions.

Solution
The number concentration of dry air molecules under near-surface conditions is approximated in Example 3.4 as $N_d = 2.55 \times 10^{19}$ molec. cm^{-3}. The molecular weight of ozone is $m = 48.0$ g mol^{-1}. Thus, from Equation 1.14, $m_{O_3} = 0.00000004 \times 48.0$ g $mol^{-1} \times 2.55 \times 10^{19}$ molec. $cm^{-3} \times 10^{12}\ \mu g\ cm^3\ g^{-1}\ m^{-3}/6.02252 \times 10^{23}$ molec. $mol^{-1} = 81.29\ \mu g$-O_3(g) m^{-3}.

1.7. Summary

In this chapter, atoms, molecules, elements, and compounds were defined. In addition, a history of the discovery of elements and compounds of atmospheric importance was given. Only a few elements, including carbon, sulfur, and certain metals, and a few solid compounds, including calcite, halite, and nitre, were identified in the ancient world. An acceleration of the discovery of elements and compounds, particularly of gases such as oxygen and nitrogen, occurred near the end of the eighteenth century. Several types of chemical reactions take place in the air, including photolysis, kinetic, thermal decomposition, isomerization, and combination reactions. The rate of a reaction depends on the reactivity and concentration of molecules. The chemical e-folding lifetime of a substance is the time required for its concentration to decrease to $1/e$ its original value and gives an indication of the reactivity of the substance. Molecules with free electrons are called free radicals and are highly reactive.

1.8. Problems

1.1. What are the main differences between gases and aerosol particles?

1.2. What compound might you expect to form on the surface of a statue made of marble or limestone (both of which contain calcium carbonate) if aqueous sulfuric acid deposits onto the statue?

1.3. Describe one experiment you could devise to isolate molecular oxygen.

1.4. What was the fundamental flaw with the theory of phlogiston?

1.5. Why did Lavoisier name oxygen as he did? Was his definition correct? Why or why not?

1.6. Is a termolecular combination reaction the result of the collision of three molecules simultaneously? Why or why not?

1.7. If the chemical e-folding lifetimes of the harmless substances A, B, and C are 1 hour, 1 week, and 1 year, respectively, and all three substances produce harmful products when they break down, which substance would you prefer to eliminate from urban air first? Why?

1.8. Match each person with a surrogate name or description of a chemical he or she discovered.

(a) Priestley	(1) "gas that is wild and dwells in out-of-the-way places"
(b) Schönbein	(2) "Poland"
(c) M. Curie	(3) "poisonous air"
(d) Baldwin	(4) "stench"
(e) Theophrastus	(5) "plaster"
(f) Paracelsus	(6) "water maker"
(g) Van Helmont	(7) "lazy gas"
(h) Balard	(8) "acid maker"
(i) Rayleigh	(9) "light bearer"
(j) D. Rutherford	(10) "to smell"

1.9. The atmosphere contained approximately 115 ppmv of anthropogenic $CO_2(g)$ in 2010. If 1 ppmv-$CO_2(g)$ is equivalent to 2184.82 Tg-C (teragrams, or 10^{12} g, of atomic carbon), calculate how many metric tonnes of quicklime [CaO(s)] were needed to fix all anthropogenic carbon dioxide into limestone that year. Assume the molecular weights of carbon, calcium, and oxygen are 12.011 g mol^{-1}, 40.078 g mol^{-1}, and 15.9994 g mol^{-1}, respectively. By what factor would the world's quicklime production need to be increased to remove all anthropogenic $CO_2(g)$ in 2010 if the 2010 quicklime production was about 2.83×10^8 tonnes-CaO(s) yr^{-1}?

1.10. Calculate the e-folding lifetime of $O(^1D)$ against the reaction $O(^1D) + N_2(g) \rightarrow O(g) + N_2(g)$ when the rate coefficient for the reaction at 298 K is $k = 2.6 \times 10^{-11}$ cm^3 molec.$^{-1}$ s^{-1} and $[N_2] = 1.9 \times 10^{19}$ molec. cm^{-3}. If the lifetime of $O(^1D)$ against a second reaction were the same as that against the first reaction, what would be the overall lifetime of $O(^1D)$ against loss by both reactions together?

1.11. Convert units of $10,670 \mu$g-CO(g) m^{-3} to ppmv-CO(g) under typical surface conditions using necessary information from Example 1.2.

Chapter 2

The Sun, the Earth, and the Evolution of the Earth's Atmosphere

Anthropogenic pollution occurs when gas and aerosol particle concentrations rise above natural, background concentrations. This chapter examines the evolution of the background atmosphere. The discussion requires a description of the sun and its origins because sunlight has affected much of the evolution of the Earth's atmosphere. The description also requires a discussion of the Earth's composition and structure because the inner Earth affects atmospheric composition through outgassing, and the crust affects atmospheric composition through exchange processes, including wind-blown soil dust, volcanic, and sea spray emissions. Earth's earliest atmosphere contained mostly hydrogen and helium. Carbon dioxide replaced these gases during the onset of the Earth's second atmosphere. Today, nitrogen and oxygen are the prevalent gases. Processes controlling the changes in atmospheric composition over time include outgassing from the Earth's interior, microbial metabolism, and atmospheric chemistry. These processes still affect the natural composition of the air today.

2.1. The Sun and Its Origin

The sun provides the energy to power the Earth. Most of the sun's energy reaching the Earth originates from the sun's surface, not from its interior. The evolution, structure, and relevant radiation emissions from the sun are discussed here.

Common theory suggests that, about 15 billion years ago (b.y.a.), all mass in the known universe was compressed into a single point, estimated to have a density of 10^9 kg m^{-3} and a temperature of 10^{12} K (Kelvin). With the **Big Bang**, this point of mass exploded, ejecting material in all directions. Aggregates of ejected material collapsed gravitationally to form the earliest stars. When temperatures in the cores of early stars reached 10 million K, **nuclear fusion** of hydrogen (H) into helium (He) and higher elements began, releasing energy that powered the stars. As early stars aged, they ultimately exploded, ejecting stellar material back into space. Table 2.1 gives the abundance of hydrogen in the universe today relative to the abundances of other interstellar elements.

About 4.6 b.y.a., some interstellar material aggregated to form a cloudy mass, the **solar nebula**, in our current solar system. The composition of the solar nebula was the same as that of 95 percent of the other stars in the universe. Gravitational collapse of the solar nebula resulted in the formation of the sun.

Today, the sun, which comprises 99.86 percent of the mass of the solar system, is divided into concentric layers, including interior and atmospheric layers. About 90 percent of the atoms in the sun are hydrogen, and 9.9 percent are helium. The remaining atoms are the other natural elements of the periodic table.

The sun's interior composition is a **plasma**, which is a fourth state of matter aside from solid, liquid, and gas, in which temperatures are so high that gas molecules break apart into positively charged atoms or molecules and negatively charged free electrons.

Table 2.1. Cosmic abundance of hydrogen relative to those of other elements

Element	Atomic mass	Abundance of H relative to element	Element	Atomic mass	Abundance of H relative to element
Hydrogen (H)	1.01	1:1	Silicon (Si)*	28.1	26,000:1
Helium (He)	4.00	14:1	Iron (Fe)*	55.8	29,000:1
Oxygen (O)	16.0	1,400:1	Sulfur (S)	32.1	53,000:1
Carbon (C)	12.0	2,300:1	Argon (Ar)	39.9	260,000:1
Neon (Ne)	20.2	10,000:1	Aluminum (Al)*	27.0	306,000:1
Nitrogen (N)	14.0	11,000:1	Calcium (Ca)*	40.1	413,000:1
Magnesium (Mg)*	24.3	24,000:1	Sodium (Na)*	23.0	433,000:1

Asterisks (*) denote rock-forming elements. All other elements vaporize more readily.
Source: Adapted from Goody (1995).

The sun's interior structure consists of a core, an intermediate interior (radiative zone), and a hydrogen convection zone (Figure 2.1). The photosphere, which is primarily gaseous, is a transition region between the sun's atmosphere and its interior. Beyond the photosphere, the sun's atmosphere consists of the chromosphere, the corona, and solar wind discharge.

The sun has an effective radius (R_p) of 696,000 km, or 109 times the radius of the Earth (6,378 km). The sun's effective radius is the distance between the center of the sun and the top of the photosphere. The mass of the sun is about 1.99×10^{30} kg, or 333,000 times the mass of the Earth (5.98×10^{24} kg). The sun's surface gravity at the top of the photosphere is about 274 m s^{-2}, or 28 times that of the Earth (9.8 m s^{-2}).

The **sun's core** lies between its center and about 0.25 R_p. Temperatures in the core reach 15 million K. At these temperatures, electrons are stripped from hydrogen atoms. The remaining nuclei (single protons) collide in intense **nuclear fusion reactions**, producing helium and releasing energy in the form of photons of radiation that power the sun. Energy from the core radiates to the **intermediate interior**, or **radiative zone**, which has a thickness of about 0.61 R_p and

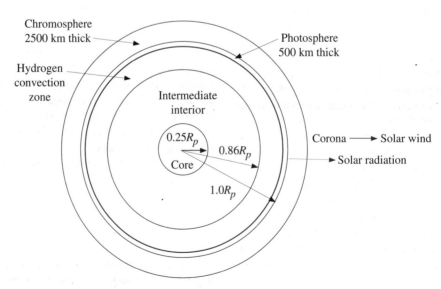

Figure 2.1. Structure of the sun.

a temperature ranging from 8 million K at its base to about 5 million K at its top. Energy transfer through the intermediate interior is also radiative. Ionized H and He atoms emit photons that are quickly absorbed by other ionized H and He atoms, which subsequently radiate photons themselves.

The next layer is the **hydrogen convection zone** (**HCZ**), a region between 0.14 R_p and 0.3 R_p in which convection of hydrogen atoms due to buoyancy takes over from radiation as the predominant mechanism of transferring energy toward the sun's surface. Temperatures at the base of the HCZ are around 5 million K; temperatures at its top are near 6,400 K. In the HCZ, a strong temperature gradient exists and the **mean free path** (average distance between collisions) between photons and hydrogen or helium atoms decreases with increasing distance from the core. A photon of radiation emitted from the core of the sun takes about 10 million years to reach the top of the HCZ.

Above the HCZ is the **photosphere** ("light sphere"), which is a relatively thin (500-km-thick) transition region between the sun's interior and its atmosphere. Temperatures in the photosphere range from 6,400 K at its base to 4,000 K at its top and average 5,785 K. The photosphere is the source of most solar energy that reaches the planets, including the Earth. Although the sun's interior is much hotter than is its photosphere, most energy produced in its interior is confined by the HCZ.

Above the photosphere lies the **chromosphere** ("color sphere"), which is a 2,500-km-thick region of hot gases. Temperatures at the base of the chromosphere are around 4,000 K. Those at the top are up to 1 million K. The name chromosphere arises because, at the high temperatures found in this region, hydrogen is energized and decays back to its ground state, emitting wavelengths of radiation in the visible part of the solar spectrum. For example, hydrogen decay results in radiation emission at 0.6563 μm, which is in the red part of the spectrum, giving the chromosphere a characteristic red coloration observed during solar eclipses.

The **corona** is the outer shell of the solar atmosphere and has an average temperature of about 1 to 2 million K. Because of the high temperature, all gases in the corona, particularly hydrogen and helium, are ionized. A low-concentration steady stream of these ions as well as electrons escapes the corona and the sun's gravitational field and propagates through space, intercepting the planets with speeds ranging from 300 to 1,000 km s^{-1}. This stream is called the **solar wind**.

Figure 2.2. Aurora Australis, as seen from Kangaroo Island, southern Australia. Photo by David Miller, National Geophysical Data Center, available from NOAA Central Library; www.photolib.noaa.gov.

The solar wind is the outer boundary of the corona and extends from the chromosphere to the outermost reaches of the solar system.

The **Earth–sun distance** (R_{es}) is about 150 million km. At the Earth, the solar wind temperature is about 200,000 K, and the number concentration of solar wind ions is a few to tens per cubic centimeter of space. As the solar wind approaches the Earth, the Earth's magnetic fields bend the path of the wind toward the poles. In the atmosphere above these regions, the ionized gases collide with air molecules, creating luminous bands of streaming, colored lights. In the Northern Hemisphere, these lights are called the **Northern Lights** or **Aurora Borealis** ("northern dawn" in Latin), and in the Southern Hemisphere, they are called the **Southern Lights** or **Aurora Australis** ("southern dawn"). Green or brownish-red colors are due to collisions of the solar wind with oxygen in the atmosphere. Red or blue colors are due to collisions with nitrogen. These lights, one of the seven natural wonders of the world, can be seen at high latitudes, such as in northern Scotland, Scandinavia, and parts of Canada in the Northern Hemisphere and in southern Australia and Argentina in the Southern Hemisphere (e.g., Figure 2.2).

2.2. Spectra of the Radiation of the Sun and the Earth

Solar radiation provides essential energy for heating the Earth and for driving the chemistry of the Earth's atmosphere. The Earth's temperatures are also controlled by infrared emissions from the Earth itself. Here, the radiation spectra of both the sun and the Earth are described.

Radiation is the emission or propagation of energy in the form of a photon or an electromagnetic wave. Whether radiation is considered a photon or a wave is still debated. A **photon** is a particle or quantum of energy that has no mass, no electric charge, and an indefinite lifetime. An **electromagnetic wave** is a disturbance traveling through a medium, such as air or space, that transfers energy from one object to another without permanently displacing the medium itself.

Because radiative energy can be transferred even in a vacuum, it is not necessary for gas molecules to be present for radiative energy transfer to occur. Thus, such transfer can occur through space, where few gas molecules exist, or through the Earth's atmosphere, where many molecules exist.

Radiation is emitted by all bodies in the universe that have a temperature above absolute zero (0 K). During emission, a body releases electromagnetic energy at different wavelengths, where a **wavelength** is the difference in distance between two adjacent peaks (or troughs) in a wave. The intensity of emission from a body varies with wavelength, temperature, and efficiency of emission. Bodies that emit radiation with perfect efficiency are called blackbodies. A **blackbody** is a body that absorbs all radiation incident upon it. No incident radiation is reflected by a blackbody. No bodies are true blackbodies, although the Earth and the sun are close, as are black carbon, platinum black, and black gold. The term blackbody was coined because good absorbers of visible radiation generally appear black. However, good absorbers of infrared radiation are not necessarily black. For example, one such absorber is white oil-based paint.

Bodies that absorb radiation incident upon them with perfect efficiency also emit radiation with perfect efficiency. The wavelength of peak intensity of emission of a blackbody is inversely proportional to the absolute temperature of the body. This law, called **Wien's displacement law**, was derived in 1893 by German physicist Wilhelm Wien (1864–1928). Wien's law states

$$\lambda_p \,(\mu m) \approx \frac{2{,}897}{T(K)} \qquad (2.1)$$

where λ_p is the wavelength (in micrometers, μm) of peak blackbody emission, and T is the absolute temperature (K) of the body. In 1911, Wien received a Nobel Prize for his discovery.

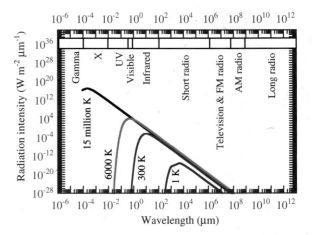

Figure 2.3. Blackbody radiation emission versus wavelength at four temperatures. Units are watts (joules of energy per second) per square meter of area per micrometer wavelength. The 15 million K spectrum represents emission from the sun's center (most of which does not penetrate to the sun's exterior). The 6,000 K spectrum represents emission from the sun's surface (photosphere) received at the top of the Earth's atmosphere (not at its surface). The 300 K spectrum represents emission from the Earth's surface. The 1 K spectrum is almost the coldest temperature possible (0 K).

Example 2.1
Calculate the peak wavelength of blackbody radiative emission for both the sun and the Earth.

Solution
The effective temperature of the sun's photosphere is 5,785 K. Thus, from Equation 2.1, the peak wavelength of the sun's emissions is about 0.5 μm. The average surface temperature of the Earth is 288 K, giving the Earth a peak emission wavelength of about 10 μm.

At any wavelength, the intensity of radiative emission from an object increases with increasing temperature. Thus, hotter bodies (e.g., the sun) emit radiation more intensely than do colder bodies (e.g., the Earth). Figure 2.3 shows radiation intensity versus wavelength for blackbodies at four temperatures. At 15 million K, a temperature at which nuclear fusion reactions occur in the sun's center, **gamma radiation** wavelengths (10^{-8} to 10^{-4} μm) and **X radiation** wavelengths (10^{-4} to 0.01 μm) are the wavelengths emitted with greatest intensity. At 6,000 K, **visible** wavelengths (0.38 to 0.75 μm) are the most intensely emitted wavelengths, although shorter **ultraviolet (UV)** wavelengths

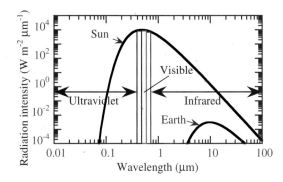

Figure 2.4. Radiation spectrum as a function of wavelength for the sun's photosphere and the Earth when both are considered blackbodies. The sun's spectrum is received at the top of the Earth's atmosphere.

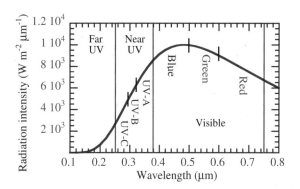

Figure 2.5. UV and visible portions of the solar spectrum. This spectrum is received at the top of the Earth's atmosphere.

(0.01 to 0.38 μm), and longer **infrared (IR)** wavelengths (0.75 to 1,000 μm) are also emitted. At 300 K, infrared wavelengths are the wavelengths emitted with greatest intensity.

Figure 2.4 focuses on the 6,000 K and 300 K spectra in Figure 2.3. These are the radiation spectra of the sun's photosphere and of the Earth, respectively. The **solar spectrum** can be divided into the UV, visible, and IR spectra. The UV spectrum can be further divided into the **far UV** (0.01 to 0.25 μm) and **near UV** (0.25 to 0.38 μm) spectra (Figure 2.5). The near UV spectrum is subdivided into **UV-A** (0.32 to 0.38 μm), **UV-B** (0.29 to 0.32 μm), and **UV-C** (0.25 to 0.29 μm) wavelength regions. The visible spectrum contains the colors of the rainbow. For simplicity, visible light is categorized as **blue** (0.38 to 0.5 μm), **green** (0.5 to 0.6 μm), or **red** (0.6 to 0.75 μm). Infrared wavelengths are partitioned into **solar-IR (near-IR)** (0.75 to 4 μm) and **thermal-IR (far-IR)** (4 to 1,000 μm) wavelengths. The intensity of the sun's emission is strongest in the visible spectrum, weaker in the the solar-IR and UV spectra, and weakest in the thermal-IR spectrum. That of the Earth's emission is strongest in the thermal-IR spectrum.

Figures 2.3 to 2.5 provide wavelength dependencies of the intensity of radiation emissions of a body at a given temperature. Integrating intensity over all wavelengths (summing the area under any of the curves) gives the total intensity of emission of a body at a given temperature. This intensity is proportional to the fourth power of the object's kelvin temperature (T) and is given by the **Stefan-Boltzmann law**, derived empirically in 1879 by Austrian physicist **Josef Stefan** (1835–1893) and theoretically in 1889 by Austrian physicist **Ludwig**

Boltzmann (1844–1906). The law states

$$F_b = \varepsilon \sigma_B T^4 \qquad (2.2)$$

where F_b is the radiation intensity (W m^{-2}), summed over all wavelengths, emitted by a body at temperature T, ε is the emissivity of the body, and $\sigma_B = 5.67 \times 10^{-8}$ W m^{-2} K^{-4} is the Stefan-Boltzmann constant. The **emissivity**, which ranges from 0 to 1, is the efficiency at which a body emits radiation in comparison with the emissivity of a blackbody, which is unity. Soil has an emissivity of 0.9 to 0.98, and water has an emissivity of 0.92 to 0.97. All the curves in Figures 2.3 to 2.5 show emission spectra for blackbodies ($\varepsilon = 1$).

Example 2.2

How does doubling the Kelvin temperature of a blackbody change the intensity of radiative emission of the body? What is the ratio of intensity of the sun's radiation compared with that of the Earth's?

Solution

From Equation 2.2, the doubling of the Kelvin temperature of a body increases its intensity of radiative emission by a factor of 16. The temperature of the sun's photosphere (5,785 K) is about twenty times that of the Earth (288 K). Assuming both are blackbodies ($\varepsilon = 1$), the intensity of the sun's radiation (63.5 million W m^{-2}) is 163,000 times that of the Earth's (390 W m^{-2}).

2.3. Primordial Evolution of the Earth and Its Atmosphere

Earth formed when rock-forming elements (identified in Table 2.1), present as gases at high temperatures

Figure 2.6. The asteroid Ida and its moon, Dactyl, taken by the Galileo spacecraft as it passed within 10,878 km of the asteroid on August 28, 1993. Available from National Space Science Data Center, http://nssdc.gsfc.nasa.gov.

vaporizing. Meteorite bombardment was intense for about 500 million years. Although the solar nebula has since cooled and most of it has been converted to solar or planetary material or has been swept away from the solar system, some planetary growth continues today, as leftover asteroids and meteorites occasionally strike the planets.

Meteorites are classified as iron, stony, or stony iron. **Iron meteorites** contain 90 to 95 percent iron, with nickel making up most of the rest. They originate from the exploded core of an earlier planet or asteroid. **Stony meteorites** by far comprise the largest number of meteorites and originate from the outer crust of an exploded planet or asteroid. Table 2.2 shows the average composition of stony meteorites, the total Earth, and the Earth's continental and oceanic crusts. The table indicates that stony meteorite composition is relatively similar to that of the total Earth, supporting the theory that stony meteorites played a role in the Earth's formation. **Stony iron meteorites** comprise only ~2 percent of all meteorites and contain roughly equal amounts of iron-nickel and other material. They originate from the middle of exploded planets or asteroids.

in the solar nebula, condensed into small solid grains as the nebula cooled. The grains grew by collision to centimeter-sized particles. Additional grains accreted onto the particles, resulting in **planetesimals**, which are small-body precursors to planet formation. Accretion of grains and particles onto planetesimals resulted in the formation of **asteroids** (Figure 2.6), which are rocky bodies 1 to 1,000 km in size that orbit the sun. Today, asteroids orbit primarily in the asteroid belt between Mars and Jupiter. Asteroids collided to form the planets.

The growth of planets was aided by the bombardment of **meteorites**, which are 1-mm- to 1-km-wide solid minerals or rocks that reach the planet's surface without

Meteorites and asteroids consist partly of **rock-forming elements** (e.g., Mg, Si, Fe, Al, Ca, Na, Ni, Cr, Mn) that condensed from the gas phase to the liquid phase before solidifying in the cooling solar nebula. They also consist of **noncondensable elements** (e.g., H, He, O, C, Ne, N, S, Ar, P), which could not condense in elemental form. How did noncondensable elements enter meteorites and asteroids, particularly as they were too light to attract to these bodies gravitationally? One theory is that noncondensable elements may have

Table 2.2. Mass percentages of major elements in stony meteorites, total Earth, and Earth's continental and oceanic crusts

Element	Stony meteorites	Total Earth	Earth's continental crust	Earth's oceanic crust
Oxygen (O)	33.24	29.50	46.6	45.4
Iron (Fe)	27.24	34.60	5.0	6.4
Silicon (Si)	17.10	15.20	27.2	22.8
Magnesium (Mg)	14.29	12.70	2.1	4.1
Sulfur (S)	1.93	1.93	0.026	0.026
Nickel (Ni)	1.64	2.39	0.075	0.075
Calcium (Ca)	1.27	1.13	3.6	8.8
Aluminum (Al)	1.22	1.09	8.1	8.7
Sodium (Na)	0.64	0.57	2.8	1.9

Source: Adapted from Cattermole and Moore (1985).

chemically reacted as gases in the solar nebula to form high-molecular-weight compounds that were condensable, although less condensable (more volatile) than rock-forming elements. When meteorites and asteroids collided with the Earth, they brought with them volatile compounds and rock-forming elements. Whereas some of the volatiles vaporized on impact, others have taken longer to vaporize and have been outgassed ever since through volcanos, fumaroles, steam wells, and geysers.

Earth's first atmosphere likely contained primarily hydrogen (H) and helium (He), the most abundant elements in the solar nebula. During the formation of the Earth, the sun was also forming. Early stars are known to blast off a large amount of gas into space. This outgassed solar material, the **solar wind**, was previously introduced as an extension of the sun's corona. During the birth of the sun, nuclear reactions in the sun that fuse hydrogen to helium were enhanced. The resulting blast increased solar wind speeds and densities to much higher values than today. This early stage of the sun is called the **T-Tauri stage** after the first star observed at this point in its evolution. The enhanced solar wind is believed to have stripped away the first atmosphere not only of the Earth, but also of all other planets in the solar system. Additional H and He were lost from the Earth's first atmosphere after escaping the Earth's gravitational field. As a result of these two loss processes (solar wind stripping and gravitational escape), the ratios of H and He to other elements in the Earth's atmosphere today are less than are the corresponding ratios in the sun.

2.3.1. Solid Earth Formation

The rock-forming elements that reached the Earth reacted to form compounds, each with different melting points, densities, and chemical reactivities. Dense and high melting point compounds, including many iron- and nickel-containing compounds, settled to the center of the Earth, called the **Earth's core**. Table 2.2 shows that the total Earth today contains more than 34 percent iron and 2 percent nickel by mass, but the Earth's **crust** (its top layer) contains less than 7 percent iron and 0.1 percent nickel by mass, supporting the contention that iron and nickel settled to the core. Low-density compounds and compounds with low melting points, including silicates of aluminum, sodium, and calcium, rose to the surface and are the most common compounds in the Earth's crust. Table 2.2 supports this hypothesis. Some moderately dense and moderately high-melting-point silicates, such as those containing magnesium or iron, settled to the Earth's **mantle**, which is a layer of Earth's interior between its crust and its core.

During the formation of the Earth's core, between 4.6 and 4.0 b.y.a., core temperatures were higher than they are today, and the only mechanism of heat escape to the surface was **conduction**, which is the transfer of energy from molecule to molecule. Because conduction is a slow process, the Earth's internal energy could not transfer to the surface and dissipate easily, so its temperature increased until the entire body became molten. In this state, the Earth's surface consisted of **magma oceans**, a hot mixture of melted rock and suspended crystals. When the Earth was molten, **convection**, the mass movement of molecules, became the predominant form of vertical energy transfer from the core to the surface. Convection occurred because temperatures in the core were hot enough for core material to expand and float to the crust, where it cooled and sank down again. This process enhanced energy dissipation from the Earth's core to space. After sufficient energy dissipation (cooling), the magma oceans solidified, creating the Earth's crust. The crust is estimated to have formed 3.8 to 4.0 b.y.a., but possibly as early as 4.2 to 4.3 b.y.a. (Crowley and North, 1991). The core cooled as well, but its outer part, the **outer core**, remains molten. Its inner part, the **inner core**, is solid.

Figure 2.7 shows temperature, density, and pressure profiles inside the Earth today. The Earth's crust extends from the topographical surface to about 10 to 75 km below continents and 8 km below the ocean floor. The crust itself contains low-density, low-melting-point silicates. The continental crust contains primarily granite, whereas the ocean crust contains primarily basalt. **Granite** is a type of rock composed mainly of quartz [$SiO_2(s)$] and potassium feldspar [$KAlSi_3O_8(s)$]. **Basalt** is a type of rock composed primarily of plagioclase feldspar [$NaAlSi_3O_3$–$CaAl_2Si_2O_8(s)$] and pyroxene (multiple compositions). The densities of both granite and basalt are about 2,800 kg m^{-3}.

Below the Earth's crust is its mantle, which consists of an upper and lower part, both made of iron-magnesium-silicate minerals. The upper mantle extends from the crust down to about 700 km. At that depth, a density gradation occurs due to a change in crystal packing. This gradation roughly defines the base of the upper mantle and the top of the lower mantle. Below 700 km, the density gradually increases to the mantle core boundary at 2,900 km.

The outer core extends from 2,900 km down to about 5,100 km. This region consists of liquid iron and nickel, although the top few hundred kilometers contain liquids and crystals. The inner core extends from 5,100 km down to the Earth's center and is solid, also consisting of iron and nickel, but packed at a higher

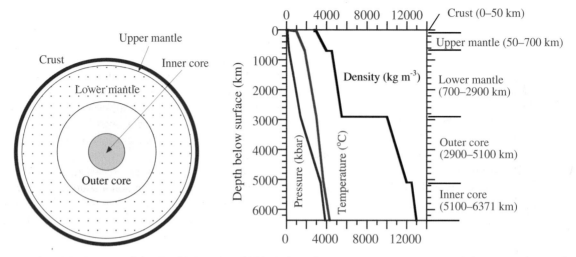

Figure 2.7. (a) Diagram of the Earth's interior. (b) Variation of pressure, temperature, and density in the Earth's interior.

density. Temperatures, densities, and pressures at the center of the Earth are estimated to be 4,300°C, 13,000 kg m^{-3}, and 3,850 kbar, respectively (1 kbar = 1,000 bar = 10^8 N m^{-2} = 10^8 Pa; for comparison, surface air pressures are about 1 bar).

2.3.2. Prebiotic Atmosphere

Earth's second atmosphere evolved as a result of outgassing from the Earth's mantle. As temperatures increased during the molten stage, hydrogen and oxygen, bound in crustal minerals as hydroxyl molecules (OH), became detached, forming the gas-phase hydroxyl radical. The hydroxyl radical then reacted with reduced gases, such as molecular hydrogen [$H_2(g)$], methane [$CH_4(g)$], ammonia [$NH_3(g)$], molecular nitrogen [$N_2(g)$], and hydrogen sulfide [$H_2S(g)$], to form oxidized gases, such as water [$H_2O(g)$], carbon monoxide [$CO(g)$], carbon dioxide [$CO_2(g)$], nitrogen dioxide [$NO_2(g)$], and sulfur dioxide [$SO_2(g)$]. As the molten rock rose to the Earth's surface during convection, oxidized and reduced gases were ejected into the air by **volcanos**, **fumaroles** (vents near volcanos), **steam wells** (vents near geothermal reservoirs), and **geysers** (springs near volcanos with intermittent ejecting of water and vapor).

After the crust and mantle solidified, outgassing continued. The resulting secondary atmosphere contained no free elemental oxygen. All oxygen was tied up in oxidized molecules. Indeed, if any free oxygen did exist, it would have been removed by chemical reaction.

Outgassed water vapor in the air condensed to form the oceans, starting around 4 b.y.a. Most outgassing occurred early on, and the new ocean was hot. Sizable oceans have been present during almost all the Earth's history (Pollack and Yung, 1980). In 1955, geochemist William W. Rubey (1898–1974) calculated that all the water in the Earth's **oceans** and atmosphere could be accounted for by the release of water vapor from volcanos that erupted throughout the Earth's history. The oceans are a critical part of today's hydrologic cycle. In this cycle, ocean water evaporates, the vapor is transported and condenses to form clouds, rain from clouds precipitates back to land or ocean surfaces, and water on land flows gravitationally back to the oceans.

2.3.3. Biotic Atmosphere before Oxygen

Table 2.3 shows an approximate timeline of important steps during the evolution of the Earth's atmosphere. Living microorganisms have been responsible for most of the changes.

Living microorganisms first developed about 3.5 b.y.a. from **amino acids**, the building blocks of life. Amino acids evolved by **abiotic synthesis**, the process by which life is created from chemical reactions and electrical discharges in the absence of oxygen. Abiotic synthesis was first hypothesized by Oparin (1938), and then demonstrated in 1953 by the American chemist Stanley Miller (1930–2007) while working in the laboratory of Harold Urey (1893–1981). Miller discharged electricity (simulating lightning) through a flask containing $H_2(g)$, $H_2O(g)$, $CH_4(g)$, $NH_3(g)$, and boiling water. He let the bubbling mixture sit for a week and, after analyzing the results, found that he had produced complex organic molecules, including the amino acids

Table 2.3. Timeline of evolution on Earth and its atmosphere

Billion years ago	Event
4.6	Formation of Earth
4.6–4.4	Earth's first atmosphere with H, He
4.3–3.8	Earth's crust formation and outgassing
4.0	$NH_3(g)$ photolysis to form $N_2(g)$ becomes important
3.5	Abiotic synthesis of amino acids
3.5	First prokaryotic bacteria
3.5	Fermenting bacteria produce $CO_2(g)$
3.5	Anoxygenic photosynthesis by sulfur cyanobacteria
3.5–2.8	Oxygenic photosynthesis by cyanobacteria
3.2	Dentrifying bacteria produce $N_2(g)$
2.9	Methanogenic bacteria produce $CH_4(g)$
2.45	Great Oxygenation Event
2.1–1.85	First eukaryotic bacteria evolve
1.8	Nitrifying bacteria convert $NH_3(g)$ to NO_3^-
1.5	Nitrogen-fixing bacteria convert $N_2(g)$ to $NH_3(g)$
0.57	First shelled invertebrates
0.43–0.5	First primitive fish
0.395–0.43	First land plants

radiative conditions. For example, one strain of bacteria found today, *Strain 121*, lives near a hydrothermal vent in the northeast Pacific Ocean and can grow at a temperature of up to 121°C. Another strain in Siberian permafrost can grow at a temperature as low as −10°C. *Ferroplasma acidarmanus* lives in acidic waters with a pH near 0 at Iron Mountain, California, and *Deinococcus radiodurans* can survive a radiation dose one thousand times higher than that lethal to humans (Cockell et al., 2007).

Living organisms today are classified according to their energy source, electron donor source, and carbon source (Table 2.4). Organisms that obtain their energy from sunlight are referred to as **phototrophs**. Those that obtain their energy from oxidation of a chemical are **chemotrophs**. Organisms that use inorganic compounds, such as carbon dioxide [$CO_2(g)$], molecular hydrogen [$H_2(g)$], hydrogen sulfide [$H_2S(g)$], water [$H_2O(g)$], the ammonium ion [NH_4^+], or the nitrite ion [NO_2^-], as electron donors are **lithotrophs**. Those that use organic compounds as electron donors are **organotrophs**. Organisms that obtain their carbon from $CO_2(g)$ are **autotrophs**. Those that obtain their carbon from organic compounds are **heterotrophs**.

Most cyanobacteria, green plants, and algae obtain their energy from sunlight, their carbon from carbon dioxide, and use water as the electron donor, and thus are **photolithotrophic autotrophs** (Figure 2.8). Some

glycine, *a-alanine*, and *b-alanine* (Miller, 1953). The amino acids could not have been produced from existing life because the boiling water and the presence of acids during the experiment would have extinguished such life. Later experiments showed that the same results could be obtained with different gases and with UV radiation as well. In all cases, much of the initial gas had to be highly reduced (Miller and Orgel, 1974).

Thus, in the prebiotic atmosphere, the amino acids required for the production of deoxyribonucleic acid (DNA) were first developed. About 3.5 b.y.a., the first microscopic cells containing DNA evolved. These **prokaryotic cells** contained a single strand of DNA but no nucleus. Today, prokaryotic microorganisms include many bacteria and blue-green algae.

Prehistoric bacteria were able to adapt to the evolving, harsh climate of the early Earth just as many bacteria today exist under extreme temperature, acidity, or

Figure 2.8. Hot sulfur springs in Lassen National Park, California. Boiling water of geothermal origin is rich in hydrogen sulfide. Photolithotrophic autotrophic bacteria thrive in the springs and oxidize the hydrogen sulfide to sulfuric acid, which dissolves the surrounding mineral and converts part of the spring into a "mud pot." The steam contains mostly water vapor, but hydrogen sulfide, sulfuric acid, and other gases are also present. Courtesy Alfred Spormann, Stanford University.

Table 2.4. Classification and examples of organisms in terms of their energy and carbon sources

Energy source	Electron donor	Carbon source	Classification	Examples
Sunlight	Inorganic compounds	Carbon dioxide	Photolithotrophic autotroph	Green plants; most algae; most cyanobacteria; purple, blue, and green sulfur bacteria
		Organic compounds	Photolithotrophic heterotroph	*Chlorella, Chlamydomonas*
	Organic compounds	Carbon dioxide	Photoorganotrophic autotroph	Some purple nonsulfur bacteria
		Organic compounds	Photoorganotrophic heterotroph	Some algae, most purple and green nonsulfur bacteria, some cyanobacteria
Chemicals	Inorganic compounds	Carbon dioxide	Chemolithotrophic autotroph	Hydrogen, colorless sulfur, methanogenic, nitrifying, iron bacteria
		Organic compounds	Chemolithotrophic heterotroph	Some colorless sulfur bacteria
	Organic compounds	Carbon dioxide	Chemoorganotrophic autotroph	*Pseudomonas oxalaticus*
		Organic compounds	Chemoorganotrophic heterotroph	Animals, most bacteria, fungi, protozoa

algae, purple and green nonsulfur bacteria, and some cyanobacteria obtain their energy from sunlight and their electrons and carbon from organic material, and thus are **photoorganotrophic heterotrophs**.

Hydrogen, colorless sulfur, methanogenic, nitrifying, and iron bacteria obtain their carbon from carbon dioxide and their energy and electrons from inorganic material, and thus are **chemolithotrophic autotrophs**. Animals, most bacteria, all fungi, and all protozoa obtain their energy, electrons, and carbon from organic material, and thus are all **chemoorganotrophic heterotrophs**. Many early prokaryotes were chemoorganotrophic heterotrophs because they obtained their energy, electrons, and carbon from organic molecules produced during abiotic synthesis.

2.3.3.1. Early Carbon Dioxide

The first major energy-producing process carried out by newly evolved prokaryotic bacteria was **fermentation**. This process developed in chemoorganotrophic heterotrophic organisms and produced **carbon dioxide** gas. One fermentation reaction is

$$C_6H_{12}O_6(aq) \rightarrow 2C_2H_5OH(aq) + 2CO_2(g) \quad (2.3)$$
$$\text{Glucose} \qquad\qquad \text{Ethanol} \qquad \text{Carbon dioxide}$$

which is exothermic (energy releasing). The energy source, glucose in this case, is only partially oxidized to carbon dioxide; thus, the reaction is inefficient. This reaction is similar to that which occurs during the production of bread, whereby fermenting fungi in yeast eat sugar added to the bread, producing ethanol and carbon dioxide. The ethanol is burned off in the oven or fire, and the carbon dioxide causes the bread to rise.

2.3.3.2. Early Methane

About 2.9 b.y.a., a new source of **methane** gas in the Earth's early atmosphere was metabolism by a new strain of bacteria called **methanogenic bacteria**. Such bacteria obtain their carbon from carbon dioxide and their energy and electrons from molecular hydrogen; thus, methanogenic bacteria are chemolithotrophic autotrophs. Their methane-producing reaction is

$$4H_2(g) + CO_2(g) \rightarrow CH_4(g) + 2H_2O(aq) \quad (2.4)$$
$$\text{Molecular hydrogen} \quad \text{Carbon dioxide} \quad \text{Methane} \quad \text{Liquid water}$$

Methanogenic bacteria use about 90 to 95 percent of carbon dioxide available to them for this process. The rest is used for synthesis of cell carbon. In Reaction 2.4, $CO_2(g)$ is reduced to $CH_4(g)$. A reaction such as this, in which cells produce energy by breaking down compounds in the absence of molecular oxygen is an **anaerobic respiration** reaction. Anaerobic ("in the absence of oxygen") respiration produces energy more efficiently than does fermentation. Even today, methanogenic bacteria are anaerobic, generally living in the intestines of cows, under rice paddies, under landfills, and in termite mounds.

In Earth's early atmosphere, methane may have had an *e*-folding lifetime as long as 10,000 years because its only loss was photolysis at very short wavelengths. Today, its lifetime is about 10 years, due to its reaction with the hydroxyl radical [OH(g)], which was hardly present in Earth's preoxygen atmosphere. The buildup of OH(g) following the buildup of oxygen caused a sudden decrease in atmospheric $CH_4(g)$, as described later in the chapter.

2.3.3.3. Early Molecular Nitrogen

When ammonia accumulated sufficiently in the air due to outgassing c. 4 b.y.a., it was photolyzed by ultraviolet sunlight to provide the major source of atmospheric **molecular nitrogen** at the time, by

$$NH_3(g) + h\nu \rightarrow \cdot\dot{N}(g) + 3\dot{H}(g) \qquad (2.5)$$

Ammonia   Atomic   Atomic
      nitrogen   hydrogen

$$\cdot\dot{N}(g) + \dot{N}(g) \xrightarrow{M} N_2(g) \qquad (2.6)$$

Atomic   Molecular
nitrogen   nitrogen

However, once atmospheric oxygen levels began to rise (2.45–0.4 b.y.a.), nitrogen production by ammonia photolysis became obsolete because oxygen absorbs the sun's ultraviolet wavelengths capable of photolyzing ammonia. Long before that, however, about 3.2 b.y.a., some anaerobic chemoorganotrophic heterotrophs developed a new mechanism of producing molecular nitrogen. In this two-step process called **denitrification**, one set of denitrifying bacteria reduce the **nitrate ion** (NO_3^-) to the **nitrite ion** (NO_2^-), and another set reduce the nitrite ion to molecular nitrogen:

$$\text{Organic compound} + NO_3^- \rightarrow CO_2(g) + NO_2^- + \cdots$$

        Nitrate   Carbon   Nitrite
        ion   dioxide   ion

$$(2.7)$$

$$\text{Organic compound} + NO_2^- \rightarrow CO_2(g) + N_2(g) + \cdots$$

        Nitrite   Carbon   Molecular
        ion   dioxide   nitrogen

$$(2.8)$$

Denitrification took over as the main source of molecular nitrogen in the air following the advent of oxygen and continues to be the main source today.

2.3.3.4. Anoxygenic Photosynthesis

Most early organisms on Earth relied on the conversion of organic or inorganic material to obtain their energy. During the microbial era, such organisms most likely lived underground or in water to avoid exposure to harmful UV radiation hitting the Earth's surface. However, about 3.5 b.y.a., certain bacteria developed the ability to obtain their energy from sunlight by a new process, **photosynthesis**. The first photosynthesizing bacteria were sulfur cyanobacteria (blue-green algae or blue-green bacteria), which are photolithotrophic autotrophs (Table 2.4). A photosynthetic reaction by blue, green, yellow, or purple sulfur cyanobacteria is

$$CO_2(g) + 2H_2S(g) + h\nu \rightarrow CH_2O(aq)$$

Carbon   Hydrogen   Carbo−
dioxide   sulfide   hydrate

$$+ H_2O(aq) + 2\dot{S}(g) \qquad (2.9)$$

Liquid   Atomic
water   sulfur

where $CH_2O(aq)$ represents a generic carbohydrate dissolved in water. Because reduced compounds [e.g., $H_2S(g)$] were not omnipresent, such bacteria flourished only in limited environments. Early **sulfur-producing photosynthesis** did not result in the production of oxygen; thus, it is referred to as **anoxygenic photosynthesis**.

2.3.4. The Oxygen Age

Oxygen-producing photosynthesis may have developed shortly after anoxygenic photosynthesis, around 3.5 to 2.8 b.y.a. However, until the onset of green plants [395 to 430 million years ago (m.y.a.)], oxygen-producing photosynthesis was carried out primarily by cyanobacteria (Figure 2.9). Prior to 2.45 b.y.a., nearly all the oxygen produced by photosynthesis was removed before it could accumulate in the air. The main removal mechanism was chemical reaction of oxygen with iron in rocks, producing **iron oxide** [$Fe_2O_3(s)$] in a process similar to rusting. Most of the oxidation occurred in ocean water, where dissolved oxygen reacted with the iron in sediments. Oxygen also reacted with sulfur to produce the sulfate ion (SO_4^{2-}), which combined further to form rock material, such as anhydrite, $CaSO_4(s)$, and gypsum, $CaSO_4–2H_2O(s)$. Today, about 58 percent

Figure 2.9. Hot spring in Yellowstone National Park, Wyoming. The hot, mineral-rich water provides ideal conditions for colored photosynthetic cyanobacteria to grow at the spring's perimeter, where the temperatures drop to about 70°C. The colors identify different photosynthetic bacteria with different temperature optima. Photo by Alfred Spormann, Stanford University.

of oxygen on Earth is bound in $Fe_2O_3(s)$ rocks, 38 percent is bound in SO_4^{2-} rocks, and 4 percent is in the air.

Around 2.45 b.y.a., rocks became saturated with oxygen, so the oxygen slowly began to accumulate in the atmosphere. Between 2.45 and 1.85 b.y.a., oxygen levels increases from near zero to 1 percent of the composition of the air. The increase in oxygen at 2.45 b.y.a. is referred to as the **Great Oxygenation Event (GOE)**. The GOE had a significant effect on Earth's climate. The increase in oxygen in the atmosphere increased the photochemical production of oxygenated chemicals, including the hydroxyl radical $[OH(g)]$ and excited atomic oxygen $[O(^1D)(g)]$. These chemicals reacted with $CH_4(g)$, a strong greenhouse gas, converting it to $CO_2(g)$, a weaker one, triggering the Huronian glaciation (Section 12.3.2.1) (Kasting and Siefert, 2002). The slight oxygen buildup also caused a mass extinction of anaerobic bacteria.

The GOE resulted in the first **ozone layer**. Oxygen plus ultraviolet radiation produces ozone high above the Earth's surface (Section 11.3), and both oxygen and ozone help shield the Earth's surface from harmful UV radiation. Molecular oxygen absorbs far UV radiation, and ozone absorbs far UV, UV-C, and a large portion of UV-B radiation. The gradually increasing abundance of oxygen and ozone in the air protected the surface of the Earth from UV radiation, allowing microorganisms to migrate to the top of the oceans and soil, ultimately helping land plants to develop 430 to 395 m.y.a.

Around 1.85 b.y.a., oxygen levels reached about 1 percent of those today. Oxygen levels stayed relatively constant at 1 percent for the next billion years, possibly because a new equilibrium had been reached between oxygen production by photosynthesis and consumption by recently evolved **eukaryotic bacteria** (Section 2.3.5). About 850 m.y.a., freshwater green algae evolved from some photosynthesizing eukaryotic organisms, increasing oxygen levels further. However, not until 430 to 395 m.y.a. did **land plants** evolve from algae, increasing oxygen levels more rapidly. Like cyanobacteria and algae, plants photosynthesized to produce oxygen. The spread of land plants resulted in oxygen levels rising to those similar to today's, where 21 of every 100 molecules in the air are molecular oxygen.

Oxygen-producing photosynthesis in plants is similar to that in bacteria. In both cases, $CO_2(g)$ and sunlight are required, and reactions occur in **chlorophylls**. Chlorophylls reside in photosynthetic membranes. In bacteria, the membranes are cell membranes; in plants and algae, photosynthetic membranes are found in **chloroplasts**.

Chlorophylls are made of **pigments**, which are organic molecules that absorb visible light. Plant and tree leaves generally contain two pigments, chlorophyll *a* and *b*, both of which absorb blue wavelengths (shorter than 500 nm) and red wavelengths (longer than 600 nm) of visible light. Chlorophyll *a* absorbs red wavelengths more efficiently than does chlorophyll *b*, and chlorophyll *b* absorbs blue wavelengths more efficiently than does chlorophyll *a*. Because neither chlorophyll absorbs in the green part of the visible spectrum (500–600 nm), chlorophyll reflects green wavelengths, giving leaves a green color. Photosynthetic bacteria generally appear purple, blue, green, or yellow, indicating that their pigments do not absorb purple, blue, or green, respectively, but do absorb other colors.

The oxygen-producing photosynthesis process in green plants is

$$6CO_2(g) + 6H_2O(aq) + h\nu \rightarrow C_6H_{12}O_6(aq)$$
$$\text{Carbon dioxide} \quad \text{Liquid water} \quad \text{Glucose}$$

$$+ 6O_2(g) \qquad\qquad (2.10)$$
$$\text{Molecular oxygen}$$

where the result, glucose, is dissolved in water in the photosynthetic membrane of the plant. The source of molecular oxygen during photosynthesis in green plants is not carbon dioxide, but water. This can be seen by first dividing Reaction 2.10 by 6, and then adding water to each side of the equation. The result is

$$CO_2(g) + 2H_2O(aq) + h\nu \rightarrow CH_2O(aq)$$

Carbon   Liquid   Carbo−
dioxide   water   hydrate

$$+ H_2O(aq) + O_2(g) \quad\quad (2.11)$$

Liquid   Molecular
water   oxygen

A comparison of Reaction 2.11 with Reaction 2.9 indicates that because the source of atomic sulfur in Reaction 2.9 is hydrogen sulfide, the analogous source of oxygen in Reaction 2.11 should be water. This was first hypothesized in 1931 by Cornelius B. Van Niel, a Dutch microbiologist working at Stanford University, and later proved to be correct experimentally with the use of isotopically labeled water.

2.3.5. Aerobic Respiration and the Oxygen Cycle

The atmospheric production of molecular oxygen and ozone 2.45 b.y.a. resulted in biological changes in organisms that shaped our present atmosphere. Most important was the development of **aerobic respiration**, which is the process by which molecular oxygen reacts with organic cell material to produce energy during cellular respiration. **Cellular respiration** is the oxidation of organic molecules in living cells.

Whereas aerobic respiration may have developed first in prokaryotes (bacteria and blue-green algae), its spread coincided with the rise of another type of organism, the **eukaryote**, about 2.1 to 1.85 b.y.a. A eukaryotic cell contains DNA surrounded by a true membrane-enclosed nucleus. This differs from a prokaryotic cell, which contains a single strand of DNA but not a nucleus. Unlike prokaryotes, many eukaryotes became multicellular. Today, the cells of all higher animals, plants, fungi, protozoa, and most algae are eukaryotic. Prokaryotic cells never evolved past the microbial stage.

Almost all eukaryotic cells respire aerobically. In fact, such cells usually switch from fermentation to aerobic respiration when oxygen concentrations reach about 1 percent of the present oxygen level (Pollack and Yung, 1980). Thus, eukaryotic cells probably developed substantially only when oxygen's atmospheric mixing ratio increased to 1 percent of its present level. This occurred about 1.85 b.y.a., after oxygen had became substantially saturated in rocks and had started to accumulate in the atmosphere.

The products of aerobic respiration are carbon dioxide and water. Aerobic respiration of glucose, a typical

Table 2.5. Sources and sinks of atmospheric molecular oxygen

Sources	Sinks
Photosynthesis by green plants and cyanobacteria	Photolysis and kinetic reaction
	Aerobic respiration
Atmospheric chemical reaction	Dissolution into ocean water
	Rusting
	Chemical reaction on soil and rock surfaces
	Fossil fuel, biofuel, and biomass burning

cell component, occurs by

$$C_6H_{12}O_6(aq) + 6O_2(g) \rightarrow 6CO_2(g) + 6H_2O(aq)$$

Glucose   Molecular   Carbon   Liquid
          oxygen   dioxide   water

$$(2.12)$$

This process produces energy more efficiently than does fermentation or anaerobic respiration. Thus, Reaction 2.12 was an evolutionary improvement.

Table 2.5 summarizes the current sources and sinks of $O_2(g)$. The primary source is photosynthesis. The major sinks are photolysis in and above the stratosphere and aerobic respiration.

2.3.6. The Nitrogen Cycle

The development of aerobic respiration hastened the evolution of organisms affecting the nitrogen cycle. This cycle centers on molecular nitrogen [$N_2(g)$], which comprises about 78 percent of total air by volume today. Figure 2.10 summarizes the major processes in the nitrogen cycle. Four of the five processes are carried out by bacteria in soils. The fifth involves nonbiological chemical reactions in the air.

The direct source of almost all molecular nitrogen in the air today is **denitrification**, the two-step process carried out by anaerobic bacteria in soils that was described by Reactions 2.7 and 2.8. Denitrification evolved around 3.2 b.y.a., prior to the buildup of oxygen. Whereas the second step of denitrification can produce either nitric oxide [$NO(g)$], nitrous oxide [$N_2O(g)$], or $N_2(g)$, $N_2(g)$ is the dominant product. $N_2(g)$ is also produced chemically from $N_2O(g)$, which can form from $NO(g)$.

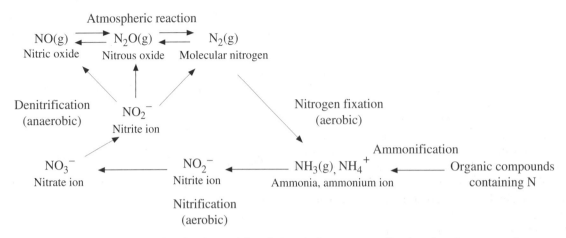

Figure 2.10. Diagram showing bacterial and chemical processes affecting the nitrogen cycle.

$N_2(g)$ is slowly removed from the air by **nitrogen fixation**. During this process, nitrogen-fixing bacteria, such as *Rhizobium*, *Azotobacter*, and *Beijerinckia*, convert $N_2(g)$ to the ammonium ion $[NH_4^+]$ dissolved in water, some of which evaporates back to the air as ammonia gas $[NH_3(g)]$. Nitrogen-fixing bacteria are prokaryotic and may be aerobic or anaerobic. Most today reside on the roots of leguminous plants.

Another source of ammonium in soils is **ammonification**, a process by which bacteria decompose organic compounds to ammonium. Today, an anthropogenic source of ammonium is fertilizer.

Ammonium is converted to nitrate in soils during a two-step process called **nitrification**. This process occurs only in aerobic environments. In the first step, nitrosofying (nitrite-forming) bacteria produce nitrite from ammonium. In the second step, nitrifying (nitrate-forming) bacteria produce nitrate from nitrite. Once nitrate is formed, the nitrogen cycle continues through the denitrification process. Nitrifying bacteria exist in most waters of moderate pH and soils.

$N_2(g)$ has few chemical sinks. Because its chemical loss was slow and because its removal by nitrogen fixation was slower than is its production by denitrification, $N_2(g)$'s concentration accumulated over time. Table 2.6 summarizes the sources and sinks of $N_2(g)$.

2.3.7. Summary of Atmospheric Evolution

Figure 2.11 summarizes the temporal evolution of $N_2(g)$, $O_2(g)$, $CO_2(g)$, and $H_2(g)$ in the Earth's second atmosphere. The atmosphere of the early Earth may have been dominated by carbon dioxide. Nitrogen gradually increased due to denitrification. Oxygen increased following the GOE 2.45 b.y.a. It reached 1 percent of its present level 1.85 b.y.a., but it did not approach the present level until after the evolution of green plants around 400 m.y.a.

Table 2.6. Sources and sinks of atmospheric molecular nitrogen

Sources	Sinks
Denitrification by bacteria	Nitrogen fixation by bacteria
Atmospheric reaction and photolysis of $N_2O(g)$	Atmospheric chemical reaction
	High-temperature combustion

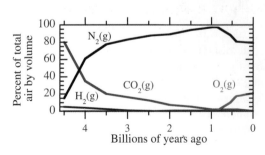

Figure 2.11. Estimated change in composition during the history of the Earth's second atmosphere. Modified from Cattermole and Moore (1985).

2.4. Summary

The sun formed from the condensation of the solar nebula about 4.6 b.y.a. Solar radiation incident on the Earth originates from the sun's photosphere. The photosphere emits radiation with an effective temperature near 6,000 K. The solar spectrum consists of UV, visible, and solar-IR wavelength regimes. The Earth formed from the same nebula as the sun. Most of the Earth's growth was due to asteroid and meteorite bombardment. The overall composition of the Earth is similar to that of stony meteorites. Dense compounds and compounds with high melting points settled to the center of the Earth. Light compounds and those with low melting points became concentrated in the crust. The first atmosphere of the Earth, which consisted of hydrogen and helium, may have been swept away by an enhanced solar wind during early nuclear explosions in the sun. The second atmosphere, which resulted from outgassing, initially consisted of carbon dioxide, water vapor, and assorted gases. When microbes first evolved, they converted carbon dioxide, ammonia, hydrogen sulfide, and organic material to methane, molecular nitrogen, sulfur dioxide, and carbon dioxide, respectively. Oxygen-producing photosynthesis led to the accumulation of oxygen and the production of ozone only after rocks had become saturated with oxygen. The presence of oxygen resulted in the evolution of aerobic respiration, which increased the efficiency of producing molecular nitrogen, the major constituent in today's atmosphere.

2.5. Problems

2.1. Explain why the Earth's core consists primarily of iron, whereas its crust consists primarily of oxygen and silicon.

2.2. Why does the Earth receive radiation from the sun as if the sun is an emitter with an effective temperature of near 6,000 K, when, in fact, the hottest temperatures of the sun are more than 15 million K?

2.3. Even though the moon is effectively the same distance from the sun as the Earth, the moon has no effective atmosphere. Why?

2.4. What is the intensity of radiation emitted by a hot desert (330 K) relative to that emitted by the stratosphere over the South Pole during July (190 K)?

2.5. What prevents the Earth from having magma oceans on its surface today, even though temperatures in the interior of the Earth exceed 4,000 K?

2.6. What peak wavelength of radiation is emitted in the center of the Earth, where the temperature is near 4,000 K?

2.7. What elements do you expect to be most abundant in soil dust particles lifted by the wind? Why?

2.8. Describe the nitrogen cycle in today's atmosphere. What would happen to molecular nitrogen production if nitrification were eliminated from the cycle? (*Hint*: Consider the processes occurring in the preoxygen atmosphere.)

2.9. Identify at least three ways in which oxygen improved the opportunity for higher life forms to develop on the Earth.

2.10. What is the advantage of aerobic respiration over fermentation?

2.11. What is the source of oxygen during photosynthesis in green plants? Explain.

2.12. If the Earth's crust contained twice as much iron as it does, would atmospheric oxygen buildup have been slowed down or sped up during the past 2.5 billion years? What might the consequence of this have been for the evolution of aerobic respiration, the ozone layer, green plants, and humans?

2.13. Explain how the Great Oxygenation Event caused an ice age.

Chapter 3

Structure and Composition of the Present-Day Atmosphere

In this chapter, the structure and composition of the present-day atmosphere are described. The structure is defined in terms of the variation of pressure, density, and temperature with height. Pressure and density are controlled by the concentrations of gases in the air, the most abundant of which are molecular nitrogen and oxygen. The temperature structure is controlled by the vertical distribution of gases that absorb UV and thermal-IR radiation. Pressure, density, and temperature are interrelated by the equation of state. Gases in the air include fixed and variable gases. In the following sections, the pressure, density, and temperature structures of the atmosphere are discussed. The main constituents of the air are then examined in terms of their sources and sinks, abundances, health effects, and importance with respect to different air pollution issues.

3.1. Air Pressure and Density Structure

Air consists of gases and particles, but the mass of air is dominated by gases. Of all gas molecules in the air, more than 99 percent are molecular nitrogen or oxygen. Thus, oxygen and nitrogen are responsible for much of the current pressure and density structure of the Earth's atmosphere. Ozone, oxygen, and nitrogen are responsible for much of the temperature structure.

Air pressure, which is the force of air exerted per unit area of underlying surface, can be calculated as the summed weight of all gas molecules between a horizontal plane and the top of the atmosphere, divided by the area of the plane. Thus, the more molecules that

are present above a plane, the greater the air pressure. Because the weight of air per unit area above a given altitude is always larger than that above any higher altitude, air pressure decreases with increasing altitude. In fact, pressure decreases exponentially with increasing altitude. **Standard sea level pressure** is 1,013 hPa (hectaPascal, where 1 hPa = 100 Pascal (Pa) = 1 millibar (mb) = 100 N m^{-2} = 100 kg m^{-1} s^{-2}). The sea level pressure at a given location and time typically differs by +10 to −20 hPa from standard sea level pressure. In a strong low-pressure system, such as at the center of a hurricane, the actual sea level pressure may be 50 to 100 hPa lower than standard sea level pressure.

Atmospheric pressure was first measured in 1643 by Italian physicist **Evangelista Torricelli** (1608–1647; Figure 3.1), an associate of **Galileo Galilei** (1564–1642). Torricelli filled a 1.2-m-long glass tube with mercury and inverted it onto a dish. He found that only a portion of the mercury flowed from the tube into the dish, and the resulting space above the mercury in the tube was devoid of air (a **vacuum**). Torricelli was the first person to record a sustained vacuum. After further observations, he suggested that the change in height of the mercury in the tube each day was caused by a change in atmospheric pressure. Air pressure balanced the pressure exerted by the column of mercury in the tube, preventing the mercury from flowing freely from the tube. Decreases in air pressure caused more mercury to flow out of the tube and the mercury level to drop. Increases in air pressure had the opposite effect. The inverted tube Torricelli used to derive this conclusion

Figure 3.1. Experiment with mercury barometer, conducted by Evangelista Torricelli (1608–1647) and directed by Blaise Pascal (1623–1662). The third person is the artist, Ernest Board. Edgar Fahs Smith Collection, University of Pennsylvania Library.

was called a **mercury barometer**, where a barometer is a device for measuring atmospheric pressure.

Example 3.1
To what height must mercury in a barometer rise to balance an atmospheric pressure of 1,000 hPa, assuming the density of mercury is 13,558 kg m^{-3}?

Solution
The pressure (kg m^{-1} s^{-2}) exerted by mercury equals the product of its density (kg m^{-3}), gravity (9.81 m s^{-2}), and the height (h, in meters) of the column of mercury. Equating this pressure with

---

the pressure exerted by air and solving for the height gives

$$h = \frac{1{,}000\,\text{hPa}}{13{,}558\,\frac{\text{kg}}{\text{m}^3} \times 9.81\,\frac{\text{m}}{\text{s}^2}} \times \frac{100\,\text{kg}\,\text{m}^{-1}\text{s}^{-2}}{1\,\text{hPa}}$$

$$= 0.752\,\text{m} = 29.6\,\text{inches},$$

indicating that a column of mercury 29.6 inches (75.2 cm) high exerts as much pressure at the Earth's surface as a 1,000-hPa column of air extending between the surface and the top of the atmosphere.

Soon after Torricelli's discovery, French mathematician **Blaise Pascal** (1623–1662) confirmed Torricelli's theory. He and his brother-in-law, Florin Périer, each carried a glass tube of mercury, inverted in a bath of mercury, up the hill of Puy-de-Dôme, France. They recorded the level of mercury at the same time at different altitudes on the hill, confirming that atmospheric pressure decreases with increasing altitude. In 1663, the Royal Society of London built its own mercury barometer based on Torricelli's model. A more advanced **aneroid barometer** was developed in 1843. The aneroid barometer measures pressure by gauging the expansion and contraction of a metal tightly sealed in a case containing no air.

Air density is the mass of air per unit volume of air. Because oxygen and nitrogen are concentrated near the Earth's surface, air density peaks near the surface. Air density decreases exponentially with increasing altitude. Figure 3.2a,b shows standard profiles of air pressure and density, respectively, and Figure 3.2a indicates that 50 percent of the atmosphere's mass lies between sea level and 5.5 km. About 99.9 percent of its mass lies below about 48 km. The Earth's radius is about 6,378 km. Thus, almost all the Earth's atmosphere lies in a layer thinner than 1 percent of the radius of the Earth.

3.2. Processes Affecting Temperature

At the Earth's surface, where air density varies only slightly in time and space, high or low temperature corresponds to how hot or cold we feel. However, **temperature** is really a measure of the average kinetic energy of an air molecule (energy giving rise to the motion of air molecules). At a given density of air, we feel hot if air molecules have a high kinetic energy because many bombard our skin, and the resulting friction against our skin converts kinetic energy from the molecules to heat that we absorb. When molecules have a low kinetic

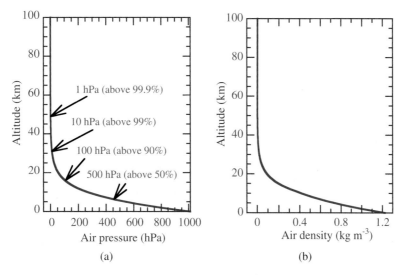

Figure 3.2. (a) Pressure and (b) density versus altitude in the Earth's lower atmosphere. The pressure diagram shows that 99.9 percent of the atmosphere lies below an altitude of about 48 km (1 hPa) and 50 percent lies below about 5.5 km (500 hPa).

energy, the frictional loss, and thus heat production, is low, so we feel cold. In the upper atmosphere, where air density is low, the temperature can be extremely high because each air molecule receives energy from intense ultraviolet radiation and thus has a high kinetic energy. However, we would feel cold if exposed to such air because the air density is low, so few air molecules would hit our skin, reducing the total heat production on our skin. In sum, temperature is a measure of hotness only at a given air density.

From gas kinetic theory, the absolute temperature (K) of air is related to kinetic energy by

$$\frac{4}{\pi} k_B T = \frac{1}{2} \bar{M} \bar{v}_a^2 \qquad (3.1)$$

where k_B is **Boltzmann's constant** (1.3807×10^{-23} kg m^2 s^{-2} K^{-1} molec^{-1}), $\bar{M}$ is the average mass of one air molecule (4.8096×10^{-26} kg molec^{-1}), and $\bar{v}_a$ is the average **thermal speed** of an air molecule (m s^{-1}). The right side of Equation 3.1 is the kinetic energy of an air molecule at its average thermal speed.

Example 3.2

What is the thermal speed of an air molecule at 200 K? At 300 K?

Solution

From Equation 3.1, the thermal speed of an air molecule is 382 m s^{-1} at 200 K and 468 m s^{-1} at 300 K.

Measurements of relative air temperature changes were first attempted in 1593 by Galileo, who devised a thermoscope to measure the expansion and contraction of air upon its heating and cooling, respectively. However, the instrument did not have a scale and was unreliable. In the mid-seventeenth century, the thermoscope was replaced by the liquid-in-glass thermometer developed in Florence, Italy. In the early eighteenth century, useful thermometer scales were developed by **Gabriel Daniel Fahrenheit** (1686–1736) of Germany and **Anders Celsius** (1701–1744) of Sweden.

The temperature at a given location and time is affected by energy transfer processes, including conduction, convection, advection, and radiation. These processes are discussed briefly here.

3.2.1. Conduction

Conduction is the transfer of energy in a medium (the conductor) from one molecule to the next in the presence of a temperature gradient. The medium, as a whole, experiences no molecular movement. Conduction occurs through soil, air, and particles. Conduction affects air temperature by transferring energy between the soil surface and the bottom molecular layers of the air. The rate of a material's conduction is determined by its **thermal conductivity** (κ, J m^{-1} s^{-1} K^{-1}), which quantifies the rate of flow of thermal energy through a material in the presence of a temperature gradient. Table 3.1 gives thermal conductivities of a few

Table 3.1. Thermal conductivities of four media

Substance	Thermal conductivity (κ) at 298.15 K (J m^{-1}s^{-1}K^{-1})
Dry air at constant pressure	0.0256
Liquid water	0.6
Clay	0.920
Dry sand	0.298

substances. It shows that liquid water, clay, and dry sand are more conductive than is dry air. Thus, energy passes through air more slowly than it passes through other materials of the same thickness, given the same temperature gradient. Clay is more conductive and dry sand is less conductive than is liquid water.

The flux of energy due to conduction (W m^{-2}) can be approximated with the **conductive heat flux equation,**

$$H_c = -\kappa \frac{\Delta T}{\Delta z} \qquad (3.2)$$

where ΔT (K) is the change in temperature over a distance Δz (m). At the ground, molecules of soil and water transfer energy by conduction to overlying molecules of air. Because the temperature gradient ($\Delta T/\Delta z$) between the surface and a thin (e.g., 1-mm) layer of air just above the surface is large, the conductive heat flux at the ground is also large. Above the ground, temperature gradients are smaller and conductive heat fluxes through the air are smaller than they are at the ground.

Example 3.3
Compare the conductive heat flux through a thin (1-mm) layer of air touching the surface if $T = 298$ K and $\Delta T = -12$ K with that through the background troposphere, where $T = 273$ K and $\Delta T/\Delta z = -6.5$ K km^{-1}.

Solution
For air, $\kappa = 0.0256$ J m^{-1} s^{-1} K^{-1}; thus, the conductive heat flux at the surface is $H_c = 307$ W m^{-2}. In the background troposphere, $H_c = 1.5 \times 10^{-4}$ W m^{-2}, which is much smaller than is the value at the surface. Thus, heat conduction through the air is important only adjacent to the ground.

3.2.2. Convection

Convection is the transfer of energy, gases, and particles by the mass movement of air, predominantly in the vertical direction. It differs from conduction in that during conduction, energy is transferred from one molecule to another, whereas during convection, energy is transferred as the molecules themselves move. Two important types of convection are forced and free.

Forced convection is an upward or downward vertical movement of air caused by mechanical means. Forced convection occurs, for example, when (1) horizontal near-surface winds converge (diverge), forcing air to rise (sink); (2) horizontal winds encounter a topographic barrier, forcing air to rise; or (3) winds blow over objects protruding from the ground, creating swirling motions of air, or **eddies**, which mix air vertically and horizontally. Objects of different size create eddies of different size. **Turbulence** is the effect of groups of eddies of different size. Turbulence from wind-generated eddies is **mechanical turbulence**.

Free convection (thermal turbulence) is a predominantly vertical motion produced by buoyancy, which occurs when the sun heats different areas of the ground differentially. Differential heating occurs because clouds or hills block the sun in some areas but not in others, or different surfaces lie at different angles relative to the sun. Over a warm, sunlit surface, conduction transfers energy from the ground to molecules of air adjacent to the ground. The warmed air above the ground rises buoyantly, producing a **thermal**. Cool air from nearby is drawn down to replace the rising air. Near the surface, the cool air heats by conduction and then rises, feeding the thermal. Free convection occurs most readily over land when the sky is cloud free and the winds are light.

3.2.3. Advection

Advection is the horizontal movement of energy, gases, and particles by the wind. Like convection, advection results in the mass movement of molecules.

3.2.4. Radiation

Radiation, first defined in Section 2.2, is the transfer of energy by electromagnetic waves or photons, which do not require a medium, such as air, for their transmission. Thus, radiative energy transfer can occur even when no atmosphere exists, such as above the moon's surface. Conduction cannot occur above the moon's

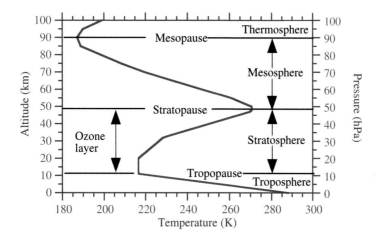

Figure 3.3. Temperature structure of the Earth's lower atmosphere, ignoring the boundary layer.

surface because too few molecules are present in the moon's atmosphere to transfer energy away from its surface. Similarly, neither convection nor advection can occur in the moon's atmosphere.

3.3. Temperature Structure of the Atmosphere

The bottom 100 km of the Earth's atmosphere, the **homosphere**, is a region in which major gases are well mixed. The homosphere is divided into four layers in which temperatures change with altitude. These are, from bottom to top, the **troposphere, stratosphere, mesosphere,** and **thermosphere**. Figure 3.3 shows an average profile of the temperature structure of the homosphere.

3.3.1. Troposphere

The troposphere is divided into the boundary layer (ignored in Figure 3.3) and the background troposphere. These regions are briefly discussed as follows.

3.3.1.1. Boundary Layer

The **boundary layer** extends from the topographical surface up to between 500 and 3,000 m above the surface. All people live in the boundary layer, so it is this region of the atmosphere in which air pollution buildup is of most concern. Pollutants emitted near the ground accumulate in the boundary layer. When pollutants escape the boundary layer, they can travel horizontally long distances before they are removed from the air by precipitation or gravitational sedimentation. The boundary layer differs from the background troposphere in that the temperature profile in the boundary layer responds to changes in ground temperatures over a period of less than an hour, whereas the temperature profile in the background troposphere responds to changes in ground temperatures over a longer period (Stull, 1988).

Figure 3.4 shows a typical temperature variation with height in the boundary layer over land during the day and night. During the day (Figure 3.4a), the boundary layer consists of a surface layer, a convective mixed layer, and an entrainment zone. The **surface layer**, which comprises the bottom 10 percent of the boundary layer, is a region of strong change of wind speed with height (wind shear). Because the boundary layer depth ranges from 500 to 3,000 m, the surface layer is about 50 to 300 m thick. Wind shear occurs in the surface layer simply because wind speeds at the ground are zero and those above the ground are not.

The **convective mixed layer** is the region of air just above the surface layer. When sunlight warms the ground during the day, some of the energy is transferred from the ground to the air just above the ground by conduction. Because the air above the ground is now warm, it rises buoyantly as a thermal. Thermals originating from the surface layer rise and gain their maximum acceleration in the convective mixed layer. As thermals rise, they displace cooler air aloft downward; thus, upward and downward motions occur, allowing air and pollutants to mix in this layer.

The top of the mixed layer is often bounded by a **temperature inversion**, which is an increase in temperature

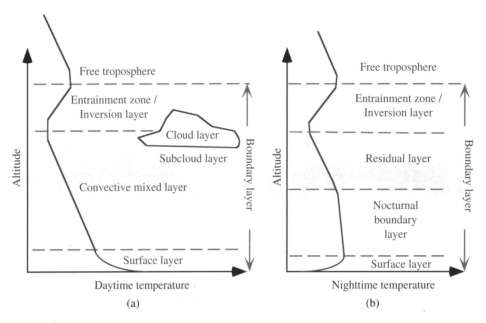

Figure 3.4. Variation of temperature with height during the (a) day and (b) night in the atmospheric boundary layer over land under a high-pressure system (Chapter 6). Adapted from Stull (1988).

with increasing height. The inversion inhibits the rise of thermals originating from the surface layer or the mixed layer. Some mixing (entrainment) between the inversion and mixed layer does occur; thus, the inversion layer is also called an **entrainment zone**. Pollutants are generally trapped beneath or within an inversion; thus, the closer the inversion is to the ground, the higher pollutant concentrations become.

Other features of the daytime boundary layer are the cloud and subcloud layers. A region in which clouds appear in the boundary layer is the **cloud layer**, and the region underneath is the **subcloud layer**.

During the night (Figure 3.4b), the ground cools radiatively, causing air temperatures to increase with increasing height from the ground, creating a surface inversion. Once the nighttime surface inversion forms, pollutants, when emitted, are confined to the surface layer.

Cooling at the top of the surface layer at night cools the bottom of the mixed layer, reducing the buoyancy and associated mixing at the base of the mixed layer. The portion of the daytime mixed layer that loses its buoyancy at night is the **nocturnal boundary layer**. The remaining portion of the mixed layer is the **residual layer**. Because thermals do not form at night, the residual layer does not undergo much change at night, except at its base. At night, the nocturnal boundary

layer thickens, eroding the residual layer base. Above the residual layer, the inversion remains.

3.3.1.2. Background Troposphere

The background **troposphere** lies between the boundary layer and the tropopause. It is a region in which, on average, the temperature decreases with increasing altitude. The average rate of temperature decrease in the background troposphere is about 6.5 K km^{-1}. The temperature decreases with increasing altitude in the background troposphere for the following reason: the ground surface receives solar energy from the sun daily, heating the ground. The ground converts that solar energy to thermal infrared (heat) radiation, which is emitted back to the atmosphere. Heat from the ground is also transferred vertically by conduction and convection. Greenhouse gases, aerosol particles, and hydrometeor particles in each layer of the atmosphere absorb the heat and reemit some of it back down toward the surface and some upward in the form of thermal-IR radiation. Thus, each successively higher layer in the atmosphere receives less heat than did the layer below. In sum, the troposphere, itself, has relatively little capacity to absorb solar energy; thus, each layer of air in the troposphere relies on heat transfer from the layer of air below it for its energy, so temperature in the troposphere decreases with increasing height.

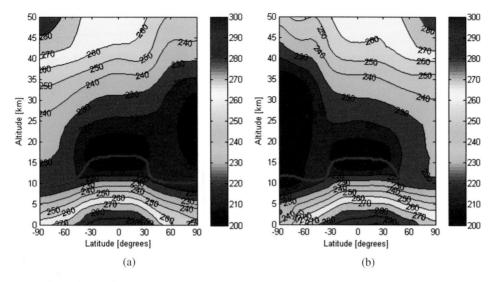

Figure 3.5. Zonally and monthly averaged temperatures between 0 and 50 km altitude for (a) December-January-February and (b) June-July-August. The thick solid line is the mean tropopause height for the period. Data for the plots were compiled from the European Center for Medium Range Weather Forecasting for the years 1969–2001; tropopause heights were determined as in Whitt et al. (2011).

The **tropopause** is the upper boundary of the troposphere. Above the tropopause base, temperatures are relatively constant with increasing height before increasing with increasing height in the stratosphere.

Figures 3.5a and 3.5b show zonally averaged temperatures for a generic December-January-February and June-July-August, respectively. A **zonally averaged** temperature is found by averaging temperatures over all longitudes at a given latitude and altitude. Tropopause heights are higher (15–18 km) over the Equator than over the poles (8–12 km). Strong vertical motions over the Equator due to buoyancy from solar heating of the surface and heat release during cloud formation raise the base of the ozone layer there. Because ozone is responsible for warming above the tropopause, pushing ozone to greater heights over the Equator increases the altitude at which warming begins. Over the poles, low temperatures cause air to sink, pushing stratospheric ozone downward as well, lowering the tropopause height.

Temperatures at the tropopause over the Equator are lower than they are over the poles. One reason is that the higher base of the ozone layer over the Equator allows tropospheric temperatures to decrease to a greater altitude over the Equator than over the poles. A second reason is that low- and midtropospheric water vapor contents are much higher over the Equator than they are over the poles. Water vapor absorbs thermal-IR radiation emitted from the Earth's surface, preventing that radiation from reaching the upper troposphere, cooling the upper troposphere over the Equator.

3.3.2. Stratosphere

The stratosphere is a large temperature inversion. The inversion is caused by ozone, which absorbs much of the sun's near-UV radiation and reemits thermal-IR radiation, heating the stratosphere. Peak stratospheric temperatures occur at the top of the stratosphere because this is the altitude at which ozone absorbs the shortest, thus most intense, UV wavelengths reaching the stratosphere (about 0.175 μm). Although the ozone concentration at the top of the stratosphere is low, each ozone molecule can absorb short UV wavelengths, increasing the average kinetic energy and, thus, temperature (through Equation 3.1) of all molecules. Short UV wavelengths do not penetrate to the lower stratosphere. Ozone densities in the stratosphere peak at 25 to 32 km.

3.3.3. Mesosphere

Temperatures decrease with increasing altitude in the **mesosphere** for the same reason they decrease in the background troposphere. Ozone densities are too low, in comparison with those of oxygen and nitrogen, for ozone absorption of UV radiation to affect the average temperature of all molecules in the mesosphere.

3.3.4. Thermosphere

In the **thermosphere**, temperatures increase with increasing altitude because $O_2(g)$ and $N_2(g)$ absorb very short, far-UV wavelengths in this region. Peak temperatures in the thermosphere range from 1,200 to 2,000 K, depending on solar activity. Air in the thermosphere would not *feel* hot to the skin because the thermosphere contains so few gas molecules. However, because each gas molecule in the thermosphere is highly energized, the average temperature is high. Because molecular oxygen and nitrogen absorb very short wavelengths in the thermosphere, such wavelengths do not penetrate to the mesosphere.

3.4. Equation of State

Pressure, density, and temperature in the atmosphere are related by the equation of state. The **equation of state** describes the relationship among pressure, volume, and absolute temperature for a real gas. The **ideal gas law** describes this relationship for an ideal gas. An ideal gas is a gas for which the product of the pressure and volume is proportional to the absolute temperature. A real gas is ideal only when intermolecular forces are small, which occurs when pressures are low enough or temperatures are high enough for the gas to be sufficiently dilute. However, under typical atmospheric temperature and pressure conditions, the ideal gas law can reasonably approximate the equation of state.

The ideal gas law is expressed as a combination of Boyle's law, Charles's law, and Avogadro's law. In 1661, **Robert Boyle** (1627–1691), an English natural philosopher and chemist (Figure 3.6), found that doubling the pressure exerted on a gas at constant temperature reduced the volume of the gas by one-half. This relationship is embodied in **Boyle's law:**

$$P \propto 1/V \text{ at constant temperature} \qquad (3.3)$$

where p is the pressure exerted on the gas (hPa) and V is the volume enclosed by the gas (m^3 or cm^3). Boyle's law describes the compressibility of a gas. When high pressure is exerted on a gas, such as in the lower atmosphere, the gas compresses until it exerts an equal pressure on its surroundings. When a gas is subject to low pressure, such as in the upper atmosphere, the gas expands until it exerts an equal pressure on its surroundings.

In 1787, French chemist **Jacques Charles** (1746–1823; Figure 3.7) found that increasing the absolute temperature of a gas at constant pressure increased the

Figure 3.6. Robert Boyle (1627–1691). Edgar Fahs Smith Collection, University of Pennsylvania Library.

volume of the gas. This relationship is embodied in **Charles's law:**

$$V \propto T \text{ at constant pressure} \qquad (3.4)$$

where T is the temperature of the gas (K). Charles's law states that, at constant pressure, the volume of a gas must decrease when its temperature decreases. Because gases change phase to liquids or solids before 0 K, Charles's law cannot be extrapolated to 0 K.

Charles is also known for his development of a balloon filled with hydrogen gas [$H_2(g)$]. On June 4, 1783, Joseph Michel (1740–1810) and **Jacques-Étienne Montgolfier** (1745–1799) launched the first untethered hot air balloon in a marketplace in Annonay, southern France. The balloon was filled with air heated by burning straw and wool under the opening of a light paper or fabric bag. Prompted by this discovery, the French Academy of Sciences asked Charles to

Figure 3.7. Jacques Alexandre Cesar Charles (1746–1823). Edgar Fahs Smith Collection, University of Pennsylvania Library.

Figure 3.8. Amedeo Avogadro (1776–1856). Edgar Fahs Smith Collection, University of Pennsylvania Library.

replicate the feat. Instead of filling his balloon with hot air, Charles filled it with $H_2(g)$, a gas fourteen times lighter than air first observed by Paracelsus and isolated by Cavendish. The balloon was launched on August 27, 1783, and flew for forty-five minutes. When it landed, the balloon, a new innovation, was ironically hacked to pieces by frightened farmers.

In 1811, **Amedeo Avogadro** (1776–1856), an Italian natural philosopher and chemist (Figure 3.8), discerned the difference between atoms and molecules. He found that molecular oxygen and nitrogen gas, believed to be single atoms at the time, were really molecules, each consisting of two atoms. He went on to hypothesize that equal volumes of all gases at the same temperature and pressure contained the same number of molecules. In other words, the volume of a gas is proportional to the number of molecules of gas present and independent of the type of gas. This relationship today is **Avogadro's law**:

$$V \propto n \text{ at constant pressure and temperature} \quad (3.5)$$

where n is the number of gas moles. The number of molecules in a mole is constant for all gases and given by **Avogadro's number**, $A = 6.02252 \times 10^{23}$ molec mol^{-1}. Avogadro did not devise this number nor was the term "mole" in the chemical vocabulary during his lifetime. In 1865, Austrian chemist **Joseph Loschmidt** (1821–1895), who isolated the first aromatic compounds, estimated the size of an air molecule and the number of molecules in a cubic centimeter of gas. Avogadro's number was devised soon after.

Combining Boyle's law, Charles's law, and Avogadro's law gives the ideal gas law or simplified equation of state as

$$p = \frac{nR^*T}{V} = \frac{nA}{V}\left(\frac{R^*}{A}\right)T = Nk_BT \quad (3.6)$$

where

$$N = \frac{nA}{V} \quad (3.7)$$

is the number concentration of gas molecules (molecules of gas per cubic meter or cubic centimeter of air), R^* is the universal gas constant (0.083145 m^3 hPa mol^{-1} K^{-1} or 8.314×10^4 cm^3 hPa mol^{-1} K^{-1}), and

$$k_B = \frac{R^*}{A} \qquad (3.8)$$

is Boltzmann's constant (1.3807×10^{-25} m^3 hPa K^{-1} molec^{-1} or 1.3807×10^{-19} cm^3 hPa K^{-1} molec^{-1}). The Appendix contains alternative units for R^* and k_B.

Example 3.4
Calculate the number concentration of air molecules in the atmosphere at standard sea-level pressure and temperature and at a pressure of 1 hPa.

Solution
At standard sea level, $p = 1{,}013$ hPa and $T = 288$ K. Thus, from Equation 3.6, $N = 2.55 \times 10^{19}$ molec cm^{-3}. From Figure 3.2a, $p = 1$ hPa occurs at 48 km. At this altitude and pressure, $T = 270$ K, as shown in Figure 3.3. Under such conditions, $N = 2.68 \times 10^{16}$ molec cm^{-3}.

Equation 3.6 can be used to relate the partial pressure exerted by a gas to its number concentration. In 1803, **John Dalton** (1766–1844), an English chemist and physicist (Figure 3.9), stated that total atmospheric pressure equals the sum of the partial pressures of the individual gases in the air. This is **Dalton's law of partial pressure**. The **partial pressure** exerted by a gas in a mixture is the pressure the gas exerts if it alone occupies the same volume as the mixture. Mathematically, the partial pressure of gas q is

$$p_q = N_q k_B T \qquad (3.9)$$

where N_q is the number concentration of the gas (molec cm^{-3}). Total atmospheric pressure is

$$p_a = \sum_q p_q. \qquad (3.10)$$

Dalton is also known for proposing the atomic theory of matter and studying color blindness (Daltonism).

Total atmospheric pressure can also be written as

$$p_a = p_d + p_v \qquad (3.11)$$

where p_d is the partial pressure exerted by dry air and p_v is the partial pressure exerted by water vapor. Dry air consists of all gases in the air, except water vapor. Together, $N_2(g)$, $O_2(g)$, $Ar(g)$, and $CO_2(g)$ constitute

Figure 3.9. John Dalton (1766–1844). Edgar Fahs Smith Collection, University of Pennsylvania Library.

99.996 percent of dry air by volume. The partial pressures of all gases aside from these four can be ignored, without much loss in accuracy, when dry air pressure is calculated. This assumption is convenient because the concentrations of most trace gases vary in time and space.

The partial pressure of dry air is related to the mass density and number concentration of dry air through the **equation of state for dry air**,

$$p_d = \rho_d R' T = N_d k_B T \qquad (3.12)$$

where ρ_d is the mass density of dry air (kg m^{-3} or g cm^{-3}) and R' is the gas constant for dry air (2.8704 m^3 hPa kg^{-1} K^{-1} or $2{,}870.3$ cm^3 hPa g^{-1} K^{-1}; alternative units are given in the Appendix), and N_d is the number concentration of dry air molecules (molec cm^{-3}). The dry air mass density, number concentration, and gas constant are further defined as

$$\rho_d = \frac{n_d m_d}{V} \quad N_d = \frac{n_d A}{V} \quad R' = \frac{R^*}{m_d} \quad (3.13)$$

respectively, where n_d is the number of moles of dry air and m_d is the molecular weight of dry air, which is a volume-weighted average of the molecular weights of

$N_2(g)$, $O_2(g)$, $Ar(g)$, and $CO_2(g)$. The standard value of m_d is 28.966 g mol^{-1}. The equation of state for water vapor is analogous to that for dry air.

Example 3.5
When $p_d = 1{,}013$ hPa and $T = 288$ K, what is the density of dry air?

Solution
From Equation 3.12, $\rho_d = 1.23$ kg m^{-3}.

The number concentration of a gas (molecules per unit volume of air) is an absolute quantity. The abundance of a gas may also be expressed in terms of a relative quantity, **volume mixing ratio**, defined as the number of gas molecules per molecule of dry air, and expressed for gas q as

$$\chi_q = \frac{N_q}{N_d} = \frac{p_q}{p_d} \qquad (3.14)$$

where N_q and p_q are the number concentration and partial pressure, respectively, of gas q. Volume mixing ratios may be multiplied by 100 and expressed as a **percentage of dry air volume**, multiplied by 10^6 and expressed in **parts per million volume** (ppmv), multiplied by 10^9 and expressed in **parts per billion volume** (ppbv), or multiplied by 10^{12} and expressed in **parts per trillion volume** (pptv).

Example 3.6
Find the number concentration and partial pressure of ozone if its volume mixing ratio is $\chi_q = 0.10$ ppmv. Assume $T = 288$ K and $p_d = 1{,}013$ hPa.

Solution
From Example 3.4, $N_d = 2.55 \times 10^{19}$ molec cm^{-3}. Thus, from Equation 3.14, the number concentration of ozone is $N_q = 0.10$ ppmv $\times 10^{-6} \times 2.55 \times 10^{19}$ molec cm$^{-3} = 2.55 \times 10^{12}$ molec cm^{-3}. From Equation 3.9, the partial pressure exerted by ozone is $p_q = 0.000101$ hPa.

3.5. Composition of the Present-Day Atmosphere

The present-day atmosphere below 100 km (the homosphere) contains only a few **well-mixed gases** that, together, comprise more than 99 percent of all gas

Table 3.2. Volume mixing ratios of well-mixed gases in the lowest 100 km of the Earth's atmosphere

Gas name	Chemical formula	Volume mixing ratio (Percent)	Volume mixing ratio (ppmv)
Molecular nitrogen	$N_2(g)$	78.08	780,000
Molecular oxygen	$O_2(g)$	20.95	209,500
Argon	$Ar(g)$	0.93	9,300
Neon	$Ne(g)$	0.0015	15
Helium	$He(g)$	0.0005	5
Krypton	$Kr(g)$	0.0001	1
Xenon	$Xe(g)$	0.000005	0.05

molecules in this region. The mixing ratios of well-mixed gases do not vary much in time or space over short time scales (hundreds to thousands of years). Nevertheless, it is the **spatially and temporally varying gases**, whose mixing ratios are small but vary in time and space, that are the most relevant to air pollution.

3.5.1. Well-Mixed Gases

Gases can become well mixed in the bottom 100 km of the atmosphere only if they (1) are long lived (have low chemical, physical, and biological loss rates) and (2) are emitted uniformly over time. Table 3.2 lists most well-mixed gases in the homosphere. Although carbon dioxide is a long-lived gas, it is not well mixed because its emission rate changes over short time scales due to human activity and because human emissions are concentrated in urban areas, giving rise to higher mixing ratios in urban areas than in surrounding rural areas. These higher mixing ratios in urban areas are referred to as **carbon dioxide domes**.

Molecular nitrogen [$N_2(g)$] and molecular oxygen [$O_2(g)$] are the most abundant well-mixed gases. At any altitude, $N_2(g)$ comprises about 78.08 percent, and $O_2(g)$ comprises about 20.95 percent and of all non-water gas molecules by volume. Although these gases have mixing ratios that are constant with increasing altitudes, they have partial pressures that decrease with increasing altitude because air pressure decreases with increasing altitude (Figure 3.2a), and $O_2(g)$ and $N_2(g)$ partial pressures are constant fractions of air pressure.

Together, $N_2(g)$ and $O_2(g)$ make up 99.03 percent of all gases in the atmosphere by volume. Argon (Ar) makes up most of the remaining 0.97 percent. Argon, the "lazy gas," is colorless and odorless. Like other noble gases, it is inert and does not react chemically.

Table 3.3. Volume mixing ratios of some spatially and temporally varying gases in three atmospheric regions

Gas name	Chemical formula	Volume mixing ratio (ppbv)		
		Clean troposphere	Polluted troposphere	Stratosphere
Inorganic				
Water vapor	$H_2O(g)$	3,000–4.0(+7)[a]	5.0(+6)–4.0(+7)[a]	3,000–6,000
Carbon dioxide	$CO_2(g)$	390,000	390,000	390,000
Carbon monoxide	$CO(g)$	40–200	2,000–10,000	10–60
Ozone	$O_3(g)$	10–100	10–350	1,000–12,000
Sulfur dioxide	$SO_2(g)$	0.02–1	1–30	0.01–1
Nitric oxide	$NO(g)$	0.005–0.1	0.05–300	0.005–10
Nitrogen dioxide	$NO_2(g)$	0.01–0.3	0.2–200	0.005–10
CFC-12	$CF_2Cl_2(g)$	0.55	0.55	0.22
Organic				
Methane	$CH_4(g)$	1,850	1,850–2,500	150–1,700
Ethane	$C_2H_6(g)$	0–2.5	1–50	–[b]
Ethene	$C_2H_4(g)$	0–1	1–30	–
Formaldehyde	$HCHO(g)$	0.1–1	1–200	–
Toluene	$C_6H_5CH_3(g)$	–	1–30	–
Xylene	$C_6H_4(CH_3)_2(g)$	–	1–30	–
Methyl chloride	$CH_3Cl(g)$	0.61	0.61	0.36

CFC, chlorofluorocarbon.
[a] 4.0(+7) means 4.0×10^7.
[b] A (–) indicates that the volume mixing ratio is negligible, on average.

Other well-mixed but inert gases present in trace concentrations include neon, helium, krypton, and xenon.

3.5.2. Spatially and Temporally Varying Gases

Gases whose volume mixing ratios change in time and space are variable gases. Table 3.3 summarizes the volume mixing ratios of some such gases in the clean troposphere, the polluted troposphere (e.g., urban areas), and the stratosphere. Many organic gases degrade chemically before they reach the stratosphere, so their mixing ratios are low in the stratosphere.

3.6. Characteristics of Selected Gases and Aerosol Particle Components

Table 3.4 lists gases and aerosol particle components relevant to each of the five major air pollution problems discussed in this book. The table indicates that each problem involves a different set of pollutants, although some pollutants are common to two or more problems.

Next, a few gases and aerosol particle components listed in Table 3.4 are discussed in terms of their relevance, abundance, sources, sinks, and health effects.

3.6.1. Water Vapor

Water vapor [$H_2O(g)$] is the most important spatially and temporally varying gas in the air. It is a natural **greenhouse gas** in that it readily absorbs thermal-IR radiation, but it also plays a vital role in the hydrologic cycle on Earth. As a natural greenhouse gas, it is much more important than carbon dioxide for maintaining a climate suitable for life on Earth. However, as a global warming agent, its anthropogenic emissions due to combustion of fossil fuels and evaporation during irrigation and power plant cooling have a much smaller impact than do emissions of other global warming agents. Nevertheless, anthropogenic water vapor emissions do affect local temperatures, stability, the relative humidity, and cloudiness, which feed back to air pollution (e.g., Jacobson, 2008a). Water vapor is not considered a health-affecting air pollutant; thus, no regulations control its mixing ratio or emission.

3.6.1.1. Sources and Sinks

Table 3.5 summarizes the sources and sinks of water vapor. The main source, evaporation from the oceans,

Table 3.4. Some gases and aerosol particle components important for specified air pollution topics

Indoor air pollution	Outdoor urban air pollution	Acid deposition	Stratospheric reduction	Global warming
Gases				
Nitrogen dioxide	Ozone	Sulfur dioxide	Ozone	Carbon dioxide
Carbon monoxide	Nitric oxide	Sulfuric acid	Nitric oxide	Methane
Formaldehyde	Nitrogen dioxide	Nitrogen dioxide	Nitric acid	Nitrous oxide
Sulfur dioxide	Carbon monoxide	Nitric acid	Hydrochloric acid	Ozone
Organic gases	Ethene	Hydrochloric acid	Chlorine nitrate	CFC-11
Radon	Formaldehyde	Carbon dioxide	Bromine nitrate	CFC-12
	Toluene		CFC-11	Water vapor
	Xylene		CFC-12	
	PAN			
Aerosol Particle Components				
Black carbon	Black carbon	Sulfate	Chloride	Black carbon
Organic matter	Organic matter	Nitrate	Sulfate	Brown carbon
Sulfate	Sulfate	Chloride	Nitrate	Other organic matter
Nitrate	Nitrate			Sulfate
Ammonium	Ammonium			Nitrate
Allergens	Soil dust			Ammonium
Asbestos	Sea spray			Soil dust
Fungal spores	Tire particles			Sea spray
Pollens	Lead			
Tobacco smoke				

CFC, chlorofluorocarbon; PAN, peroxyacetyl nitrate.

accounts for approximately 85 percent of water vapor. The primary anthropogenic sources of water vapor are

Table 3.5. Sources and sinks of atmospheric water vapor

Sources	Sinks
Evaporation from oceans, lakes, rivers, and soil	Condensation to liquid water in clouds
Sublimation from sea ice and snow	Vapor deposition to ice crystals in clouds
Transpiration from plant leaves	Deposition to oceans, sea ice, snow, and soils
Atmospheric chemical reaction	Atmospheric chemical reaction
Evaporation during power plant industrial water cooling	
Evaporation during irrigation	
Fossil fuel, biofuel, and biomass burning combustion	

evaporation of water used to cool coal, nuclear, natural gas, and biofuel power plants and industrial facilities; evaporation of water upon crop irrigation; and burning of fossil fuels, biofuels, and outdoor biomass, which produces water vapor as a combustion product. The anthropogenic emission rate of water vapor is about 1/8,000th its natural emission rate.

3.6.1.2. Mixing Ratios

The mixing ratio of water vapor varies with location and time but is physically limited to no more than 4 to 5 percent of total air by its **saturation mixing ratio**, which is the maximum water vapor the air can hold at a given temperature before the water vapor condenses on the surfaces of aerosol particles as a liquid. When temperatures are low, such as over the poles and in the stratosphere, saturation mixing ratios are low (<0.1 percent of total air), and water vapor readily deposits as ice or condenses as liquid water. When temperatures are high, such as over the Equator, saturation mixing ratios are high (>3 percent of total air), and liquid water may evaporate readily to the gas phase.

3.6.1.3. Health Effects

Water vapor has no direct harmful effects on humans. Liquid water in aerosol particles indirectly causes health problems when pollutant gases dissolve in it or other solid particles attach to it. Small particles can be inhaled, causing health problems.

3.6.2. Carbon Dioxide

Carbon dioxide [$CO_2(g)$] is a colorless, odorless, natural greenhouse gas that is currently also responsible for almost half of the higher global temperatures due to global warming. Although it does not directly impact health at typical mixing ratios found in outdoor air, its anthropogenic emissions over cities produce **carbon dioxide domes** that increase local temperatures and water vapor, both of which increase the concentrations of other pollutants that damage health (Jacobson, 2010a) (Section 12.5.6). $CO_2(g)$ also acidifies rain slightly because it is a weak acid that dissolves in rainwater (Section 10.2.1), but such acidity does not cause environmental damage. $CO_2(g)$ also plays a role in stratospheric ozone depletion because its global warming in the troposphere enhances global cooling of the stratosphere, and stratospheric cooling damages the ozone layer (Section 12.5.7). Ambient mixing ratios of carbon dioxide are not regulated in any country. $CO_2(g)$ emission controls are the subject of an ongoing effort by the international community to reduce global warming.

3.6.2.1. Sources and Sinks

Table 3.6 lists the major sources and sinks of $CO_2(g)$. Carbon dioxide is produced during several bacterial metabolism processes, including fermentation (Reaction 2.3) and denitrification (Reactions 2.7 and 2.8), and aerobic respiration in plant and animal cells (Reaction 2.12). Other sources of $CO_2(g)$ consist of evaporation from the oceans, chemical oxidation of carbon monoxide and organic gases, volcanic outgassing, fossil fuel combustion, biofuel combustion, and biomass burning (Figure 3.10). The single largest source of $CO_2(g)$ is bacterial decomposition of dead organic matter. Indoor sources of $CO_2(g)$ include human exhalation and combustion of gas, kerosene, fuel oil, wood, and coal.

$CO_2(g)$ is removed from the air by oxygenic photosynthesis (Reaction 2.10), dissolution in surface water (e.g., ocean, lake, river, stream water); deposition to sea ice, snow, soil, vegetation, and structures; and chemical weathering.

Table 3.6. Sources and sinks of atmospheric carbon dioxide

Sources	Sinks
Bacterial fermentation	Oxygen-producing photosynthesis
Bacterial anaerobic respiration	Sulfur-producing photosynthesis
Bacterial aerobic respiration	Bacterial anaerobic respiration
Plant, animal, fungus, and protozoa aerobic respiration	Dissolution in surface water
Evaporation from oceans	Deposition to sea ice, snow, soil, vegetation, and structures
Atmospheric chemical reaction	Chemical weathering of carbonate rocks
Volcanic outgassing	
Biomass burning	Photolysis in upper atmosphere
Fossil fuel and biofuel combustion	
Cement production	

Like water vapor, carbon dioxide is a greenhouse gas. Unlike water vapor, $CO_2(g)$ has few chemical loss processes in the gas phase. Its main chemical loss is photolysis to $CO(g)$ in the upper stratosphere and mesosphere, but the $CO_2(g)$ lifetime against loss by this process alone is about 200 years. Its data-constrained *e*-folding lifetime, from emission to removal, due to all loss processes, ranges from 30 to 50 years. An important removal mechanism of $CO_2(g)$ is its dissolution in surface water. Dissolution occurs by the reversible

Figure 3.10. Vegetation fire. Emissions from the fire include gases (e.g., carbon dioxide, carbon monoxide, nitric oxide, organics) and aerosol particles (e.g., soot, organic matter). © Eliasgomez/Dreamstime.com.

(denoted by the double arrows) reaction

$$CO_2(g) \rightleftharpoons CO_2(aq) \tag{3.15}$$

Gaseous carbon dioxide   Dissolved carbon dioxide

followed by the rapid combination of $CO_2(aq)$ with water to form carbonic acid [$H_2CO_3(aq)$] and the dissociation of carbonic acid to the **hydrogen ion** [H^+], the **bicarbonate ion** [HCO_3^-], or the **carbonate ion** [CO_3^{2-}] by the reversible reactions

$$CO_2(aq) + H_2O(aq) \rightleftharpoons H_2CO_3(aq)$$

Dissolved carbon dioxide   Liquid water   Dissolved carbonic acid

$$\rightleftharpoons H^+ + HCO_3^- \rightleftharpoons 2H^+ + CO_3^{2-} \tag{3.16}$$

Hydrogen ion   Bicarbonate ion   Hydrogen ion   Carbonate ion

Ocean water is alkaline (or basic), the opposite of acidic (Section 5.3.2.3), with a pH $\approx$ 8.1. Under such conditions, nearly all dissolved $CO_2(g)$ dissociates to the bicarbonate ion, and a small fraction dissociates to the carbonate ion. Certain organisms in the ocean are able to synthesize the carbonate ion with the calcium ion [Ca^{2+}] to form calcium carbonate [$CaCO_3(s)$, calcite] shells by

$$Ca^{2+} + CO_3^{2-} \rightarrow CaCO_3(s) \tag{3.17}$$

Calcium ion   Carbonate ion   Calcium carbonate

When shelled organisms die, they sink to the bottom of the ocean, where they are ultimately buried and their shells are turned into calcite rock.

Another removal process of $CO_2(g)$ from the air is **chemical weathering**, which is the breakdown and reformation of rocks and minerals at the atomic and molecular level by chemical reaction. One chemical weathering reaction is

$$CaSiO_3(s) + CO_2(g) \rightleftharpoons CaCO_3(s) + SiO_2(s) \tag{3.18}$$

Generic calcium silicate   Carbon dioxide   Calcium carbonate (calcite)   Silicon dioxide (quartz)

in which calcium-bearing silicate rocks react with $CO_2(g)$ to form calcium carbonate rock and quartz rock [$SiO_2(s)$]. At high temperatures, such as in the Earth's mantle, the reverse reaction also occurs, releasing $CO_2(g)$, which is expelled to the air by volcanic eruptions.

Another chemical weathering reaction involves carbon dioxide and calcite rock. During this process, $CO_2(g)$ enters surface water or groundwater by Reaction 3.16 and forms carbonic acid [$H_2CO_3(aq)$] by Reaction 3.17. The acid reacts with calcite, producing the calcium ion and the bicarbonate ion by

$$CaCO_3(s) + CO_2(g) + H_2O(aq)$$

Calcium carbonate   Gaseous carbon dioxide   Liquid water

$$\rightleftharpoons CaCO_3(s) + H_2CO_3(aq) \rightleftharpoons Ca^{2+} + 2HCO_3^-$$

Calcium carbonate   Carbonic acid   Calcium ion   Biocarbonate ion

$$\tag{3.19}$$

Because Reaction 3.19 is reversible, it can proceed either to the right or left. When the partial pressure of $CO_2(g)$ is high, the reaction proceeds to the right, breaking down calcite, removing $CO_2(g)$, and producing Ca^{2+}. Within soils, root and microorganism respiration and organic matter decomposition cause the partial pressure of $CO_2(g)$ to be about 10 to 100 times that in the atmosphere (Brook et al., 1983). Thus, calcite is broken down, and $CO_2(g)$ is removed more readily within soils than at soil surfaces. Dissolved calcium ultimately flows with runoff back to the oceans, where some of it is stored and the rest of it is converted to shell material.

3.6.2.2. Mixing Ratios

Figure 3.11 shows that outdoor $CO_2(g)$ mixing ratios have increased steadily since 1958 at the Mauna Loa Observatory, Hawaii. Average global $CO_2(g)$ mixing ratios have increased from approximately 275 ppmv in the mid-1700s to approximately 393 ppmv in 2011. The yearly increases are due to increased $CO_2(g)$ emission from fossil fuel combustion and permanent deforestation resulting from biomass burning.

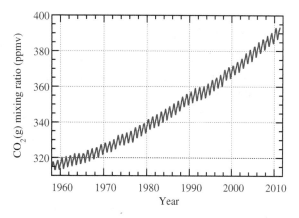

Figure 3.11. Yearly and seasonal fluctuations in carbon dioxide mixing ratio at Mauna Loa Observatory, Hawaii (19.4795°N, 155.603°W) between 1958 and 2011. From Mauna Loa Data Center (2011).

The seasonal fluctuation in $CO_2(g)$ is due to the seasonal cycle of photosynthesis and bacterial decomposition of **annual plants** (plants that germinate, flower, and die during a year), **deciduous trees** (trees that lose their leaves seasonally), and **coniferous trees** (trees with cones, needles, or scalelike leaves that lose their needles annually but are mostly evergreen). Such vegetation generally grows at high latitudes, outside the tropics. When annual plants, deciduous trees, and coniferous trees grow during the spring and summer, photosynthesis removes $CO_2(g)$ from the air, increasing plant mass. When such plants die during the autumn and winter, their decomposition by bacteria adds $CO_2(g)$ to the air, reducing plant mass. Because the Northern Hemisphere contains much more land area and plant mass than does the Southern Hemisphere, $CO_2(g)$ reductions are dominated by photosynthesis during the Northern Hemisphere spring and summer. In fact, the seasonal reductions in Figure 3.11 always start in May and end in October. Because tropical plants are mostly **evergreen** (maintaining their leaves during all seasons), they contribute less to the seasonal $CO_2(g)$ cycle than does vegetation at higher latitudes.

Typical indoor mixing ratios of $CO_2(g)$ are 700 to 2,000 ppmv but can exceed 3,000 ppmv when unvented appliances are used (Arashidani et al., 1996).

3.6.2.3. Equation for Estimating Mixing Ratio

For many types of analyses, it is useful to estimate the past, current, or future globally averaged mixing ratio of $CO_2(g)$. Here, a simple yet relatively accurate equation is presented that depends only on the emission rate, overall lifetime against removal, and preindustrial mixing ratio of $CO_2(g)$.

The **mixing ratio** (χ, ppmv) of $CO_2(g)$ can be expressed as $\chi = \chi_b + \chi_a$, where χ_b is the background or **preindustrial mixing ratio** prior to 1750 and χ_a is the **anthropogenic mixing ratio**, which is the mixing ratio due to anthropogenic emissions since 1750. The year 1750 precedes the **Industrial Revolution** (Section 4.1.3). Because the preindustrial mixing ratio was relatively constant ($\chi_b = 275$ ppmv) averaged over hundreds of years before 1750, preindustrial emission rates of $CO_2(g)$ are in relative equilibrium with preindustrial mixing ratios, so a time-dependent expression for χ_b is not needed for analyses on time scales of several hundred years. However, because anthropogenic emissions have been increasing over time, they are not in equilibrium with

χ_a, so χ_a must be calculated in a time-dependent manner.

The rate of change of the anthropogenic mixing ratio (χ_a, ppmv) of $CO_2(g)$ or any other gas with a constant emission source (E, ppmv yr^{-1}) and a specified e-folding lifetime (τ, years) against overall removal from the air can be calculated over time (t, years) by solving

$$\frac{d\chi_a(t)}{dt} = E - \frac{\chi_a(t)}{\tau} \tag{3.20}$$

where $\chi_a(t)$ is the anthropogenic mixing ratio expressed as a function of time. To determine the $CO_2(g)$ lifetime used in this equation, which varies in time slowly, it is necessary first to rearrange the equation as

$$\tau = \frac{\chi_a(t)}{E - d\chi_a(t)/dt} \tag{3.21}$$

The lifetime can be determined from data by using the $CO_2(g)$ mixing ratio $\chi_a(t)$ and its slope over time from Figure 3.11 (assuming $\chi_b = 275$ ppmv) and an annually varying anthropogenic emission rate consisting of a fossil fuel emission rate from Figure 12.11, plus a permanent deforestation emission rate of 1,500 to 2,700 Tg-C yr^{-1}. Because the lifetime determined from Equation 3.21 is constrained by data, it is referred to as the **data-constrained e-folding lifetime of $CO_2(g)$**. The data-constrained lifetime represents the overall lifetime against $CO_2(g)$ loss by all processes in Table 3.6. Whereas each individual loss process, acting on its own, would cause $CO_2(g)$ to have a different lifetime, the overall lifetime accounts for the aggregate of individual process lifetimes in a manner similar to Equation 1.13. However, unlike Equation 1.13, the data-constrained lifetime is obtained directly from observations rather than estimates. The data-constrained lifetime varies over time, accounting for the fact that as conditions change on the Earth, the lifetime changes as well.

Figure 3.12 shows the data-constrained e-folding lifetime of $CO_2(g)$ since 1960 calculated in this manner. The figure indicates that the lifetime has slowly increased but has historically ranged from around 30 to 50 years, with a rough mean of $\tau \sim 40$ years. The increase in lifetime with time is due in part to the fact that $CO_2(g)$ solubility in ocean water decreases with increasing temperature. Because ocean temperatures have been increasing, the relative rate of atmospheric $CO_2(g)$ dissolution into ocean water has been

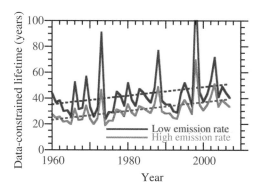

Figure 3.12. Data-constrained e-folding lifetime of carbon dioxide from Equation 3.21. The mean lifetime is about $\tau = 40$ years, with a rough range of 30–50 years. The dashed lines are linear fits to the data. The two curves correspond to low and high carbon emission rates from permanent deforestation added to constant levels of fossil fuel emissions in each case. Updated from Jacobson (2005c).

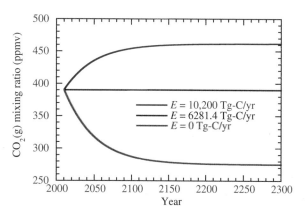

Figure 3.13. Time-dependent change in $CO_2(g)$ mixing ratio from Equation 3.22 under three different emission scenarios: (a) with 2010 emissions from fossil fuels plus permanent deforestation of 10,200 Tg-C yr^{-1}, (b) with an emission rate derived from $E = \chi_a(0)/\tau = 6,281.4$ Tg-C yr^{-1} that gives a constant mixing ratio of $CO_2(g)$, and (c) with zero emissions. Other conditions are $\chi_b = 275$ ppmv, $\chi_a(0) = 115$ ppmv, and $\tau = 40$ years.

decreasing. In addition, continued permanent deforestation since 1960 has reduced the vegetation mass available for $CO_2(g)$ uptake by photosynthesis. Both factors have contributed to a greater residence time of $CO_2(g)$ in the air since 1960.

The exact solution to the change in mixing ratio over time from Equation 3.20 is

$$\chi_a(t) = \chi_a(0)e^{-t/\tau} + \tau E(1 - e^{-t/\tau}) \quad (3.22)$$

where $\chi_a(0) = \chi(0) - \chi_b$ is the initial anthropogenic mixing ratio (ppmv) at time $t = 0$, which corresponds to a base year of interest. For example, if 2010 is selected as the base year ($t = 0$), and if the total mixing ratio of $CO_2(g)$ in 2010 is $\chi(0) = 390$ ppmv, then $\chi_a(0) = 390 - 275 = 115$ ppmv. Because emission rates are generally in units of Tg-C/yr, whereas Equation 3.22 requires emission units of ppmv-$CO_2(g)$/yr, the constant **2,184.82 Tg-C/ppmv-$CO_2(g)$** is used to convert units for the equation.

The solution in Equation 3.22 suggests that, in the absence of emissions, the mixing ratio of $CO_2(g)$ will decay exponentially to $1/e$ its initial value when $t = \tau$. The equation also suggests that, in steady state ($t \to \infty$), the mixing ratio approaches $\chi_a(\infty) = \tau E$. Once Equation 3.22 is solved, the new total mixing ratio of $CO_2(g)$ is $\chi(t) = \chi_b + \chi_a(t)$.

Figure 3.13 illustrates how $CO_2(g)$ changes according to Equation 3.22 under three constant emission scenarios: (a) one with 2010 emissions; (b) one with emissions reduced to $E = \chi(0)/\tau$, which forces the anthropogenic mixing ratio of $CO_2(g)$ over time to equal a constant initial anthropogenic mixing ratio; and (c) one with zero emissions. In all cases, the lifetime of $CO_2(g)$ is held to 40 years. In cases (a) and (c), the atmospheric mixing ratio is not initially in equilibrium with its emissions, so the atmospheric mixing ratio increases and decreases, respectively, over time to reach equilibrium.

Case (a) indicates that, even if E were held constant at the 2010 level, the $CO_2(g)$ atmospheric mixing ratio would increase for many years. Today, the $CO_2(g)$ emission rate is increasing, not staying constant. As such, the $CO_2(g)$ mixing ratio will rise more in reality than it will with the highest emission rate scenario in Figure 3.13. Cases (b) and (c) suggest that the only way to reduce the ambient mixing ratio of $CO_2(g)$ is to reduce emissions from 10,200 Tg-C yr^{-1} to less than 6,281 Tg-C yr^{-1}.

As illustrated in Example 3.8, it will be necessary to reduce 2030 emissions by about 64 percent to stabilize $CO_2(g)$ to 360 ppmv after 2030. The emission reduction in 2030 represents an 80 percent reduction of 2010 anthropogenic emissions. As such, any effort to stabilize $CO_2(g)$ at 360 ppmv requires at least an 80 percent reduction of 2010 anthropogenic emissions to account for anticipated future growth in emissions.

Example 3.7

For an emission rate of 8,400 Tg-C yr^{-1} from fossil fuels and 1,800 Tg-C yr^{-1} from permanent deforestation, calculate the equilibrium mixing ratio of $CO_2(g)$ in the atmosphere assuming an overall lifetime of 40 years.

Solution

The total anthropogenic emission rate of 10,200 Tg-C yr^{-1} is first converted to ppmv-$CO_2(g)$/yr with 2,184.82 Tg-C/ppmv-$CO_2(g)$, giving $E = 4.67$ ppmv-$CO_2(g)$ yr^{-1}. Substituting this emission rate into $\chi_a(\infty) = \tau E$ gives 187 ppmv. Adding this to $\chi_b = 275$ ppmv gives the steady-state total mixing ratio, $\chi(\infty) = 462$ ppmv $CO_2(g)$.

Example 3.8

In 2030, the projected fossil fuel plus permanent deforestation emission rate for the world is expected to be 12,800 Tg-C yr^{-1}. What percent reduction in 2030 anthropogenic emissions is needed to stabilize the atmospheric mixing ratio of $CO_2(g)$ to 360 ppmv? What percent of 2010 anthropogenic emissions does this reduction represent?

Solution

The emission rate needed to stabilize $CO_2(g)$ at a total mixing ratio of $\chi = 360$ ppmv is $E = \chi_a(\infty)/\tau = (360 - 275 \text{ ppmv})/40 \text{ years} = 2.125$ ppmv yr^{-1}. Multiplying this by 2,184.82 Tg-C/ppmv-$CO_2(g)$ gives 4,642.74 Tg-C yr^{-1}. The 2030 emission rate reduction needed is, therefore, 8,160 Tg-C yr^{-1}, or a 63.7 percent reduction. This reduction represents 80 percent of the 2010 emission rate of 10,200 Tg-C yr^{-1}.

Table 3.7. Storage reservoirs of carbon in Earth's atmosphere, oceans, sediments, and land in 2011

Location and form of carbon	Gigatonnes of carbon
Atmosphere	
Gas and particulate carbon	859
Surface oceans	
Live organic carbon	5
Dead organic carbon	30
Bicarbonate ion	500
Deep oceans	
Dead organic carbon	3,000
Bicarbonate ion	40,000
Ocean sediments	
Dead organic carbon	10,000,000
Land/ocean sediments	
Carbonate rock	60,000,000
Land	
Live organic carbon	800
Dead organic carbon	2,000

nearest carbon-containing rival, $CH_4(g)$. The atmospheric mass of carbon pales in comparison with the mass of carbon in other reservoirs, particularly the deep oceans, ocean sediments, and carbonate rocks. Table 3.7 shows the relative abundance of carbon in each reservoir.

Exchanges of carbon among the reservoirs include exchanges between the surface ocean (0–60 m below sea level) and deep ocean (below the surface ocean) by up- and down-welling of water. Exchanges also occur between the deep ocean and sediments by gravitational sinking and burial of dead organic matter and shell material, between the sediments and atmosphere by volcanism, between the land and atmosphere by oxygen-producing photosynthesis and bacterial metabolism, and between the surface ocean and atmosphere by evaporation and dissolution.

3.6.2.4. Carbon Reservoirs

At 393 ppmv $CO_2(g)$ in 2011, the atmosphere contained about 859 gigatonnes (GT) of carbon (1 ppmv-$CO_2(g)$ = 2,184.82 Tg-C; 1 GT = 10^9 tonnes = 10^{15} g = 1,000 Tg). Almost all carbon in the air is in its most oxidized form, $CO_2(g)$, but some is in its most reduced form, methane [$CH_4(g)$], and in many inorganic and organic gas and particle components. The mass of carbon in airborne $CO_2(g)$ is more than 200 times that of its

3.6.2.5. Health Effects

Carbon dioxide mixing ratios must be higher than 15,000 ppmv to affect human respiration. Mixing ratios higher than 30,000 ppmv cause headaches, dizziness, or nausea (Schwarzberg, 1993). In indoor air, carbon dioxide mixing ratios may build up enough to cause discomfort, but levels above 15,000 ppmv are rare. Outdoor mixing ratios of carbon dioxide are almost always too low to cause noticeable direct health impacts. One exception was an anomalous event at **Lake Nyos, Cameroon**. This crater lake sits above a pocket of magma that leaks carbon dioxide into the lake water. On August 21, 1986, a landside triggered a large release of carbon dioxide that asphyxiated 1,700 people and 3,500 animals in nearby towns and villages.

Although direct health effects of $CO_2(g)$ have been insignificant beyond the event at Lake Nyos, local or global increases in $CO_2(g)$ increase temperatures and water vapor. Both higher temperatures and higher water vapor independently increase ozone and particulate matter, as discussed in Section 12.5.6. Ozone and particulate matter both harm human health.

3.6.3. Carbon Monoxide

Carbon monoxide [$CO(g)$] is a tasteless, colorless, and odorless gas. Although $CO(g)$ is the most abundantly emitted spatially and temporally varying gas aside from $H_2O(g)$ and $CO_2(g)$, it plays a lesser role in ozone formation in urban areas than do many organic gases. In the background troposphere, however, it plays a relatively larger role. $CO(g)$ is a minor greenhouse gas because it absorbs some thermal-IR radiation. Its emission and oxidation to $CO_2(g)$ also affect global climate. $CO(g)$ is not important with respect to stratospheric ozone reduction or acid deposition. However, it is an important component of urban and indoor air pollution because it has harmful short-term health effects. It is one of six pollutants called **criteria air pollutants** (Section 8.1.6) for which U.S. National Ambient Air Quality Standards (NAAQS) were set by the U.S. Environmental Protection Agency (U.S. EPA) under the 1970 U.S. Clean Air Act Amendments (CAAA70). $CO(g)$ is regulated in most countries of the world today (Section 8.2).

3.6.3.1. Sources and Sinks

Table 3.8 summarizes the sources and sinks of $CO(g)$. A major source of $CO(g)$ is incomplete combustion during fossil fuel and biofuel combustion. $CO(g)$ emission sources include wildfires, biomass burning, nontransportation combustion, some industrial processes, and biological activity. Indoor sources of $CO(g)$ include water heaters, coal and gas heaters, and gas stoves. The major sink of $CO(g)$ is chemical conversion to $CO_2(g)$. $CO(g)$ is also lost by deposition to sea ice, snow, soil, vegetation, and structures and dissolution in surface water. Because it is relatively insoluble, its dissolution rate is low.

Table 3.8. Sources and sinks of atmospheric carbon monoxide

Sources	Sinks
Fossil fuel and biofuel combustion	Atmospheric chemical reaction to carbon dioxide
Biomass burning	Dissolution in surface water
Atmospheric chemical reaction	Deposition to sea ice, snow, soil, vegetation, and structures
Plants and biological activity in oceans	

In 2008, about 70 million metric tonnes of $CO(g)$ were emitted from anthropogenic sources in the United States (Figure 3.14). The largest source (73 percent) was on-road plus nonroad transportation. Carbon monoxide emissions decreased in the United States between 1970 and 2008 by about 62 percent, despite a large increase in the number of vehicles. The reason was due primarily to the development and mandatory use of the catalytic converter in motor vehicles (Chapter 8).

3.6.3.2. Mixing Ratios

Mixing ratios of $CO(g)$ in polluted urban air away from freeways are typically 2 to 10 ppmv. On freeways and in traffic tunnels, they can exceed 100 ppmv. In indoor air, hourly average mixing ratios can reach 6 to 12 ppmv when a gas stove is turned on (Samet et al., 1987). In the absence of indoor sources, $CO(g)$ indoor mixing ratios are usually less than are those outdoors (Jones,

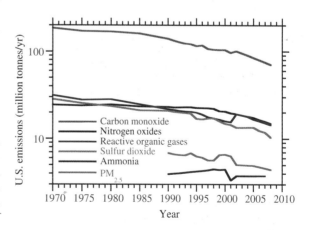

Figure 3.14. U.S. anthropogenic emissions by pollutant 1970–2008. Nitrogen oxides are $NO_x(g) = NO(g) + NO_2(g)$. $PM_{2.5}$ is particulate matter $\leq 2.5\ \mu m$ in diameter. Data from U.S. EPA (2011a).

1999). In the background troposphere, CO(g) mixing ratios generally vary from 50 to 150 ppbv.

3.6.3.3. Health Effects

Exposure to 300 ppmv of CO(g) for one hour causes a headache. Exposure to 700 ppmv for one hour causes death. From 1999 to 2004, CO(g) poisoning caused about 500 unintentional U.S. deaths per year. Many of these deaths were due to indoor CO(g) buildup from leaking indoor combustion heaters combined with poor ventilation. CO(g) poisons by dissolving in blood and replacing oxygen as an attachment to hemoglobin [Hb(aq)], an iron-containing compound. The conversion of $O_2Hb(aq)$ to $COHb(aq)$ (carboxyhemoglobin) causes suffocation. CO(g) can also interfere with $O_2(g)$ diffusion in cellular mitochondria and with intracellular oxidation (Gold, 1992). For the most part, the effects of CO(g) are reversible once exposure to CO(g) is reduced. However, following acute exposure, individuals may still express neurological or psychological symptoms for weeks or months, especially if they become temporarily unconscious (Choi, 1983).

3.6.4. Methane

Methane [$CH_4(g)$] is the most reduced form of carbon in the air. It is also the simplest and most abundant hydrocarbon and organic gas. Methane is a greenhouse gas that absorbs thermal-IR radiation twenty-five times more efficiently, molecule for molecule, than does $CO_2(g)$, but mixing ratios of carbon dioxide are much higher than are those of methane. Because its e-folding lifetime against chemical breakdown is so long (8–12 years), methane hardly enhances ozone formation in photochemical smog. In contrast, because it builds up in the background troposphere and breaks down slowly over time, it is an important contributor to global tropospheric ozone. In the stratosphere, methane has little effect on the ozone layer because other sources of ozone are much larger, but its chemical decomposition provides one of the few sources of stratospheric water vapor.

3.6.4.1. Sources and Sinks

Table 3.9 summarizes the sources and sinks of atmospheric methane. Methane is produced in anaerobic environments, where methanogenic bacteria consume organic material and excrete methane (Reaction 2.4). Ripe anaerobic environments include rice paddies (Figure 3.15); landfills; wetlands; and the digestive tracts of cattle, sheep, and termites. Methane is also

Table 3.9. Sources and sinks of atmospheric methane

Sources	Sinks
Methanogenic bacteria	Atmospheric chemical reaction
Natural gas leaks during fossil fuel mining and transport	Dissolution in surface water
Leaks from deep ocean and from under permafrost	Deposition to sea ice, snow, soil, vegetation, and structures
Fossil fuel, biomass, and biofuel combustion	Methanotrophic bacteria
Atmospheric chemical reaction	

Figure 3.15. Rice paddies, such as this one in the town of Sapa, Vietnam, produce not only an important source of food, but also methane gas. © Juliengrondin/Dreamstime.com.

produced in the ground from the decomposition of fossilized carbon. The resulting **natural gas**, which contains more than 90 percent pure methane, often leaks to the air or is harnessed and used for energy. Methane also leaks from storage under permafrost and in the deep ocean as temperatures rise (Section 12.3.2.4). Methane is also produced during biomass burning, fossil fuel and biofuel combustion, and atmospheric chemical reactions. Its sinks include chemical reaction; dissolution into surface water; deposition to sea ice, snow, soil, vegetation, and structures; and consumption by methanotrophic bacteria. The *e*-folding lifetime of tropospheric methane against loss by chemical reaction, its major loss process, is 8 to 12 years, which is slow in comparison with the lifetimes of other organic gases. Because methane is relatively insoluble, its dissolution rate in water is low. Approximately 80 percent of the methane in the air today is biogenic in origin; the rest originates from fuel combustion and natural gas leaks.

3.6.4.2. Mixing Ratios

Methane's average mixing ratio in the troposphere is near 1.85 ppmv, which is an increase from about 0.8 ppmv in the mid-1800s (Ethridge et al., 1992). Its tropospheric mixing ratio increased steadily due to increased biomass burning, fossil fuel and biofuel combustion, fertilizer use, and landfill development until the 1990s, when they leveled off through much of the 2000s. They began to increase again in 2007 (Figure 12.8b). Mixing ratios of methane are relatively constant with height in the troposphere, but decrease in the stratosphere and above due to its increase in chemical loss with increasing height. At 25 km, methane's mixing ratio is about half that in the troposphere.

3.6.4.3. Health Effects

Methane has no direct harmful human health effects at typical outdoor or indoor mixing ratios.

3.6.5. Ozone

Ozone [$O_3(g)$] is a relatively colorless gas at typical mixing ratios. It appears faintly purple when its mixing ratio is high because it weakly absorbs green wavelengths of visible light and transmits red and blue, which combine to form purple. Ozone exhibits an odor when its mixing ratio exceeds 0.02 ppmv. In urban smog or indoors, it is an air pollutant that harms humans, animals, plants, and materials. In the stratosphere, its absorption of UV radiation provides a protective shield for life on Earth. Although ozone is considered to be

Table 3.10. Sources and sinks of atmospheric ozone

Sources	Sinks
Chemical reaction of $O(g)$ with $O_2(g)$	Photolysis
	Atmospheric reaction in gas phase and on surfaces
	Dissolution in surface water
	Deposition to sea ice, snow, soil, vegetation, and structures

"good" when it is in the stratosphere and "bad" when it is in the boundary layer, ozone molecules are the same in both cases. In the United States, it is one of the six criteria air pollutants whose outdoor mixing ratios require control under CAAA70. It is also regulated in most countries.

3.6.5.1. Sources and Sinks

Table 3.10 summarizes the sources and sinks of ozone. Ozone is not emitted. Its only source in the air is chemical production. Sinks of ozone include chemical reaction; dissolution in surface water; and deposition to sea ice, snow, soil, vegetation, and structures. Because ozone is relatively insoluble, its dissolution rate is relatively low.

3.6.5.2. Mixing Ratios

In the background troposphere, ozone mixing ratios are 20 to 40 ppbv near sea level and 30 to 70 ppbv at higher altitudes. In urban air, ozone mixing ratios today range from <10 ppbv at night to 550 ppbv during afternoons in the most polluted cities of the world, with typical values of 70 to 150 ppbv during moderately polluted afternoons. Indoor ozone mixing ratios are almost always less than are those outdoors. In the stratosphere, peak ozone mixing ratios are 10 to 12 ppmv.

3.6.5.3. Health Effects

Symptoms of ozone health problems start with headache (>150 ppbv) and chest pain (>250 ppbv), and then advance to shortness of breath, cough, and breathing discomfort (>300 ppbv). Ozone decreases lung function for people who exercise steadily for more than an hour while exposed to >300 ppbv. Over the long term, ozone causes respiratory illness, the aging of lung tissue, and lung disease, all of which can be fatal. The elderly, sick, asthmatics, those with chronic bronchitis, those with emphysema, and young children are most susceptible to the corrosive effects of ozone on the respiratory tract and lungs. Above 100 ppbv, ozone affects animals by increasing their susceptibility to bacterial infection. It also interferes with the growth of plants and trees and deteriorates organic materials, such as rubber, textile dyes and fibers, and some paints and coatings (U.S. EPA, 1978). Ozone increases plant and tree stress and their susceptibility to disease, infestation, and death.

3.6.6. Sulfur Dioxide

Sulfur dioxide [$SO_2(g)$] is a colorless gas that exhibits a taste above 0.3 ppmv and a strong odor above 0.5 ppmv. $SO_2(g)$ is a precursor to sulfuric acid [$H_2SO_4(aq)$], an aerosol particle component that affects acid deposition, global climate, and the global ozone layer. $SO_2(g)$ is one of the six air pollutants for which NAAQS standards are set by the U.S. EPA under CAAA70. $SO_2(g)$ is regulated in most countries.

3.6.6.1. Sources and Sinks
Table 3.11 summarizes the major sources and sinks of $SO_2(g)$. Some sources include coal-fired power plants, vehicles, and volcanoes. $SO_2(g)$ is also produced chemically in the air from biologically produced dimethylsulfide [$DMS(g)$] and hydrogen sulfide [$H_2S(g)$]. $SO_2(g)$ is removed by chemical reaction; dissolution in water (as it is relatively soluble); and deposition to sea ice, snow, soil, vegetation, and structures. Anthropogenic $SO_2(g)$ emissions decreased in the United States between 1970 and 2008 by about 63 percent (Figure 3.14). In 2008, its total emissions were ~10.4 million tonnes.

3.6.6.2. Mixing Ratios
In the background troposphere, $SO_2(g)$ mixing ratios range from 10 pptv to 1 ppbv. In polluted air, they range

Table 3.11. Sources and sinks of atmospheric sulfur dioxide

Sources	Sinks
Oxidation of dimethylsulfide	Atmospheric chemical reaction to produce sulfuric acid
Volcanic emission	
Fossil fuel combustion	Dissolution in cloud drops and surface water
Mineral ore processing	
Chemical manufacturing	Deposition to sea ice, snow, soil, vegetation, and structures

from 1 to 30 ppbv. $SO_2(g)$ levels are usually lower indoors than outdoors. The indoor-to-outdoor ratio of $SO_2(g)$ is typically 0.1:1 to 0.6:1 in buildings without indoor sources (Jones, 1999). Indoor mixing ratios can range from 30 to 57 ppbv in homes equipped with a kerosene heater or a gas stove (Leaderer et al., 1993).

3.6.6.3. Health Effects
Because $SO_2(g)$ is soluble, it dissolves and oxidizes in the mucous membranes of the nose and respiratory tract to sulfurous acid [$H_2SO_3(aq)$], burning the respiratory tract upon inhalation. Particles containing sulfuric acid [$H_2SO_4(aq)$] are also soluble, but their deposition rates into the respiratory tract depend on particle size (Maroni et al., 1995). High concentrations of $SO_2(g)$ and $H_2SO_4(aq)$ can harm the lungs (Islam and Ulmer, 1979). Bronchiolar constrictions and respiratory infections can occur at mixing ratios greater than 1.5 ppmv. Long-term exposure to $SO_2(g)$ is associated with impaired lung function and other respiratory ailments (Qin et al., 1993). People exposed to open coal fires emitting $SO_2(g)$ are likely to suffer from breathlessness and wheezing more than are those not exposed to such fires (Burr et al., 1981). $SO_2(g)$ can cause death at mixing ratios of 400 to 500 ppmv and above.

3.6.7. Nitric Oxide

Nitric oxide [$NO(g)$] is a colorless gas and a free radical. It is important because it is a precursor to tropospheric ozone, nitric acid [$HNO_3(g)$], and particulate nitrate [NO_3^-]. Whereas $NO(g)$ does not directly affect acid deposition, nitric acid does. $NO(g)$ does not affect climate directly either, but ozone and nitrate particulate matter produced from it do. Natural $NO(g)$ reduces ozone in the upper stratosphere. Emissions of $NO(g)$ from jets that fly in the stratosphere also reduce stratospheric ozone. Outdoor levels of $NO(g)$ are not regulated in any country.

3.6.7.1. Sources and Sinks
Table 3.12 summarizes the sources and sinks of $NO(g)$. $NO(g)$ is emitted by microorganisms in soils and plants during denitrification, and it is produced by lightning, combustion, and chemical reaction. Combustion sources include fossil fuel, biofuel, and biomass burning. The primary sink of $NO(g)$ is atmospheric chemical reaction. Figure 3.14 shows that, in 2008, 14.8 million tonnes of anthropogenic $NO_x(g)$ [= $NO(g) + NO_2(g)$] were emitted in the United States. $NO(g)$ emissions represent about 5 to 15 percent those of $NO_x(g)$. About 58

Table 3.12. Sources and sinks of atmospheric nitric oxide

Sources	Sinks
Denitrification in soils and plants	Atmospheric chemical reaction
Lightning	Dissolution in surface water
Fossil fuel and biofuel combustion	Deposition to sea ice, snow, soil, vegetation, and structures
Biomass burning	
Photolysis and kinetic chemical reaction	

Table 3.13. Sources and sinks of atmospheric nitrogen dioxide

Sources	Sinks
Atmospheric chemical reaction	Atmospheric chemical reaction
Fossil fuel and biofuel combustion	Dissolution in surface water
Biomass burning	Deposition to sea ice, snow, soil, vegetation, and structures

percent of $NO_x(g)$ emissions were from transportation sources, whereas much of the rest were from electric power and industrial production. Between 1970 and 2008, $NO_x(g)$ emissions decreased 39.2 percent in the United States.

3.6.7.2. Mixing Ratios

A typical sea-level mixing ratio of $NO(g)$ in the background troposphere is 5 pptv. In the upper troposphere, $NO(g)$ mixing ratios are 20 to 60 pptv. In urban regions, $NO(g)$ mixing ratios reach 0.1 ppmv in the early morning but may decrease to zero by midmorning due to reaction with ozone.

3.6.7.3. Health Effects

Nitric oxide has no harmful human health effects at typical outdoor or indoor mixing ratios.

3.6.8. Nitrogen Dioxide

Nitrogen dioxide [$NO_2(g)$] is a brown gas with a strong odor. It absorbs short (blue and green) wavelengths of visible radiation, transmitting the remaining green and all red wavelengths, causing $NO_2(g)$ to appear brown. $NO_2(g)$ is an intermediary between $NO(g)$ emission and $O_3(g)$ formation. It is also an immediate precursor to nitric acid, a component of acid deposition. Natural $NO_2(g)$, like natural $NO(g)$, reduces ozone in the upper stratosphere. $NO_2(g)$ is one of the six criteria air pollutants for which outdoor standards are set by the U.S. EPA under CAAA70. It is regulated in many countries.

3.6.8.1. Sources and Sinks

Table 3.13 summarizes sources and sinks of $NO_2(g)$. Its major source is oxidation of $NO(g)$. Minor sources are fossil fuel and biofuel combustion and biomass burning. During combustion or burning, $NO_2(g)$ emissions are about 5 to 15 percent those of total $NO_x(g)$. Indoor sources of $NO_2(g)$ include gas appliances, kerosene heaters, wood-burning stoves, other biofuel burning for heating and cooking, and cigarettes. Sinks of $NO_2(g)$ include photolysis, chemical reaction, dissolution in surface water, and deposition to ground surfaces. $NO_2(g)$ is relatively insoluble in water.

3.6.8.2. Mixing Ratios

Mixing ratios of $NO_2(g)$ near sea level in the background troposphere range from 10 to 50 pptv. In the upper troposphere, they range from 30 to 70 pptv, and in urban regions, from 50 to 250 ppbv. Outdoors, $NO_2(g)$ reaches its peak mixing ratio during midmorning because sunlight breaks down most $NO_2(g)$ past midmorning. In homes with gas cooking stoves or unvented gas space heaters, weekly average $NO_2(g)$ mixing ratios range from 20 to 50 ppbv, although peak mixing ratios may reach 400 to 1,000 ppbv (Spengler, 1993; Jones, 1999).

3.6.8.3. Health Effects

Although exposure to high mixing ratios of $NO_2(g)$ harms the lungs and increases respiratory infections (Frampton et al., 1991), epidemiologic evidence suggests that exposure to typical mixing ratios has little health impact. Children and asthmatics are more susceptible to illness associated with high $NO_2(g)$ mixing ratios than are adults (Li et al., 1994). Pilotto et al. (1997) found that levels of $NO_2(g)$ greater than 80 ppbv resulted in more sore throats, colds, and absences from school. Goldstein et al. (1988) found that exposure to 300 to 800 ppbv $NO_2(g)$ in kitchens reduced lung capacity by about 10 percent. $NO_2(g)$ may trigger asthma by damaging or irritating and sensitizing the lungs, making people more susceptible to allergic response to indoor allergens (Jones, 1999). At mixing ratios much higher than normal indoor or outdoor conditions, $NO_2(g)$ can result in acute bronchitis (25–100 ppmv) or death (150 ppmv).

3.6.9. Lead

Lead [Pb(s)] is a gray-white, solid heavy metal with a low melting point that is present in air pollution as an aerosol particle component. It is soft, malleable, a poor conductor of electricity, and resistant to corrosion. It was first regulated as a criteria air pollutant in the United States in 1977. Many countries now regulate the emission and outdoor concentration of lead. Nevertheless, lead poisoning, particularly in children, is still a pervasive problem in many countries of the world.

3.6.9.1. Sources and Sinks

Table 3.14 summarizes the sources and sinks of atmospheric lead. Lead is emitted during combustion of vehicle fuel containing a lead additive, manufacture of lead acid batteries, crushing of lead ore, condensation of lead fumes from lead ore smelting, solid waste disposal, uplift of lead-containing soils, and crustal weathering of lead ore. Between the 1920s and 1970s, the largest source of atmospheric lead was vehicle combustion.

In December 1921, General Motors researcher **Thomas J. Midgley, Jr.** (1889–1944) (Figure 3.16) discovered that adding tetraethyl lead to fuel reduced automobile engine knock, increased octane levels, and increased engine power and efficiency. Midgley later discovered chlorofluorocarbons (CFCs), the precursors to stratospheric ozone destruction (Section 11.5.1).

Table 3.14. Sources and sinks of atmospheric lead

Sources	Sinks
Leaded fuel combustion in transportation vehicles	Deposition to oceans, sea ice, snow, soil, and vegetation
Lead acid battery manufacturing	Inhalation
Lead ore crushing and smelting	
Dust from soils contaminated with lead-based paint	
Solid waste disposal	
Crustal physical weathering	

Figure 3.16. Thomas J. Midgley, Jr. (1889–1944). Inventor of leaded gasoline and chlorofluorocarbons (CFCs), Midgley was born in Beaver Falls, Pennsylvania, in 1889. He grew up in Dayton and Columbus, Ohio, and graduated from Cornell University with a degree in mechanical engineering in 1911. In 1916, he joined the Dayton Engineering Laboratories Company (DELCO) as a researcher. DELCO became the main research laboratory for General Motors in 1919. In 1921, Midgley invented leaded gasoline, which he named ethyl. In 1923, he became vice president of the Ethyl Gasoline Corporation, a subsidiary of General Motors and Standard Oil. In 1924, he was forced to step down due to management problems. He returned to research on synthetic rubber at the Thomas and Hochwalt Laboratory in Dayton, Ohio, with funding from General Motors. In 1928, Midgley and two assistants invented CFCs as a substitute refrigerant for ammonia. Midgley moved on to became vice president of Kinetic Chemicals, Inc. (1930); director and vice president of the Ethyl-Dow Chemical Company (1933); and director and vice president of the Ohio State University Research Foundation (1940–1944). In 1940, he became afflicted with polio, causing him to lose a leg and design a system of ropes to pull himself out of bed. On November 2, 1944, he died of strangulation in the rope system, possibly by suicide. Edgar Fahs Smith Collection, University of Pennsylvania Library.

Although Midgley also found that ethanol/benzene blends reduced knock in engines, he chose to advocate for tetraethyl lead, and it was first marketed in 1923 under the name **ethyl gasoline**. That year, Midgley and three other General Motors laboratory employees experienced lead poisoning. Despite his personal experience and warnings sent to him from leading experts on the poisonous effects of lead, Midgley countered,

> The exhaust does not contain enough lead to worry about, but no one knows what legislation might come into existence fostered by competition and fanatical health cranks. (Kovarik, 1999)

Between September 1923 and April 1925, 17 workers at du Pont, General Motors, and Standard Oil died and 149 were injured due to lead poisoning during the processing of leaded gasoline. Five of the workers died in October 1924 at a Standard Oil of New Jersey refinery after they became suddenly insane from the cumulative exposure to high concentrations of tetraethyl lead. Despite the deaths and public outcry, Midgley continued to defend his additive. In a paper presented at the American Chemical Society conference in April 1925, he stated that

> tetraethyl lead is the only material available which can bring about these (antiknock) results, which are of vital importance to the continued economic use by the general public of all automotive equipment, and unless a grave and inescapable hazard exists in the manufacture of tetraethyl lead, its abandonment cannot be justified. (Midgley, 1925b)

Midgley's claim about the lack of antiknock alternatives contradicted his own work with ethanol/benzene blends, iron carbonyl, and other mixes that prevented knock.

In May 1925, the U.S. Surgeon General (head of the Public Health Service) put together a committee to study the health effects of tetraethyl lead. The Surgeon General argued that because no regulatory precedent existed, the committee would have to find striking evidence of serious and immediate harm for action to be taken against lead (Kovarik, 1999). Based on measurements that showed lead contents in fecal pellets of typical drivers and garage workers lower than those of lead industry workers, and based on the observations that drivers and garage workers had not experienced direct lead poisoning, the Surgeon General concluded that there were "no grounds for prohibiting the use of Ethyl gasoline" (U.S. Public Health Service, 1925). He did caution that further studies should be carried out.

Despite the caution, more studies were not carried out for thirty years, and effective opposition to the use of leaded gasoline ended.

By the mid-1930s, 90 percent of U.S. gasoline was leaded. Industrial backing of lead became so strong that, in 1936, the U.S. Federal Trade Commission issued a restraining order forbidding commercial criticism of tetraethyl lead, stating that it is

> . . . entirely safe to the health of (motorists) and to the public in general when used as a motor fuel, and is not a narcotic in its effect, a poisonous dope, or dangerous to the life or health of a customer, purchaser, user or the general public. (Federal Trade Commission, 1936)

Only in 1959 did the Public Health Service reinvestigate the issue of tetraethyl lead. At that time, it found it

> . . . regrettable that the investigations recommended by the Surgeon General's Committee in 1926 were not carried out by the Public Health Service. (U.S. Public Health Service, 1959)

Despite the concern, tetraethyl lead was not regulated as a pollutant in the United States until 1977. In 1975, the catalytic converter, which reduced emissions of carbon monoxide, hydrocarbons, and, eventually, oxides of nitrogen from cars, was invented. Because lead deactivates the catalyst in the catalytic converter, vehicles using catalytic converters could run only on unleaded fuel. Thus, the use of the catalytic converter in new cars inadvertently provided a convenient method to phase out the use of lead. The regulation of lead as a criteria air pollutant in the United States in 1977 due to its health effects also hastened the phase-out of lead as a gasoline additive.

Between 1970 and 1997, total lead emissions in the United States decreased from 199,000 to 3,600 tonnes per year, with only 13.3 percent of total lead emissions in 1997 coming from transportation (U.S. EPA, 1998). Today, the largest emission sources of lead in the United States are lead ore crushing and smelting and battery manufacturing. Since the 1980s, leaded gasoline has been phased out in many countries, although it is still an additive to gasoline in several others.

3.6.9.2. Concentrations

The U.S. National Ambient Air Quality Standard for lead is 1.5 μg m^{-3}, averaged over a calendar quarter. Ambient concentrations of lead between 1980 and 2009 decreased from about 1.4 to 0.1 μg m^{-3} (U.S. EPA, 2011c). The highest concentrations of lead are

Table 3.15. Selected toxic compounds and their major sources and health effects

Compound	Source	Health effects
Benzene	Gasoline combustion, solvents, tobacco smoke	Respiratory irritation, dizziness, headache, nausea, chromosome aberrations, leukemia, produces ozone
Styrene	Plastic and resin production, clothing, building materials	Eye and throat irritation, carcinogenic, produces ozone
Toluene	Gasoline combustion, biofuel and biomass burning, petroleum refining, detergent production, painting, building materials	Skin and eye irritation, fatigue, nausea, confusion, fetal toxicity, anemia, liver damage, dysfunction of central nervous system, coma, death, produces ozone
Xylene	Gasoline combustion, lacquers, glues	Eye, nose, and throat irritation; liver and nerve damage; produces ozone
1,3-Butadiene	Manufacture of synthetic rubber, combustion of fossil fuels, tobacco smoke	Eye, nose, and throat irritation; central nervous system damage; cancer; produces ozone
Acetone	Nail polish and paint remover, cleaning solvent	Nose and throat irritation, dizziness, produces ozone
Methyethylketone	Solvent in paints, adhesives, and cosmetics	Headaches, vision reduction, memory loss
Methylene chloride	Solvent, paint stripper, degreaser	Skin irritation, heart and nervous system disorders, carcinogenic
Vinyl chloride	Polyvinylchloride (PVC), plastics, building materials	Liver, brain, and lung cancer; mutagenic

Sources: Turco (1997); U.S. EPA (1998); Rushton and Cameron (1999).

now found near lead ore smelters and battery manufacturing plants.

3.6.9.3. Health Effects

Health effects of lead were known by the early Romans. Marcus Vitruvius Pollio, a Roman engineer, stated in the first century BC,

> We can take example by the workers in lead who have complexions affected by pallor. For when, in casting, the lead receives the current of air, the fumes from it occupy the members of the body, and burning them thereon, rob the limbs of the virtues of the blood. Therefore it seems that water should not be brought in lead pipes if we desire to have it wholesome. (Kovarik, 1998)

Lead, which accumulates in bones, soft tissue, and blood, can affect the kidneys, liver, and nervous system. Severe effects of lead poisoning include mental retardation, behavior disorders, and neurologic impairment. A disease associated with lead accumulation is **plumbism**. Symptoms at various stages include abdominal pains, a black line near the base of the gums, paralysis, loss of nerve function, dizziness, blindness, deafness, coma, and death. Low doses of lead have been linked to nervous system damage in fetuses and young children, resulting in learning deficits and low IQs. Lead may also contribute to high blood pressure and heart disease (U.S. EPA, 1998).

3.6.10. Hazardous Organic Compounds

The 1990 Clean Air Act Amendments (CAAA90) required that the U.S. EPA develop emission standards for each of 189 hazardous air pollutants that were believed to pose a risk of cancer, birth defects, or environmental or ecological damage. Toxic compounds are released into the air from area sources (e.g., buildings, industrial complexes), stationary point sources (e.g., smokestacks), and mobile sources. Whereas most toxics do not affect concentrations of ozone, the main component of photochemical smog, many others do. Table 3.15 identifies selected hazardous organics, their sources, and their effects on health and ozone levels.

3.7. Summary

In this chapter, the structure and composition of the present-day atmosphere were discussed. Air pressure, density, and temperature are interrelated by the equation of state. Pressure and density decrease exponentially with increasing altitude throughout the atmosphere. Temperature decreases with increasing altitude in the troposphere and mesosphere but increases with increasing altitude in the stratosphere and thermosphere. The temperature is affected by energy transfer processes, including conduction, convection, advection, and radiation. The troposphere is divided into the boundary layer and background troposphere. The daytime boundary layer is often characterized by a surface layer, a convective mixed layer, and an elevated inversion layer. The nighttime boundary layer is often characterized by a surface layer, a nocturnal boundary layer, a residual layer, and an elevated inversion layer. Pollutants emitted from the surface are initially confined to the boundary layer. Total air pressure consists of the sum of the partial pressure of each gas in the air. The air consist of well-mixed gases, such as molecular nitrogen, molecular oxygen, and argon, and spatially and temporally varying gases, such as water vapor, carbon dioxide, methane, ozone, and nitric oxide. Some spatially and temporally varying gases have adverse health effects; others affect radiation transfer through the air. Most are chemically reactive.

3.8. Problems

3.1. How tall must a column of liquid water with a density of 1,000 kg m^{-3} be to balance an atmospheric pressure of 1,000 hPa?

3.2. What do the balloon of Charles and an air parcel experiencing free convection have in common?

3.3. Calculate the conductive heat flux through 1 cm of clay soil if the temperature at the top of the soil is 283 K and that at the bottom is 284 K. How does this flux compare with the fluxes from Example 3.3?

3.4. If $T = 295$ K at 1 mm above the ground and the conductive heat flux is $H_c = 250$ W m^{-2}, estimate the air temperature at the ground.

3.5. If oxygen, nitrogen, and ozone did not absorb UV radiation, what would you expect the temperature profile in the atmosphere to look like between the surface and top of the thermosphere?

3.6. If temperatures in the middle of the sun are 15 million K and those on the Earth are near 300 K, what is the relative ratio of the thermal speed of a hydrogen atom in the middle of the sun to that on the Earth? (Assume the mass of one hydrogen atom is 1.66×10^{-27} kg molec^{-1}.) What are the actual speeds in both cases? How long would a hydrogen atom take to travel from the center of the sun to the Earth if it could escape the sun and if no energy losses occurred during its journey?

3.7. Do convection, conduction, or advection occur in the moon's atmosphere? Why or why not?

3.8. In the absence of an elevated inversion layer in Figure 3.4a, do you expect pollutant concentrations to build up or decrease? Why?

3.9. Why does the lower part of the daytime convective mixed layer lose its buoyancy at night? How does the loss of buoyancy in this region affect concentrations of pollutants emitted at night?

3.10. Why are the coldest temperatures in the lowest 50 km on Earth generally at the tropical tropopause?

3.11. Why does the tropopause height decrease with increasing latitude?

3.12. According to the equation of state, if temperature increases with increasing height and pressure decreases with increasing height in the stratosphere, how must density change with increasing height?

3.13. If $N_q = 1.5 \times 10^{12}$ molec cm^{-3} for ozone gas, $T = 285$ K, and $p_d = 980$ hPa, find the volume mixing ratio and partial pressure of ozone.

3.14. If carbon dioxide were not removed from the oceans by shell production and sedimentation, what would happen to its atmospheric mixing ratios?

3.15. When the carbon dioxide mixing ratio in the atmosphere increases, what happens to the concentration of dissolved carbon dioxide in the ocean?

3.16. Show and explain a possible chemical weathering process between carbon dioxide and magnesite [$MgCO_3(s)$].

3.17. Explain why the volume mixing ratio of oxygen is constant but its number concentration decreases exponentially with altitude in the bottom 100 km of the atmosphere.

3.18. Calculate the percent reduction in $CO_2(g)$

emissions needed to stabilize $CO_2(g)$ at a total mixing ratio of $\chi = 360$ ppmv, assuming a 2010 emission rate of 10,200 Tg-C yr^{-1} (from fossil fuel and permanent deforestation sources) and an e-folding lifetime for carbon dioxide of $\tau = 40$ years.

3.19. In 1990, the total $CO_2(g)$ emission rate for the world was $\sim$7,660 Tg-C yr^{-1}. To what steady-state mixing ratio does this emission rate correspond? Is this number lower or higher than the mixing ratio needed to stabilize climate, which is 360 ppmv?

Chapter 4

Urban Air Pollution

Anthropogenic air pollution has affected the environment since the development of the first human communities. Today, pollution arises due to the burning of wood, vegetation, coal, natural gas, oil, gasoline, kerosene, diesel, liquid biofuels, waste, and chemicals. Two general types of urban-scale pollution were identified in the twentieth century: London-type smog and photochemical smog. The former results from the burning of coal and other raw materials in the presence of a fog or strong temperature inversion, and the latter results from the emission of hydrocarbons and oxides of nitrogen in the presence of sunlight. In most places, urban air pollution consists of a combination of the two. In this chapter, urban outdoor gas-phase air pollution is discussed in terms of its history, early regulation, and chemistry.

4.1. History and Early Regulation of Outdoor Urban Air Pollution

Before the twentieth century, air pollution was treated as a regulatory or legal problem rather than a scientific problem. Because regulations were often weak or not enforced and health effects of air pollution were not well understood, pollution problems were rarely mitigated. In this section, a brief history of air pollution and its regulation until the 1940s is discussed.

4.1.1. Before 1200: Metal Smelting and Wood Burning

Available data suggest that air pollution in ancient Greece and the Roman Empire was likely severe due both to metal production and wood burning. From 500 BC to 300 AD, lead concentrations in Greenland ice cores were four times their natural values. These concentrations were due to the transport of pollution from Greek and Roman lead and silver mining and smelting operations (Hong et al., 1994). Lead emissions in ancient Greece occurred predominantly from smelting during the production of silver coins. Each gram of silver resulted in ~300 g of lead by-product. Lead emissions rose further during the Roman Empire because the Romans used lead in cookware, pipes, face powders, rouges, and paints. Cumulative lead pollution from smelting reaching Greenland between 500 BC and 300 AD was about 15 percent of that from leaded gasoline worldwide between 1930 and 1990.

Whereas the Romans mined up to 80,000 to 100,000 tonnes/yr of lead, they also mined 15,000 tonnes/yr of copper, 10,000 tonnes/yr of zinc, and 2 tonnes/yr of mercury (Nriagu, 1996). Ores containing these metals were smelted in open fires, producing pollution that was transported locally and long distances. For example, as evidenced by Greenland ice core data, the smelting of copper to produce coins during the Roman Empire and in China during the Song Dynasty (960–1279) increased atmospheric copper concentrations (Hong et al., 1996).

However, pollution in Rome was due to both metal smelting and wood burning. The Roman poet Horace noted thousands of wood-burning fires (Hughes, 1994) and the blackening of buildings (Brimblecombe, 1999). Air pollution events caused by emissions under strong inversions in Rome were called "heavy heavens" (Hughes, 1994).

Air pollution control in ancient Greece and Rome was relatively weak. In ancient Greece, town leaders were responsible for keeping sources of odors outside of town. In Rome, air pollution resulted in civil lawsuits.

During the fall of the Roman Empire, lead smelter emissions declined significantly. Such emissions rebounded, however, c. 1000 AD, following the discovery of lead and silver mines in Germany, Austria, Hungary, and other parts of central and eastern Europe (Hong et al., 1994). The discovery of these mines resulted in the movement of lead pollution from southern to central Europe. Wood burning, though, continued throughout the decline of the Roman Empire in all population centers of the world.

4.1.2. 1200–1700: Quicklime Production and Coal Burning

In London during the Middle Ages, a new source of pollution was the production of quicklime [CaO(s), calcium oxide] to form a building material. Quicklime was produced by the heating of limestone [which contains calcium carbonate, $CaCO_3(s)$] in kilns with oak brushwood as the primary fuel source. Quicklime was then mixed with water to produce a cement, slaked lime [calcium hydroxide, $Ca(OH)_2(s)$], a building material. This process released organic gases, nitric oxide, carbon dioxide, and organic particulate matter into the air.

Sea coal was introduced to London by 1228 and gradually replaced the use of wood as a fuel in lime kilns and forges. Wood shortages may have led to a surge in sea coal use by the mid-1200s. The burning of sea coal resulted in the release of sulfur dioxide, carbon dioxide, nitric oxide, soot, and particulate organic matter. Coal merchants in London worked on Sea Coal Lane, and they would sell their coal to limeburners on nearby Limeburner's Lane (Brimblecombe, 1987). The ratio of coal burned per forge to that burned per lime kiln may have been 1:1,000.

The pollution in London due to the burning of sea coal became sufficiently severe that a commission was ordered by King Edward I in 1285 to study and remedy the situation. The commission met for several years, and finally, in 1306, the king banned the use of coal in lime kilns. The punishment was "grievous ransom," which may have meant fines and furnace confiscation (Brimblecombe, 1987). One person may have been condemned to death because he violated the law three times

(Haagen-Smit, 1950). However, by 1329, the ban had either been lifted or lost its effect.

Between the thirteenth and eighteenth centuries, the use of sea coal and charcoal increased in England. Coal was used not only in lime kilns and forges, but also in glass furnaces, brick furnaces, breweries, and home heating. One of the early writers on air pollution was **John Evelyn** (1620–1706), who wrote *Fumifugium* or *The Inconveniencie of the Aer and the Smoake of London Dissipated* in 1661. He explained how smoke in London was responsible for the fouling of churches, palaces, clothes, furnishings, paintings, rain, dew, water, and plants. He blamed "Brewers, Diers, Limeburners, Salt and Sope-boylers" for the problems.

4.1.3. 1700–1840: The Steam Engine

Air quality in Great Britain (the union of England, Scotland, and Wales) increased by orders of magnitude in severity during the eighteenth century due to the invention of the **steam engine**, a machine that burned coal to produce mechanical energy. The idea for the steam engine originated with the French-born English physicist **Denis Papin** (1647–1712), who invented the pressure cooker in 1679 while working with Robert Boyle. In this device, water was boiled under a closed lid. The addition of steam (water vapor at high temperature) to the air in the cooker increased the total air pressure exerted on the cooker's lid from within. Papin noticed that the high pressure pushed the lid up. The phenomenon gave him the idea that steam could be used to push a piston up in a cylinder, and the movement of the cylinder could be used to do work. Although he designed a model of such a cylinder-and-piston steam engine in 1690, Papin never built one.

Capitalizing on the idea of Papin, **Thomas Savery** (1650–1715), an English engineer, patented the first practical steam engine in 1698. It replaced horses as a source of energy to pump water out of coal mines. The engine worked when water was boiled in a boiler to produce vapor that was transferred to a steam chamber. A pipe from the steam chamber to the water source in the mine was then opened. Liquid water was sprayed on the hot vapor in the steam chamber to recondense the vapor, creating a vacuum that sucked the water from the mine into the steam chamber. The pipe from the chamber to the mine was then closed, and another pipe from the chamber to outside the mine was opened. Finally, the boiler was fired up again to produce more water vapor to force the liquid water from the steam chamber out of the mine through the second pipe.

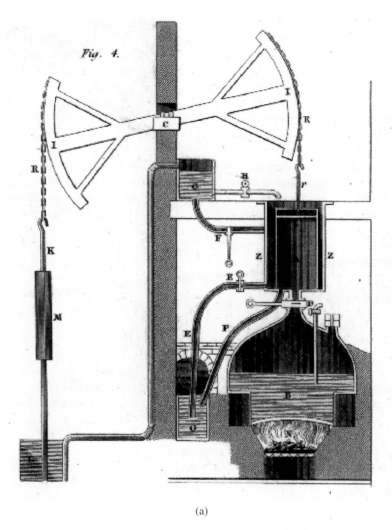

Figure 4.1. (a) Newcomen steam engine. From Brewster (1832), www.uh.edu/engines/epi69.htm and www.uh.edu/engines/watt2.gif.

In 1712, **Thomas Newcomen** (1663–1729), an English engineer, developed a more efficient steam engine than Savery by introducing a piston. Water evaporated into the steam chamber under the piston pushed the piston and an attached lever up (Figure 4.1a), producing mechanical energy. Liquid water was then squirted into the steam chamber to recondense the vapor, creating a vacuum that pulled the piston and lever down.

Because Savery had a patent on the steam engine, Newcomen was forced to enter into partnership with Savery to market the Newcomen engine. Newcomen's engine was used to pump water out of mines and to power waterwheels. Steam engines in the early eighteenth century were inefficient, capturing only 1 percent of their input raw energy (McNeill, 2000). Because coal

mines were not located in cities, early steam engines did not contribute much to urban pollution.

In 1763, Scottish engineer and inventor **James Watt** (1736–1819) was given a Newcomen steam engine to repair. He found an inefficiency with the original engine in that it allowed evaporation and condensation to occur in the same chamber. The squirting of cold water to condense vapor and create a vacuum for pulling the piston down resulted in more heat required to produce steam to push the piston back up if both operations occurred in the same chamber rather than in different chambers. Watt overcame this shortcoming by developing two chambers: one in which condensation occurred, and the second in which evaporation occurred. The condensation chamber stayed cold due to the squirting of cold water into it, whereas the evaporation chamber

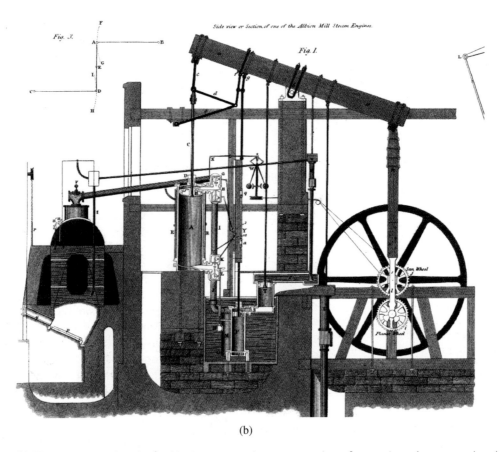

Figure 4.1 (b) Watt steam engine. In the Newcomen engine, evaporation of water into the steam chamber under the piston pushed the piston up. Liquid water squirted into the chamber recondensed the vapor, creating a vacuum to pull the piston back down. In the Watt engine, evaporation and condensation occurred in separate chambers, and motions were circular instead of up and down. From Brewster (1832), www.uh.edu/engines/epi69.htm and www.uh.edu/engines/watt2.gif.

stayed warm due to the boiling of water under it. In 1769, Watt patented this revised steam engine, which had double the efficiency of the previous one.

Watt made further modifications until 1800, including an engine in which the steam was supplied to both sides of the piston and an engine in which motions were circular instead of up and down (Figure 4.1b). Watt's engines were used not only to pump water out of mines, but also to provide energy for paper, iron, flour, cotton, and steel mills; distilleries; canals; waterworks; and locomotives. For many of these uses, steam engines were located in urban areas, thus increasing urban air pollution. Pollution became particularly severe because, although Watt had improved the steam engine, it still captured only 5 percent of the energy it used by 1800 (McNeill, 2000). Because the steam engine was a large, centralized source of energy, it was responsible for the shift from the artisan shop to the factory system of

industrial production during the **Industrial Revolution** of 1760 to 1880 (Rosenberg and Birdzell, 1986).

In the nineteenth century, the steam engine was used not only in Great Britain, but also in many other countries, providing a new source of energy and pollution. The steam engine played a large part in a hundredfold global increase in coal combustion between 1800 and 1900. Industries centered on coal burning arose in the United States, Belgium, Germany, Russia, Japan, India, South Africa, and Australia, among other nations.

Pollution problems in Great Britain worsened not only because of steam engine emissions, but also because of coal combustion in furnaces and boilers and chemical combustion in factories. Between 1800 and 1900, the death rate due to air pollution in Great Britain may have been four to seven times that worldwide (Clapp, 1994).

4.1.4. Regulation in the United Kingdom: 1840–1930

Pollution due to industrial growth in the United Kingdom (UK; the 1803 union of Great Britain and Northern Ireland) grew significantly enough so that, by the 1840s, legislation was considered to control it. In 1843, a committee was first set up in London to obtain information about pollution from furnaces and heated steam boilers. Bills were subsequently brought before Parliament in 1843 and 1845 to limit emissions, but they were defeated. A third bill was withdrawn in 1846. In 1845, the Railway Clauses Consolidated Act, which required railway engines to consume their own smoke, was enacted. The act stated that

> . . . every locomotive steam-engine to be used on a railway shall, if it use coal or other similar fuel emitting smoke, be constructed on the principle of consuming, and so as to consume its own smoke; and if any engine be not so constructed, the company or party using such engine shall forfeit 5 pounds for every day during which such engine shall be used on the railway . . . (In *Manchester, Sheffield and Lincolnshire Rail Co. v Wood*, 29 Law J. Rep. M.C. 29)

In 1846, a public health bill passed with a clause discussing the reduction of smoke emissions from furnaces and boilers. The clause was removed following pressure by industry. Additional bills failed in 1849 and 1850.

In 1851, an emission clause passed in a sewer bill for the city of London, and this clause was enforced through citations. In 1853, a Smoke Nuisance Abatement (Metropolis) Act also passed through Parliament. This law gave power to the Home Office to appoint an inspector who would work with the police to reduce smoke from furnaces in London and from steam vessels near London Bridge. This law was enforced only after several years of delay. In 1863, Parliament passed the British Alkali Act, which reduced emissions of hydrochloric acid gas formed during soap production. Air pollution legislation also appeared in the Sanitary Acts of 1858 and 1866, the Public Health Acts of 1875 and 1891, and the Smoke Abatement Act of 1926.

4.1.5. Regulation in the United States: 1869–1940

Most pollution in the United States in the nineteenth century resulted from the burning of coal for manufacturing, home heating, and transportation and wood for heating and cooking. Early pollution regulations were set not by state or national agencies, but by municipalities. The first clean air law in the United States may have been a 1869 ordinance by the city of **Pittsburgh** outlawing the burning of soft coal in locomotives within the city limits. This law was not enforced. In 1881, **Cincinnati** passed a law requiring smoke reductions and the appointment of a smoke inspector. Again, the ordinance was not enforced, although the three primary causes of death in Cincinnati in 1886 – tuberculosis, pneumonia, and bronchitis – were all lung related (Stradling, 1999). In 1881, a smoke reduction law passed in **Chicago**. Although this law had the support of the judiciary, it had little effect.

St. Louis may have been the first city to pass potentially effective legislation. In 1893, the city council first passed a law forbidding the emission of "dense black or thick gray smoke" and then a second law creating a commission to appoint an inspector and examine smoke-related issues. However, the ordinance was overturned by the Missouri State Supreme Court in 1897. The Court stated that the ordinance exceeded the "power of the city under its charter" and was "wholly unreasonable" (Stradling, 1999). Nevertheless, by 1920, air pollution ordinances existed in 175 municipalities; by 1940, this number increased to 200 (Heinsohn and Kabel, 1999).

In 1910, **Massachusetts** became the first state to regulate air pollution by enacting smoke control laws for the city of Boston. The first federal involvement in air pollution was probably the creation of an **Office of Air Pollution** by the **Department of Interior's Bureau of Mines** in the early 1900s. This office, the purpose of which was to control emission of coal smoke, was relatively inactive and was thus eliminated shortly thereafter.

Although early regulations in the United Kingdom and United States did not reduce pollution, they led to pollution control technologies, such as technologies for recycling chlorine from soda ash factory emissions and the electrostatic precipitator, used for reducing particle emissions from smokestacks. Inventions unrelated to air pollution relocated some pollution problems. The advent of the electric motor in the twentieth century, for example, centralized sources of combustion at electric utilities, reducing air pollution at individual factories that had relied on energy from the steam engine.

4.1.6. London-Type Smog

In 1905, the term **smog** was first used to describe the combination of smoke and fog that was visible in several cities throughout Great Britain by **Harold Antoine Des Voeux**, a member of the Coal Smoke Abatement Society in London. The term spread after Des Voeux presented a report at the Manchester Conference of the Smoke Abatement League of Great Britain in 1911, describing smog events in the autumn of 1909 in Glasgow and Edinburgh, Scotland, that killed more than 1,000 people.

The smoke in smog at the time was due to emissions from the burning of coal and other raw materials. Coal was combusted to generate energy, and raw materials were burned to produce chemicals, particularly **soda ash** [$Na_2CO_3(s)$], used in consumable products, such as soap, detergents, cleansers, paper, glass, and dyes. To produce soda ash, many materials, including charcoal, elemental sulfur, potassium nitrate, sodium chloride, and calcium carbonate were burned, emitting soot, sulfuric acid, nitric acid, hydrochloric acid, calcium sulfide, and hydrogen sulfide, among other compounds. Emissions from soda ash factories were added to the landscape of the United Kingdom and France starting in the early 1800s.

Today, pollution resulting from coal and chemical combustion smoke in the presence of fog or a low-lying temperature inversion is referred to as **London-type smog**. Some of the chemistry associated with London-type smog is described in Chapter 10. Several deadly London-type smog events have occurred in the nineteenth and twentieth centuries. These episodes, discussed briefly next, provided motivation for modern-day air pollution regulation.

Figure 4.2. Daytime darkness during the 1952 smog event along the Strand in London. Library of Congress, Prints and Photographs Division, NYWT&S Collection, reproduction no. LC-USZ62-114381.

4.1.6.1. London, United Kingdom

Several London-type smog events were recorded in London in the nineteenth and twentieth centuries. These include events in December 1873 (270–700 more deaths than the average death rate for this period), January 1880 (700–1,100 excess deaths), December 1892 (1,000 excess deaths), November 1948 (300 excess deaths), **December 1952** (4,000 excess deaths), January 1956 (480 excess deaths), December 1957 (300–800 excess deaths), and December 1962 (340–700 excess deaths) (Brimblecombe, 1987).

The worst of these episodes occurred in December 1952, with 4,000 excess deaths. Although excess deaths occurred in every age group, the greatest number occurred for those older than forty-five. People with a history of heart or respiratory problems comprised 80 percent of those who died. During the episodes, temperature inversions coupled with fog and heavy emissions

of pollutants, particularly from combustion of coal and other raw materials, were blamed for the disasters. During the 1952 episode, the peak mixing ratio of $SO_2(g)$ and peak concentration of particulate matter were estimated to be 1.4 ppmv and 4,460 $\mu g\ m^{-3}$, respectively. (This compares with 24-hour federal standards in the United States in 2011 of 0.14 ppmv for $SO_2(g)$ and 150 $\mu g\ m^{-3}$ for particulate matter less than 10 μm in diameter.) The particle and fog cover was so heavy during the event that the streets of London were dark at noon, and it was necessary for buses to be guided by lantern light (Figure 4.2).

4.1.6.2. Meuse Valley, Belgium

In December 1930, a five-day fog event in the presence of a strong temperature inversion and heavy emissions of $SO_2(g)$ from coal burning resulted in 63 deaths and 6,000 illnesses, mostly during the last two days of the pollution episode. The majority of those who died were

(a)

(b)

(c)

Figure 4.3. Panoramic views of (a) Reading, Pennsylvania, c. 1909, photo by O. Conneaut, available from Library of Congress, Prints and Photographs Division, Washington, DC; (b) Youngstown, Ohio, c. 1910, photo by O. Conneaut, available from Library of Congress Prints and Photographs Division, Washington, DC; and (c) Indiana Steel Co.'s big mills, Gary, Indiana, c. 1912, photo by Crose Photo Co., available from Library of Congress Prints and Photographs Division, Washington, DC.

elderly with previous heart or lung disease. Symptoms of the victims included chest pain, cough, shortness of breath, and eye irritation.

4.1.6.3. Donora, Pennsylvania, United States

Pittsburgh, Pennsylvania, is located near large coal deposits and major river arteries. In 1758, coal was first burned in Pittsburgh to produce energy for iron and glass manufacturing. By 1865, half of all glass and 40 percent of all iron in the United States were produced in Allegheny County, where Pittsburgh is located (McNeill, 2000). In the late 1800s, the county emerged as a major industrial consumer of coal, particularly for steel. In 1875, Andrew Carnegie opened the Edgar Thomson Works in Braddock, Pennsylvania, introducing a source of high-volume steel to Allegheny County.

Steel manufacturing spread to many other cities in the region, including Reading, Pennsylvania; Youngstown, Ohio; and Gary, Indiana. These cities burned coal not only for steel manufacturing, but also for iron manufacturing and railway transportation. Figure 4.3 shows that the burning of coal in Reading, Youngstown, and Gary darkened the skies of these cities.

Uncontrolled burning of coal in Allegheny County continued through 1941, when the first strong smoke abatement laws were passed in Pittsburgh. The 1941 laws, however, were suspended until the end of World War II, and even then resulted in only minor improvements in air quality due to a lack of enforcement.

Donora, Pennsylvania, is a town south of Pittsburgh along the Monongahela River. Between October 26 and 31, 1948, heavy emissions of soot and sulfur dioxide

Figure 4.4. Noontime photograph of Donora, Pennsylvania, on October 29, 1948, during a deadly smog event. Copyright Photo Archive/Pittsburgh Post-Gazette, 2001. All rights reserved. Reprinted with permission.

from steel mills and metal fumes from the Donora Zinc Works, a zinc smelter, under a strong temperature inversion resulted in the deaths of 20 people and 10 dogs, as well as the respiratory illness of 7,000 of the town's 14,000 human residents. Of the 20 people who died, 14 had a known heart or lung problem. Both young and old people became ill, and most of the illnesses arose by the third day. Symptoms included cough, sore throat, chest constriction, shortness of breath, eye irritation, nausea, and vomiting. A sample of 229 dogs and 165 cats after the event found that 15.5 percent of dogs and 7.3 percent of cats suffered illness (Catcott, 1955). The smog event darkened the city during peak daylight hours (Figure 4.4).

During the 1940s and 1950s, the darkening of cities throughout the United States and Europe during the day due to factory smoke and vehicle exhaust was not unusual (e.g., Figure 4.5). This unabated air pollution shortened the lives of city dwellers by several years, contributed to millions of illnesses and hundreds of thousands of premature deaths per year, as well as a general feeling of helplessness among those trapped under the pollution.

4.1.7. Photochemical Smog

Although short, deadly air pollution episodes have attracted public attention, persistent pollution problems in sunny regions gained additional notoriety in the twentieth century. Most prominent was a layer of pollution that formed almost daily over Los Angeles, California. In the early twentieth century, this layer was caused by a combination of directly emitted smoke (London-type smog) and chemically formed pollution called **photochemical smog**. Sources of the smoke were primarily particles from fuel combustion in factories and power plants and from the open burning of waste. Sources of the chemically formed pollution were primarily gases from factory and automobile emissions. In 1903, the factory smoke was so thick that one day, residents of Los Angeles believed that they were observing an eclipse of the sun (South Coast Air Quality Management District (SCAQMD), 2011). Between 1905 and 1912, regulations controlling smoke emissions were adopted by the Los Angeles City Council. However, such regulations were of little benefit, as illustrated by Figure 4.6, which shows clouds of dark smoke billowing out of stacks and traveling across the city.

As automobile use increased, the relative fraction of photochemical versus London-type smog in Los Angeles increased between the 1910s and 1940s. From 1939 to 1943, visibility in Los Angeles declined precipitously. On July 26, 1943, a plume of pollution engulfed downtown Los Angeles, reducing visibility to three blocks. Even after a local Southern California Gas Company plant suspected of releasing butadiene gas was shut

(a)   (b)

Figure 4.5. (a) Photograph of Fifth Avenue, Pittsburgh, Pennsylvania, at 11:00 AM on November 5, 1945, Library of Congress, Prints and Photographs Division, Washington, DC, reproduction no. LC-USZ62-76932. (b) Smog in New York City, November 20, 1953, Library of Congress, Prints and Photographs Division, Washington, DC, NYWT&S Collection, reproduction no. LC-USZ62-114346.

down, the pollution event continued, suggesting that the pollution was not from the plant.

In 1945, the Los Angeles County Board of Supervisors banned the emission of dense smoke and designed an office called the Director of Air Pollution Control. The city of Los Angeles mandated emission controls during the same year. Also in 1945, Los Angeles County Health Officer **H. O. Swartout** suggested that pollution in Los Angeles originated not only from smokestacks, but also from locomotives, diesel trucks, backyard incinerators, lumber mills, city dumps, and automobiles. In 1946, an air pollution expert from St. Louis,

Raymond R. Tucker, was hired by the *Los Angeles Times* to suggest methods of ameliorating air pollution problems in Los Angeles. Tucker proposed twenty-three methods and suggested that a countywide air pollution agency be set up to enforce air pollution regulations (SCAQMD, 2011).

In the face of opposition from oil companies and the Los Angeles Chamber of Commerce, the Los Angeles County Board of Supervisors drafted legislation to be submitted to the state of California that would allow counties throughout the state to set up unified air pollution control districts. The legislation was supported

Figure 4.6. Panoramic view of Los Angeles, California, taken from Third and Olive Streets, December 3, 1909. Photo by Chas. Z. Bailey, available from Library of Congress, Prints and Photographs Division, Washington, DC.

(a)

(b)

Figure 4.7. (a) Dozens of open waste incineration sites, such as this, checkered Los Angeles until the late 1940s, when garbage collection and regulated incineration began. Photo: August 2, 1945. (b) A patrolman tells a new resident of Los Angeles, on June 25, 1960, that burning waste in a backyard incinerator was banned effective October 1, 1957 and to extinguish the fire immediately. Photos courtesy Los Angeles Public Library, *Los Angeles Herald Examiner* Photo Collection.

by the League of California Cities, who believed that air pollution could be regulated more effectively at the county level rather than at the city level. The bill passed 73 to 1 in the California State Legislature and 20 to 0 in the State Assembly; it was signed by Governor Earl Warren on June 10, 1947. On October 14, 1947, the Board of Supervisors created the first regional air pollution control agency in the United States, the Los Angeles Air Pollution Control District. On December 30, 1947, the district issued its first mandate, requiring major industrial emitters to obtain emission permits. In the late 1940s and early 1950s, the district further regulated open burning in garbage dumps (e.g., Figure 4.7a), emission of sulfur dioxide from refineries, and emission from industrial gasoline storage tanks (1953). In 1954, it banned the use of 300,000 backyard incinerators in Los Angeles, effective October 1, 1957 (Figure 4.7b). Nevertheless, smog problems in Los Angeles persisted, resulting in serious respiratory and cardiovascular disease and death, eye irritation, and general discomfort (Figure 4.8).

In 1950, 1957, and 1957, Orange, Riverside, and San Bernardino counties, respectively, established their own air pollution control districts. These districts merged with the Los Angeles district in 1977 to form the South Coast Air Quality Management District (SCAQMD), which currently controls air pollution in the four-county Los Angeles region.

The chemistry of photochemical smog was first elucidated by **Arie Haagen-Smit** (1900–1977), a Dutch professor of biochemistry at the California Institute of Technology (Figure 4.9). In 1948, Haagen-Smit began studying plants damaged by smog. In 1950, he found that plants exposed to ozone sealed in a chamber exhibited the same type of damage as plants exposed to outdoor smog (e.g., Figure 4.10). Symptoms of damage included the silvering or bronzing of the underside of the leaves of spinach, sugar beets, and endives, and the bleaching of alfalfa and oats. About fifty other chemicals in the air tested did not produce the same damage, indicating that ozone was present in the smog and was causing the damage (Haagen-Smit, 1950).

Haagen-Smit also found that ozone caused eye irritation, damage to materials, and respiratory problems. Meanwhile, tire manufacturers discovered that their rubber tires deteriorated faster in Los Angeles than in other parts of the United States. Bradley and Haagen-Smit (1951) determined that bent rubber could crack within seven minutes when exposed to 100 ppbv ozone, whereas such cracking took thirty to sixty minutes when exposed to 20 to 30 ppbv ozone. Thus, ozone also damaged tires and other rubber products.

(a)

(b)

Figure 4.8. (a) A man wearing a gas mask and women fighting eye irritation walk on a sidewalk in smoggy downtown Los Angeles on September 19, 1958. By Art Worden. (b) Smog soaks the Los Angeles Basin on August 22, 1964. By Mike Sergieff. Photos courtesy Los Angeles Public Library, *Los Angeles Herald Examiner* Collection.

Figure 4.9. Arie Haagen-Smit (1900–1977). Courtesy the Archives, California Institute of Technology.

In 1952, Haagen-Smit discovered the mechanism of ozone formation in smog. In the laboratory, he produced ozone from oxides of nitrogen and reactive organic gases in the presence of sunlight. He suggested that ozone and these precursors were the main constituents of Los Angeles photochemical smog.

On the discovery of the sources of ozone in smog, oil companies and business leaders argued that the ozone

Figure 4.10. Smog damage to sugar beets. Courtesy South Coast Air Quality Management District, www.aqmd.gov.

Figure 4.11. Photochemical smog in Los Angeles, California, on July 23, 2000. The smog hides the high-rise buildings in downtown Los Angeles and the mountains in the background. Photo by Mark Z. Jacobson.

originated from the stratosphere. Subsequent measurements showed that ozone levels were low at nearby Catalina Island, proving that ozone in Los Angeles was local in origin.

Photochemical smog has persisted to this day in the Los Angeles Basin (e.g., Figure 4.11). It has also been observed in most cities of the world. Notable sites of photochemical smog include Mexico City, Santiago, Tokyo, Beijing, Calcutta, Johannesburg, and Athens. Unlike London-type smog, photochemical smog does not require smoke or fog for its production. Both London-type and photochemical smog are exacerbated by a strong temperature inversion (Section 3.3.1.1).

In the following sections, gas chemistry of background tropospheric air and photochemical smog are discussed. Regulatory efforts to control smog since the 1940s are summarized in Chapter 8.

4.2. Gas-Phase Chemistry of the Background Troposphere

Today, no region of the global atmosphere is unaffected by anthropogenic pollution. Nevertheless, the background troposphere is cleaner than are urban areas, and to understand photochemical smog, it is useful to examine the gas-phase chemistry of the background troposphere. The background troposphere is affected by inorganic, light organic, and a few heavy organic gases. The sources of many gases in the background troposphere are natural and include microorganisms, vegetation, volcanos, lightning, and fires (which may be natural or anthropogenic). The major heavy organic gas in the background troposphere is isoprene, a hemiterpene emitted by vegetation. Because rural areas can have a lot of vegetation, isoprene emissions occur over many land areas of the world. The background troposphere also receives inorganic and light organic gases emitted anthropogenically from urban regions. However, most heavy organic gases emitted in urban air, such as toluene and xylene, break down chemically over hours to a few days; thus, these gases do not reach the background troposphere, even though their breakdown products do. In the following subsections, inorganic and light organic chemical pathways important in the background troposphere are described.

4.2.1. Photostationary-State Ozone Concentration

In the background troposphere, the ozone [O_3(g)] mixing ratio is controlled primarily by a set of three reactions involving itself, nitric oxide [NO(g)], and nitrogen dioxide [NO_2(g)]. These reactions are

$$\dot{N}O(g) + O_3(g) \rightarrow \dot{N}O_2(g) + O_2(g) \quad (4.1)$$
Nitric   Ozone   Nitrogen   Molecular
oxide   dioxide   oxygen

$$\dot{N}O_2(g) + h\nu \rightarrow \dot{N}O(g) + \cdot\dot{O}(g) \quad \lambda < 420\,nm \quad (4.2)$$
Nitrogen   Nitric   Atomic
dioxide   oxide   oxygen

$$\cdot\dot{O}(g) + O_2(g) \overset{M}{\rightarrow} O_3(g) \quad (4.3)$$
Atomic   Molecular   Ozone
oxygen   Oxygen

Background tropospheric mixing ratios of O_3(g) (20–60 ppbv) are much higher than are those of NO(g) (1–60 pptv) or NO_2(g) (5–70 pptv). Because the mixing ratio of NO(g) is much lower than is that of O_3(g), Reaction 4.1 does not deplete ozone during the day or night in background tropospheric air. In urban air, Reaction 4.1

can deplete local ozone at night because NO(g) mixing ratios at night may exceed those of O_3(g).

If k_1 (cm^3 molec^{-1} s^{-1}) is the rate coefficient of Reaction 4.1 and J (s^{-1}) is the photolysis rate coefficient of Reaction 4.2, the volume mixing ratio of ozone can be calculated from these two reactions as

$$\chi_{O_3(g)} = \frac{J}{N_d k_1} \frac{\chi_{NO_2(g)}}{\chi_{NO(g)}} \quad (4.4)$$

where χ is volume mixing ratio (molecule of gas per molecule of dry air) and N_d is the concentration of dry air (molecules of dry air per cubic centimeter). This equation is called the **photostationary-state relationship**. The equation does not state that ozone is affected by only NO(g) and NO_2(g). Indeed, other reactions affect ozone, including ozone photolysis. Instead, Equation 4.4 says that the ozone mixing ratio is a function of the ratio of the NO_2(g) to NO(g) ratio. Many other reactions can affect the NO_2(g):NO(g) ratio, thereby affecting ozone.

Example 4.1
Find the photostationary-state mixing ratio of O_3(g) at midday when $p_d = 1{,}013$ hPa, $T = 298$ K, $J \approx 0.01$ s^{-1}, $k_1 = 1.8 \times 10^{-14}$ cm^3 molec^{-1} s^{-1}, $\chi_{NO(g)} = 5$ pptv, and $\chi_{NO_2(g)} = 10$ pptv (typical free-tropospheric mixing ratios).

Solution
From Equation 3.12, $N_d = 2.46 \times 10^{19}$ molec cm^{-3}. Substituting this into Equation 4.4 gives $\chi_{O_3(g)} = 45.2$ ppbv, which is a typical free-tropospheric ozone mixing ratio.

4.2.2. Daytime Removal of Nitrogen Oxides

During the day, NO_2(g) is removed slowly from the photostationary-state cycle by the reaction

$$\dot{N}O_2(g) + \dot{O}H(g) \overset{M}{\rightarrow} HNO_3(g) \quad (4.5)$$
Nitrogen   Hydroxyl   Nitric
dioxide   radical   acid

Although HNO_3(g) photolyzes back to NO_2(g) + OH(g), its e-folding lifetime against photolysis is 15 to 80 days, depending on the day of the year and the latitude. Because this lifetime is fairly long, HNO_3(g) serves as a sink for **nitrogen oxides** [NO_x(g) = NO(g) + NO_2(g)] in the short term. In addition, because HNO_3(g) is soluble, much of it dissolves in cloud drops or aerosol particles before it photolyzes back to NO_2(g).

Reaction 4.5 requires the presence of the **hydroxyl radical** [OH(g)], an oxidizing agent that decomposes (scavenges) many gases. Given enough time, OH(g) breaks down every organic gas and most inorganic gases in the air.

The daytime average OH(g) concentration in the clean background troposphere usually ranges from 2×10^5 to 3×10^6 molec cm^{-3}. In urban air, OH(g) concentrations typically range from 1×10^6 to 1×10^7 molec cm^{-3}. The primary free-tropospheric source of OH(g) is the pathway

$$O_3(g) + h\nu \rightarrow O_2(g) + \cdot\dot{O}(^1D)(g) \quad \lambda < 310\,\text{nm}$$

Ozone   Molecular   Excited
          Oxygen   atomic
                   oxygen

(4.6)

$$\cdot\dot{O}(^1D)(g) + H_2O(g) \rightarrow 2\dot{O}H(g) \qquad (4.7)$$

Excited   Water   Hydroxyl
atomic   vapor   radical
oxygen

Thus, OH(g) concentrations in the background troposphere depend on ozone and water vapor contents.

4.2.3. Nighttime Nitrogen Chemistry

During the night, Reaction 4.2 (nitrogen dioxide photolysis) shuts off, eliminating the major chemical sources of O(g) and NO(g). Because O(g) is necessary for the formation of ozone, ozone production also shuts down at night. Thus, at night, neither O(g), NO(g), nor O_3(g) is produced chemically. However, NO(g) can be emitted at night. When this occurs, it destroys ozone by Reaction 4.1, causing a local loss in ozone.

Because NO_2(g) photolysis shuts off at night, NO_2(g) becomes available to produce NO_3(g), N_2O_5(g), and HNO_3(aq) at night by the sequence

$$\dot{N}O_2(g) + O_3(g) \rightarrow \dot{N}O_3(g) + O_2(g) \qquad (4.8)$$

Nitrogen   Ozone   Nitrate   Molecular
dioxide   radical   oxygen

$$\dot{N}O_2(g) + \dot{N}O_3(g) \overset{M}{\rightleftharpoons} N_2O_5(g) \qquad (4.9)$$

Nitrogen   Nitrate   Dinitrogen
dioxide   radical   pentoxide

$$N_2O_5(g) + H_2O(aq) \rightarrow 2HNO_3(aq) \qquad (4.10)$$

Dinitrogen   Liquid   Dissolved
pentoxide   water   nitric acid

Reaction 4.10 occurs on aerosol or hydrometeor particle surfaces. During the morning, sunlight breaks down NO_3(g) within seconds, so NO_3(g) is not a source of pollution during the day. Because N_2O_5(g) forms from NO_3(g) and decomposes thermally within seconds at high temperatures by the reverse of Reaction 4.9, N_2O_5(g) is also unimportant during the day.

4.2.4. Ozone Production from Carbon Monoxide

Daytime ozone production in the background troposphere is enhanced by carbon monoxide [CO(g)], methane [CH_4(g)], and certain nonmethane organic gases. CO(g) produces ozone by

$$CO(g) + \dot{O}H(g) \rightarrow CO_2(g) + \dot{H}(g) \qquad (4.11)$$

Carbon   Hydroxyl   Carbon   Atomic
monoxide   radical   dioxide   hydrogen

$$\dot{H}(g) + O_2(g) \overset{M}{\rightarrow} H\dot{O}_2(g) \qquad (4.12)$$

Atomic   Molecular   Hydroperoxy
hydrogen   oxygen   radical

$$\dot{N}O(g) + H\dot{O}_2(g) \rightarrow \dot{N}O_2(g) + \dot{O}H(g) \qquad (4.13)$$

Nitric   Hydroperoxy   Nitrogen   Hydroxyl
oxide   radical   dioxide   radical

$$\dot{N}O_2(g) + h\nu \rightarrow \dot{N}O(g) + \cdot\dot{O}(g) \quad \lambda < 420\,\text{nm} \quad (4.14)$$

Nitrogen   Nitric   Atomic
dioxide   oxide   oxygen

$$\cdot\dot{O}(g) + O_2(g) \overset{M}{\rightarrow} O_3(g) \qquad (4.15)$$

Atomic   Molecular   Ozone
oxygen   oxygen

Because the e-folding lifetime of CO(g) against breakdown by OH(g) in the background troposphere is 28 to 110 days, the sequence does not affect the photostationary-state relationship among O_3(g), NO(g), and NO_2(g) much, except to increase the NO_2(g):NO(g) ratio slightly, thus increasing ozone. Reaction 4.11 also produces carbon dioxide, and Reaction 4.12 is almost instantaneous.

4.2.5. Ozone Production from Methane

Methane [CH_4(g)], with a mixing ratio of 1.85 ppmv, is the most abundant organic gas in the Earth's atmosphere. Its free-tropospheric e-folding lifetime against chemical destruction is eight to twelve years. This long lifetime has enabled it to mix uniformly up to the tropopause. Above the tropopause, its mixing ratio decreases due to chemical reaction and photolysis. Methane's most important tropospheric loss mechanism is its reaction with the hydroxyl radical. The methane loss pathway produces ozone by increasing the NO_2(g):NO(g) ratio, but the incremental quantity of ozone produced is small compared with the pure photostationary quantity of ozone. The methane oxidation sequence producing ozone is

$$CH_4(g) + \dot{O}H(g) \rightarrow \dot{C}H_3(g) + H_2O(g) \qquad (4.16)$$

Methane   Hydroxyl   Methyl   Water
           radical   radical   vapor

$$\dot{C}H_3(g) + O_2(g) \xrightarrow{M} CH_3\dot{O}_2(g) \quad (4.17)$$

Methyl radical · , Molecular oxygen, Methylperoxy radical

$$\dot{N}O(g) + CH_3\dot{O}_2(g) \rightarrow \dot{N}O_2(g) + CH_3\dot{O}(g) \quad (4.18)$$

Nitric oxide, Methylperoxy radical, Nitrogen dioxide, Methoxy radical

$$\dot{N}O_2(g) + h\nu \rightarrow \dot{N}O(g) + \cdot\dot{O}(g) \quad \lambda < 420\,nm \quad (4.19)$$

Nitrogen dioxide, Nitric oxide, Atomic oxygen

$$\cdot\dot{O}(g) + O_2(g) \xrightarrow{M} O_3(g) \quad (4.20)$$

Atomic oxygen, Molecular oxygen, Ozone

In the first reaction, OH(g) **abstracts** (removes) a hydrogen atom from methane, producing the **methyl radical** [$CH_3(g)$] and water. In the stratosphere, this reaction is an important source of water vapor. As with Reaction 4.12, Reaction 4.17 is fast. The remainder of the sequence is similar to the remainder of the carbon monoxide sequence, except that the **methylperoxy radical** [$CH_3O_2(g)$] converts NO(g) to $NO_2(g)$, whereas in the carbon monoxide sequence, the **hydroperoxy radical** [$HO_2(g)$] performs the conversion.

4.2.6. Ozone Production from Formaldehyde

An important by-product of the methane oxidation pathway is **formaldehyde** [$HCHO(g)$]. Formaldehyde is a colorless gas with a strong odor at mixing ratios higher than 0.05 to 1.0 ppmv. It is the most abundant aldehyde in the air and moderately soluble in water. Aside from gas-phase chemical reaction, the most important source of formaldehyde is off-gassing from plywood, resins, adhesives, carpeting, particleboard, fiberboard, and other building materials (Hines et al., 1993). Mixing ratios of formaldehyde in urban air are generally less than 0.1 ppmv (Maroni et al., 1995). Indoor mixing ratios range from 0.07 to 1.9 ppmv and typically exceed outdoor mixing ratios (Anderson et al., 1975; Jones, 1999).

Because formaldehyde is moderately soluble in water, it dissolves readily in the upper respiratory tract. Below mixing ratios of 0.05 ppmv, formaldehyde causes no known health problems. Above this level, it causes the following effects: 0.05 to 1.5 ppmv, neurophysiologic effects; 0.01 to 2.0 ppmv, eye irritation; 0.1 to 25 ppmv, irritation of the upper airway; 5 to 30 ppmv, irritation of the lower airway and pulmonary problems; 50 to 100 ppmv, pulmonary edema, inflammation, and pneumonia; and greater than 100 ppmv, coma or death (Hines et al., 1993; Jones, 1999). Formaldehyde is but one of many eye irritants in photochemical smog. Compounds

that cause eyes to swell, redden, and tear are **lachrymators**. The methoxy radical from Reaction 4.18 produces formaldehyde by

$$CH_3\dot{O}(g) + O_2(g) \rightarrow HCHO(g) + H\dot{O}_2(g) \quad (4.21)$$

Methoxy radical, Molecular oxygen, Formaldehyde, Hydroperoxy radical

The *e*-folding lifetime of $CH_3O(g)$ against destruction by $O_2(g)$ is ~0.0001 seconds. Once formaldehyde forms, it produces ozone precursors by

$$HCHO(g) + h\nu \rightarrow \begin{cases} H\dot{C}O(g) + \dot{H}(g) & \lambda < 334\,nm \\ \text{Formyl radical} \quad \text{Atomic hydrogen} \\ CO(g) + H_2(g) & \lambda < 370\,nm \\ \text{Carbon monoxide} \quad \text{Molecular hydrogen} \end{cases} \quad (4.22)$$

Formaldehyde

$$HCHO(g) + \dot{O}H(g) \rightarrow H\dot{C}O(g) + H_2O(g) \quad (4.23)$$

Formaldehyde, Hydroxyl radical, Formyl radical, Water vapor

$$H\dot{C}O(g) + O_2(g) \rightarrow CO(g) + H\dot{O}_2(g) \quad (4.24)$$

Formyl radical, Molecular oxygen, Carbon monoxide, Hydroperoxy radical

$$\dot{H}(g) + O_2(g) \xrightarrow{M} H\dot{O}_2(g) \quad (4.25)$$

Atomic hydrogen, Molecular oxygen, Hydroperoxy radical

CO(g) forms ozone through Reactions 4.11 to 4.15, and $HO_2(g)$ forms ozone through Reactions 4.13 to 4.15.

4.2.7. Ozone Production from Ethane

The most concentrated nonmethane hydrocarbons in the background troposphere are **ethane** [$C_2H_6(g)$] and **propane** [$C_3H_8(g)$]. Background tropospheric mixing ratios are 0 to 2.5 ppbv for ethane and 0 to 1.0 ppbv for propane. These hydrocarbons originate substantially from anthropogenic sources and have relatively long lifetimes against photochemical destruction, so they persist after leaving urban areas. The *e*-folding lifetime of ethane against chemical destruction is about twenty-three to ninety-three days, and that of propane is five to twenty-one days. The primary oxidant of ethane and propane is OH(g). In both reactions, OH(g) initiates the breakdown. The sequence of reactions, with respect to ethane, is

$$C_2H_6(g) + \dot{O}H(g) \rightarrow \dot{C}_2H_5(g) + H_2O(g) \quad (4.26)$$

Ethane, Hydroxyl radical, Ethyl radical, Water vapor

$$\dot{C}_2H_5(g) + O_2(g) \xrightarrow{M} C_2H_5\dot{O}_2(g) \quad (4.27)$$

Ethyl radical, Molecular oxygen, Ethylperoxy radical

$$\dot{N}O(g) + C_2H_5\dot{O}_2(g) \rightarrow \dot{N}O_2(g) + C_2H_5\dot{O}(g)$$

Nitric oxide, Ethylperoxy radical, Nitrogen dioxide, Ethoxy radical

$$(4.28)$$

$$\dot{N}O_2(g) + h\nu \rightarrow \dot{N}O(g) + \cdot\dot{O}(g) \quad \lambda < 420\,nm$$

Nitrogen   Nitric   Atomic
dioxide   oxide   oxygen

(4.29)

$$\cdot\dot{O}(g) + O_2(g) \xrightarrow{M} O_3(g) \quad (4.30)$$

Atomic   Molecular   Ozone
oxygen   oxygen

4.2.8. Ozone and Peroxyacetyl Nitrate Production from Acetaldehyde

An important by-product of ethane oxidation is **acetaldehyde** [$CH_3CH(=O)(g)$], produced from the ethoxy radical formed in Reaction 4.28. The reaction producing acetaldehyde is

$$C_2H_5\dot{O}(g) + O_2(g) \rightarrow CH_3CH(=O) + H\dot{O}_2(g)$$

Ethoxy   Molecular   Acetaldehyde   Hydroperoxy
radical   oxygen   radical

(4.31)

This reaction is relatively instantaneous. Acetaldehyde is a precursor to **peroxyacetyl nitrate** (PAN), a daytime component of the background troposphere that was discovered during laboratory experiments of photochemical smog formation (Stephens et al., 1956). The only source of PAN is chemical reaction in the presence of sunlight. As such, its mixing ratio peaks during the afternoon, at the same time that ozone mixing ratios peak. Like formaldehyde, PAN is an eye irritant and lachrymator, but it does not have other serious human health effects. It does damage plants by discoloring their leaves. Mixing ratios of PAN in clean air are typically 2 to 100 pptv. Those in rural air downwind of urban sites are up to 1 ppbv. Polluted air mixing ratios are generally not higher than 35 ppbv, with typical values of 10 to 20 ppbv. PAN is not an important constituent of air at night or in regions of heavy cloudiness. The reaction pathway producing PAN is

$$CH_3CH(=O)(g) + O\dot{H}(g) \rightarrow CH_3\dot{C}(=O)(g) + H_2O(g)$$

Acetaldehyde   Hydroxyl   Acetyl   Water
              radical   radical   vapor

(4.32)

$$CH_3\dot{C}(=O)(g) + O_2(g) \xrightarrow{M} CH_3C(=O)\dot{O}_2(g)$$

Acetyl   Molecular   Peroxyacetyl
radical   oxygen   radical

(4.33)

$$CH_3C(=O)\dot{O}_2(g) + \dot{N}O_2(g) \xrightleftharpoons{M} CH_3C(=O)O_2NO_2(g)$$

Peroxyacetyl   Nitrogen   Peroxyacetyl nitrate
radical   dioxide   (PAN)   (4.34)

The last reaction in this sequence is reversible and strongly temperature dependent. At 300 K and at surface pressure, PAN's *e*-folding lifetime against thermal

decomposition by the reverse of Reaction 4.34 is about twenty-five minutes. At 280 K, its lifetime increases to thirteen hours. As such, PAN's mixing ratios decrease with increasing temperature.

Conversely, when PAN rises vertically to above the boundary layer, where temperatures are low, it persists longer than at the surface and can be blown long distances by the wind. When it comes back to the surface, it can dissociate again, releasing ozone-forming pollutants far from urban areas.

Acetaldehyde also produces ozone. The peroxyacetyl radical in Reaction 4.34, for example, converts NO(g) to $NO_2(g)$ by

$$CH_3C(=O)\dot{O}_2(g) + \dot{N}O(g) \rightarrow CH_3C(=O)\dot{O}(g)$$

Peroxyacetyl   Nitric   Acetyloxy
radical   oxide   radical

$$+ \dot{N}O_2(g) \quad (4.35)$$

Nitrogen
dioxide

$NO_2(g)$ forms $O_3(g)$ through Reactions 4.2 and 4.3. A second mechanism of ozone formation from acetaldehyde is through photolysis:

$$CH_3CHO(g) + h\nu \rightarrow \dot{C}H_3(g) + H\dot{C}O(g) \quad (4.36)$$

Acetaldehyde   Methyl   Formyl
               radical   radical

The methyl radical from this reaction forms ozone through Reactions 4.17 to 4.20. The formyl radical forms ozone through Reaction 4.24, followed by Reactions 4.13 to 4.15.

4.3. Chemistry of Photochemical Smog

Photochemical smog is a soup of gases and aerosol particles. Some of the substances in smog are emitted, whereas others form chemically or physically in the air from precursor gases or particles. In this section, the gas-phase components of smog are discussed. Aerosol particles in smog are discussed in Chapter 5.

Photochemical smog differs from background air in two ways. First, smog contains more high-molecular-weight organics, particularly aromatic compounds, than does background air. Because most high-molecular-weight and complex organic compounds break down quickly in urban air, they are unable to survive long enough to disperse to the background troposphere. Second, the mixing ratios of nitrogen oxides and organic gases are higher in polluted air than in background air, causing mixing ratios of ozone to be higher in urban air than in background air.

Photochemical smog involves reactions among **nitrogen oxides** [$NO_x(g) = NO(g) + NO_2(g)$] and **reactive organic gases** (ROGs, total organic gases minus methane) in the presence of sunlight. The most recognized gas-phase by-product of smog reactions is ozone because ozone has harmful health effects (Section 3.6.5) and is an indicator of the presence of other pollutants.

On a typical day, ozone forms following the emission of NO(g) and ROGs. Emitted pollutants are called **primary pollutants**. Pollutants, such as ozone, that form chemically or physically in the air are called **secondary pollutants**.

Primary pollutant ROGs are broken down by chemical reaction into **peroxy radicals**, denoted by $RO_2(g)$. Peroxy radicals and NO(g) form secondary pollutant ozone by the following sequence:

$$\dot{N}O(g) + R\dot{O}_2(g) \rightarrow \dot{N}O_2(g) + R\dot{O}(g) \quad (4.37)$$

Nitric oxide · Organic peroxy radical · Nitrogen dioxide · Organic oxy radical

$$\dot{N}O(g) + O_3(g) \rightarrow \dot{N}O_2(g) + O_2(g) \quad (4.38)$$

Nitric oxide · Ozone · Nitrogen dioxide · Molecular oxygen

$$\dot{N}O_2(g) + h\nu \rightarrow \dot{N}O(g) + \cdot\dot{O}(g) \quad \lambda < 420\,nm$$

Nitrogen dioxide · Nitric oxide · Atomic oxygen

$$(4.39)$$

$$\cdot\dot{O}(g) + O_2(g) \xrightarrow{M} O_3(g) \quad (4.40)$$

Atomic oxygen · Molecular oxygen · Ozone

Figure 4.12 shows ozone mixing ratios resulting from different initial mixtures of $NO_x(g)$ and ROGs. This plot is called an **ozone isopleth**. The figure shows that, for low mixing ratios of $NO_x(g)$, ozone mixing ratios are relatively insensitive to the quantity of ROGs. For high $NO_x(g)$, an increase in ROGs increases ozone.

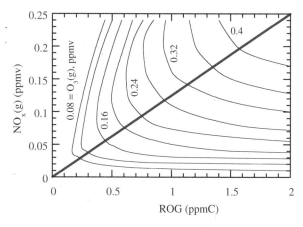

Figure 4.12. Peak ozone mixing ratios resulting from different initial mixing ratios of $NO_x(g)$ and reactive organic gases (ROGs). The ROG:$NO_x(g)$ ratio along the line through zero is 8:1. Adapted from Finlayson-Pitts and Pitts (1999).

The plot also shows that, for low ROGs, increases in $NO_x(g)$ above 0.05 ppmv decrease ozone. For high ROGs, increases in $NO_x(g)$ always increase ozone.

The plot is useful for regulatory control of ozone. If ROG mixing ratios are high (e.g., 2 ppmC) and $NO_x(g)$ mixing ratios are moderate (e.g., 0.06 ppmv), the plot indicates that the most effective way to reduce ozone is to reduce $NO_x(g)$. Reducing ROGs under these conditions has little effect on ozone. If ROG mixing ratios are low (e.g., 0.7 ppmC) and $NO_x(g)$ mixing ratios are high (e.g., 0.2 ppmv), the most effective way to reduce ozone is to reduce ROGs. Reducing $NO_x(g)$ under these conditions actually increases ozone. In many polluted urban areas, the ROG:$NO_x(g)$ ratio is lower than 8:1, indicating that limiting ROG emissions should be the most effective method of controlling ozone. However, because ozone mixing ratios depend not only on chemistry, but also on meteorology, deposition, and gas-to-particle conversion, such a conclusion is not always clear cut.

Figure 4.13 shows the evolution of NO(g), $NO_2(g)$, and $O_3(g)$ during one day at two locations – central Los Angeles and San Bernardino – in the Los Angeles Basin. In the basin, a daily sea breeze transfers primary pollutants [NO(g) and ROGs], emitted on the west side of the basin (i.e., central Los Angeles) to the east side of the basin (i.e., San Bernardino), where they arrive as secondary pollutants [$O_3(g)$ and PAN]. Whereas NO(g) mixing ratios peak on the west side of Los Angeles, as shown in Figure 4.13a, $O_3(g)$ mixing ratios peak on the east side, as shown in Figure 4.13b. Thus, the west side of the basin is a **source region**, and the east side is a **receptor region** of photochemical smog.

A significant difference between ozone production in urban air and the background troposphere is that peroxy radicals convert NO(g) to $NO_2(g)$ in urban air but less so in clean air. As a result, the $NO_2(g):NO(g)$ ratio is much higher in urban air than clean air. Reaction 4.4 suggests that an increase in the $NO_2(g):NO(g)$ ratio increases ozone levels above the pure photostationary-state ozone level.

4.3.1. Emissions of Photochemical Smog Precursors

Major gases emitted in urban air include nitrogen oxides, ROGs, carbon monoxide [CO(g)], and sulfur oxides [$SO_x(g) = SO_2(g) + SO_3(g)$]. Of these, $NO_x(g)$ and ROGs are the main precursors of photochemical smog. Sources of CO(g), $SO_x(g)$, and $NO_x(g)$ are primarily incomplete combustion. Sources of ROGs

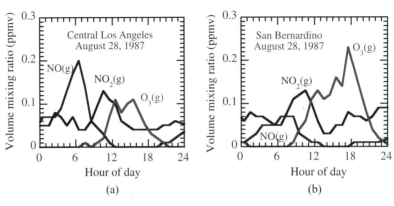

Figure 4.13. Evolution of NO(g), NO₂(g), and O₃(g) mixing ratios in (a) central Los Angeles and (b) San Bernardino on August 28, 1987. Central Los Angeles is closer to the coast than is San Bernardino. A sea breeze sends primary pollutants, such as NO(g), from the west side of the Los Angeles Basin (i.e., central Los Angeles) toward the east side (i.e., San Bernardino). As the pollutants travel, organic peroxy radicals convert NO(g) to NO₂(g). Photolysis of NO₂(g) produces atomic oxygen, which forms ozone, a secondary pollutant.

include evaporative emissions and incomplete combustion. Evaporative emissions originate from oil fields, refineries, gas stations, automobiles, petroleum product production, solvent production and use, and painting, among other sources. Incomplete combustion occurs in power plants, factories, homes, incinerators, and automobiles.

Table 4.1 shows the rates of primary pollutant gas emissions in the Los Angeles Basin and California for several species and groups of chemicals in 2005. Carbon dioxide is the most abundantly emitted anthropogenic gas. It is not reactive chemically, but its local emissions result in the formation of **carbon dioxide domes** over Los Angeles and other cities that warm the air locally. The warmer air evaporates water, and both the water and higher temperature increase the rate of ozone

formation in polluted air (Section 12.5.6). Carbon monoxide is the second most abundantly emitted gas. It increases ozone slightly in urban air but is more important for producing ozone in the background troposphere. Methane is the most abundantly emitted organic gas in urban areas; however, due to its long chemical lifetime, it has little direct chemical impact on local ozone production. However, it can enhance local ozone by forming a **methane dome**, which is similar to a carbon dioxide dome. Methane domes enhance local temperatures and thus local ozone.

Of the ROGs, hexane, butanol, pentane, toluene, xylene, butane, ethane, ethene, various aldehydes, and benzene are generally emitted in the greatest abundance. NO$_x$(g) emission rates are five to ten times higher than are SO$_x$(g) emission rates, indicating that Los Angeles and California are more susceptible to nitric acid deposition than to sulfuric acid deposition problems (Chapter 10). Sulfur emission rates in California are low relative to those in many other regions worldwide. Ammonia emission rates are lower than are sulfur emission rates in California, but they are still high enough to produce ammonium sulfate and ammonium nitrate aerosol particles (Chapter 5).

Table 4.2 shows the approximate percentage emission of several gases by source category in Los Angeles. Emissions originate from point, area, and mobile sources. A **point source** is an individual pollutant source, such as a smokestack, fixed in space. A **mobile source** is a moving individual pollutant source, such as the exhaust of a motor vehicle or an airplane. An **area source** is an area, such as a city block, agricultural

Table 4.1. Gas-phase emissions (tonnes/day) in Los Angeles Basin and California, 2005

Substance	Los Angeles	California
Carbon dioxide [CO₂(g)]	693,000	1,450,000
Carbon monoxide [CO(g)]	4,740	13,900
Methane [CH₄(g)]	479	1,975
Total ROGs	1,070	3,240
Total NO$_x$(g) as NO₂(g)	1,460	3,840
Total SO$_x$(g) as SO₂(g)	162	781
Ammonia [NH₃(g)]	70	553

ROG, reactive organic gas.
Source: Jacobson and Ginnebaugh (2010).

Table 4.2. Percentage emission of several gases by source category in Los Angeles Basin, 2005

Source category	CO(g)	ROG	NO$_x$(g)	SO$_x$(g)	NH$_3$(g)
Stationary	11	67	17	38	74
Mobile	89	33	83	62	26
TOTAL	100	100	100	100	100

Source: U.S. EPA (2011b).

field, or industrial facility, over which many fixed pollutant sources aside from point source smokestacks exist. Together, point and area sources are **stationary sources**. Table 4.2 shows that carbon monoxide and nitrogen oxides originate predominantly from mobile sources. The thermal combustion reaction in automobiles that produces nitric oxide at a high temperature is

$$N_2(g) + O_2(g) \xrightarrow{\text{High temperature}} 2\dot{N}O(g) \quad (4.41)$$

Molecular nitrogen   Molecular oxygen   Nitric oxide

Table 4.2 shows that mobile sources emit less ROG and ammonia than do stationary sources. In the 1980s, the distribution of ROGs between stationary and mobile sources was roughly equal. The decrease in mobile source ROGs since then is due primarily to the catalytic converter, a control device required in all new U.S. automobiles since the 1970s (Section 8.1.7). Because Los Angeles has few coal-burning sources, most of its SO$_x$(g) emissions are from mobile sources.

4.3.2. Reactive Organic Gas Breakdown Processes

Once ROGs are emitted, they are broken down chemically into free radicals. Six major processes break down ROGs – photolysis and reaction with OH(g), HO$_2$(g), O(g), NO$_3$(g), and O$_3$(g). OH(g) and O(g) are present during the day only because they are short lived and require photolysis for their production. NO$_3$(g) is present during the night only because it photolyzes quickly during the day. O$_3$(g) and HO$_2$(g) may be present during both day and night.

OH(g) is produced in urban air by some of the same reactions that produce it in the background troposphere. An early morning source of OH(g) in urban air is photolysis of **nitrous acid** [HONO(g)]. HONO(g) may be emitted by automobiles; thus, it is more abundant in urban air than in the background troposphere. Midmorning sources of OH(g) in urban air are aldehyde photolysis and oxidation. The major afternoon source of OH(g) in urban air is ozone photolysis. In sum, the three major reaction mechanisms that produce the hydroxyl radical in urban air are

Early morning source

$$HONO(g) + h\nu \rightarrow \dot{O}H(g) + \dot{N}O(g) \quad \lambda < 400\,\text{nm}$$

Nitrous acid   Hydroxyl radical   Nitric oxide

$$(4.42)$$

Midmorning source

$$HCHO(g) + h\nu \rightarrow H\dot{C}O(g) + \dot{H}(g) \quad \lambda < 334\,\text{nm}$$

Formaldehyde   Formyl radical   Atomic hydrogen

$$(4.43)$$

$$\dot{H}(g) + \dot{O}_2(g) \xrightarrow{M} H\dot{O}_2(g) \quad (4.44)$$

Atomic hydrogen   Molecular oxygen   Hydroperoxy radical

$$H\dot{C}O(g) + O_2(g) \rightarrow CO(g) + H\dot{O}_2(g) \quad (4.45)$$

Formyl radical   Molecular oxygen   Carbon Monoxide   Hydroperoxy radical

$$\dot{N}O(g) + H\dot{O}_2(g) \rightarrow \dot{N}O_2(g) + \dot{O}H(g) \quad (4.46)$$

Nitric oxide   Hydroperoxy radical   Nitrogen dioxide   Hydroxyl radical

Afternoon source

$$O_3(g) + h\nu \rightarrow O_2(g) + \cdot\dot{O}(^1D)(g) \quad \lambda < 310\,\text{nm}$$

Ozone   Molecular oxygen   Excited atomic oxygen

$$(4.47)$$

$$\cdot\dot{O}(^1D)(g) + H_2O(g) \rightarrow 2\dot{O}H(g) \quad (4.48)$$

Excited atomic oxygen   Water vapor   Hydroxyl radical

ROGs emitted in urban air include alkanes, alkenes, alkynes, aldehydes, ketones, alcohols, aromatics, and hemiterpenes. Table 4.3 shows lifetimes of these ROGs against breakdown by six processes. Photolysis breaks down aldehydes and ketones; OH(g) breaks down all eight groups during the day; HO$_2$(g) breaks down aldehydes during both day and night; O(g) breaks down alkenes and terpenes during the day; NO$_3$(g) breaks down alkanes, alkenes, aldehydes, aromatics, and terpenes during the night; and O$_3$(g) breaks down alkenes and terpenes during both day and night.

ROG chemical breakdown produces radicals that lead to ozone formation. The ozone-forming potential of an ROG is a function not only of the ROG's emission rate, but also of its reactivity. The **reactivity** of a gas is the ozone-forming potential of the gas for given emission rates of the gas and NO$_x$(g). For example, 1 kg of emitted formaldehyde can form two orders of magnitude more ozone than 1 kg of emitted ethane under typical urban conditions (Russell et al., 1995). Knowing both

Table 4.3. Estimated e-folding lifetimes of reactive organic gases representing alkanes, alkenes, alkynes, aldehydes, ketones, alcohols, aromatics, and hemiterpenes, respectively, against photolysis and oxidation by gases at specified concentrations in urban air

ROG species	Photolysis	Lifetime in polluted urban air at sea level				
		$OH(g)$ 5×10^6 molec cm^{-3}	$HO_2(g)$ 2×10^9 molec cm^{-3}	$O(g)$ 8×10^4 molec cm^{-3}	$NO_3(g)$ 1×10^{10} molec cm^{-3}	$O_3(g)$ 5×10^{12} molec cm^{-3}
n-Butane	–	22 h	1,000 y	18 y	29 d	650 y
trans-2-Butene	–	52 m	4 y	6.3 d	4 m	17 m
Acetylene	–	3.0 d	–	2.5 y	–	200 d
Formaldehyde	7 h	6.0 h	1.8 h	2.5 y	2.0 d	3,200 y
Acetone	23 d	9.6 d	–	–	–	–
Ethanol	–	19 h	–	–	–	–
Toluene	–	9.0 h	–	6 y	33 d	200 d
Isoprene	–	34 m	–	4 d	5 m	4.6 h

Lifetimes were obtained from rate and photolysis coefficient data. Gas concentrations are typical (but not necessarily average) values for each region. Units: m, minutes; h, hours; d, days; y, years; –, insignificant loss. ROG, reactive organic gas.

the reactivity and emission rate of an ROG is important for prioritizing which organics to control first in an urban area.

Table 4.4 shows the most important ROGs in Los Angeles during the summer of 1987, in terms of a combination of abundance and reactive ability to form ozone. The table shows that m- and p-xylene, both aromatic hydrocarbons, were the most important gases in terms of generating ozone. Although alkanes are emitted in greater abundance than are other organics, they are less reactive in producing ozone than are aromatics, alkenes, or aldehydes.

In the following subsections, photochemical smog processes involving the chemical breakdown of organic gases to produce ozone are discussed.

Table 4.4. Ranking of most important species in terms of chemical reactivity and abundance during Southern California air quality study, summer 1987

1. m- and p-Xylene	6. i-Pentane
2. Ethene	7. Propene
3. Acetaldehyde	8. o-Xylene
4. Toluene	9. Butane
5. Formaldehyde	10. Methylcyclopentane

Ranking was determined by multiplying the weight fraction of each organic present in the atmosphere by a species-specific reactivity scaling factor developed by Carter (1991).
Source: Lurmann et al. (1992).

4.3.3. Ozone Production from Alkanes

Table 4.4 shows that i-pentane and butane are important alkanes in terms of concentration and reactivity in producing ozone in Los Angeles air. As in the background troposphere, the main pathway of alkane decomposition in urban air is OH(g) attack. Alkane concentrations are not affected much by photolysis or reaction with $O_3(g)$, $HO_2(g)$, or $NO_3(g)$ (Table 4.3). Of all alkanes, methane is the least reactive and the least important with respect to urban air pollution. Methane is more important with respect to background tropospheric and stratospheric chemistry. The oxidation pathways of methane were given in Reactions 4.16 to 4.20, and those of ethane were shown in Reactions 4.26 to 4.30.

4.3.4. Ozone Production from Alkenes

Table 4.4 shows that alkenes, such as ethene and propene, are important ozone precursors in photochemical smog. Mixing ratios of ethene and propene in polluted air reach 1 to 30 ppbv. Table 4.3 indicates that alkenes react most rapidly with OH(g), $O_3(g)$, and $NO_3(g)$. In the following subsections, the first two of these reaction pathways are discussed.

4.3.4.1. Alkene Reaction with the Hydroxyl Radical

When ethene reacts with the hydroxyl radical, the radical substitutes into ethene's double bond to produce an **ethanyl radical** in an OH(g) **addition** process. The ethanyl radical then reacts to produce $NO_2(g)$. The sequence is

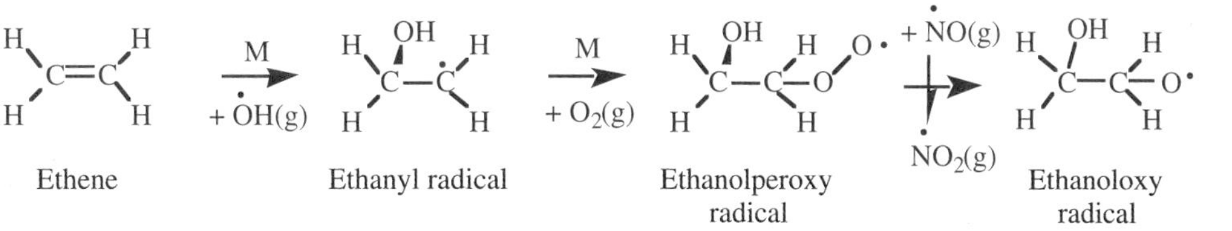

(4.49)

Ethene   Ethanyl radical   Ethanolperoxy radical   Ethanoloxy radical

$NO_2(g)$ produces ozone by Reactions 4.2 and 4.3. The **ethanoloxy radical**, a by-product of ethene oxidation,

The **excited Criegee biradical** isomerizes, and its product, excited formic acid, thermally decomposes by

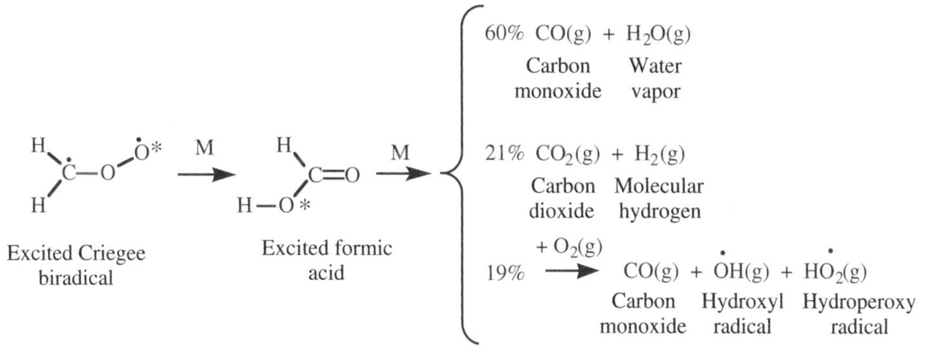

$$
\begin{cases}
60\% \; CO(g) + H_2O(g) \\
\quad \text{Carbon} \quad \text{Water} \\
\quad \text{monoxide} \quad \text{vapor} \\[4pt]
21\% \; CO_2(g) + H_2(g) \\
\quad \text{Carbon} \quad \text{Molecular} \\
\quad \text{dioxide} \quad \text{hydrogen} \\[4pt]
\quad\quad + O_2(g) \\
19\% \longrightarrow \; CO(g) + \dot{O}H(g) + H\dot{O}_2(g) \\
\quad\quad \text{Carbon} \quad \text{Hydroxyl} \quad \text{Hydroperoxy} \\
\quad\quad \text{monoxide} \quad \text{radical} \quad \text{radical}
\end{cases}
$$

(4.52)

goes on to produce formaldehyde and **glycol aldehyde** [$HOCH_2CHO(g)$], both of which contribute to further ozone formation. The formaldehyde ozone process is described in Section 4.2.6.

4.3.4.2. Alkene Reaction with Ozone

When ethene reacts with ozone, the ozone substitutes into ethene's double bond to form an unstable **ethene molozonide**. The molozonide decomposes to products that are also unstable. The reaction sequence of ethene with ozone is

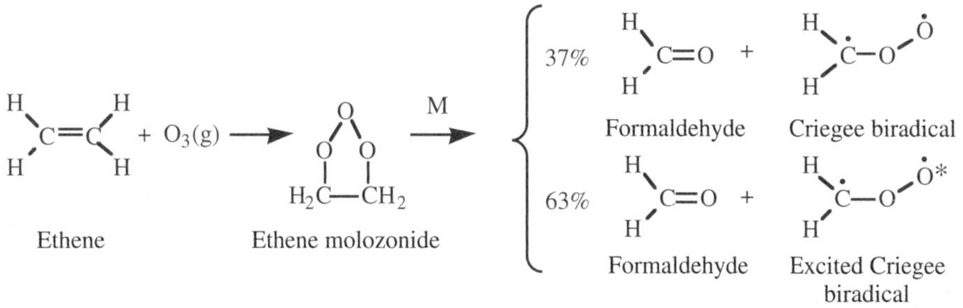

In sum, ozone attack on ethene produces $HCHO(g)$, $HO_2(g)$, $CO(g)$, and $NO_2(g)$. These gases not only reform the original ozone lost, but also produce new ozone.

4.3.5. Ozone Production from Aromatics

Toluene [$C_6H_5CH_3(g)$] originates from the combustion of gasoline, diesel, biofuels, and outdoor biomass; petroleum refining; detergent production; and offgassing from paints, solvents, and building materials.

(4.50)

Ethene   Ethene molozonide

37%   Formaldehyde   Criegee biradical

63%   Formaldehyde   Excited Criegee biradical

Formaldehyde produces ozone as in Section 4.2.6. The **Criegee biradical** from Reaction 4.50 forms $NO_2(g)$ by

$$H_2\dot{C}O\dot{O}(g) + \dot{N}O(g) \rightarrow HCHO(g) + \dot{N}O_2(g) \quad (4.51)$$

Criegee biradical   Nitric oxide   Formaldehyde   Nitrogen dioxide

After methane, it is the second most abundantly emitted organic gas in Los Angeles air and the fourth most important gas in terms of abundance and chemical reactivity (Table 4.4). Mixing ratios of toluene in polluted air range from 1 to 30 ppbv (Table 3.3). Table 4.3 shows that toluene is decomposed almost exclusively by OH(g). The hydroxyl radical breaks down toluene by abstraction and addition. The respective pathways are

4.3.6. Ozone Production from Terpenes

The background troposphere and urban areas are affected by biogenic emissions of isoprene and other terpenes. **Biogenic emissions** are produced from bio-

$$(4.53)$$

The benzylperoxy radical, formed from the abstraction pathway, converts NO(g) to NO_2(g). It also results in the formation of **benzaldehyde** [C_6H_5CHO(g)], which, like formaldehyde and acetaldehyde, decomposes to form ozone. The toluene-hydroxyl radical adduct, which is also a peroxy radical, converts NO(g) to NO_2(g). Cresol reacts with OH(g) to form the methylphenylperoxy radical [$C_6H_5CH_3O_2$(g)], which converts NO(g) to NO_2(g), resulting in O_3(g) formation.

The most important organic gas producing ozone in urban air is **xylene** [$C_6H_4(CH_3)_2$(g)] (Table 4.4), which is present in gasoline, lacquers, and glues. Its mixing ratios in polluted air range from 1 to 30 ppbv (Table 3.3). As with toluene oxidation, xylene oxidation is primarily through reaction with OH(g). Oxidation of xylene by OH(g) produces peroxy radicals, which convert NO(g) to NO_2(g), resulting in ozone formation.

logical sources, such as plants, trees, algae, bacteria, and animals. Strictly speaking, **terpenes** are hydrocarbons that have the formula $C_{10}H_{16}$. Loosely speaking, they are a class of compounds that include hemiterpenes [C_5H_8(g)] such as **isoprene**; monoterpenes [$C_{10}H_{16}$(g)] such as α-**pinene**, β-**pinene**, and **d-limonene**; sesquiterpenes [$C_{15}H_{24}$(g)]; and diterpenes [$C_{20}H_{32}$(g)]. Isoprene is emitted by sycamore, oak, aspen spruce, willow, balsam, and poplar trees; α-pinene is emitted by pines, firs, cypress, spruce, and hemlock trees; β-pinene is emitted by loblolly pine, spruce, redwood, and California black sage trees; and d-limonene is emitted by loblolly pine, eucalyptus, and California black sage trees, and by lemon fruit.

Table 4.3 shows that OH(g), O_3(g), and NO_3(g) decompose isoprene. The reaction pathways of isoprene with OH(g) produce at least six peroxy radicals. The pathways are

$$(4.54)$$

Isoprene peroxy radicals

(Paulson and Seinfeld, 1992). The *e*-folding lifetime of isoprene against reaction with OH(g) is about thirty minutes when [OH] = 5.0 × 10⁶ molec cm⁻³. All six peroxy radicals convert NO(g) to NO₂(g). The second and fifth radicals also create **methacrolein** and **methylvinylketone** by

background levels. In less vegetated areas and in areas where anthropogenic emissions are large, such as in Los Angeles, they account for only 3 to 8 percent of ozone above background levels.

$$(4.55)$$

Isoprene peroxy radical   Methacrolein   Formaldehyde

$$(4.56)$$

Isoprene peroxy radical   Methylvinylketone   Formaldehyde

respectively. The NO₂(g) from these reactions produces ozone. Methacrolein and methylvinylketone react further with OH(g) and O₃(g) to form products that convert NO(g) to NO₂(g), resulting in additional ozone.

The isoprene ozone reaction, not shown, is slower than is the isoprene hydroxyl-radical reaction. Products of the isoprene ozone reaction include methacrolein, methylvinylketone, the criegee biradical, and formaldehyde, all of which reproduce ozone lost in the reaction and create additional ozone.

In cities near forests, such as Atlanta, Georgia, terpenes can account for up to 40 percent of ozone above

4.3.7. Ozone Production from Alcohols

Alcohols, which can be distilled from corn, grapes, potatoes, sugarcane, molasses, and artichokes, among other farm products, have been used as an engine fuel since April 1, 1826, when Orford, New Hampshire native **Samuel Morey** (1762–1843) patented the first internal combustion engine. His engine ran on **ethanol** [C₂H₅OH(g)] and turpentine. In September 1829, his engine was used to power a 5.8-m-long boat up the Connecticut River at seven to eight miles per hour.

Although alcohols were used in later prototype engines, they became relatively expensive in the United States due to a federal tax placed on alcohol following the Civil War of 1860 to 1865.

On August 27, 1859, **Edwin Laurentine Drake** (1819–1880) discovered oil after using a steam engine to power a drill through 21 m of rock in Titusville, Pennsylvania. This discovery is considered the beginning of the oil industry. Oil was soon refined to produce gasoline. The lower cost of gasoline in comparison with that of ethanol resulted in the comparatively greater use of gasoline than ethanol in early U.S. and European automobiles. The first practical gasoline-powered engine, which ran on illuminating gas, was constructed by **Étienne Lenoir** (1822–1900) of France in 1860. In 1862, he built an automobile powered by this engine. In 1876, German **Nikolaus Otto** (1832–1891) developed the first four-stroke internal combustion engine. In 1885, **Karl Benz** (1844–1929) of Germany designed and built the first practical automobile powered by an internal combustion engine. The same year, **Gottlieb Daimler** (1834–1900) of Germany patented the first successful high-speed internal combustion engine and developed a carburetor that allowed the use of gasoline as a fuel. In 1893, **J. Frank Duryea** (1869–1967) and **Charles E. Duryea** (1861–1938) produced the first successful gasoline-powered vehicle in the United States. In 1896, **Henry Ford** (1863–1947) completed his first successful automobile in Detroit, Michigan.

Whereas the United States had large oil reserves to draw on, France and Germany had few oil reserves and thus used ethanol as a fuel in automobiles to a greater extent. In 1906, 10 percent of engines at Otto Gas Engine Works in Germany ran on ethanol (Kovarik, 1998). The same year, the United States repealed the federal tax on ethanol, making it more competitive with gasoline. Soon after, however, oil fields in Texas were discovered, leading to a reduction in gasoline prices and the near death of the alcohol fuel industry.

Yet, the alcohol fuel industry managed to survive. From the 1920s to the 2000s, every industrialized country, except the United States, marketed blends of ethyl alcohol with gasoline in greater than nontrivial quantities. In the 1920s, I. G. Farben, a German firm, discovered a process to make synthetic **methanol** [$CH_3OH(g)$] from coal. As Hitler prepared for war in 1937, production of alcohol as a fuel in Germany increased to about 52 million gallons per year (Egloff, 1940). Nevertheless, alcohol may never have represented more than 5 percent of the total fuel use in Europe in the 1930s (Egloff, 1940).

When methanol is burned as a fuel, its major by-products are unburned methanol and the carcinogen formaldehyde. In the atmosphere, unburned methanol oxidizes to formaldehyde and ozone. Table 4.3 indicates that the only important chemical loss process of alcohols in the air is through reaction with the hydroxyl radical. The reaction of methanol with OH(g) is

$$(4.57)$$

Methanol lost from these reactions has an e-folding lifetime of 71 days when $[OH] = 5.0 \times 10^6$ molec cm^{-3}; thus, the reaction is not rapid. The organic product of the first reaction is formaldehyde, and that of the second reaction is the methoxy radical, which produces formaldehyde by Reaction 4.21. Formaldehyde is an ozone precursor.

In the 1970s, Brazil began a national effort to ensure that all gasoline sold contained ethanol (Section 8.2.14). In the United States, gasoline prices have always been much lower than alcohol fuel prices, inhibiting the popularity of alcohol as an alternative to gasoline. When ethanol is burned as a fuel, its major by-products are unburned ethanol and acetaldehyde, a carcinogen. Atmospheric oxidation of unburned ethanol results in more acetaldehyde (a precursor to PAN) and ozone through

$$(4.58)$$

Ethanol lost from the most probable (middle) reaction has an e-folding lifetime of about 19 hours when $[OH] = 5.0 \times 10^6$ molec cm^{-3}. Acetaldehyde, formed from the middle reaction, produces PAN and ozone. The relatively long lifetime of ethanol oxidation to acetaldehyde suggests that unburned ethanol from ethanol-fueled vehicles may be a free-tropospheric and urban source of near-surface ozone (Jacobson, 2007).

Table 4.5. Percent difference in emissions of several chemicals between E85 and gasoline near room temperature and at low temperature

Substance	22°C	−7°C
Nitrogen oxides	−38	−21
Carbon monoxide	+1	+94
Nonmethane hydrocarbons	+14	+133
Benzene	−65	−15
1,3-Butadiene	−66	−0.3
Acetaldehyde	+4,500	+8,200
Formaldehyde	+125	+204

Sources: Ginnebaugh et al. (2010), citing data from Westerholm et al. (2008) for nitrogen oxides and carbon monoxide averaged over two vehicles, a Saab 9-5 biopower and a Volvo V50 flex fuel vehicle run on E85 versus E5, and data from Whitney and Fernandez (2007) for the remaining emissions averaged over three vehicles, a 2007 Chevrolet Silverado, a 2006 Lincoln Town Car, and a 2006 Dodge Stratus, each run on E85 at 22°C but E70 at −7°C.

4.3.8. Ethanol versus Gasoline Effects on Air Pollution and Health

Current ethanol fuel blends used in vehicles range from E6 (6 percent ethanol fuel, 94 percent gasoline) to E100 (100 percent ethanol fuel). However, in the United States and many countries, 100 percent ethanol fuel really contains 95 percent ethanol and 5 percent gasoline added as a **denaturant**, which is a poisonous or untasteful chemical added to a fuel to prevent people from drinking it.

Many vehicles today are designed for the use of **E85**, which is effectively 81 percent ethanol and 19 percent gasoline due to the presence of the denaturant in the fuel. An important question is whether the use of E85 or other blends of ethanol improve or exacerbate air pollution and global warming relative to gasoline vehicles or other vehicles, such as battery electric or hydrogen fuel cell vehicles. These issues are discussed at length in Chapter 13; however, a comparison of the air pollution and health effects of vehicles powered by E85 versus gasoline is given here.

The primary by-products of E85 combustion in a vehicle include unburned ethanol, acetaldehyde, formaldehyde, methane, nitrogen oxides, and carbon monoxide. Table 4.5 compares the percent change in emissions of several important pollutants between E85 and gasoline vehicles from data at two temperatures. Among the chemicals, benzene, 1,3-butadiene,

acetaldehyde, and formaldehyde are the major carcinogens of concern from both gasoline and E85 exhaust. Generally, at room temperature, E85 increases acetaldehyde significantly and formaldehyde to a lesser extent but reduces benzene and 1,3-butadiene. When the population distribution and actual emissions at room temperature throughout the United States are accounted for, the net effect of cancer risks due to E85 and gasoline are similar (Jacobson, 2007). However, Table 4.5 shows that at low temperature, emissions of all four carcinogens increase for E85 relative to gasoline (Ginnebaugh et al., 2010). Because emissions occur in reality across the entire temperature range rather than at room temperature alone, E85 increases the overall cancer risk compared with gasoline. However, the cancer risks of both gasoline and E85 are much higher than are those from battery electric or hydrogen fuel cell vehicles (Chapter 13) and much lower than the ozone mortality effects associated with either gasoline or E85 (discussed next).

Table 4.5 shows that E85 decreases nitrogen oxide but increases organic gas emissions relative to gasoline. Much of the additional organic gas emissions due to E85 is in the form of unburned ethanol and acetaldehyde. The upper triangle of the ozone isopleth in Figure 4.12 indicates that, when $NO_x(g)$ is high relative to ROGs (e.g., in Los Angeles or along the east coast of the United States), both a decrease in $NO_x(g)$ and an increase in ROGs independently increase ozone. This suggests that a conversion from gasoline to E85 should increase ozone in the Los Angeles Basin. Figure 4.14a supports this supposition with a computer model simulation for 2020, in which all gasoline vehicles are converted to E85 vehicles and emissions are at room temperature in both cases. Such a conversion increases ozone in the basin, a result consistent with expectations from the isopleth. The conversion to E85 also increases ozone-related mortality, hospitalization, and asthma by about 9 percent in Los Angeles. If cold temperature emissions, which apply at night and early morning and in the winter, are considered, ozone increases further due to E85.

For region where forests emit significant ROGs, such as in the southeast United States, the ratio of ROGs to $NO_x(g)$ is high, and ozone is governed by the lower triangle of the isopleth in Figure 4.12. In such cases, a decrease in $NO_x(g)$ (which occurs with E85 vs. gasoline) should decrease ozone, and an increase in ROGs should have little impact on ozone. Indeed, the computer simulation results in Figure 4.14b indicate that a conversion to E85 might decrease ozone slightly in the southeast United States. However, most populated areas

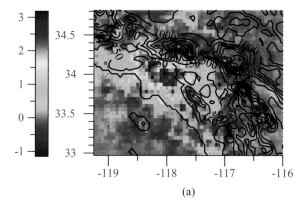

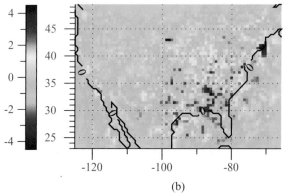

Figure 4.14. Modeled difference in the August 24-hour average near-surface ozone mixing ratio in (a) Los Angeles and (b) the United States when all gasoline vehicles in 2020 are converted to E85 vehicles. The increase in population-weighted ozone (population in a model grid cell multiplied by ozone in the cell, summed over all cells, then divided by total population) is +1.4 ppbv in the Los Angeles simulation and +0.27 ppbv in the U.S. simulation. From Jacobson (2007).

of the United States fall in the upper triangle of the isopleths, where ozone increases due to E85. As such, a conversion to E85 across the entire United States increases the overall number of air pollution–related deaths in Figure 4.14b by 4 percent (Jacobson, 2007). Because gasoline and E85 both cause significant mortality, neither is an ideal option. Instead, a conversion to battery electric or hydrogen fuel cell vehicles, where the raw energy originates from clean, renewable sources, eliminates nearly all air pollution mortality from vehicles (Chapter 13).

4.4. Pollutant Removal

Severe air pollution episodes generally last from a few days to more than a week, depending on the meteorol-

ogy. During a pollution episode, air is usually confined, resulting in the buildup of concentrations over successive days. The major removal processes of pollutant gases during and following an episode are chemical reaction in the air, dissolution in falling raindrops or ocean water, chemical reaction with the ground and other surfaces, and transport (vertical and horizontal) to the larger scale.

Organic gases emitted in polluted air ultimately break down chemically to carbon dioxide and water. In the case of aromatics and other heavy organics, the initial breakdown steps are relatively fast. In the case of many simpler, lighter organic gases, the initial breakdown steps are often slow. Organic gases are also removed by dissolution in rainwater and ocean surfaces, chemical reaction with the ground, conversion to aerosol particle constituents, and transport to the background troposphere. Oxides of nitrogen often evolve chemically to nitric acid, which converts to particulate matter or deposits to the soil or surface water. Oxides of nitrogen also react with organic gases to form organic nitrate gases. Such gases decompose, convert to particulate matter, are rained out, or deposit to the ocean and ground.

4.5. Summary

Urban air pollution has been a problem since the beginning of civilization. Prior to the twentieth century, most air pollution problems arose from the burning of wood, coal, and other raw materials without emission controls. Such burning resulted not only in smoky cities, but also in health problems. In the early and mid-twentieth century, severe London-type smog events occurred during which emissions coupled with a fog or a strong temperature inversion were responsible for several fatal episodes. Increased use of the automobile in the 1900s increased emissions of nitrogen oxides and ROGs. In the presence of sunlight, these chemicals produce ozone, PAN, and a host of other products, giving rise to photochemical smog. Smog initiates when ROGs photolyze or are oxidized by $OH(g)$, $HO_2(g)$, $NO_3(g)$, $O_3(g)$, or $O(g)$ to produce organic radicals. The radicals convert $NO(g)$ to $NO_2(g)$, which photolyzes to $O(g)$, which reacts with $O_2(g)$ to form $O_3(g)$. The most important ROGs in urban air are aromatics, alkenes, and aldehydes. Although alkanes are emitted in greater abundance than the other organics, alkanes are less reactive and longer lived than the others. Most organic gases are destroyed in urban air, but long-lived organics, particularly methane, ethane, and propane, are transported to the background troposphere, where they decay and produce ozone. Carbon monoxide is another abundantly emitted gas in urban air that escapes to the background troposphere because of its long lifetime. $CO(g)$ not only produces ozone in the background troposphere, but is also a chemical source of $CO_2(g)$. In the background troposphere, the concentrations of ozone, nitric oxide, and nitrogen dioxide are strongly coupled through the photostationary-state relationship. In polluted urban air, organics convert more $NO(g)$ to $NO_2(g)$, increasing the $NO_2(g):NO(g)$ ratio and thus ozone compared with the photostationary-state ozone level. The use of ethanol as a fuel may increase ozone production relative to gasoline, particularly at low temperature. However, both ethanol and gasoline, as well as other internal combustion fuels, cause severe health problems. Such problems can be eliminated most effectively by switching away from internal combustion, as discussed in Chapter 13.

4.6. Problems

4.1. Calculate the photostationary-state mixing ratio of ozone when $p_d = 1,013$ hPa, $T = 298$ K, $J \approx 0.01$ s^{-1}, $k_1 \approx 1.8 \times 10^{-14}$ cm^3 molec^{-1} s^{-1}, $\chi_{NO(g)} = 180$ ppbv, and $\chi_{NO_2(g)} = 76$ ppbv. Perform the same calculation under the same conditions, except $\chi_{NO(g)} = 9$ ppbv and $\chi_{NO_2(g)} = 37$ ppbv.

(a) Ignoring the constant temperature and photolysis coefficient, which of the two cases do you believe represents afternoon conditions? Why?

(b) If the $NO(g)$ and $NO_2(g)$ mixing ratios were measured in urban air, and ROG mixing ratios were much higher in the morning than in the afternoon, do you believe that the morning or afternoon ozone mixing ratio calculated by the photostationary-state relationship would be closer to the actual mixing ratio of ozone? Why?

4.2. Explain why the photostationary-state relationship is a useful relationship for background tropospheric air but less useful for urban air.

4.3. Why does ozone not form at night?

4.4. Why does the hydroxyl radical not form at night?

4.5. Why are nighttime ozone mixing ratios always nonzero in the background troposphere but sometimes zero in urban areas?

4.6. If nighttime ozone mixing ratios in one location are zero and in another nearby location are nonzero, what do you believe is the reason for the difference?

4.7. If ozone mixing ratios are 0.16 ppmv and ROG mixing ratios are 1.5 ppmC, what is the best regulatory method of reducing ozone, if only the effects of $NO_x(g)$ and ROGs on ozone are considered?

4.8. If $NO_x(g)$ mixing ratios are 0.2 ppmv and ROG mixing ratios are 0.3 ppmC, what is the best regulatory method of reducing ozone if only the effects of $NO_x(g)$ and ROGs on ozone are considered?

4.9. If $NO_x(g)$ mixing ratios are 0.05 ppmv and ROG mixing ratios are 1 ppmC, what would be the resulting ozone mixing ratio if ROGs were increased by 1 ppmC and $NO_x(g)$ were increased by 0.1 ppmv?

4.10. In Figure 4.13, why are ozone mixing ratios high and nitric oxide mixing ratios low in San Bernardino, whereas the reverse is true in central Los Angeles?

4.11. Why don't aromatic gases, emitted in urban air, reach the stratosphere to produce ozone?

4.12. What are the two fundamental differences between ozone production in the background troposphere and in urban air?

4.13. If the hydroxyl radical did not break down ROGs, would aromatics still be important smog producers? What about aldehydes? Explain.

4.14. Why is $CO(g)$, the most abundantly emitted gas in urban air, not an important smog producer?

4.15. Why should PAN mixing ratios peak at about the same time as ozone mixing ratios during the day?

4.16. In terms of chemical lifetimes and by-products, what are some of the costs and benefits of methanol and ethanol as alternative fuels?

4.17. Write out a chemical mechanism for the production of ozone from propane [$C_3H_8(g)$] oxidation by $OH(g)$.

4.18. Write out a chemical mechanism showing how benzaldehyde [$C_6H_5CHO(g)$] can form ozone.

4.19. Explain the main differences between the effects of ethanol (E85) and gasoline on ozone production when the ROG:$NO_x(g)$ ratio in background air is (a) low (<8:1) and (b) high. Explain your answer in terms of an ozone isopleth.

Chapter 5

Aerosol Particles in the Polluted and Global Atmosphere

Although air pollution regulations have historically focused on gases, aerosol particles cause more health problems and visibility degradation than do gases. Particles smaller than 2.5 μm in diameter cause the most severe health problems. They enter the atmosphere by emissions and nucleation. In the air, their number concentrations and sizes change by coagulation, condensation, chemistry, water uptake, rainout, washout, sedimentation, dry deposition, and transport. Particle concentration, size, and morphology affect the radiative energy balance in both urban air and the global atmosphere. In this chapter, compositions, concentrations, sources, transformation processes, sinks, and health effects of aerosol particles are discussed. The effects of aerosol particles on visibility are described in Chapter 7, and regulations related to particles are given in Chapter 8.

5.1. Size Distributions

Aerosol and hydrometeor particles (defined in Section 1.1.2) are characterized by their size distribution and composition. A **size distribution** is the variation in particle number concentration or mass concentration, for example, with particle size. **Number concentration** is the number of particles of a given size per unit volume of air, whereas **mass (or volume) concentration** is the mass (or volume) of particles of a given size per unit volume of air.

Aerosol particle sizes range from 1 nm to 10 mm in diameter, and thus span seven orders of magnitude.

Aerosol particles smaller than 100 nm in diameter are **nanoparticles**, also known as **ultrafine particles**. Those between 100 and 2,500 nm in diameter are **fine particles**. Particles larger than 2,500 nm are **coarse particles**.

The sum of all aerosol particles smaller than 2.5 μm (2,500 nm) in diameter is referred to as **PM₂.₅**. The sum of all particles smaller than 10 μm in diameter is referred to as **PM₁₀**. As such, PM₁₀ includes the contribution of PM₂.₅. **PM₁** is the sum of all particles smaller than 1 μm in diameter. Such particles are also referred to as **submicron particles**. Conversely, **supermicron particles** are those larger than 1 μm in diameter.

Hydrometeor particles generally range from 5 μm to 8 mm in diameter, although hail can reach 115 mm in diameter.

Table 5.1 compares typical number concentrations and mass concentrations in different size ranges of gases, aerosol particles, and hydrometeor particles under lower tropospheric conditions. It indicates that number and mass concentrations of gas molecules are much larger than are those of particles. The number concentration of aerosol particles decreases with increasing particle size. The number concentration of hydrometeor particles is typically less than is that of aerosol particles, but the mass concentration of hydrometeor particles always exceeds that of aerosol particles.

Aerosol particle size distributions can be divided into **modes**, which are regions of the size spectrum space in which distinct peaks in concentration occur. Usually, each mode is described analytically with a

Table 5.1. Characteristics of gases, aerosol particles, and hydrometeor particles

	Diameter range (μm)	Number concentration (molecules or particles cm^{-3})	Mass concentration (μg m^{-3})
Gas molecules	0.0005	2.45×10^{19}	1.2×10^9
Aerosol particles			
Small	<0.1	10^3–10^6	<1
Medium	0.1–2.5	1–104	<250
Large	2.5–8,000	<1–10	<500
Hydrometeor particles			
Fog drops	5–20	1–500	10^4–10^6
Cloud drops	10–200	1–1,000	10^4–10^7
Drizzle	200–1,000	0.01–1	10^5–10^7
Raindrops	1,000–8,000	0.001–0.01	10^5–10^7
Hail	5,000–115,000	0.0001–0.001	10^5–10^7

Data are for typical lower tropospheric conditions.

lognormal function, which is a bell curve distribution on a log-log scale, as shown in Figure 5.1a. Figure 5.1b shows the same distribution on a log-linear scale.

Aerosol particle size distributions with one, two, three, four, and even five modes are called unimodal, bimodal, trimodal, quadramodal, or pentamodal, respectively. Distributions may include one or two nucleation modes, one or two accumulation modes, and a coarse mode. A **nucleation mode** is a mode with mean diameter smaller than 0.1 μm (100 nm). Thus, nucleation mode particles are nanoparticles or ultrafine particles. The nucleation mode contains newly nucleated particles (those formed directly from the gas phase) or small emitted particles. Nucleated particles are between 1 and 3 nm in diameter because they consist of a small cluster of gas molecules that have changed phase to a liquid.

Two nucleation modes often exist in the case of vehicle emissions. One mode has a peak smaller than 10 nm

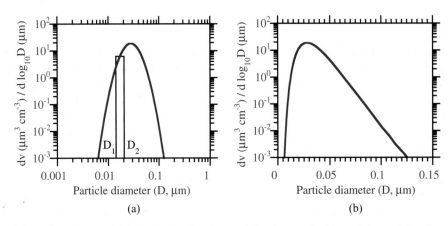

Figure 5.1. (a) A lognormal particle volume distribution on a log-log scale. The incremental volume concentration (dv, μm^3 cm^{-3}-air) of material between any two diameters (D_1 and D_2, μm) is estimated by multiplying the average value from the curve between the two diameters by d $\log_{10}D = \log_{10}D_2 - \log_{10}D_1$. Thus, for example, the volume concentration between diameters D_1 and D_2 is approximately 6 μm^3 cm^{-3} μm$^{-1} \times 0.15$ μm = 0.9 μm^3 cm^{-3}. (b) The lognormal curve shown in (a), drawn on a log-linear scale.

in diameter and contains liquid organic material (often lubricating oil and unburned fuel oil) and sulfuric acid. The second mode often peaks between 20 and 100 nm in diameter and contains primarily aggregates of black carbon spherules coated with lubricating oil, unburned fuel oil, and sulfuric acid. Small nucleated or emitted particles increase in size by coagulation (collision and coalescence of particles) and growth (condensation of gases onto particles). Only a few gases, such as sulfuric acid, water, and some heavy organic gases, among others, condense onto particles as they age in urban and background air. Molecular oxygen and nitrogen, which comprise the bulk of gas in the air, do not condense.

Growth and coagulation move nucleation mode particles into the **accumulation mode**, where diameters are 0.1 to 2.5 μm (100–2,500 nm). Some of these particles are removed by rain, but they are too light to fall out of the air by **sedimentation** (dropping by their own weight against the force of drag). The accumulation mode sometimes consists of two submodes with mean diameters near 200 nm and 500 to 700 nm (Hering and Friedlander, 1982; John et al., 1989), corresponding to newer and aged particles, respectively. The accumulation mode is important for two reasons. First, unlike larger particles, accumulation mode particles are likely to affect health by penetrating deep into the lungs. Second, accumulation mode particles are close in size to the peak wavelengths of visible light and, as a result, affect visibility (Chapter 7). Particles in the nucleation and accumulation modes together are fine particles.

The **coarse mode** consists of particles larger than 2.5 μm in diameter. These particles originate from windblown dust, sea spray, volcanos, plants, fossil fuel combustion, tire erosion, and other sources. Coarse mode particles are generally heavy enough to sediment out rapidly within hours to days. The emission sources and deposition sinks of fine particles differ from those of coarse mode particles. Fine particles usually do not grow by condensation to much larger than 1 μm, indicating that coarse mode particles originate primarily from emissions.

In general, the nucleation mode has the highest number concentration of aerosol particles, the accumulation mode has the highest surface area concentration, and the coarse mode has the highest volume (or mass) concentration. Figure 5.2 shows a quadramodal distribution, fitted from data at Claremont, California, for the morning of August 27, 1987. All four modes (one nucleation mode, two subaccumulation modes, and one coarse particle mode) are most noticeable in the number concentration distribution. The nucleation mode is marginally

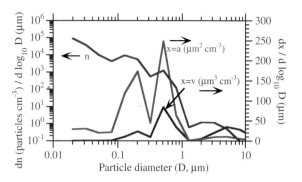

Figure 5.2. Number (n, particles cm^{-3}), area (a, μm^2 cm^{-3}), and volume (v, μm^3 cm^{-3}) concentration size distribution of particles at Claremont, California, on the morning of August 27, 1987. Sixteen model size bins and four lognormal modes were used to simulate the distribution (Jacobson, 1997). The distribution does not include the smallest nucleation mode particles (those less than 20 nm in diameter).

noticeable in the area concentration distribution and invisible in the volume concentration distribution.

5.2. Sources and Compositions of New Particles

New aerosol particles originate from two sources: emissions and homogeneous nucleation. Emitted particles are called **primary particles**. Particles produced by homogenous nucleation, a gas-to-particle conversion process, are called **secondary particles**. Primary particles may originate from point, area, or mobile sources.

5.2.1. Emissions

Aerosol particle emission sources may be natural or anthropogenic. Natural emission sources include sea spray, soil dust, volcanos, natural biomass fires, and biological materials. Major anthropogenic sources include fugitive dust emissions (dust from road paving, passenger and agricultural vehicles, and building construction/demolition), fossil fuel and biofuel combustion, anthropogenic biomass burning, industrial emissions, and tire erosion.

In the United States, about 4.4 million tonnes of anthropogenic particles smaller than 2.5 μm in diameter were emitted in 2008 (Figure 3.14). On a global scale, more than half of all particle emissions are anthropogenic in origin.

Table 5.2 summarizes the natural and anthropogenic sources of the major components present in aerosol

Table 5.2. Nontrivial sources of major components of aerosol particles

	Sea spray emissions	Soil dust emissions	Volcanic emissions	Biomass and biofuel burning	Fossil fuel combustion for transportation and energy	Fossil fuel and metal combustion for industrial processes
Black carbon (C)			✓	✓	✓	✓
Organic matter (C,H,O,N)	✓	✓	✓	✓	✓	✓
Ammonium (NH_4^+)				✓	✓	✓
Sodium (Na^+)	✓	✓	✓	✓		✓
Calcium (Ca^{2+})	✓	✓	✓	✓		✓
Magnesium (Mg^{2+})	✓	✓	✓	✓		✓
Potassium (K^+)	✓	✓	✓	✓		✓
Sulfate (SO_4^{2-})	✓	✓	✓	✓	✓	✓
Nitrate (NO_3^-)				✓		✓
Chloride (Cl^-)	✓	✓	✓	✓		✓
Silicon (Si)		✓	✓		✓	✓
Aluminum (Al)		✓	✓		✓	✓
Iron (Fe)		✓	✓	✓	✓	✓

particles. These sources are discussed in more detail later in the chapter.

5.2.1.1. Sea Spray Emissions

The most abundant natural aerosol particles in terms of mass present in the atmosphere at a given time are sea spray drops. **Sea spray** forms when winds and waves force air bubbles to burst at the sea surface (Woodcock, 1953). Sea spray is emitted primarily in the coarse mode of the particle size distribution. Winds also tear off wave crests to form larger drops, called **spume drops**, but these drops fall back to the ocean quickly. Sea spray initially contains all the components of sea water. About 96.8 percent of sea spray weight is water and 3.2 percent is sea salt, most of which is sodium chloride. Table 5.3 shows the relative composition of the major constituents in sea water. The chlorine-to-sodium mass ratio in sea water, obtained from the table, is about 1.8 to 1. Although organic carbon comprises only a small percent of total sea spray mass worldwide, it comprises an increasing percentage of particle mass with decreasing size in phytoplankton blooms (Facchini et al., 2008).

When sea water is emitted as sea spray or spume drops, the chlorine-to-sodium mass ratio, originally 1.8:1, sometimes decreases because the chlorine is removed by sea spray acidification (e.g., Eriksson, 1960; Duce, 1969; Hitchcock et al., 1980). **Sea spray acidification** occurs when sulfuric or nitric acid enters

Table 5.3. Mass percentages of major constituents in sea water

Constituent	Mass percentages in sea water	Constituent	Mass percentages in sea water
Water	96.78	Sulfur	0.0876
Sodium	1.05	Calcium	0.0398
Chlorine	1.88	Potassium	0.0386
Magnesium	0.125	Carbon	0.0027

The concentration of these constituents, including water, in sea water is 1,033,234 mg/L.
Source: Lide (1998).

a sea spray drop and forces chloride to evaporate as hydrochloric acid [HCl(g)]. Some sea spray drops lose all their chloride in the presence of nitric or sulfuric acid.

The size of a sea spray drop is also affected by dehydration. **Dehydration** (loss of water) occurs when water from a drop evaporates due to a decrease in the relative humidity between the air just above the ocean surface and that a few meters higher. Dehydration increases the concentration of solute in a drop.

5.2.1.2. Soil Dust and Fugitive Dust Emissions

Soil is the natural, unconsolidated mineral and organic matter lying above bedrock on the surface of the Earth. A **mineral** is a natural, homogeneous, inorganic, solid substance with a characteristic chemical composition, crystalline structure, color, and hardness. Minerals in soil originate from the breakdown of rocks. Organic matter originates from the decay of dead plants and animals. **Rocks** are consolidated or unconsolidated aggregates of minerals or biological debris.

Three generic rock families exist: sedimentary, igneous, and metamorphic. **Sedimentary rocks** cover about 75 percent of the Earth's surface; they form primarily on land and on the floors of lakes and seas by the slow layer-by-layer deposition and cementation of carbonates, sulfates, chlorides, and shell and existing rock fragments. An example of a sedimentary rock is **chalk**, made of skeletons of microorganisms. **Igneous rocks** ("rocks from fire" in Greek) form by the cooling of magma, which is a molten, silica-rich mixture of liquid and crystals. Igneous rocks can form either on the Earth's surface following a volcanic eruption or beneath the surface when magma is present. **Granite** is one example of an igneous rock. **Metamorphic rocks** (rocks that "change in form") result from the structural transformation of existing rocks due to high temperatures and pressures in the Earth's interior. During metamorphosis, no rock melting or change in chemical composition occurs; elements merely rearrange themselves to form new mineral structures that are stable in the new environment. **Marble** is an example of a metamorphic rock.

5.2.1.2.1. Breakdown of Rocks to Soil Material.

Breakdown of rocks to soil material occurs primarily by physical weathering. **Physical weathering** is the disintegration of rocks and minerals by processes that do not involve chemical reactions. Disintegration may occur when a stress is applied to a rock, causing it to break into blocks or sheets of different size and, ultimately, into fine soil minerals. Stresses arise when rocks are subjected to high pressure by soil or other rocks lying above. Stresses also arise when rocks freeze and then thaw, or when saline solutions enter cracks and cause rocks to disintegrate or fracture. Salts have higher thermal expansion coefficients than do rocks; thus, when temperatures increase, salts within rock fractures expand, forcing the rock to open and break apart. One source of salt for rock disintegration is sea spray transported from the oceans. Another source is desert salt (from deposits) transported by winds. Physical weathering of rocks on the Earth's surface can occur by their constant exposure to winds or running water.

Another process that causes some rocks to break down and others to reform is **chemical weathering**. Reaction 3.20, which showed calcium carbonate breaking down upon reaction with dissolved carbonic acid, was a chemical weathering reaction. Another chemical weathering reaction is

$$CaSO_4\text{–}2H_2O(s) \rightleftharpoons Ca^{2+} + SO_4^{2-} + 2H_2O(aq) \quad (5.1)$$

$$\underset{\substack{\text{Calcium sulfate} \\ \text{dihydrate (gypsum)}}}{} \quad \underset{\substack{\text{Calcium} \\ \text{ion}}}{} \quad \underset{\substack{\text{Sulfate} \\ \text{ion}}}{} \quad \underset{\substack{\text{Liquid} \\ \text{water}}}{}$$

which occurs when gypsum dissolves in water. Alternatively, when high sulfate and calcium ion concentrations are present in surface or groundwater, Reaction 5.1 can proceed to the left, producing gypsum. Gypsum forms when sulfate-containing particles from the air dissolve in stream water containing calcium ions. Crystalline gypsum can also form when sea spray drops, which contain calcium, collide with volcanic aerosol particles, which contain sulfate.

5.2.1.2.2. Types of Minerals in Soil Dust.

Table 2.2 shows the most abundant elements in the Earth's continental crust. Soil dust particles, emitted from the top layer of the crust, contain minerals made of these elements and organic matter. Minerals in soil include quartz, feldspars, hematite, calcite, dolomite, gypsum, epsomite, kaolinite, illite, smectite, vermiculite, and chlorite.

Pure **quartz** [$SiO_2(s)$] is a clear, colorless mineral that is resistant to chemical weathering. The Greeks believed that it was frozen water. Its name originates from the Saxon word *querkluftertz*, which means "cross-veined ore."

Feldspars, which comprise at least 50 percent of the rocks on the Earth's surface, are by far the most abundant minerals on Earth. The name feldspar originates from the Swedish words *feld* ("field") and *spar* (the name of a mineral commonly found overlying granite). Two common types of feldspars are potassium feldspar [$KAlSi_3O_8(s)$] and **plagioclase feldspar** [$NaAlSi_3O_3–CaAl_2Si_2O_8(s)$].

Hematite [$Fe_2O_3(s)$] (Greek for "bloodlike stone") is an oxide mineral because it includes a metallic element bonded with oxygen. The iron within it causes it to appear reddish-brown (Figure 1.4).

Calcite [$CaCO_3(s)$] and **dolomite** [$CaMg(CO_3)_2(s)$] are carbonate minerals. The name calcite is derived from the word **calcspar**, which was derived from the Greek word for limestone, *khálix*. Dolomite is similar in form to calcite. It was named after French geologist and mineralogist Silvain de Dolomieu (1750–1801).

Gypsum [$CaSO_4–2H_2O(s)$] and **epsomite** [$MgSO_4–7H_2O(s)$] are two of only a handful of sulfate-containing minerals. Gypsum, which ranges from colorless to white, was named after the Greek word *gypsos* ("plaster"). Epsomite was named after the location – Epsom, England – where it was first found. As seen in Table 2.2, sulfur is not an abundant component of the Earth's crust.

Kaolinite, illite, smectite, vermiculite, and chlorite are all **clays**, which are odorous minerals resulting from the weathering of rocks. Clays are usually soft, compact, and composed of aggregates of small crystals. The major components of clays are oxygen, silicon, aluminum, iron, and magnesium. **Kaolinite** [$Al_4Si_4O_{10}(OH)_8(s)$ in pure form] was named after the Chinese word *kauling* ("high ridge"), which is the name of a hill near Jauchu Fa, where clay for the manufacture of porcelain was dug. Early porcelain clay was called kaolinite. **Illite** is a mineral group name, originating from the state name for Illinois, where many illite minerals were first studied. **Smectite** is a mineral group name, originating from *smectis*, a Greek name for "fuller's earth." **Vermiculite** is a mineral group name that originates from the Latin word *vermiculari* ("to breed worms"), which refers to the mineral's ability to exude wormlike structures when rapidly heated. **Chlorite** is a mineral group name, originating from the Greek word for green, which is the common color of this group of clays. In addition to containing minerals, soils contain organic matter, such as plant litter or animal tissue broken down by bacteria.

Soil dust, which consists of the minerals and organic material making up soil, is lifted into the air by winds. The extent of lifting depends on the wind speed and particle mass. Most mass of soil dust lifted into the air is in particles larger than 1 μm in diameter; thus, soil dust particles are predominantly coarse mode particles. Those larger than 10 μm fall out quite rapidly, but those between 1 and 10 μm can stay in the air for days to

Table 5.4. Time for particles to fall 1 km in the atmosphere by sedimentation under near-surface conditions

Particle diameter (μm)	Time to fall 1 km
0.02	228 years
0.10	36 years
1	328 days
10	3.6 days
100	1.1 hours
1,000	4 minutes
5,000	1.8 minutes

weeks or longer, depending on the height to which they are originally lifted. Table 5.4 shows the time required for particles of different diameters to fall 1 km in the air by sedimentation. The table indicates that particles 1 μm in diameter take 328 days to fall 1 km, whereas those 10 μm in diameter take 3.6 days to fall 1 km. Although most soil dust particles are not initially lofted more than 1 km in the air, many are, and these particles can travel long distances before falling to the ground.

Source regions of soil dust on a global scale include deserts (e.g., the Sahara in North Africa; Gobi in Mongolia; Great Basin, Mojave, and Sonora in the southwestern United States; the deserts of Australia) and regions where foliage has been cleared by biomass burning and plowing. Figure 5.3 shows a satellite image of

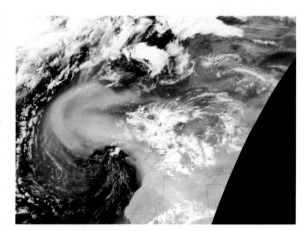

Figure 5.3. Dust storms originating from northwest Africa and Portugal/Spain, captured by SeaWiFS satellite, February 28, 2000. Courtesy SeaWiFS Project, NASA/Goddard Space Flight Center and GeoEye, Inc.

soil dust plumes originating from the Sahara Desert. Soil dust also enters the air when off-road and agricultural vehicles drive over loose soil or when on-road vehicles resuspend soil dust. Additional dust is resuspended during construction or demolition. Fifty percent of all soil dust emitted may be due to anthropogenic activities (Tegen et al., 1996).

5.2.1.3. Volcanic Eruptions

The word **volcano** originates from the ancient Roman god of fire *Vulcan*, after whom the Romans named an active volcano, *Vulcano*. Today, more than 500 volcanos are active on the Earth's surface. Volcanos result from the sudden release of gases dissolved in magma, which contains 1 to 4 percent gas by mass. Water vapor is the most abundant gas in magma, comprising 50 to 80 percent of its mass. Figure 5.4, showing emissions from the Sarychev Volcano, indicates that some of the fast-rising water vapor (in the form of steam) from the volcano expands and cools to form a cloud on

Figure 5.4. Eruption of Sarychev Volcano, Kuril Islands, northeast of Japan on June 12, 2009. The satellite image from the International Space Station indicates a cloud forming on top of the plume due to condensation of rapidly rising and cooling hot water vapor. The volcano punched a hole through a stratiform cloud deck due to the heat emanating out from the eruption. Image courtesy of Earth Sciences and Image Analysis Laboratory, NASA Johnson Space Center, ISS020-E-9048.JPG, http://eol.jsc.nasa.gov.

top of the plume. Volcanic magma also contains carbon dioxide [$CO_2(g)$], sulfur dioxide [$SO_2(g)$], carbonyl sulfide [$OCS(g)$], and molecular nitrogen [$N_2(g)$]. Lesser gases include carbon monoxide [$CO(g)$], molecular hydrogen [$H_2(g)$], molecular sulfur [$S_2(g)$], hydrochloric acid [$HCl(g)$], molecular chlorine [$Cl_2(g)$], and molecular fluorine [$F_2(g)$].

Volcanos emit particles that contain the elements of the Earth's mantle (e.g., Table 2.2). The most abundant particle components in volcanic eruptions are silicate minerals (minerals containing Si). Emitted volcanic particles range in diameter from smaller than 0.1 μm to larger than 100 μm. As seen in Table 5.4, particles 100 μm in diameter take 1.1 hours to fall 1 km by sedimentation. The only volcanic particles that survive more than a few months before falling to the ground are those smaller than 4 μm. Such particles require at least 23 days to fall 1 km. Larger volcanic particles fall to the ground more quickly. Volcanic particles of most sizes are removed more readily by rain than by sedimentation and contain the components listed in Table 5.2.

Volcanic gases such as carbonyl sulfide [$OCS(g)$] and sulfur dioxide [$SO_2(g)$] are sources of new particles. When $OCS(g)$ is injected volcanically into the stratosphere, some of it photolyzes, and its products react to form $SO_2(g)$, which oxidizes to gas-phase sulfuric acid [$H_2SO_4(g)$], which then nucleates to form new sulfuric acid–water aerosol particles. A layer of such particles, called the **Junge layer**, has formed in the stratosphere by this mechanism (Junge, 1961). The average diameter of these particles is 0.14 μm. Sulfuric acid is the dominant particle constituent, aside from liquid water, in the stratosphere and upper troposphere. More than 97 percent of particles in the lower stratosphere and 91 to 94 percent of particles in the upper troposphere contain oxygen and sulfur in detectable quantities (Sheridan et al., 1994).

5.2.1.4. Biomass Burning

Biomass burning is the burning of evergreen forests, deciduous forests, woodlands, grassland, and agricultural land, either to clear land for other use, stimulate grass growth, manage forest growth, or satisfy a ritual. Biomass fires are responsible for a large portion of particle emissions. Such fires may be natural or anthropogenic in origin. Figure 5.5 shows a savannah fire in Kenya.

Biomass burning produces gases, such as $CO_2(g)$, $CO(g)$, $CH_4(g)$, $NO_x(g)$, and ROGs, and particles,

Figure 5.5. Savannah fire at Masai Mara, Kenya.
© Eric Isselee/Dreamstime.com.

such as ash, plant fibers, and soil dust (Figure 5.6), as well as primary organic matter and soot. **Ash** is the primarily inorganic solid or liquid residue left after biomass burning, but may also contain organic compounds oxidized to different degrees. **Primary organic matter** (POM) consists of carbon- and hydrogen-based compounds and often contains oxygen (O), nitrogen (N), etc., as well. It is usually grey (e.g., Figure 5.5), yellow, or brown. **Soot** contains spherules of **black carbon** (BC) (carbon atoms bonded to each other in a sphere) aggregated together and coated by POM. The coating is generally in the form of aliphatic hydrocarbons, polycyclic aromatic hydrocarbons (PAHs), and small amounts of O and N (Chang et al., 1982; Reid and Hobbs, 1998; Fang et al., 1999) The ratio of BC to POM produced in smoke depends on temperature. High-temperature flames produce more BC than POM, whereas low-temperature flames (e.g., in smoldering biomass) produce less. Because BC is black, the more BC produced, the blacker the smoke (Figure 5.7).

Vegetation contains low concentrations of metals, including titanium (Ti), manganese (Mn), zinc (Zn), lead (Pb), cadmium (Cd), copper (Cu), cobalt (Co), antimony (Sb), arsenic (As), nickel (Ni), and chromium (Cr). These substances vaporize during burning and then quickly recondense onto soot or ash particles. Young smoke also contains K^+, Ca^{2+}, Mg^{2+}, Na^+, NH_4^+, Cl^-, NO_3^-, and SO_4^{2-} (Andreae et al., 1998;

Ferek et al., 1998; Reid et al., 1998). Young smoke particles are found in the upper nucleation mode and lower accumulation mode.

5.2.1.5. Fossil Fuel Combustion

Fossil fuel combustion is an anthropogenic emission source. Fossil fuels that produce aerosol particles include coal, oil, natural gas, gasoline, kerosene, and diesel.

Coal is a combustible brown-to-black carbonaceous sedimentary rock formed by compaction of partially decomposed plant material. Metamorphosis of plant material to black coal goes through several stages, from unconsolidated brown-black **peat** to consolidated brown-black **peat coal**, to brown-black **lignite coal**, to dark brown-to-black **bituminous (soft) coal**, to black **anthracite (hard) coal**. Countries with the largest bituminous and anthracite coal reserves include the United States (22.6 percent of worldwide reserves in 2008), Russia (14.4 percent), China (12.6 percent), Australia (8.9 percent), India (7.0 percent), and Germany (4.7 percent) (World Energy Council, 2010). The United States doubled coal production between 1923 and 1998. Even at the present rate of coal recovery, U.S. reserves in 2010 may last 245 years, although those in China may last only 38 years and those worldwide, 119 years.

Oil (or **petroleum**) is a natural greasy, viscous, combustible liquid (at room temperature) that is insoluble in water. It forms from the geological scale decomposition of plants and animals and is made of hydrocarbon complexes. Oil is found in the Earth's continental and ocean crusts.

Natural gas is a colorless, flammable gas made primarily of methane and other hydrocarbon gases that is often found near petroleum deposits. As such, natural gas is mined along with petroleum.

Gasoline is a volatile mixture of liquid hydrocarbons derived by refining petroleum.

Kerosene is a combustible, oily, water-white liquid with a strong odor that is distilled from petroleum and used as a fuel and a solvent.

Diesel fuel is a combustible liquid distilled from petroleum after kerosene. Diesel fuel is named after **Rudolf Diesel** (1858–1913), the German automotive engineer who designed and built the first engine to run on diesel fuel.

Particle components emitted during the combustion of fossil fuels include soot (BC and POM) (Figure 5.8), POM alone, sulfate (SO_4^{2-}), metals, and **fly ash**. Most of the soot, metals, and sulfate are in particles smaller

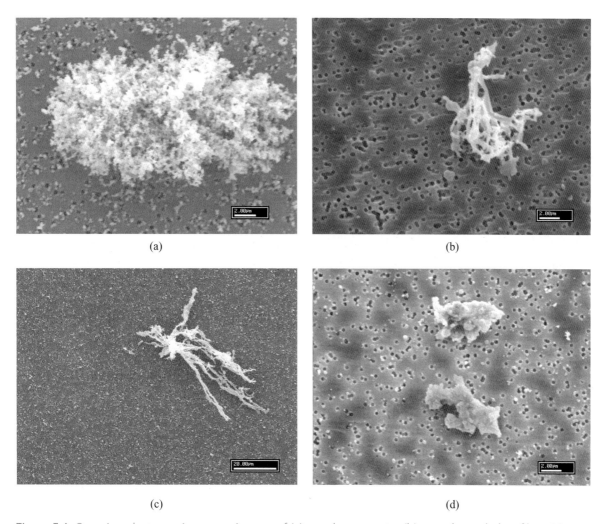

Figure 5.6. Scanning electron microscopy images of (a) an ash aggregate, (b) a combusted plant fiber, (c) an elongated ash particle, and (d) soil dust particles collected from forest fire emissions in Brazil by Reid and Hobbs (1998).

Figure 5.7. Grass fire, February 28, 2009. © Fairiegoodmother/Dreamstime.com.

than 1 μm in diameter. Fly ash is generally found in particles larger than 1 μm in diameter. POM is emitted in both size ranges.

Coal combustion results in the emission of soot, POM, sulfate, and fly ash. The fly ash consists of oxygen, silicon, aluminum, iron, calcium, and magnesium in the form of quartz, hematite, gypsum, and clays. Combustion in gasoline engines usually results in the emission of submicron POM, sulfate, some soot, and elemental silicon, iron, zinc, and sulfur. Combustion in diesel engines usually results in the emission of these components, plus more soot and ammonium.

Diesel-powered vehicles without particle control devices (**aftertreatment technologies**) emit 10 to 100 times more particle mass than do gasoline-powered

(a)   (b)   (c)

Figure 5.8. Soot emissions from (a) natural gas flaring, Sisak, Croatia, June 29, 2009, © Srecko Petrovic/ Dreamstime.com; (b) a diesel locomotive, © Kenneth Sponsler/Dreamstime.com; and (c) a diesel ship, © Efired/Dreamstime.com.

vehicles. With aftertreatment, the ratio is reduced significantly, but diesel vehicles still emit 2 to 20 times more particle mass than do gasoline vehicles. In the United States, 99 percent of heavy-duty trucks and buses but only 0.1 percent of light-duty vehicles use diesel fuel. In Europe, 99 percent of heavy-duty trucks and buses and over half of light-duty vehicles use diesel. In many European countries, the largest sources of soot is diesel engines.

Most soot from fossil fuels originates from coal, diesel fuel, and jet fuel combustion. The smallest emitted particles in diesel exhaust (<15 nm in diameter) are generally **semivolatile** (evaporating readily), containing primarily unburned fuel (C_{15}–C_{23} organics), unburned lubricating oil (C_{15}–C_{36} organics), and sulfate, and form during dilution and cooling of exhaust (e.g., Kittelson, 1998; Sakurai et al., 2003). These particles generally do not contain BC because the diameter of a single **BC spherule** is 5 to 50 nm with a mean of 25 to 35 nm, and particles with BC contain tens to hundreds of spherules aggregated together. As such, **BC aggregates** generally occur only in particles 15 to 200 nm in diameter. BC aggregates are coated by polycyclic aromatic hydrocarbons (PAHs), unburned fuel oil, and unburned lubricating oil (e.g., Steiner et al., 1992; Sakurai et al., 2003). The PAHs include nitro-PAHs (PAHs with nitrogen-containing functional

groups), which may be harmful. Gasoline vehicles emit primarily semivolatile compounds with at least one peak of 20- to 30-nm diameter and a larger peak smaller than 10 nm.

The lower-molecular-weight ($<C_{24}$) semivolatile organic components of both small and large diesel exhaust particles readily evaporate upon dilution of the exhaust, leaving the higher-molecular-weight organics and BC in the particles. The resulting shrinkage of these particles increases their rate of diffusion through the air, and thus their rate of collision and coalescence with each other to larger size (Jacobson et al., 2005).

5.2.1.6. Industrial Sources

Many industrial processes involve the burning of fossil fuels together with metals. As such, industrial combustion emits soot, sulfate, fly ash, and metals. Fly ash from industrial processes often contains $Fe_2O_3(s)$, $Fe_3O_4(s)$, $Al_2O_3(s)$, $SiO_2(s)$, and various carbonaceous compounds that have been oxidized to different degrees (Greenberg et al., 1978). Fly ash is emitted primarily in the coarse mode (particles greater than 2.5 μm in diameter). Metals are emitted during high-temperature industrial processes, such as waste incineration, smelting, cement kilning, and power plant combustion. In such cases, heavy metals vaporize at high temperature, and then recondense onto soot and fly ash particles that

Table 5.5. Metals present in fly ash of different industrial origin

Source	Metals present in fly ash
Smelters	Fe, Cd, Zn
Oil-fired power plants	V, Ni, Fe
Coal-fired power plants	Fe, Zn, Pb, V, Mn, Cr, Cu, Ni, As, Co, Cd, Sb, Hg
Municipal waste incineration	Zn, Fe, Hg, Pb, Sn, As, Cd, Co, Cu, Mn, Ni, Sb
Open hearth furnaces at steel mills	Fe, Zn, Cr, Cu, Mn, Ni, Pb

Sources: Henry and Knapp (1980); Schroeder et al. (1987); Pooley and Mille (1999); Ghio and Samet (1999).

are emitted simultaneously. Table 5.5 lists some metals present in fly ash of different industrial origin.

Of the metals emitted into the air industrially, iron is by far the most abundant. Lead, a criteria air pollutant, is emitted industrially from lead ore smelting, lead acid battery manufacturing, lead ore crushing, and solid waste disposal.

5.2.1.7. Miscellaneous Sources

Additional particle types in the air include tire rubber particles, pollen, spores, bacteria, viruses, plant debris, and meteoric debris. **Tire rubber particles** are emitted due to the constant erosion of a tire at the tire–road interface. Such particles are generally larger than 2.5 μm in diameter. **Pollens**, **spores**, **bacteria**, **viruses**, and **plant debris** are biological particles lifted by the wind. They often serve as sites on which cloud drops and ice crystals form. A stratosphere source of new particles is **meteoric debris**. Most meteorites disintegrate before they fall to an altitude of 80 km. Those that reach the stratosphere contain iron (Fe), titanium (Ti), and aluminum (Al), among other elements. The net contribution of meteorites to particles in the stratosphere is small (Sheridan et al., 1994).

5.2.2. Homogeneous Nucleation

Aside from emissions, homogeneous nucleation is the only source of new particles in the air. **Homogeneous nucleation** is a process by which gas molecules aggregate to form clusters that change phase to a liquid or solid. If the radius of the cluster reaches a critical size (generally fifteen to twenty molecules), the cluster becomes a stable new particle and can grow further.

Homogeneous nucleation occurs without the aid of an existing surface. Another type of nucleation is **heterogeneous nucleation**, which occurs when a cluster forms on a preexisting particle surface. Thus, it does not result in new particles. If background aerosol particles are present, a nucleating gas is more likely to nucleate heterogeneously than homogeneously. Thus, homogeneous nucleation is likely to occur either in relatively clean air or if concentrations of a nucleating gas become very high. In addition, gases that nucleate must be condensable. Conditions for condensation are discussed in Section 5.3.2.1. Homogeneous or heterogeneous nucleation must occur before a particle can grow by condensation or vapor deposition, processes discussed later in the chapter.

Homogeneous and heterogeneous nucleation are homomolecular, binary, or ternary. **Homomolecular nucleation** occurs when molecules of only one gas species nucleate, **binary nucleation** occurs when molecules of two gas species nucleate, and **ternary nucleation** occurs when molecules of three gas species nucleate.

Some gases that undergo homomolecular nucleation include high-molecular-weight organic gas oxidation products of toluene, xylene, alkylbenzenes, alkanes, alkenes, and terpenes. In addition, the chemical reaction of ammonia gas plus hydrochloric acid gas produces ammonium chloride gas, which can homogeneously nucleate to form solid ammonium chloride if concentrations are sufficiently high and the relative humidity is sufficiently low (Seinfeld and Pandis, 2006). Homogenous nucleation of water vapor does not occur under typical atmospheric conditions. Water nucleation is always heterogeneous. Indeed, a cloud drop in the atmosphere forms when water nucleate heterogeneously on one aerosol particles, and additional water then rapidly condenses on the embryo. Aerosol particles that can become cloud drops following heterogeneous nucleation and condensation are called **cloud condensation nuclei** (CCN).

Sulfuric acid and water are the most common gases that undergo binary nucleation. Newly formed homogeneously nucleated sulfuric acid–water particles are typically 1 to 3 nm in diameter. In the remote atmosphere (e.g., over the ocean), homogenous nucleation events can produce more than 10^4 particles cm^{-3} in this size range over a short period.

If ammonia gas is present, it can combine with sulfuric acid and water in a ternary nucleation event.

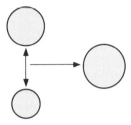

Figure 5.9. Schematic showing coagulation. When two particles collide, they may coalesce to form one large particle, thereby reducing the number concentration but conserving the volume concentration of particles.

Sulfuric acid–water–ammonia are more likely to homogeneously nucleate than sulfuric acid–water if all other conditions are equal.

5.3. Processes Affecting Particle Size

Once in the air, particles increase in size by coagulation and growth. These processes are discussed in the following subsections.

5.3.1. Coagulation

Coagulation occurs when two particles collide and stick together, or coalesce (Figure 5.9), reducing the number concentration but conserving the volume of the sum of all particles in the air. It can occur between two small particles, a small and a large particle, or two large particles.

Coagulation involves the simultaneous collision and coalescence among many particles of different size. Particles of a given **volume** v (cm^3 per particle) are produced by collision and coalescence of particles of any smaller volume $\bar{v}$ with particles of volume $v - \bar{v}$. However, particles of volume v can collide and coalesce with particles of any size to form larger particles. The **basic coagulation equation** that describes the change in the number concentration (n_v, number of particles per cubic centimeter of air) of particles of volume v due to these coagulation production and loss processes is

$$\frac{\partial n_v}{\partial t} = \frac{1}{2}\int_0^v \beta_{v-\bar{v},\bar{v}} n_{v-\bar{v}} n_{\bar{v}} d\bar{v} - n_v \int_0^\infty \beta_{v,\bar{v}} n_{\bar{v}} d\bar{v} \quad (5.2)$$

where β is the **coagulation rate coefficient** (kernel) of two colliding particles (cm^3 particle^{-1} s^{-1}). Equation 5.2 states that the change in number concentration with time of particles of volume v equals the rate at which

particles of volume $v - \bar{v}$ coagulate with particles of volume $\bar{v}$ minus the rate at which particles of volume v are lost due to coagulation with particles of all sizes. The first integral in Equation 5.2 is multiplied by one-half to eliminate double counting of production terms.

Example 5.1
Estimate the concentration of 0.2-μm-diameter particles after 10 seconds of coagulation if 10^5 particles cm^{-3} of this size exist initially, no other particles exist, and the coagulation rate coefficient is 6×10^{-9} cm^3 particle^{-1} s^{-1}. In this case, what is the resulting concentration of new particles formed, assuming that they are formed only from the 0.2-μm-diameter particles and none of the new particles is lost?

Solution
Because particles smaller than 0.2 μm in diameter do not exist in this case, no new particles of this size can form; thus, only the rightmost term in Equation 5.2 needs to be considered. A simple discretization of the resulting equation gives $n_{t+h} = n_t - h\beta_t n_t n_t$, where t is the initial time, h is the time step, and the volume subscripts were removed and replaced by time subscripts. Based on the initial conditions provided, the loss of 0.2-μm particles over $h = 10$ s is $h\beta_t n_t n_t = 600$ particles cm^{-3}, giving the new concentration of 0.2-μm particles as 99,400 particles cm^{-3}. Because each coagulation interaction resulted in the loss of two smaller particles to produce one larger particle, the number of new, larger particles formed must equal half the number of particles lost, thus 300 particles cm^{-3}.

Several important physical processes that drive particles to collide and stick together (coalesce), thereby increasing the coagulation rate coefficient, are Brownian motion, gravitational collection, and van der Waals forces.

Brownian motion is the random movement of particles suspended in a fluid. Coagulation due to Brownian motion is the process by which particles diffuse, collide, and coalesce due to random motion. When two particles collide due to Brownian motion, they may or may not stick together, depending on the efficiency of coalescence, which, in turn, depends on particle shape, composition, and surface characteristics. Because the kinetic energy of a small particle is small relative to that of a large particle at a given temperature, the likelihood that bounce-off occurs when small particles collide is low; thus, such particle collisions usually result in coalescence (Pruppacher and Klett, 1997).

A second important mechanism causing coagulation is **gravitational collection**. When two particles of different size fall, the larger one may catch up and collide with the smaller one. The kinetic energy of the larger particle is higher, increasing the chance that collision will result in a bounce-off rather than a coalescence; thus, not all collisions during gravitational collection result in coalescence. Gravitational collection is an important mechanism for producing raindrops.

Van der Waals forces are weak dipole–dipole attractions caused by brief, local charge fluctuations in nonpolar molecules having no net charge. That is, uncharged particles experience random charge fluctuations that cause one part of the particle to experience a brief positive charge and the other part to experience a brief negative charge, so that the particle still exhibits no net charge. When a particle experiencing a brief charge fluctuation approaches another, the first induces a charge of the opposite sign on the closest end of the second particle. The opposite charge between the two particles causes an attraction, enhancing the rate of coagulation between the particles. Van der Waals forces enhance the rate of coagulation of small particles, particularly of particles smaller than 50 nm in diameter.

Brownian motion affects coagulation significantly when at least one of two colliding particles is small. When both particles are large (but not exactly the same size), gravitational collection is the dominant coagulation process. For small nanoparticles (<15 nm in diameter), van der Waals forces can increase the rate of coagulation compared with Brownian motion alone by a factor of five or more.

Outside of clouds, small aerosol particles are affected more by coagulation than are large aerosol particles because noncloudy air contains many more small particles than large particles, and coagulation rates depend a lot on particle number. In urban regions, coagulation affects the number concentration of aerosol particles primarily smaller than 0.2 μm (200 nm) in size over the course of a day. Figure 5.10 shows results from a model calculation of the change in the number and volume concentration of particles in polluted urban air over a 24-hour period. Whereas the number concentration of small particles was affected, changes in the volume concentration size distribution were affected less.

Over the ocean, coagulation is an important mechanism by which sea spray drops become internally mixed with other aerosol constituents, such as soil dust particles. Andreae et al. (1986), for example, found that

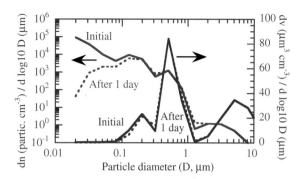

Figure 5.10. Modeled change in aerosol number and volume concentrations at Claremont over a 24-h period when coagulation alone was considered. Number concentration is shown in red; volume concentration, in blue. From Jacobson (1997).

80 to 90 percent of silicate particles over the equatorial Pacific Ocean between Ecuador and Hawaii contained sea spray constituents. Murphy et al. (1998) discovered that almost all aerosol particles larger than 0.13 μm in the boundary layer in a remote South Pacific site contained sea spray components. Pósfai et al. (1999) found that almost all soot particles in the North Atlantic contained sulfate. The internal mixing of aerosols by coagulation is supported by model simulations that show that on a global scale, about half the increase in size of soot particles following their emissions may be due to coagulation with nonsoot particles, such as sulfate, organic matter, sea spray, and soil, whereas the rest may be due to growth processes (Jacobson, 2001b). Thus, although coagulation does not affect the number concentration of large particles very much, it does affect the composition and mixing state of particles of all sizes (Jacobson, 2002b).

5.3.2. Growth Processes

Coagulation is a process that involves only particles, whereas condensation/evaporation; vapor deposition/sublimation; dissolution, dissociation, and hydration; and gas-aerosol chemical reaction are gas-to-particle conversion processes. These processes are discussed in the following subsections.

5.3.2.1. Condensation/Evaporation

Condensation and evaporation occur only after homogeneous or heterogeneous nucleation. On a nucleated liquid surface, gas molecules continuously **condense** (change state from gas to liquid) and liquid molecules continuously **evaporate** (change state from liquid

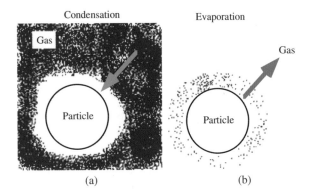

Condensation   Evaporation

Gas

Gas

Particle

Particle

(a)   (b)

Figure 5.11. (a) Condensation occurs when the partial pressure of a gas away from a particle surface (represented by the thick cloud of gas away from the surface) exceeds the saturation vapor pressure (SVP) of the gas over the surface (represented by the thin cloud of gas near the surface). (b) Evaporation occurs when the SVP exceeds the partial pressure of the gas. The schematics are not to scale.

to gas). The partial pressure of the gas immediately over the particle's surface is called the gas's **saturation vapor pressure** (SVP). If the partial pressure of the gas away from the surface increases above the SVP over the surface, excess molecules diffuse to the surface (Figure 5.11a) and condense. If the gas's partial pressure decreases below the SVP, gas molecules over the surface diffuse away from the surface (Figure 5.11b), and liquid molecules on the surface evaporate to maintain saturation over the surface. In sum, *if the ambient partial pressure of a gas exceeds the gas's SVP, condensation occurs.* If the ambient partial pressure of the gas falls below the gas's SVP, evaporation occurs. Thus, the lower its SVP, the more likely a gas is to condense.

The most abundant condensing gas in the air is water vapor. Figure 5.12a shows the SVP of water over a liquid surface versus temperature. It indicates that the SVP of water vapor increases superlinearly with increasing temperature. This rule applies to any gas. The consequence is that *gases evaporate faster with increasing temperature and condense faster with decreasing temperature.*

Figure 5.12b shows the SVP of water over a liquid surface and an ice surface at temperatures below 0°C. Because water vapor's partial pressure cannot exceed its SVP without the excess vapor condensing, the SVP is effectively the maximum possible partial pressure of water vapor in the air at a given temperature. Near the poles, where temperatures are below 0°C, the SVP can be as low as 0.0003 percent of sea level air pressure. Near the Equator, where temperatures are close to 30°C, the SVP can increase to 4 percent or more of sea level air pressure.

> **Example 5.2**
> Determine the maximum partial pressure and percentage water vapor in the atmosphere at 0°C and 30°C.
>
> **Solution**
> From Figure 5.12a, the SVP and, therefore, the maximum partial pressure of water vapor at 0°C and 30°C are 6.1 and 42.5 hPa, respectively. Because sea level dry air pressure is 1,013 hPa, water vapor comprises no more than 0.6 and 4.2 percent of total air by volume, respectively, at these two temperatures.

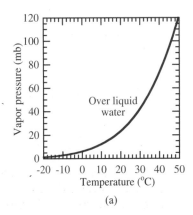

(a)

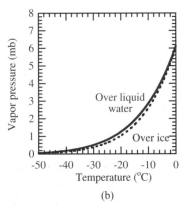

(b)

Figure 5.12. Saturation vapor pressure over (a) liquid water versus temperature and (b) liquid water and ice versus temperature.

During condensation, latent heat is released, warming the surface of a drop. This warming immediately increases the SVP over the drop surface, as illustrated in Figure 5.12. The warming also creates a temperature gradient between the drop surface and the air around it, causing energy to flow from the surface back to the cooler air, slightly reducing the warming of the drop surface. Nevertheless, the SVP increase due to the warmer surface slows the rate of vapor transfer to the surface. Vapor transfer is also slowed by the fact that drop growth depletes vapor away from the drop, decreasing the partial pressure over time.

These physical processes are described by Equation 5.3, which gives the rate of change in volume (v, cm^3 per particle) of a single liquid drop due to condensation or evaporation as

$$\frac{dv}{dt} = \frac{4\pi r D(p - p_s)}{\dfrac{DL_e \rho p_s}{\kappa T}\left(\dfrac{L_e m}{R^* T} - 1\right) + \dfrac{R^* T \rho}{m}} \quad (5.3)$$

where r is drop radius (cm), D is the diffusion coefficient of vapor through the air (cm^2 s^{-1}), p is the partial pressure of vapor away from the surface (hPa), p_s is the SVP over the surface (hPa), L_e is the latent heat of evaporation (J g^{-1} = 10^4 cm^3 hPa g^{-1}), ρ is the liquid density of the drop (g cm^{-3}), κ is the thermal conductivity of air (J cm^{-1} s^{-1} K^{-1}; Section 3.2.1), T is the drop surface temperature, m is the molecular weight of the vapor (g mol^{-1}), and R^* is the universal gas constant (cm^3 hPa mol^{-1} K^{-1}). A **diffusion coefficient** is the rate of transfer of a gas through a bath of air molecules arising from random motion and redirection upon collision with the air molecules. The **latent heat of evaporation** is the energy added to a liquid to evaporate it and equals the energy released to the air by a vapor to condense it. For water vapor, $D = 0.211$ cm^2 s^{-1}, $L_e = 2{,}501$ J g^{-1}, $m = 18.015$ g mol^{-1}, and $\rho = 1$ g cm^{-3}.

Equation 5.3 indicates that the rate of change of drop volume is proportional to the gradient between the partial pressure and SVP, and to the diffusion coefficient of the vapor through air. Thus, the greater the partial pressure relative to the SVP, the faster the growth rate of the drop. Similarly, the greater the diffusion coefficient, the greater the growth rate. If the SVP exceeds the partial pressure, the drop shrinks (liquid evaporates).

Equation 5.3 also shows that the greater the latent heat of evaporation, the slower the drop growth rate. In other words, if more heat is released during condensation, less vapor can condense because the heat feeds back to increase the SVP. Conversely, the greater the thermal conductivity of air, the faster the growth

rate because heat released during condensation transfers away from the drop surface faster, diminishing the warming upon condensation and the increase in SVP.

Example 5.3
Calculate the percent increase in the growth rate of single-particle volume from Equation 5.3 if the radius doubles, $p_s = 1$ hPa, the partial pressure (p) increases from 1.01 p_s to 1.02 p_s, and all other parameters stay constant.

Solution
The ratio of the final to initial pressure gradient from the example is $(1.02p_s - p_s)/(1.01p_s - p_s) = 2$. Multiplying this by a factor of 2 to account for the doubling of particle radius gives a final growth rate 4 times the initial value, for an increase of 300 percent. Thus, the growth rate of a single particle increases with increasing particle size and with increasing difference between the partial pressure and saturation vapor pressure.

The **relative humidity** (RH) is the partial pressure of water vapor (p) divided by the SVP of water over a flat, pure liquid water surface (p_s), all multiplied by 100 percent. When the relative humidity exceeds 100 percent, $p > p_s$, and water vapor condenses onto CCN to form **cloud drops**. When the RH drops below 100 percent, $p < p_s$, and liquid water on a CCN surface evaporates, leaving the residual aerosol particle.

Example 5.4
If the partial pressure of water vapor is 20 hPa and the temperature is 30°C, what is the relative humidity?

Solution
From Figure 5.12a, the SVP is 42.5 hPa. Therefore, the relative humidity is 100 percent × 20 hPa/42.5 hPa = 47 percent.

Sulfuric acid gas, which has a low SVP, also condenses onto particles. Once condensed, sulfuric acid rarely evaporates because its SVP is so low. Sulfuric acid condenses primarily onto accumulation mode particles because the accumulation mode has a larger surface area concentration (surface area per volume of air) than do other modes.

Some other condensable gases (with low SVPs) include high-molecular–weight organic gases, such as certain products of toluene, xylene, alkylbenzene, alkane, alkene, and biogenic hydrocarbon oxidation (Pandis et al., 1992).

5.3.2.1.1. Vapor Deposition/Sublimation.

Vapor deposition is the process by which gas diffuses to an aerosol particle surface and deposits on the surface as a solid. Water vapor deposition to ice is the most common type of vapor deposition in the atmosphere. It occurs in clouds only at subfreezing temperatures (below 0°C) and when the partial pressure of water exceeds the SVP of water over ice. Figure 5.12b shows the SVP of water vapor over ice at subfreezing temperatures. The reverse of water vapor deposition is **sublimation**, the conversion of ice to water vapor. Deposition of ammonium chloride gas or ammonium nitrate gas to the solid phase can also occur when the relative humidity is sufficiently low. At high relative humidity, though, these gases dissolve in liquid water within aerosol particles.

5.3.2.1.2. Dissolution, Dissociation, and Hydration.

Dissolution is the process by which a gas, suspended over an aerosol particle surface, diffuses to and dissolves in a liquid on the surface. The liquid in which the gas dissolves is a **solvent**. In aerosol and hydrometeor particles, liquid water is most often the solvent. Any gas, liquid, or solid that dissolves in a solvent is a **solute**. One or more solutes plus the solvent comprise a **solution**. The ability of a gas to dissolve in water depends on the **solubility** of the gas, which is the maximum amount of a gas that can dissolve in a given amount of solvent at a given temperature.

In a solution, dissolved molecules may **dissociate** (break into simpler components, namely, ions). Positive ions, such as H^+, Na^+, K^+, Ca^{2+}, and Mg^{2+}, are **cations**. Negative ions, such as OH^-, Cl^-, NO_3^-, HSO_4^-, SO_4^{2-}, HCO_3^-, and CO_3^{2-}, are **anions**. The dissociation process is reversible, meaning that ions can reform a dissolved molecule. Substances that undergo partial or complete dissociation in solution are **electrolytes**. The degree of dissociation of an electrolyte depends on the acidity of the solution, the strength of the electrolyte, and the concentrations of ions in solution.

The **acidity** of a solution is a measure of the concentration of **hydrogen ions** (protons or H^+ ions)

in solution. Acidity is measured in terms of **pH**, where

$$pH = -\log_{10}[H^+] \tag{5.4}$$

$[H^+]$ is the **molarity** of H^+ (moles of H^+ per liter of solution). The more acidic a solution, the higher the molarity of H^+ and the lower the pH. The pH scale (Figure 10.3) ranges from less than 0 (highly acidic) to greater than 14 (highly basic or alkaline). In pure water, the only source of H^+ is

$$\underset{\substack{\text{Liquid} \\ \text{water}}}{H_2O(aq)} \rightleftharpoons \underset{\substack{\text{Hydrogen} \\ \text{ion}}}{H^+} + \underset{\substack{\text{Hydroxide} \\ \text{ion}}}{OH^-} \tag{5.5}$$

where OH^- is the **hydroxide ion** and arrows in both directions indicate that the reaction is reversible. Because the product $[H^+][OH^-]$ must equal 10^{-14} mol^2 L^{-2}, and $[H^+]$ must equal $[OH^-]$ to balance charge, the pH of pure water is 7 ($[H^+] = 10^{-7}$ mol L^{-1}).

Acids are substances that, when added to a solution, dissociate, increasing the molarity of H^+. The more H^+ added, the stronger the acid and the lower the pH. Common acids include sulfuric [$H_2SO_4(aq)$], hydrochloric [$HCl(aq)$], nitric [$HNO_3(aq)$], and carbonic [$H_2CO_3(aq)$] acids. When the pH is low (<2), $HCl(aq)$, $HNO_3(aq)$, and $H_2SO_4(aq)$ dissociate readily, whereas $H_2CO_3(aq)$ does not. The former acids are **strong acids**, and the latter acid is a **weak acid**.

Bases (alkalis) are substances that, when added to a solution, remove H^+, increasing pH. Some bases include ammonia [$NH_3(aq)$] and slaked lime [$Ca(OH)_2(aq)$].

When anions, cations, or certain undissociated molecules are dissolved in water, the water can bond to the ion in a process called **hydration**. Several water molecules can hydrate to each ion. Hydration, which increases the liquid water content of aerosol particles, is important when the relative humidity is less than 100 percent. The higher the sub–100-percent relative humidity and the greater the quantity of solute in solution, the greater the liquid water content of aerosol particles due to hydration. At relative humidities greater than 100 percent, the volume of water added to a particle by hydration is small compared with that added by water vapor condensation.

Next, dissolution and reaction of some strong acids and a base in aerosol particles are discussed. The discussion extends to cloud drops and rain in Chapter 10.

5.3.2.1.3. Hydrochloric Acid.

Gas-phase **hydrochloric acid** [$HCl(g)$] is abundant over the ocean, where it originates from sea spray and sea water evaporation. Over land, it is emitted anthropogenically during coal combustion. If $HCl(g)$ becomes supersaturated in the gas phase (if its partial pressure exceeds its saturation vapor pressure), $HCl(g)$ dissolves into water-containing particles and dissociates by the reversible process

$$\underset{\substack{\text{Hydrochloric} \\ \text{acid gas}}}{HCl(g)} \rightleftharpoons \underset{\substack{\text{Dissolved} \\ \text{hydrochloric acid}}}{HCl(aq)} \rightleftharpoons \underset{\substack{\text{Hydrogen} \\ \text{ion}}}{H^+} + \underset{\substack{\text{Chloride} \\ \text{ion}}}{Cl^-} \quad (5.6)$$

Dissociation of $HCl(aq)$ is complete as long as the pH exceeds −6, which almost always occurs. The pH of fresh sea spray drops, which are primarily in the coarse particle mode, ranges from +7 to +9; however, the pH of such drops decreases during dehydration and sea spray acidification. **Sea spray acidification**, briefly discussed in Section 5.2.1.1, occurs when nitric acid or sulfuric acid enters a particle and dissociates, adding H^+ to the solution and forcing Cl^- to reassociate with H^+ and evaporate as $HCl(g)$. A net sea spray acidification process involving nitric acid is

$$\underset{\substack{\text{Nitric} \\ \text{acid gas}}}{HNO_3(g)} + \underset{\substack{\text{Chloride} \\ \text{ion}}}{Cl^-} \rightleftharpoons \underset{\substack{\text{Hydrochloric} \\ \text{acid gas}}}{HCl(g)} + \underset{\substack{\text{Nitrate} \\ \text{ion}}}{NO_3^-} \quad (5.7)$$

Sea spray acidification is most severe along coastal regions near pollution sources and can result in a depletion of chloride ions from sea spray drops. Figure 5.13 shows the effect of sea spray acidification and the measured composition of aerosol particles 3.3 to 6.3 μm in diameter at Riverside, California, about 60 km from the Pacific Ocean. Sodium in the particles originated from the ocean. In clean air over the ocean, the mass ratio of chloride to sodium is typically 1.8:1. Figure 5.13 further shows that, over Riverside, the ratio was about 0.18:1, one-tenth the clean air ratio. The fact that Riverside particles contained a lot of nitrate and sulfate suggests that acidification by these ions was responsible for the near depletion of chloride in the particles.

5.3.2.1.4. Nitric Acid.
Gas-phase **nitric acid** [$HNO_3(g)$] forms from the chemical oxidation of nitrogen dioxide (Reaction 4.5). Because emitted aerosol particles generally do not contain nitric acid, nitric acid usually enters aerosol particles from the gas phase. The process is

$$\underset{\substack{\text{Nitric} \\ \text{acid gas}}}{HNO_3(g)} \rightleftharpoons \underset{\substack{\text{Dissolved} \\ \text{nitric acid}}}{HNO_3(aq)} \rightleftharpoons \underset{\substack{\text{Hydrogen} \\ \text{ion}}}{H^+} + \underset{\substack{\text{Nitrate} \\ \text{ion}}}{NO_3^-} \quad (5.8)$$

Nitric acid is a strong acid because it dissociates when the pH exceeds −1.

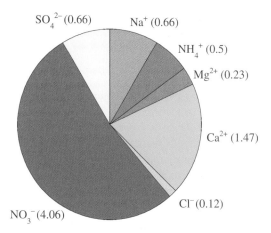

Figure 5.13. Example of sea spray and soil dust acidification. The pie chart shows measured mass concentration (μg m^{-3}, in parentheses) of inorganic ions summed over particles with diameters between about 3.3 and 6.3 μm on August 29, 1987, from 05:00 to 08:30 PST, at Riverside, California. Chloride associated with the sodium in sea spray and carbonate associated with the calcium in soil dust particles were most likely displaced to the gas phase by the addition of nitrate and sulfate to the particles. Data from the seventh stage of eight-stage impactor measurements by John et al. (1990).

When nitric acid dissolves in sea spray drops containing chloride, it displaces the chloride to the gas phase by sea spray acidification, as shown in Reaction 5.7. Similarly, when nitric acid dissolves in soil particle solutions containing calcium carbonate and water, it causes calcium carbonate to dissociate and the carbonate to reform carbon dioxide gas, which evaporates. The process, called **soil dust acidification**, is described by

$$\underset{\substack{\text{Calcium} \\ \text{carbonate}}}{CaCO_3(s)} + \underset{\substack{\text{Nitric} \\ \text{acid gas}}}{2HNO_3(g)} \rightleftharpoons \underset{\substack{\text{Calcium} \\ \text{ion}}}{Ca^{2+}} + \underset{\substack{\text{Nitrate} \\ \text{ion}}}{2NO_3^-}$$
$$+ \underset{\substack{\text{Carbon} \\ \text{dioxide gas}}}{CO_2(g)} + \underset{\substack{\text{Liquid} \\ \text{water}}}{H_2O(aq)} \quad (5.9)$$

(e.g., Dentener et al., 1996; Hayami and Carmichael, 1997; Tabazadeh et al., 1998; Jacobson, 1999c). The net result of this process is that nitrate ions build up in soil dust particles that contain calcite. A similar result occurs when nitric acid gas is exposed to soil dust particles that contain magnesite [$MgCO_3(s)$]. Because nitric acid readily enters soil dust and sea spray particles during acidification and these particles are primarily in the coarse mode, nitrate is usually in the coarse mode. The high coarse mode nitrate concentration in Figure 5.13 was most likely due to acidification of sea spray and soil dust particles.

5.3.2.1.5. Sulfuric Acid.

Gas-phase **sulfuric acid** [$H_2SO_4(g)$] is condensable due to its low saturation vapor pressure. Once it condenses, it does not readily evaporate, so it is **involatile**. As sulfuric acid condenses, water vapor molecules simultaneously hydrate to it. Thus, condensation of sulfuric acid produces a solution of sulfuric acid and water, even if a solution did not preexist.

Once condensed irreversibly, sulfuric acid dissociates reversibly. Condensation and dissociation are represented by

$$\underset{\substack{\text{Sulfuric} \\ \text{acid gas}}}{H_2SO_4(g)} \rightarrow \underset{\substack{\text{Dissolved} \\ \text{sulfuric acid}}}{H_2SO_4(aq)} \rightleftharpoons \underset{\substack{\text{Hydrogen} \\ \text{ion}}}{H^+} + \underset{\substack{\text{Bisulfate} \\ \text{ion}}}{HSO_4^-}$$

$$\rightleftharpoons \underset{\substack{\text{Hydrogen} \\ \text{ion}}}{2H^+} + \underset{\substack{\text{Sulfate} \\ \text{ion}}}{SO_4^{2-}} \quad (5.10)$$

The first dissociation [producing the **bisulfate ion** (HSO_4^-)] occurs when the pH exceeds −3, so sulfuric acid is a strong acid. The second dissociation [producing the **sulfate ion** (SO_4^{2-})] occurs when the pH exceeds +2, so the bisulfate ion is also a strong acid.

Condensation of sulfuric acid occurs most readily over the particle size mode with the most surface area, which is the accumulation mode. When sulfuric acid condenses on coarse mode sea spray drops, it displaces the chloride ion as hydrochloric acid to the gas phase. When it condenses on soil dust particles, sulfuric acid displaces the carbonate ion as carbon dioxide to the gas phase. In a competition with nitrate, sulfuric acid also displaces the nitrate ion as nitric acid to the gas phase.

5.3.2.1.6. Ammonia.

Ammonia gas [$NH_3(g)$] is emitted during bacterial metabolism in domestic and wild animals and their waste, humans, fertilizers, natural soil, and the oceans. It is also emitted during biomass burning (forest fires, savannah and grassland burning, outdoor agricultural waste burning), biofuel burning, and fossil fuel combustion. Figure 5.14 summarizes the relative source contributions to worldwide ammonia emissions in 2005.

When ammonia dissolves in water, it combines with the hydrogen ion to form the **ammonium ion**

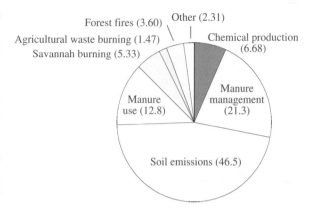

Figure 5.14. Percentage of the 2005 worldwide ammonia gas emissions of 48.389 Tg-NH_3(g) from different sources. "Other" includes electricity and heat production (0.21 percent), other energy industries (0.023 percent), manufacturing industries and construction (0.31 percent), road transport (0.95 percent), residential (0.086 percent), fugitive dust emissions from solid fuels (0.0047 percent), mineral production (0.077 percent), grassland fires (0.61 percent), and other waste handling (0.037 percent). Data from EC-JRC/PBL (2010).

(NH_4^+) by

$$\underset{\substack{\text{Ammonia} \\ \text{gas}}}{NH_3(g)} \rightleftharpoons \underset{\substack{\text{Dissolved} \\ \text{ammonia}}}{NH_3(aq)} \quad (5.11)$$

$$\underset{\substack{\text{Dissolved} \\ \text{ammonia}}}{NH_3(aq)} + \underset{\substack{\text{Hydrogen} \\ \text{ion}}}{H^+} \rightleftharpoons \underset{\substack{\text{Ammonium} \\ \text{ion}}}{NH_4^+} \quad (5.12)$$

Because the ammonium ion is positively charged, it primarily enters particles that have an abundance of negatively charged ions (anions) in order to maintain charge balance. Particles that have an abundance of anions are acidic and thus have a low pH. Acidic particles generally contain sulfate, nitrate, or chloride, which are all anions. Particles that are basic (have a high pH), often contain significant amounts of sodium, potassium, calcium, or magnesium, which are all cations. *Thus, the ammonium ion generally enters particles that have an abundance of anions but not particles that have an abundance of cations.* Because sea spray and soil particle solutions contain high concentrations of cations, and thus have a high pH, ammonia rarely enters these particles. An exception is when high concentrations of sulfate or nitrate ions are also present (e.g., Figure 5.13).

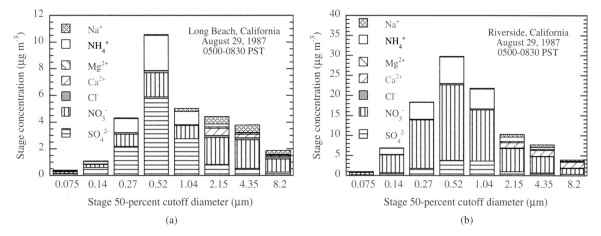

Figure 5.15. Measured concentrations of inorganic aerosol particle components versus particle diameter at (a) Long Beach and (b) Riverside, California, on the morning of August 29, 1987. Data were obtained by John et al. (1990) with an eight-stage Berner impactor.

Ammonia is frequently present in particles containing sulfate, and such particles are usually present in the accumulation mode. When nitrate ion concentrations in the accumulation mode are high, the nitrate is also balanced by ammonium ions. Figure 5.15 illustrates this point by showing measured aerosol particle compositions versus size at Long Beach and Riverside, California. Long Beach is a coastal site, and Riverside is located 60 km inland in the Los Angeles Basin. At Long Beach, sulfate dominated, but nitrate was present in the accumulation mode. At Riverside, nitrate dominated, but some sulfate was present in the accumulation mode. Ammonium ions balanced charge with nitrate and sulfate ions in the accumulation mode at both Long Beach and Riverside.

5.3.2.1.7. Solid Precipitation.

When their concentrations in aerosol particle solution are high, ions may precipitate to form **solid electrolytes**. Indeed, many soils of the world have formed from the deposition of solid minerals originating from aerosol particle solutions. **Precipitation** is the formation of an insoluble solid compound due to the buildup in concentration of dissolved ions in a solution. Solids can be suspended throughout a solution but are not part of the solution. If the water content of a solution suddenly increases, solid electrolytes often dissociate back to ions. Solid electrolytes generally do not form in cloud drops because these drops are too dilute. For a similar reason, solid formation is often inhibited in aerosol particles when the relative humidity is high.

The most abundant sulfate-containing electrolyte in aerosol particles on a global scale may be **gypsum** [$CaSO_4–2H_2O(s)$] (Jacobson, 2001a), which can form at any relative humidity below 98 percent. Gypsum forms when calcium and sulfate react in sea spray drops or soil dust particles; thus, it is present primarily in the coarse mode. **Ammonium sulfate** [$(NH_4)_2SO_4(s)$], which forms in accumulation mode particles, may be less abundant than is gypsum, but it is more important than gypsum in terms of its effects on visibility because the accumulation mode affects radiative fields more than does the coarse mode.

In urban regions, where nitrate production and ammonia gas emissions are high, concentrations of solid **ammonium nitrate** [$NH_4NO_3(s)$] often build up. In Figure 5.15, for example, some of the ammonium and nitrate in accumulation mode particles probably formed ammonium nitrate crystals. Ammonium nitrate, in either liquid or solid form, is considered to be one of the major causes of visibility reduction in Los Angeles smog. The major sources of the nitrate ion is nitric acid formed from fossil fuel emissions of $NO_x(g)$. The major source of the ammonium ion is ammonia emitted by cattle, such as in the feedlots of Chino, just east of Los Angeles.

5.3.3. Removal Processes

Aerosol particles are removed from the air by rainout, washout, sedimentation, and dry deposition. **Rainout** occurs when an aerosol particle CCN activates to form a liquid cloud drop or ice crystal and the drop or crystal coagulates with other cloud drops or crystals to become rain or graupel, which falls to the surface, removing the aerosol particle inclusion. **Washout** occurs when growing or falling precipitation particles coagulate with aerosol particles that are either interstitially between cloud drops or in the clear sky below clouds that the rain passes through. When the precipitation falls to the surface, it brings the aerosol particles with it. Together, rainout and washout are the most important mechanisms removing aerosol particles globally (Jacobson, 2010b). Because rain clouds occur only in the troposphere, rainout is not a process by which stratospheric particles are removed. However, rainout is an effective removal process for volcanic particles.

Sedimentation is the sinking of particles to lower altitudes by their own weight against the force of drag imposed by the air. **Dry deposition** is a process by which gases and aerosol particles are carried by molecular diffusion; turbulent diffusion; or winds to the surface of trees, grass, rocks, the ocean, buildings, or roads, and then rest on, bond to, or react with the surface.

Sedimentation and dry deposition are important removal processes for very large particles over short periods and small particles over long periods, but not so important relative to rainout or washout for small particles over short periods. Table 5.4 indicates that particles less than 0.5 μm in diameter stay in the air several years before sedimenting even 1 km by their own weight. For these and smaller particles, sedimentation is only a long-term removal process. Gases also sediment, but their weights are so small that their sedimentation velocities are negligible. A typical gas molecule has a diameter of 0.5 to 1 nm. Such diameters result in gases falling only 1 to 3 km per 10,000 years.

If small particles are near the ground, dry deposition can usually remove them more efficiently than can sedimentation. Dry deposition is more efficient for removing particles than gases because particles are heavier than are gases. As such, particles fall and tend to stay on a surface more readily than do gases, unless wind speeds are high. Gases, especially if they are chemically unreactive, are more likely to be resuspended into the air.

5.4. Summary of the Composition of Aerosol Particles

The composition of aerosol particles varies with particle size and location. Some generalities about composition are summarized as follows:

- Newly nucleated aerosol particles usually contain sulfate ions and water, although they may also contain ammonium ions.
- Biomass, biofuel, and fossil fuel combustion produce primarily small accumulation mode particles, but coagulation and gas-to-particle conversion move these particles to the middle and high accumulation modes. Coagulation also moves some particles to the coarse mode.
- Metals that evaporate during industrial emissions recondense, primarily onto accumulation mode soot particles and coarse mode fly ash particles. The metal emitted in greatest abundance is usually iron.
- Sea spray and soil particles are primarily in the coarse mode.
- When sulfuric acid condenses, it usually condenses onto accumulation mode particles because these aerosol particles have more surface area, when averaged over all particles in the mode, than do nucleation or coarse mode particles.
- Once in accumulation mode particles, sulfuric acid dissociates primarily to sulfate ions [SO_4^{2-}]. To maintain charge balance, ammonia gas [$NH_3(g)$] dissolves and dissociates in such particles, producing ammonium ions [NH_4^+], the major cation in accumulation mode particles. Thus, ammonium and sulfate ions often coexist in accumulation mode particles.
- Because sulfuric acid has a lower SVP and a greater solubility than does nitric acid, nitric acid is inhibited from entering those accumulation mode particles that already contain sulfuric acid.
- Nitric acid tends to dissolve in coarse mode particles that contain cations (and have a high pH). It displaces the chloride ion in sea spray drops to hydrochloric acid gas and the carbonate ion in soil dust particles to carbon dioxide gas during sea spray acidification and soil dust acidification, respectively. Sulfuric acid also displaces chloride ions and carbonate ions during acidification.

Table 5.6 summarizes the predominant components and their sources in the nucleation, accumulation, and coarse particle modes.

5.5. Aerosol Particle Morphology and Shape

The morphologies (structures) and shapes of aerosol particles vary with composition. The older an aerosol particle, the greater the number of layers and attachments the particle is likely to have. If the aerosol particle

Table 5.6. Dominant sources and components of nucleation, accumulation, and coarse mode particles

Nucleation mode	Accumulation mode	Coarse mode
Homogeneous nucleation $H_2O(aq)$, SO_4^{2-}, NH_4^+	Industrial emissions BC, POM, Fe, Al, S, P, Mn, Zn, Pb, Ba, Sr, V, Cd, Cu, Co, Hg, Sb, As, Sn, Ni, Cr, $H_2O(aq)$, NH_4^+, Na^+, Ca^{2+}, K^+, SO_4^{2-}, NO_3^-, Cl^-, CO_3^{2-}	Sea spray emissions $H_2O(aq)$, Na^+, Ca^{2+}, Mg^{2+}, K^+, Cl^-, SO_4^{2-}, Br^-, POM
Fossil fuel emissions BC, POM, SO_4^{2-}, Fe, Zn	Fossil fuel emissions BC, POM, SO_4^{2-}, Fe, Zn	Soil dust emissions Si, Al, Fe, Ti, P, Mn, Co, Ni, Cr, Na^+, Ca^{2+}, Mg^{2+}, K^+, SO_4^{2-}, Cl^-, CO_3^{2-}, POM
Biomass- and biofuel-burning emissions BC, POM, K^+, Na^+, Ca^{2+}, Mg^{2+}, SO_4^{2-}, NO_3^-, Cl^-, Fe, Mn, Zn, Pb, V, Cd, Cu, Co, Sb, As, Ni, Cr	Biomass- and biofuel-burning emissions BC, POM, K^+, Na^+, Ca^{2+}, Mg^{2+}, SO_4^{2-}, NO_3^-, Cl^-, Fe, Mn, Zn, Pb, V, Cd, Cu, Co, Sb, As, Ni, Cr	Biomass- and biofuel-burning ash, industrial fly ash, tire particle emissions, pollen, spores, bacteria, viruses
Condensation/dissolution $H_2O(aq)$, SO_4^{2-}, NH_4^+, POM	Condensation/dissolution $H_2O(aq)$, SO_4^{2-}, NH_4^+, POM	Condensation/dissolution $H_2O(aq)$, NO_3^-
	Coagulation of all components from nucleation mode	Coagulation of all components from smaller modes

is hygroscopic, it absorbs liquid water at high relative humidity and becomes spherical. If ions are present and the relative humidity decreases, crystals may form within the particle. Some observed aerosol particles are flat, others are globular or contain layers, and still others are fibrous.

Of particular interest is the morphology and shape of soot particles, which contain BC, POM, O, N, and H. They also have important optical effects. The only atmospheric source of soot is emissions. The main combustion sources of soot are fossil fuel, biofuel, and biomass burning. An emitted soot particle is irregularly shaped and mostly solid, containing from 30 to 2,000 graphitic spherules aggregated with random orientation by collision during combustion (Katrlnak et al., 1993). Figure 5.16b shows an example of a soot aggregate.

Once emitted, soot particles can coagulate or grow. Because soot particles are porous and have a large surface area, they serve as sites on which condensation occurs. Although BC in soot is hydrophobic, some organics in soot attract water, in which inorganic gases dissolve (Andrews and Larson, 1993). Evidence of soot growth is abundant because traffic tunnel studies (Venkataraman et al., 1994) and test vehicle

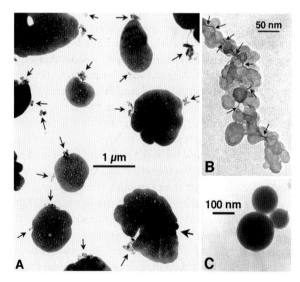

Figure 5.16. Transmission electron microscopy images of (a) ammonium sulfate particles containing soot (arrows point to soot inclusions), (b) a chainlike soot aggregate, and (c) fly ash spheres consisting of amorphous silica collected from a polluted marine boundary layer in the North Atlantic Ocean. From Pósfai et al. (1999).

studies (European Automobile Manufacturers Association (ACEA), 1999; Maricq et al., 1999) indicate that most fossil fuel BC is emitted in particles smaller than 100 nm in diameter, but ambient measurements in Los Angeles, the Grand Canyon, Glen Canyon, Chicago, Lake Michigan, Vienna, and the North Sea show that accumulation mode BC often exceeds emissions mode BC (McMurry and Zhang, 1989; Hitzenberger and Puxbaum, 1993; Venkataraman and Friedlander, 1994; Berner et al., 1996; Offenberg and Baker, 2000). The most likely way that ambient BC redistributes so dramatically is by coagulation and growth. Similarly, the measured mean number diameter of biomass-burning smoke less than four minutes old is 100 to 130 nm (Reid and Hobbs, 1998); yet the mass of such aerosol particles increases by 20 to 40 percent during aging, with one-third to one-half the growth occurring within hours after emissions (Reid et al., 1998).

Transmission electron microscopy (TEM) images support the theory that soot particles can become coated once emitted. Katrlnak et al. (1992, 1993) show TEM images of soot from fossil fuel sources coated with sulfate or nitrate. Martins et al. (1998) show a TEM image of a coated biomass-burning soot particle, and Pósfai et al. (1999) show TEM images (Figure 5.16a) of North Atlantic soot particles coated by ammonium sulfate. They found that internally mixed soot and sulfate appeared to comprise a large fraction of aerosol particles in the troposphere. Almost all soot particles found in the North Atlantic contained sulfate. Strawa et al. (1999) took scanning electron microscopy (SEM) images of black carbon particles in the Arctic stratosphere. One such image is reproduced in Figure 5.17. The rounded edges of the particle seem to indicate that the particle is coated. Katrlnak et al. (1993) report that rounded grains on black carbon aggregates indicate a coating. As soot aggregates become coated, they compress into a more spherical shape (Schnaiter et al., 2003; Wentzel et al., 2003). As such, particle shapes change over time in the air.

In sum, whereas emitted soot particles are relatively distinct, or **externally mixed** from other aerosol particles, soot particles typically coagulate or grow to become **internally mixed** with other particle components. Although soot becomes internally mixed, it does not become "well mixed" (diluted) in an internal mixture because soot consists of a solid aggregate of many graphite spherules. Thus, soot is a distinct component, generally a core, in a mixed particle.

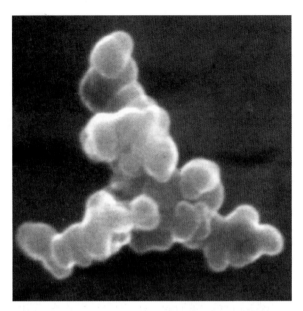

Figure 5.17. Scanning electron microscopy image of a coated soot particle from the Arctic stratosphere. From Strawa et al. (1999).

5.6. Health Effects of Aerosol Particles

Aerosol particles contain a variety of hazardous inorganic and organic substances. Some hazardous organic substances include benzene, polychlorinated biphenyls, and polycyclic aromatic hydrocarbons (PAHs). Hazardous inorganic substances include metals and sulfur compounds. Metals cause lung injury, bronchioconstriction, and increased incidence of infection (Ghio and Samet, 1999). Particles smaller than 10 μm in diameter (**PM_{10}**) have been correlated with asthma and chronic obstructive pulmonary disease (MacNee and Donaldson, 1999).

With respect to outdoor air, some studies found that there may be **no low threshold for PM_{10}–related health problems** (Pope et al., 1995). Because most mass of PM_{10} is not hazardous, damage from PM_{10} may be due primarily to small particles, particularly **ultrafine particles**, which are particles smaller than 100 nm in diameter. Such particles may be toxic to the lungs, even when the particles contain components that are not toxic when present in larger particles (MacNee and Donaldson, 1999).

Studies in the 1970s found a link between cardiopulmonary disease and high concentrations of aerosol particles and sulfur oxides. Subsequent studies found a link between low concentrations of aerosol particles and health. A review by Pope (2000) concluded that short-term (acute) increases of 10 μg m^{-3} PM$_{10}$ were associated with a 0.5 to 1.5 percent increase in daily mortality, higher hospitalization and health care visits for respiratory and cardiovascular disease, and enhanced outbreaks of asthma and coughing. Increased death rates usually occurred within 1 to 5 days following an air pollution episode. Long-term exposures to 5 μg m^{-3} of particles smaller than 2.5 μm in diameter (PM$_{2.5}$) above background levels resulted in a variety of cardiopulmonary problems, including increased mortality, increased disease, and decreased lung function in adults and children (Pope and Dockery, 1999). Pope et al. (2002) found a mean increased risk of mortality due to long-term exposure to PM$_{2.5}$ of 4 percent per 10 μg m^{-3}, or 0.004 per μg m^{-3}.

Small particles (PM$_{2.5}$) result in more respiratory illness and premature death than do larger aerosol particles (Özkatnak and Thurston, 1987; U.S. EPA, 1996). One six-city, 16-year study concluded that people living in areas where aerosol particle concentrations were lower than even the U.S. federal PM$_{10}$ standard had a lifespan 2 years shorter than people living in cleaner air (Dockery et al., 1993). Air pollution was correlated with death from lung cancer and cardiopulmonary disease. Fine particles, including sulfates, were correlated with mortality. A more recent study found that each 10 μg m^{-3} of PM$_{2.5}$ reduces life expectancy by 5 to 10 months (Pope et al., 2009).

A review of health studies concluded that ambient *PM$_{2.5}$ increases premature mortality, hospital admissions, and emergency room visits for cardiovascular and respiratory disease* and development of chronic respiratory disease (U.S. EPA, 2009b). Although some individual chemicals in PM$_{2.5}$ are likely to be responsible for most of the health damage, isolating health effects of such chemicals has been difficult to date. Some studies, however, discovered a link between short-term exposure of black carbon, for example, and cardiovascular effects (e.g., U.S. EPA, 2009b; Mordukhovich et al., 2009).

Figure 5.18 illustrates the health damage to the lungs of a teenage nonsmoker resulting from severe air pollution, primarily particulate matter, in Los Angeles in the 1970s. Air pollution levels at the time were equivalent to smoking two packs of cigarettes per day. Although air pollution levels in the United States and much of Europe have improved significantly since then, pollution levels in most developing countries of the world are similar to those found in Los Angeles in the 1970s (Chapter 8).

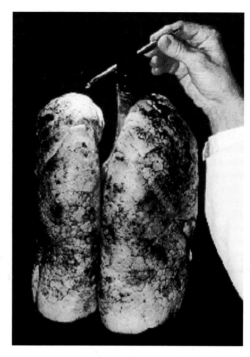

Figure 5.18. Lungs of a nonsmoking teenager living in Los Angeles who died accidentally in the 1970s. South Coast Air Quality Management District, www.aqmd.gov.

Living in some places, such as Linfen, China, in 2010, was equivalent to smoking three packs of cigarettes per day.

5.7. Quantifying the Health Effects of Particles or Gases

Health effect rates (y) (e.g., deaths/yr, cancers/yr, hospitalizations/yr) due to a particle or gas pollutant can be quantified with

$$y = y_0 P \left(1 - \exp\left[-\beta \times \max\left(x - x_{th},\ 0\right)\right]\right) \quad (5.13)$$

where x is the average concentration or mixing ratio of the pollutant, x_{th} is the threshold concentration or mixing ratio below which no health effect occurs, β is the fractional increase in risk of the health effect per unit x, y_0 is the baseline health effect rate per unit population, and P is the population. For example, in the United States, the all-cause death rate is approximately $y_0 = 833$ per 100,000 population per year. For ozone, the threshold mixing ratio above which short-term health effects occur is $x_{th} = 35$ ppbv, and for PM$_{2.5}$, it is 0 μg/m^3. A compilation of studies suggests that the increased risk of death due to short-term exposure to ozone is ~0.0004 per ppbv above the threshold (Ostro et al., 2006). Although health effects of $PM_{2.5}$ vary for different chemical components within $PM_{2.5}$, almost all epidemiological studies correlating particle changes with health use ambient $PM_{2.5}$ measurements to derive such correlations. The increased risk of mortality due to long-term exposure to $PM_{2.5}$ may be ~0.004 per $\mu g\ m^{-3}$ (Pope et al., 2002).

Example 5.5
Calculate the number of United States deaths per year due to short-term ozone exposure if the entire population of 300 million were exposed to 40 ppbv.

Solution
Substituting $y_0 = 0.00833$, $P = 300,000,000$, $\beta = 0.0004$ per ppbv, $x = 40$ ppbv, and $x_{th} = 35$ ppbv into Equation 5.13 gives $y \approx 5,000$ additional deaths/yr.

5.8. Summary

Aerosol particles appear in a variety of shapes and compositions and vary in size from a few gas molecules to the size of a raindrop. Natural sources of aerosol particles include sea spray uplift, soil dust uplift, volcanic eruptions, natural biomass burning, meteoric debris, and wind-driven emissions of pollen, spores, bacteria, viruses, and plant debris. Anthropogenic sources include fugitive dust emissions; biomass, biofuel and fossil fuel combustion; and industrial sources. Aerosol particle size distributions generally contain three to five modes, including one or two subnucleation modes, one or two subaccumulation modes, and a coarse particle mode. Homogeneous nucleation and emissions from fossil fuel, biofuel, and biomass burning dominate the nucleation mode. Emissions of sea spray, natural soil dust, fugitive soil dust, pollen, spores, and bacteria dominate the coarse mode. Aerosol particles coagulate and grow by condensation or dissolution from the nucleation mode to the accumulation mode. Chemistry within aerosol particles and between gases and aerosol particles affects growth. Growth does not move accumulation mode particles to the coarse mode, except when water vapor grows onto aerosol particles to form cloud drops. The main removal processes of aerosol particles from the atmosphere are rainout, washout, sedimentation, and dry deposition. Aerosol particles are responsible for a variety of health problems that persist worldwide.

5.9. Problems

5.1. Why do accumulation mode aerosol particles not grow readily into the coarse mode?

5.2. Why do accumulation mode particles generally contain more sulfate than do coarse mode particles?

5.3. On a global scale, why is most chloride observed in the coarse mode?

5.4. Why is most ammonium found in accumulation mode particles?

5.5. Write an equilibrium reaction showing nitric acid gas reacting with magnesite, a solid. In what particle size mode should this reaction most likely occur?

5.6. Why is the carbonate ion not abundant in aerosol particles?

5.7. Why might a sea spray drop over midocean lose all its chloride when it reaches the coast?

5.8. Why is more nitrate than sulfate generally observed in particles containing soil minerals?

5.9. Why is coagulation not an important process for moving particle mass from the lower to the upper accumulation mode?

5.10. Estimate the concentration of 0.01-μm-diameter particles after 10 s of coagulation if 10^6 particles cm^{-3} of this size exist initially, no other particles exist, and the coagulation rate coefficient is $10^{-8}\ cm^3$ particle^{-1} s^{-1}. What is the resulting concentration of new particles formed assuming that they are formed only from the 0.01-μm particles and none of the new particles is lost?

5.11. Calculate the percent change in the growth rate of single-particle volume from Equation 5.3 if the diffusion coefficient is cut in half, $p_s = 10^{-6}$ hPa, the partial pressure (p) is increased from $2p_s$ to $4p_s$, particle density is decreased by a factor of two, and all other parameters remain constant. *Hint*: For this small value of p_s, the left side of the denominator in Equation 5.3 can be assumed to be zero relative to the right side.

5.12. Visibility is affected primarily by particles with diameter close to the wavelength of visible light, 0.5 μm. Which particle mode does this correspond to, and which three particle components in this mode do you think affect visibility in the background troposphere the most?

5.13. Particles smaller than 2.5 μm in diameter affect human health more than do larger particles. Identify five chemicals that you might expect to see in high concentrations in these particles in polluted air. Why did you pick these chemicals?

5.14. Calculate the additional number of U.S. deaths per year due to long-term PM$_{2.5}$ exposure if the entire population of 300 million were exposed to 12 instead of 10 μg m^{-3}.

Chapter 6

Effects of Meteorology on Air Pollution

The concentrations of gases and aerosol particles in the air are affected by winds, temperatures, vertical temperature profiles, clouds, and the relative humidity. These meteorological parameters are influenced by large- and small-scale weather systems. Large-scale weather systems are controlled by vast regions of high and low pressure, whereas small-scale weather systems are controlled by ground temperatures, soil moisture, and small-scale variations in pressure. The first section of the chapter examines the forces acting on air, and the second section examines how forces combine to form winds. In the third section, the way in which radiation coupled with forces and the rotation of the Earth generates the global circulation of the atmosphere is discussed. Sections four and five discuss characteristics of the two major types of large-scale pressure systems. In the sixth section, the effects of such pressure systems on air pollution are addressed. The last section focuses on the effects of local meteorology on air pollution.

6.1. Forces

Winds arise due to forces acting on the air. The four major forces – pressure gradient force, apparent Coriolis force, friction force, and apparent centrifugal force – are described in this section.

6.1.1. Pressure Gradient Force

When air pressure is high in one location and low nearby, air moves from high to low pressure. The force causing this motion is the **pressure gradient force** (PGF), which is proportional to the difference in pressure divided by the distance between the two locations and always acts from high to low pressure.

6.1.2. Apparent Coriolis Force

When air is in motion over a rotating Earth, it appears to an observer fixed in space to accelerate to the right in the Northern Hemisphere and to the left in the Southern Hemisphere by the **apparent Coriolis force** (ACoF; Figure 6.1). The ACoF is not a real force; rather, it is an acceleration that arises when the Earth rotates under a body (air in this case) in motion. The ACoF is zero at the Equator, maximum at the poles, zero for bodies at rest, proportional to the speed of the air, and always acts 90 degrees to the right (left) of the moving body in the Northern (Southern) Hemisphere.

Figure 6.1 illustrates the ACoF. If the Earth did not rotate, an object thrown from point A directly north would be received at point B, along the same longitude as point A. Because the Earth rotates, objects thrown to the north have a west-to-east velocity equal to that of the Earth's rotation rate at the latitude from which they originate. The Earth's rotation rate near the Equator (low latitude) is greater than that near the poles (high latitudes); thus, objects thrown from low latitudes have a greater west-to-east velocity than does the Earth below them when they reach a high latitude. For example, an object thrown from point A to point B in the north will end up at point C, instead of at point B′ by the time the person at point A reaches point A′, and the object

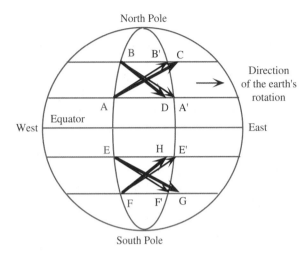

Figure 6.1. Example of the apparent Coriolis force (ACoF), described in the text. Thin arrows represent intended paths, and thick arrows are actual paths.

will appear as if it has been deflected to the right (from point B′ to point C). Similarly, an object thrown from point B to point A in the south will end up at point D, instead of at point A′, by the time the person at point B reaches point B′. The Coriolis effect, therefore, appears to deflect moving bodies to the right in the Northern Hemisphere and to the left in the Southern Hemisphere.

6.1.3. Friction Force

A third force that acts on moving air, the **friction force** (FF), is important near the surface only. The FF slows the wind. Its magnitude is proportional to the wind speed, and it acts in exactly the opposite direction from the wind. The rougher the surface, the greater the FF. The FF over oceans and flat deserts is small, whereas that over forests and buildings is large.

6.1.4. Apparent Centrifugal Force

A fourth force, which also acts on moving air, is the **apparent centrifugal force** (ACfF). This force is another fictitious force; it arises when an object rotates around an axis. The apparent force is directed outward, away from the axis of rotation. When a passenger in a car rounds a curve, for example, a viewer traveling with a passenger sees the passenger being pulled outward, away from the axis of rotation, by this force. In contrast, a viewer fixed in space sees the passenger accelerating inward due to a **centripetal acceleration**, which is equal in magnitude to, but opposite in direction from, the apparent centrifugal force.

6.2. Winds

The major forces acting on the air in the horizontal are the PGF, ACoF, FF, and ACfF. In the vertical, the major forces are the upward-directed vertical pressure gradient force and the downward-directed force of gravity. These forces drive winds. Examples of horizontal winds arising from force balances are given next.

6.2.1. Geostrophic Wind

The type of wind involving the least number of forces is the **geostrophic** ("Earth-turning") **wind**. This wind is driven by only the pressure gradient force and the apparent Coriolis force. It occurs above the boundary layer, where surface friction is negligible, and along straight isobars, where the apparent centrifugal force is negligible. An **isobar** is a line of constant pressure drawn on a constant altitude map. Suppose a horizontal pressure gradient, represented by two parallel isobars, exists, such as at the top of Figure 6.2. The PGF causes still air to move from high to low pressure. As the air moves, the ACoF deflects the air to the right in the Northern Hemisphere. The ACoF continues deflecting the air until the ACoF exactly balances the magnitude and is in the opposite direction from the PGF. This condition is referred to as **geostrophic balance**. Figure 6.2 shows that the resulting wind (the geostrophic wind) flows parallel to the isobars. The closer the isobars are together, the faster the geostrophic wind. In reality, geostrophic balance occurs following a process called **geostrophic adjustment**, during which the wind overshoots and then undershoots its ultimate path in an oscillatory fashion. In the Southern Hemisphere, the ACoF deflects

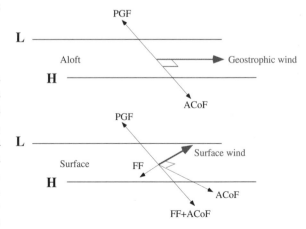

Figure 6.2. Forces acting to produce winds aloft and at the surface in the Northern Hemisphere.

moving air to the left, so the geostrophic wind flows in the opposite direction from that shown at the top of Figure 6.2.

6.2.2. Surface Winds along Straight Isobars

When isobars are straight near the surface, the friction force also affects the equilibrium wind speed and direction. The bottom of Figure 6.2 shows the wind direction that results from a balance among the PGF, ACoF, and FF at the surface in the Northern Hemisphere. Friction, which acts in the opposite direction from the wind, slows the wind. Because the magnitude of the ACoF is proportional to the wind speed, a reduction in wind speed reduces the magnitude of the ACoF. Because the sum of the FF and the ACoF must balance the magnitude of and be in the opposite direction from the PGF, the equilibrium wind direction shifts toward low pressure. On average, surface friction turns winds 15 to 45 degrees toward low pressure, with lower values corresponding to smooth surfaces and higher values corresponding to rough surfaces. As such, if the wind is at your back in the Northern Hemisphere, turn 15 to 45 degrees clockwise, and low pressure will be on your left.

In the Southern Hemisphere, surface winds are also angled toward low pressure, but symmetrically in the opposite direction from the wind shown in the bottom of Figure 6.2. Thus, if the wind is at your back in the Southern Hemisphere, turn 15 to 45 degrees counterclockwise, and low pressure will be on your right.

6.2.3. Gradient Wind

When centers of low and high pressure (relative to pressures in nearby regions at the same altitude) appear above the boundary layer, wind speed and direction are controlled by three forces, the PGF, ACoF, and ACfF. The resulting wind is called the **gradient wind**, which is a circular wind around centers of low or high pressure aloft.

Gradient winds aloft flow counterclockwise around a center of low pressure in the Northern Hemisphere. An easy rule for determining the direction of flow in the Northern Hemisphere is the **left-hand rule**. Point your left thumb down (for low pressure) and follow your fingers to determine the direction of the gradient wind around a low-pressure center aloft. For a high-pressure center, point your left thumb up (for high pressure) and follow your fingers. They will indicate that the gradient wind flows clockwise around high-pressure centers aloft in the Northern Hemisphere.

Figure 6.3a illustrates the forces acting on the air and the resulting gradient wind around a low-pressure center in the Northern Hemisphere. The PGF must point toward the center of the low, and the ACfF must point away from it. However, the ACfF cannot, on its own, balance the magnitude of the PGF because any moving wind has an ACoF associated with it that needs to be counted in the force balance. The PGF must be balanced in magnitude and opposite in direction from the sum of the ACfF and the ACoF. Because the wind must point 90 degrees to the left of the ACoF (or the ACoF points

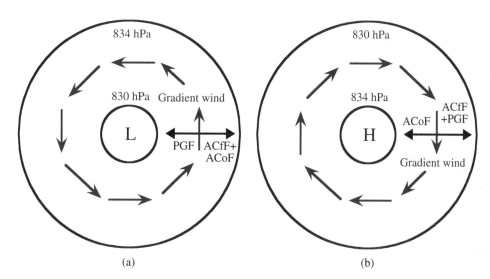

Figure 6.3. Gradient winds (in green) around a center of (a) low and (b) high pressure in the Northern Hemisphere and the forces (in blue) affecting them. The pressures given represent those along the circular isobars (in red).

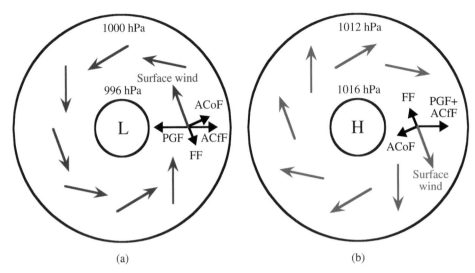

Figure 6.4. Surface winds (in green) around centers of (a) low and (b) high pressure in the Northern Hemisphere and the forces (in blue) affecting them. The pressures given represent those along the circular isobars (in red).

90 degrees to the right of the wind), the resulting wind must flow counterclockwise around the center of low pressure.

Around a high-pressure center aloft in the Northern Hemisphere, the PGF and ACfF point opposite from the center of high. To balance the sum of these two forces, the ACoF must point toward the center of high. Because the wind is always pointed 90 degrees to the left of the ACoF, the resulting gradient wind must flow clockwise around the center of high (Figure 6.3b).

In the Southern Hemisphere, the **right-hand rule** is used. It indicates that the gradient winds flow clockwise around low-pressure centers and counterclockwise around high-pressure centers, thus opposite in direction from their respective flows around lows and highs in the Northern Hemisphere.

6.2.4. Surface Winds along Curved Isobars

The large-scale circulation of the wind around a surface low-pressure center is called a **cyclone**, and that around a surface high-pressure center is called an **anticyclone**. **Cyclonic flow** is flow around a low-pressure center (either at the surface or aloft), and **anticyclonic flow** is flow around a high-pressure center (either at the surface or aloft).

Near the Earth's surface, the friction force slows and turns the wind toward the center of low pressure in a cyclone and away from the center of high pressure in an anticyclone. The flow of air into a center of low

pressure is called **convergence**, whereas the flow of air away from a center of high pressure is **divergence**. Air converges into surface centers of low pressure and diverges from surface centers of high pressure in both hemispheres.

In the Northern Hemisphere, surface winds converge while flowing counterclockwise around the center of low pressure and diverge while flowing clockwise around the center of high pressure (left-hand rule). Figure 6.4 shows the force balances and resulting winds in the presence of a surface (a) low-pressure system and (b) high-pressure system in the Northern Hemisphere.

In the low-pressure case, the PGF is balanced by the sum of the ACoF, ACfF, and FF. The resulting wind converges counterclockwise into the center of the low. The converging air rises, and the rising air expands and cools. If sufficient cooling occurs, clouds form. As such, surface cyclones are frequently associated with stormy weather.

In the high-pressure case, the sum of the PGF and ACfF is balanced by the sum of the FF and ACoF. The resulting wind diverges clockwise out of the center of the high. The diverging air pulls more air downward in the center of the high. The descending air compresses and warms, potentially evaporating clouds. As such, surface high-pressure centers, or anticyclones, are often associated with sunny skies and warm weather.

In the Southern Hemisphere, surface winds converge while flowing clockwise around the center of

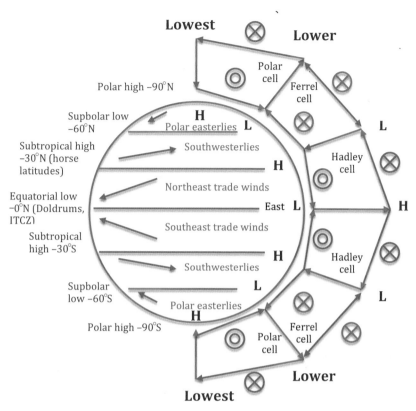

Figure 6.5. Diagram of the three major circulation cells, the predominant surface pressure systems, and the predominant surface wind systems on the Earth. H at the surface equals high pressure. H aloft is high altitude at a given pressure above the surface, which is similar to saying it is high pressure at a given altitude above the surface. L at surface equals low pressure. L aloft is low altitude at a given pressure above the surface (or low pressure at a given altitude). "Lower" equals lower altitude at the same pressure as L aloft (or lower pressure at the same altitude). "Lowest" equals lower altitude than "Lower" at the same pressure as L aloft (or lower pressure at the same altitude). Thus, elevated high and low altitudes are relative to other altitudes at the same pressure. This is similar to saying that elevated high and low pressures are relative to other pressures at the same altitude. The circles with xs denote winds going into the page (west to east). The circles with embedded circles denote winds coming out of the page (east to west).

low pressure and diverge while flowing counterclockwise around the center of high pressure (right-hand rule).

6.3. Global Circulation of the Atmosphere

Air pollution is affected by winds, winds are affected by large-scale pressure systems, and large-scale pressure systems are affected by the global circulation of the atmosphere. Figure 6.5 shows features of the global circulation, including the major circulation cells, the belts of low and high pressure, and the predominant wind directions.

Winds have a west–east (**zonal**), south–north (**meridional**), and vertical component. The three circulation

cells in each hemisphere shown in Figure 6.5 represent the meridional and vertical components of the Earth's winds, averaged zonally (over all longitudes) and over a long time period. The cells are symmetric about the Equator and extend up to the tropopause (Section 3.3.1.2), which is near 18 km altitude over the Equator and near 8 km altitude over the poles.

Two cells, called **Hadley cells**, extend from 0°N to 30°N and 0°S to 30°S latitude, respectively. These cells were named after **George Hadley** (1685–1768), an English physicist and meteorologist, who, in 1735, first proposed the cells in his paper, "Concerning the Cause of the General Trade Winds," which was presented to the Royal Society of London. Hadley's original cells, however, extended between the Equator and the poles.

Figure 6.6. William Ferrel (1817–1891). National Oceanic and Atmospheric Administration Central Library.

In 1855, **William Ferrel** (1817–1891; Figure 6.6), an American school teacher, meteorologist, and oceanographer, published an article in the *Nashville Journal of Medicine* pointing out that Hadley's one-cell model did not fit observations so well as did the three-cell model shown in Figure 6.5. In 1860, Ferrel went on to publish a collection of papers showing the first application of mathematical theory to fluid motions on a rotating Earth. Today, the middle cell in the three-cell model is called the **Ferrel cell**. A Ferrel cell extends from 30°N to 60°N and from 30°S to 60°S, whereas a **polar cell** extends from 60 to 90 degrees in each hemisphere as well.

6.3.1. Equatorial Low-Pressure Belt

Circulation in the three cells is controlled by heating at the Equator, cooling at the poles, and the rotation of the Earth. In the Hadley cells, air rises over the Equator because the sun heats this region intensely. Much of the heating occurs over water, some of which evaporates. As air containing water vapor rises, the air expands and cools, and the water vapor recondenses to form clouds of great vertical extent. Condensation of water vapor releases latent heat, providing the air with more buoyancy. Over the Equator, the air can rise up to about 18 km before it is decelerated by the stratospheric inversion. Once the air reaches the tropopause, it cannot rise much farther, so it diverges to the north and south. At the surface on the Equator, air is drawn in horizontally to replace the rising air. As long as divergence aloft exceeds convergence at the surface, surface air pressure decreases and the altitude of a given air pressure aloft increases over the Equator (relative to altitudes at the same pressure level aloft but at other latitudes). The surface low-pressure belt at the Equator is called the **equatorial low-pressure belt**. Because pressure gradients are weak, winds are light, and the weather is often rainy over equatorial waters, this region is also called the **doldrums**.

6.3.2. Winds Aloft in the Hadley Cells

As air diverges toward the north in the elevated part of the Northern Hemisphere Hadley cell, the ACoF force deflects much of it to the right (to the east), giving rise to **westerly winds** aloft (winds are generally named after the direction that they originate from). Westerly winds aloft in the Northern Hemisphere Hadley cell increase in magnitude with increasing distance from the Equator until they meet equatorward-moving air from the Ferrel cell at 30°N, the **subtropical front**. The front is a region of sharp temperature contrast. The winds at the front are strongest at the tropopause, where they are called the **subtropical jet stream**. Winds aloft in the Southern Hemisphere Hadley cell are also westerly and culminate in a tropopause subtropical jet stream at 30°S.

6.3.3. Subtropical High-Pressure Belts

As air converges at the subtropical fronts at 30°N and 30°S, much of it descends. Air is then drawn in horizontally aloft to replace the descending air. As long as inflow aloft exceeds outflow at the surface, surface air pressure builds up. The surface high-pressure belts at 30°N and 30°S are called **subtropical high-pressure belts**. Because descending air compresses and warms evaporating clouds, and because pressure gradients are relatively weak around high-pressure centers, surface high-pressure systems are characterized by sunny skies and light winds. Sunny skies and the lack of rainfall at 30°N and 30°S are two reasons why many deserts of the world are located at these latitudes. The light winds forced some ships sailing at 30°N to lighten their cargo, the heaviest and most dispensable component of which was often horses. Thus, the 30°N latitude band was also known as the **horse latitudes**.

6.3.4. Trade Winds

At the surface at 30°N and 30°S, descending air diverges both equatorward and poleward. Most of the air moving equatorward is deflected by the ACoF to the right (toward the west) in the Northern Hemisphere and to the left (toward the west) in the Southern Hemisphere, except that friction reduces the extent of ACoF turning. The resulting winds in the Northern Hemisphere are called the **northeast trade winds** because they originate from the northeast (Figure 6.5). Those in the Southern Hemisphere are called the **southeast trade winds** because they originate from the southeast. Sailors from Europe have used the northeast trades to speed their voyages westward since the fifteenth century. The trade winds are consistent winds. The northeast and southeast trade winds converge at the **Intertropical Convergence Zone** (ITCZ; Figure 6.5), which moves north of the Equator in the Northern Hemisphere summer and south of the Equator in the Southern Hemisphere summer, generally following the direction of the sun. At the ITCZ, air convergence and surface heating lead to the rising arm of the Hadley cells.

6.3.5. Subpolar Low-Pressure Belts

As surface air moves poleward in the Ferrel cells, the ACoF turns it toward the right (east) in the Northern Hemisphere and left (east) in the Southern Hemisphere. However, surface friction reduces the extent of turning, so that near-surface winds at **midlatitudes** (30°N to 60°N and 30°S to 60°S) are generally westerly to southwesterly (from the west or southwest) in the Northern Hemisphere and westerly to northwesterly in the Southern Hemisphere. In both hemispheres, poleward-moving near-surface air in the Ferrel cell meets equatorward-moving air from the polar cell at the **polar front**, which is a region of sharp temperature contrast between these two cells. Converging air at the surface front rises and diverges aloft, reducing surface air pressure and increasing air pressure aloft relative to pressures at other latitudes. The surface low-pressure regions at 60°N and 60°S are called **subpolar low-pressure belts**. Regions of rising air and surface low pressure are associated with storms. Thus, the intersection of the Ferrell and polar cells is associated with stormy weather. Unlike at the Equator, surface pressure gradients and winds at the polar front are relatively strong. West–east wind speeds also increase with increasing height at the polar fronts. At the tropopause in each hemisphere, they culminate in the **polar front jet streams**. Although the subtropical jet streams do not meander to the north or south over great distances, the polar front jet streams do. Their predominant direction is still from west to east.

6.3.6. Westerly Winds Aloft at Midlatitudes

One might expect air in the elevated portion of the Ferrel cell to move equatorward and for the ACoF to deflect such air to the west, creating easterly winds (from east to west) aloft at midlatitudes in both hemispheres. In fact, winds aloft in the Ferrel cell are generally westerly (from west to east), although they meander between south and north. *Part of the reason for the westerly winds aloft in the Ferrel cell is that heights of constant pressure (or pressures at a constant height) decrease between the Equator and the poles in the upper troposphere*, as shown in Figure 6.5. Winds tend to start flowing down the height (or pressure) gradient, toward the poles in the Ferrel cell aloft. The Coriolis force then acts on this moving air, turning it toward the east in both hemispheres, creating westerly winds aloft in both hemispheres. The reason heights of constant pressure (or pressures at a constant height) decline from equator to pole is that temperatures transition from warm to cold between the Equator and pole. Warm air rises and cold air sinks; thus, near the Equator, the rising air pushes up the height of a constant pressure level (or increases the pressure at a constant height), and near the pole, sinking air pushes down the height of a constant pressure level (or decreases the pressure at a constant height).

The second reason for westerly winds aloft in the Ferrel cell relates to the presence of centers of low and high pressure. Descending air at 30°N and 30°S creates centers of surface high pressure, and rising air at 60°N and 60°S creates bands or centers of surface low pressure. Figure 6.7b shows an example of surface high- and low-pressure centers over the Pacific Ocean in the Northern Hemisphere between 10°N and 80°N. Surface winds moving around a Northern Hemisphere surface high-pressure center travel clockwise (diverging away from the center of the high), and surface winds moving around a surface low-pressure center travel counterclockwise (converging into the center of the low) (Sections 6.2.3 and 6.2.4). Indeed, these characteristics are seen in Figure 6.7b, which shows winds traveling clockwise around the highs and counterclockwise around the lows. The positions of the highs and lows

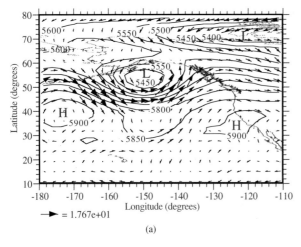

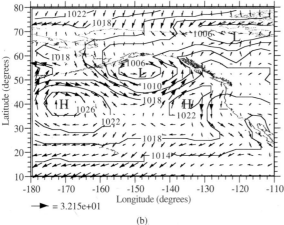

Figure 6.7. Maps of (a) 500-hPa height contours (m) and wind vectors (m s^{-1}) (b) sea level pressure contours (hPa) and near-surface wind vectors (m s^{-1}) obtained from National Centers for Environmental Prediction (NCEP, 2000) for August 3, 1990, at 12 GMT for the northern Pacific Ocean. Height contours on the constant pressure map (a) are analogous to isobars on a constant height map; thus, high (low) heights in map (a) correspond to high (low) pressures on a constant height map. The surface low-pressure system at −148°W, 53°N in (b) is the Aleutian low. The surface high-pressure system at −134°W, 42°N in (b) is the Pacific high. The arrow below each map gives the scale of the wind speed arrow in m s^{-1}.

Figure 6.7a shows an elevated map corresponding to the surface map in Figure 6.7b. The elevated map shows height contours on a surface of constant pressure (500 hPa) and winds traveling around centers of low and high heights. Height contours on a constant pressure graph are analogous to isobars on a constant altitude graph; thus, high (low) heights in Figure 6.7a correspond to high (low) pressures on a constant height graph. The lows aloft lie slightly to the west of the surface lows. The figure indicates that winds traveling around the highs and lows aloft connect, resulting in sinusoidal west-to-east flow around the globe. *Thus, high- and low-pressure systems aloft are responsible for accelerating winds aloft in the Ferrel cell from west to east (in the westerly direction).*

6.3.7. Polar Easterlies

Air moving poleward aloft in the polar cells is turned toward the east in both hemispheres by the ACoF, causing elevated winds in the polar cells to be westerly. At the poles, cold air aloft descends, increasing surface air pressure. The surface high-pressure regions are called **polar highs**. Air at the polar surface diverges equatorward. The ACoF turns this air toward the west. Friction is weak over the Arctic because polar surfaces are either snow or ice, and sea ice is relatively flat; therefore, surface winds are relatively easterly. The Antarctic is a continent with high mountains and rough surfaces, so the surface winds are turned more toward low pressure by friction and thus are more southeasterly (coming from the southeast). Nevertheless, in both hemispheres, the resulting surface winds in the polar cells are called **polar easterlies** (Figure 6.5).

6.4. Semipermanent Pressure Systems

The subtropical high-pressure belts in the Northern and Southern Hemispheres are dominated by surface high-pressure centers over the oceans. These high-pressure centers are called **semipermanent surface high-pressure centers** because they are usually visible on a sea-level map most of the year. These pressure systems tend to move northward in the Northern Hemisphere summer and southward in the winter. On average, they are centered near 30°N or 30°S. In the Northern Hemisphere, the two semipermanent surface high-pressure systems are the **Pacific high** (in the Pacific Ocean) and the **Bermuda-Azores high** (in the Atlantic Ocean). In Figure 6.7b, the Pacific high is the surface high-pressure center at −134°W, 42°N. In the Southern Hemisphere, semipermanent high-pressure systems are

in Figure 6.7b create a near-surface west-to-east flow that meanders sinusoidally around the globe. The flow created by these highs and lows is consistent with the expectation that near-surface winds in the Ferrel cell are predominantly westerly.

located at 30°S in the South Pacific, South Atlantic, and Indian Oceans.

The subpolar low-pressure belt in the Northern Hemisphere is dominated by **semipermanent surface low-pressure centers**. The subpolar low-pressure belt in the Southern Hemisphere is dominated by a band of low pressure. In the Northern Hemisphere, the two semipermanent low-pressure centers are the **Aleutian low** (in the Pacific Ocean) and the **Icelandic low** (in the Atlantic Ocean). These pressure systems tend to move north in the Northern Hemisphere summer and south in the winter, but generally stay between 40°N and 65°N. In Figure 6.7b, the Aleutian low is the surface low-pressure center at −148°W, 53°N.

6.5. Thermal Pressure Systems

Although semipermanent surface high- and low-pressure centers exist over the oceans all year, thermal surface high- and low-pressure systems form over land seasonally. Thermal pressure systems form in response to surface heating and cooling, which depend on properties of soil and water, such as specific heat. **Specific heat** (J kg^{-1} K^{-1}) is the energy required to increase the temperature of 1 g of a substance 1 K. Soil has a lower specific heat than does water, as shown in Table 6.1. During the day, the addition of the same amount of sunlight increases the temperature of soil more than it increases water temperature. During the night, the release of the same amount of thermal-IR energy decreases soil temperature more than it decreases water temperature. As such, land heats during the day and cools during the night more than does water. Similarly, land heats during the summer and cools during the winter more than does water.

Specific heat varies not only between land and water, but also between different soil types, as shown in Table 6.1. Because sand has a lower specific heat than does clay, sandy soil heats to a greater extent than does clayey soil during the day and summer. The preferential heating

Table 6.1. Specific heats of four media at 298.15 K

Substance	Specific heat (J kg^{-1} K^{-1})
Dry air at constant pressure	1,004.67
Liquid water	4,185.5
Clay	1,360
Dry sand	827

of sand over clay and the preferential heating of land over water are important factors giving rise to thermal low-pressure centers.

When a region of soil warms, the air above the soil warms, rises, and diverges horizontally aloft, creating low pressure at the surface. Because the low-pressure system forms by heating, it is called a **thermal low-pressure system**. Thermal low-pressure centers form in the summer over deserts and other sunny areas (e.g., the Mojave Desert in Southern California, the plateau of Iran, the north of India). These areas are all located near 30°N, the same latitude as the semipermanent highs. Descending air at 30°N due to the highs helps form the thermal lows by evaporating clouds and thus clearing the skies. The descending air also keeps the thermal lows shallow (air rising in the thermal lows diverges horizontally at a low altitude). In some cases, such as over the Mojave Desert, the thermal lows do not produce clouds due to the lack of water vapor and shallowness of the low. In other cases, such as over the Indian continent in the summer, rising air in a low sucks in warm, moist surface air horizontally from the ocean, producing heavy rainfall. The strong sea breeze due to this thermal low is the summer monsoon, where a **monsoon** is a seasonal wind caused by a strong seasonal variation in temperature between land and water.

During the winter, when temperatures decrease over land, air densities increase, causing air to descend. Air from aloft is drawn in horizontally to replace the descending air, building up surface air pressure. The resulting surface high-pressure system is a **thermal high-pressure system**. In the Northern Hemisphere, two thermal high-pressure systems are the **Siberian high** (over Siberia) and the **Canadian high** (over the Rocky Mountains between Canada and the United States). As with thermal low-pressure system, the thermal high-pressure systems are often shallow.

6.6. Effects of Large-Scale Pressure Systems on Air Pollution

Semipermanent and thermal pressure systems affect air pollution. Table 6.2 compares characteristics of such pressure systems, including their effects on pollution. Semipermanent low-pressure systems are associated with cloudy skies, stormy weather, fast surface winds, and low penetrations of solar radiation to the surface. Thermal low-pressure systems, which are often shallow and occur in desert areas, may or may not produce clouds. Air rises in both types of low-pressure systems, dispersing near-surface pollution upward. When clouds

Table 6.2. Summary of characteristics of Northern Hemisphere surface low- and high-pressure systems

Characteristic	Surface low-pressure systems		Surface high-pressure systems	
	Semipermanent	Thermal	Semipermanent	Thermal
Latitude range	45–65°N	25–45°N	25–45°N	45–65°N
Surface pressure gradients	Strong	Variable	Weak	Variable
Surface wind speeds	Fast	Fast/variable	Slow	Slow/variable
Surface wind directions	Converging, counterclockwise	Converging, counterclockwise	Diverging, clockwise	Diverging, clockwise
Vertical air motions	Upward	Upward	Downward	Downward
Cloud cover	Cloudy	Cloudy or cloud free	Cloud free, sunny	Cloud free
Surface solar radiation	Low	Low or high	High	High
Storm formation?	Yes	Sometimes	No	No
Effect on air pollution	Reduces	Reduces	Enhances	Enhances

form in low-pressure systems, they block sunlight that would otherwise drive photochemical reactions, reducing pollution further. If clouds do not form, significant solar radiation, including UV radiation, reaches the surface, heating the surface and driving photochemical reactions.

Surface high-pressure systems are characterized by relatively slow surface winds, sinking air, cloudfree skies, and high penetrations of solar radiation to the surface. In such pressure systems, air sinks, confining near-surface pollution. Slow near-surface winds associated with high-pressure systems also prevent horizontal dispersion of pollutants, and the cloudfree skies caused by the pressure systems maximize the sunlight available to drive photochemical smog formation. In sum, the major effects of pressure systems on pollution are through vertical pollutant transfer, horizontal pollutant transfer, and cloud cover. Each of these effects is discussed in turn.

6.6.1. Vertical Pollutant Transport

Pressure systems affect vertical air motions and, therefore, pollutant dispersion by forced and free convection (Section 3.2.2). In a semipermanent low-pressure system, for example, near-surface winds converge and rise, dispersing near-surface pollutants upward. In a semipermanent high-pressure system, winds aloft converge and sink, confining near-surface pollutants. Both cases illustrate forced convection. In thermal low-pressure systems, surface warming causes near-surface air to become buoyant and rise. In thermal high-pressure systems, surface cooling causes near-surface

air to become negatively buoyant and sink or stagnate. Both cases illustrate free convection. To understand better how free convection in thermal pressure systems and forced convection in semipermanent pressure systems affect pollutant dispersion, it is necessary to discuss adiabatic processes and atmospheric stability.

6.6.1.1. Adiabatic and Environmental Lapse Rates

Whether air rises or sinks buoyantly in a thermal pressure system depends on atmospheric stability, which depends on adiabatic and environmental lapse rates. These terms are discussed next.

Imagine a balloon filled with air. The air pressure inside the balloon exactly equals the air pressure outside the balloon; otherwise, the balloon would continue to expand or contract. Also imagine that no energy (e.g., solar or thermal-IR energy or latent heat energy created by condensation of water vapor) can enter or leave the balloon, but that the balloon's membrane is flexible enough for it to expand and contract due to changes in air pressure outside the balloon. Suppose now that the balloon rises. Because air pressure always decreases with increasing altitude, the balloon must rise into decreasing air pressure. For the air pressure inside the balloon to decrease to the air pressure outside the balloon, the balloon must now expand, increasing in volume. This type of expansion, caused by a change in air pressure alone, is called an **adiabatic expansion**. Solar heating and latent heat release are **diabatic heating processes** and do not contribute to an adiabatic expansion.

During an adiabatic expansion, kinetic energy of air molecules is converted to work to expand the air. Because temperature is proportional to the kinetic energy of air molecules (Equation 3.1), an adiabatic expansion cools the air. In sum, rising air expands and expanding air cools; thus, rising air cools during an adiabatic expansion. The rate of cooling during an adiabatic expansion near the surface of the Earth is approximately 9.8 K or °C per kilometer increase in altitude. This rate is called the **dry** or **unsaturated adiabatic lapse rate** (Γ_d). Lapse rates are opposite in sign to changes in temperature with height; thus, a positive lapse rate indicates that temperature decreases with increasing height.

If the balloon in our example rises in air saturated with water vapor, the resulting adiabatic expansion cools the air and decreases the saturation vapor pressure of water in the balloon, causing the relative humidity to increase to more than 100 percent and water vapor to condense to form cloud drops, releasing latent heat. The rate of temperature increase with increasing height due to this latent heat release in saturated air parcels is typically 4°C km^{-1}, but it increases to 8°C km^{-1} in the tropics. Subtracting this latent heat release rate from the dry adiabatic lapse rate gives a net lapse rate during cloud formation of between 6 and 2°C km^{-1}. This rate is called the **wet, saturated, or pseudoadiabatic lapse rate** (Γ_w). It is the negative rate of change of temperature with increasing altitude during an adiabatic expansion in which condensation also occurs. The wet adiabatic lapse rate is applicable only in clouds.

The opposite of an adiabatic expansion is an **adiabatic compression**, which occurs when a balloon sinks adiabatically from low to high pressure. The compression is due to the increased air pressure around the balloon. During an adiabatic compression, work is converted to kinetic energy (which is proportional to temperature), warming the air in the balloon.

Although dry and wet adiabatic lapse rates describe the extent of cooling of a balloon rising adiabatically, the **environmental lapse rate** describes the actual change in air temperature with altitude in the environment outside a balloon. It is defined as

$$\Gamma_e = -\frac{\Delta T}{\Delta z} \qquad (6.1)$$

where $\Delta T / \Delta z$ is the actual change in air temperature with a change in altitude. An increasing temperature with increasing altitude gives a negative environmental lapse rate.

> **Example 6.1**
> If the observed temperature cools 14°C between the ground and 2 km above the ground, what is the environmental lapse rate? Is it larger or smaller than the dry adiabatic lapse rate?
>
> **Solution**
> The environmental lapse rate in this example is $\Gamma_e = +7°C$ km^{-1}, which is less than the dry adiabatic lapse rate of $\Gamma_d = +9.8°C$ km^{-1}.

6.6.1.2. Stability

One purpose of examining dry, wet, and environmental lapse rates is to determine the stability of the air, where the **stability** is a measure of whether pollutants emitted will convectively rise and disperse or build up in concentration near the surface.

Figure 6.8 illustrates the concept of stability. When a parcel of air (the balloon, in our example) that is unsaturated with respect to water is displaced vertically, it rises, expands, and cools dry adiabatically (along the dashed line). If the environmental temperature profile is stable (right thick line), the rising parcel is cooler and more dense than is the air in the environment around it at every altitude. As a result of its lack of buoyancy, the parcel sinks, compresses, and warms until its temperature (and density) equals that of the air around it. In reality, the parcel overshoots its original altitude as it descends, but eventually comes back to the original altitude in an oscillatory manner. In sum, a parcel of pollution at the temperature of the environment that is accelerated vertically in **stable air** returns to its original altitude and temperature. Stable air is associated with near-surface pollution buildup because pollutants perturbed vertically in stable air cannot rise and disperse.

In **unstable air**, an unsaturated parcel that is perturbed vertically continues to accelerate in the direction

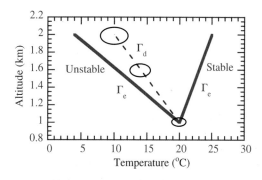

Figure 6.8. Stability and instability in unsaturated air, as described in the text.

of the perturbation. Unstable air is associated with near-surface pollutant cleansing. If the environmental temperature profile is unstable (left thick line in Figure 6.8), a parcel rising adiabatically (along the dashed line) is warmer and less dense than is the environment around it at every altitude, and the parcel continues to accelerate. The parcel stops accelerating only when it encounters air with the same temperature (and density) as the parcel. This occurs when the parcel reaches a layer with a new environmental lapse rate.

In neutral air (when the dry and environmental lapse rates are equal), an unsaturated parcel that is perturbed vertically neither accelerates nor decelerates, but continues along the direction of its initial perturbation at a constant velocity. Neutral air results in pollution dilution slower than in unstable air but faster than in stable air.

Whether unsaturated air is stable or unstable can be determined by comparing the dry adiabatic lapse rate with the environmental lapse rate. Symbolically, the stability criteria are

$$\Gamma_e \begin{cases} > \Gamma_d & \text{dry unstable} \\ = \Gamma_d & \text{dry neutral} \\ < \Gamma_d & \text{dry stable} \end{cases} \quad (6.2)$$

If the air is saturated, such as in a cloud, the wet adiabatic lapse rate is used to determine stability. In such a case, the stability criteria are

$$\Gamma_e \begin{cases} > \Gamma_w & \text{wet unstable} \\ = \Gamma_w & \text{wet neutral} \\ < \Gamma_w & \text{wet stable} \end{cases} \quad (6.3)$$

Although stability at any point in space and time depends on Γ_d or Γ_w, but not both, generalized stability criteria for all temperature profiles are often summarized as follows:

$$\begin{cases} \Gamma_e > \Gamma_d & \text{absolutely unstable} \\ \Gamma_e = \Gamma_d & \text{dry neutral} \\ \Gamma_d > \Gamma_e > \Gamma_w & \text{conditionally unstable} \quad (6.4) \\ \Gamma_e = \Gamma_w & \text{wet neutral} \\ \Gamma_e < \Gamma_w & \text{absolutely stable} \end{cases}$$

These conditions indicate that when $\Gamma_e > \Gamma_d$, the air is **absolutely unstable** or unstable, regardless of whether the air is saturated or unsaturated. Conversely, if $\Gamma_e < \Gamma_w$, the air is **absolutely stable** or stable, regardless of whether the air is saturated or unsaturated. If the air is **conditionally unstable**, stability depends on whether the air is saturated. Figure 6.9 illustrates the stability criteria in Equation 6.4.

Example 6.2

Given the environmental lapse rate from Example 6.1, determine the stability class of the atmosphere.

Solution

The environmental lapse rate in the example was $\Gamma_e = +7°C \text{ km}^{-1}$. Because the wet adiabatic lapse rate ranges from $\Gamma_w = +2$ to $+6°C \text{ km}^{-1}$, the atmosphere in this example is conditionally unstable (Equation 6.4).

In thermal low-pressure systems, sunlight warms the surface. The surface energy is conducted to the air, warming the lower boundary layer and decreasing the stability of the boundary layer. When the temperature profile near the surface becomes unstable, convective thermals rise buoyantly from the surface, carrying pollution with them.

In thermal high-pressure systems, radiative cooling of the surface stabilizes the temperature profile, preventing near-surface air and pollutants from rising. In many cases, air near the surface becomes so stable that a temperature inversion forms, further inhibiting vertical pollution dispersion. Inversions are discussed next.

6.6.1.3. Temperature Inversions

The stable environmental profile in Figure 6.8 and Profile 4 in Figure 6.9 are both **temperature inversions**,

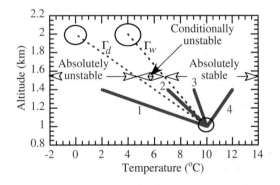

Figure 6.9. Stability criteria for unsaturated and saturated air. If air is saturated, the environmental lapse rate is compared with the wet adiabatic lapse rate to determine stability. Environmental lapse rates 3 and 4 are stable, and 1 and 2 are unstable, with respect to saturated air. Environmental lapse rates 2, 3, and 4 are stable, and 1 is unstable, with respect to unsaturated air. A rising or sinking air parcel follows the Γ_d line when the air is unsaturated and the Γ_w line when the air is saturated.

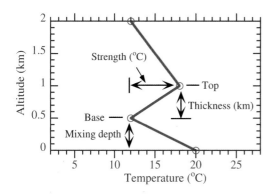

Figure 6.10. Characteristics of a temperature inversion.

which are increases in air temperature with increasing height. Inversions are always stable, but a stable temperature profile is not necessarily an inversion (e.g., Profile 3 in Figure 6.9). Inversions are important because they trap pollution near the surface to a greater extent than does a stable temperature profile that is not an inversion.

An inversion is characterized by its strength, thickness, top/base height, and top/base temperatures. The **inversion strength** is the difference between the temperature at the inversion top and that at its base. The **inversion thickness** is the difference between the inversion's top and base heights. The **inversion base height** is the height from the ground to the bottom of the inversion. It is also called the **mixing depth** because it is the estimated height to which pollutants released from the surface mix. In reality, pollutants often mix into the inversion layer itself. Inversion layer characteristics are illustrated in Figure 6.10.

Stable air and inversions, in particular, trap pollutants, preventing them from dispersing into the background troposphere and causing pollutant concentrations to build up near the surface. Figure 6.11 illustrates how such trapping occurs. It shows two air parcels with different initial temperatures released at the surface under an inversion. Suppose the parcels represent exhaust plumes that are initially warmer than the environment. Due to their buoyancy, both parcels rise, expand, and cool at the dry adiabatic lapse rate of near 10°C km^{-1}. The parcel released at 20°C rises and cools until its temperature approaches that of the air around it. At that point, the parcel decelerates and then comes to rest after oscillating around its final altitude. The path of this parcel illustrates how an inversion traps pollutants emitted from the surface. It also illustrates that pollutants often penetrate into the inversion layer. The parcel released at 30°C also rises and cools,

but it passes easily through the inversion layer. Because the background troposphere above the inversion is stable, the parcel ultimately comes to rest above the inversion.

Inversions form whenever a parcel of air becomes colder than the air above it. Common inversion types include the radiation inversion, the large-scale subsidence inversion, the marine inversion, the frontal inversion, and the small-scale subsidence inversion. These are discussed next.

6.6.1.4. Radiation Inversion

Radiation (nocturnal) inversions occur nightly as land cools by emitting thermal-IR radiation. During the day, land also emits thermal-IR radiation, but this loss is exceeded by a gain in solar radiation. At night, thermal-IR emissions cool the ground, which in turn cools molecular layers of air above the ground relative to air aloft, creating an inversion. The strength of a radiation inversion is maximized during long, calm, cloud-free nights when the air is dry. Long nights maximize the time during which thermal-IR cooling occurs, calm nights minimize downward turbulent mixing of energy, cloudfree nights minimize absorption of thermal-IR energy by cloud drops, and dry air minimizes absorption of thermal-IR energy by water vapor. The morning temperature profile in Figure 6.12 shows a radiation inversion. Radiation inversions also form in the winter during the day in regions that are not exposed to much

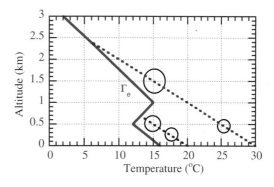

Figure 6.11. Schematic of pollutants trapped by an inversion and stable air. The parcel of air released at 20°C rises at the unsaturated adiabatic lapse rate of 10°C km^{-1} until its temperature equals that of the environment. This parcel is trapped by the inversion. The parcel of air released at 30°C also rises at 10°C km^{-1}. It escapes the inversion, but stops rising in stable free-tropospheric air, where the environmental lapse rate is 6.5°C km^{-1}.

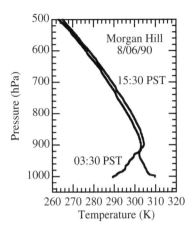

Figure 6.12. Observed temperature profiles in the early morning and late afternoon at Morgan Hill, California, on August 6, 1990. The morning sounding shows a radiation inversion coupled with a large-scale subsidence inversion. The afternoon sounding shows a large-scale subsidence inversion.

sunlight. They do not form regularly over the ocean because ocean water cools only slightly at night.

6.6.1.5. Large-Scale Subsidence Inversion

A **large-scale subsidence inversion** occurs within a surface high-pressure system. In such a system, air descends, compressing and warming adiabatically. When a layer of air descends adiabatically, the entire layer becomes more stable, often to the point that an inversion forms. The descension of air and creation of an inversion in a high-pressure system both contribute to near-surface pollution buildup.

Figure 6.13, illustrating the formation of a subsidence inversion, shows a 1.37-km-thick layer of air based at 3 km. At this altitude, the pressure thickness of the layer is 114 hPa. The initial temperature profile of the layer is stable, but not an inversion. As the layer descends, both its top and bottom compress and warm adiabatically at the rate of about $10°C$ km^{-1}. Although the pressure thickness of the layer remains constant during descension to conserve mass, the height thickness of the layer decreases with decreasing altitude (due to compression). Thus, at the surface, the pressure thickness of the layer is still 114 hPa, but the height thickness is 1 km. Figure 6.13 shows that the final temperature profile in the layer is more stable than is the initial profile. In fact, the sinking of air created an inversion. Conversely, the lifting of an unsaturated layer of air decreases the layer's stability.

Over land at night, air in a large-scale pressure system can descend and warm on top of near-surface air that has been cooled radiatively, creating a combined radiation/large-scale subsidence inversion. The morning inversion in Figure 6.12 shows such a case, and the afternoon profile shows the contribution of the large-scale subsidence inversion. The subsidence inversion in Figure 6.12 was due to the Pacific high, well known for producing large-scale subsidence inversions. Such inversions are present 85 to 95 percent of the days of the year in Los Angeles, which is one reason this city has historically had severe air pollution problems.

Figure 6.12 indicates that the radiation inversion eroded between morning and afternoon, so that all that remained was the large-scale subsidence inversion. Figure 6.14 illustrates this process, showing that, at 03:00, a strong radiation/large-scale subsidence inversion exists. At 06:00, after the sun rises, the ground and near-surface air heat sufficiently to chip away at the bottom of the inversion (Figure 6.14a). The chipping continues until late afternoon, when the inversion base height (mixing depth) reaches a maximum (Figure 6.14b). As the sun goes down, the surface temperature decreases, and the inversion base height decreases again. If the ground in Figure 6.14a were heated to $35°C$ instead of $30°C$ in late afternoon, the inversion would disappear. The elimination of an inversion due to surface heating is called **popping the inversion**.

A large-scale subsidence inversion serves as a lid to confine pollution beneath it, as shown in Figure 6.15. The mixing depth in which pollution is trapped is generally thickest in the late afternoon and thinnest during the night and early morning. Mixing ratios of primary pollutants, such as CO(g), NO(g), primary ROGs, and primary aerosol particles, usually peak during the morning, when mixing depths are low and rush-hour traffic emission rates are high. Although emission rates are high during afternoon rush hours, afternoon mixing

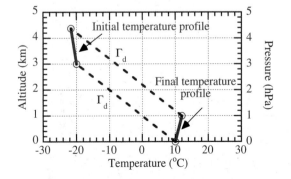

Figure 6.13. Formation of a subsidence inversion in sinking unsaturated air, as described in the text.

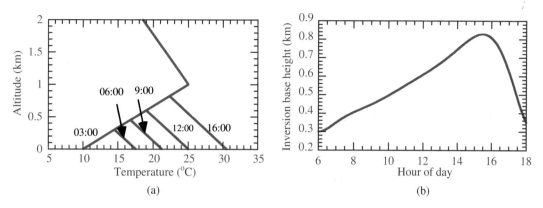

Figure 6.14. (a) Schematic showing the gradual chipping away of a morning radiation/large-scale subsidence inversion to produce an afternoon large-scale subsidence inversion. (b) Change in the inversion base height during the day corresponding to the chipping-away process in (a).

depths are also high, so primary pollutants become more diluted, decreasing their mixing ratios. Afternoon rush hours are also spread over more hours than are morning rush hours. Despite thick afternoon mixing depths, some secondary pollutants, such as $O_3(g)$, PAN(g), and secondary (aged) aerosol particles, reach their peak mixing ratios in the afternoon, when their chemical formation rates peak.

Figure 6.16 illustrates the seasonal variation of the afternoon inversion profile in Los Angeles. During the winter, the large-scale subsidence inversion is often weak because the center of the Pacific high is far from Los Angeles and temperatures aloft in Los Angeles are low. During the summer, the inversion is often strong because the center of the Pacific high is close to Los Angeles and temperatures aloft are high. During the spring, the inversion is of medium strength because the center of the Pacific high is a medium distance from Los Angeles. The inversion base height may be higher in spring than in summer, as illustrated in Figure 6.16.

Figure 6.15. Photograph of pollution trapped under a large-scale subsidence inversion in Los Angeles, California, on the afternoon of July 23, 2000. Photo by Mark Z. Jacobson.

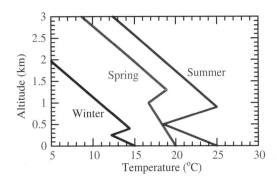

Figure 6.16. Schematic showing seasonal variation of afternoon inversions in Los Angeles.

6.6.1.6. Marine Inversion

A marine inversion occurs over coastal areas. During the day, sunlight heats land faster than it heats water. The rising air over land decreases near-surface air pressure, creating a sea–land pressure gradient that draws air inland from the sea. The resulting breeze between sea and land is the **sea breeze**. As cool, marine air moves inland during a sea breeze, it forces warm, inland air to rise, creating warm air over cold air and a **marine inversion**, which contributes to the inversion strength dominated by the large-scale subsidence inversion in Los Angeles.

6.6.1.7. Small-Scale Subsidence Inversion

As air flows down a mountain slope, it compresses and warms adiabatically, just as in a large-scale subsidence inversion. When the air compresses and warms on top of cool air, a **small-scale subsidence inversion** forms. In Los Angeles, air sometimes flows from east to west over the San Bernardino Mountains and into the Los Angeles Basin. Such air compresses and warms as it descends the mountain onto cool marine air below it, creating an inversion.

6.6.1.8. Frontal Inversion

A **cold front** is the leading edge of a cold **air mass** and the boundary between a cold air mass and a warm air mass. An air mass is a large body of air with similar temperature and moisture characteristics. Cold fronts and warm fronts form around low-pressure centers, particularly in the cyclones that predominate at 60°N. Cold fronts rotate counterclockwise around a surface cyclone. At a cold front, cold, dense air acts as a wedge and forces air in the warm air mass to rise, creating warm air over cold air and a **frontal inversion**. Because frontal inversions occur in low-pressure systems, where air generally rises and clouds form, frontal inversions are not usually associated with air pollution buildup.

6.6.2. Horizontal Pollutant Transport

Large-scale pressure systems affect the direction and speed of winds, which, in turn, affect pollutant transport. In this subsection, large-scale winds and their effects on pollution are examined. Small-scale winds are discussed in Section 6.7.4.

6.6.2.1. Effects of Wind Speeds on Pollutants

Winds around a surface low-pressure systems are generally faster than are those around a surface high-pressure systems, partly because pressure gradients in a low-pressure system are generally stronger than are those in a high-pressure system. Strong winds tend to clear out chemically produced pollution faster than do weak winds, but strong winds also resuspend more soil dust and other aerosol particles from the ground than do weak winds. Most soil dust resuspension occurs when the soil is bare, as illustrated in Figure 6.17.

6.6.2.2. Effects of Wind Direction on Pollutants

Large-scale pressure systems redirect air pollution. When a surface low-pressure center in the Northern Hemisphere is located to the west of a location, it produces southerly to southwesterly winds at the location, as seen in Figure 6.4a. Similarly, surface high-pressure centers located to the west of a location produces northwesterly to northerly winds at the location, as seen in

Figure 6.17. Dust storm approaching Spearman, Texas, on April 14, 1935. *Mon. Wea. Rev., 63*, 148, 1935, available from the National Oceanic and Atmospheric Administration Central Library; www.photolib.noaa.gov.

Figure 6.4b. In summers, the Pacific high is sometimes located to the northwest of the San Francisco Bay Area. Under such conditions, air traveling clockwise around the high passes through the Sacramento Valley, where temperatures are hot and pollutant concentrations are high, into the Bay Area, where temperatures are usually cooler and pollution also exists. The transport of hot, polluted air from the Sacramento Valley increases temperatures and pollution levels in the Bay Area.

6.6.2.3. Santa Ana Winds

In autumn and winter, the Canadian high-pressure system forms over the Great Basin, an elevated plateau, due to cold surface temperatures. Winds around the high flow clockwise down the Rocky Mountains, compressing and warming adiabatically. The winds then travel through Utah, New Mexico, Nevada, and into Southern California. As the winds travel over the Mojave Desert, they heat further and pick up desert soil dust. The winds then reach the San Gabriel and San Bernardino Mountains, which enclose the Los Angeles Basin. As winds are compressed through Cajon Pass and Banning Pass, two passes into the basin, they speed up to conserve momentum (mass times velocity). The fast winds resuspend soil dust. The winds, called the **Santa Ana winds**, then enter the Los Angeles Basin, bringing dust with them. If the winds are strong, they overpower the sea breeze (which blows from ocean to land), clearing pollution out of the basin to the ocean. Warm, dry, strong Santa Ana winds are also responsible for spreading brush fires in the basin. When the Santa Ana winds are weak, they are countered by the sea breeze, resulting in stagnation. Some of the heaviest air pollution events in the Los Angeles Basin occur under weak Santa Ana conditions. In such cases, pollution builds up over a period of several days. Santa Ana winds are generally strongest between October and December.

6.6.2.4. Long-Range Transport of Air Pollutants

Winds carry air pollution, sometimes over long distances. The chimney was developed centuries ago, not only to lift pollution above the ground, but also to take advantage of winds aloft that disperse pollution horizontally. Chimneys exacerbate pollution downwind of the point of emission. Starting in the seventeenth century, for example, chimney emissions from the Besshi copper mine and smelter on Shikoku Island, Japan, vented pollution to agricultural fields downwind. In the eighteenth century, smoke from soda ash factory chimneys in France and Great Britain devastated nearby countrysides (Chapter 10). In the 1930s to 1950s, a smelter in Trail, British Columbia, Canada, released pollution that traveled to Washington State in the United States. This last case is an example of **transboundary air pollution**, which occurs when pollution crosses political boundaries.

Sulfur dioxide emitted from tall smokestacks is often carried long distances before it deposits to the ground as sulfuric acid. Because most anthropogenic $SO_2(g)$ is emitted in midlatitudes, where the prevailing near-surface winds are southwesterly and the prevailing elevated winds are westerly, $SO_2(g)$ is transported to the northeast or east. If it is emitted at a high enough altitude, $SO_2(g)$ can travel hundreds to thousands of kilometers. The second largest smokestack in the world (after the 419.7-m stack in the GRES-2 coal power plant in Ekibastuz, Kazakhstan) is located in **Sudbury**, Ontario, Canada, north of Lake Huron. This nickel smelting stack, which is 380 m tall, was built in 1972 and designed to carry $SO_2(g)$ and $NO_x(g)$ emissions far from the local region. Although the stack reduced local pollution levels substantially, it increased pollution in all directions within 240 km of the plant, including to the United States, causing devastation, particularly in the prevailing wind direction.

Long-range transport affects not only pollutants emitted from stacks, but also photochemical smog closer to the ground. In 1987, Wisconsin believed that high mixing ratios of ozone there were exacerbated by ozone transport from Illinois and Indiana. Wisconsin then filed a lawsuit to force Illinois and Indiana to control pollutant emissions better. The lawsuit led to a settlement mandating a study of ozone transport pathways (Gerritson, 1993).

Pollutants travel long distances along many other well-documented pathways. Pollutants from New York City, for example, travel to Mount Washington, New Hampshire. Pollutants travel along the **BoWash corridor** between Boston and Washington, DC. Pollutants from the northeast United States travel to the clean north Atlantic Ocean (Liu et al., 1987; Dickerson et al., 1995; Moody et al., 1996; Levy et al., 1997; Prados et al., 1999). Pollutants from Los Angeles travel northward to Santa Barbara, northeastward to the San Joaquin Valley, southward to San Diego, and eastward to the Mojave Desert. Such pollutants have also been traced to the Grand Canyon, Arizona (Poulos and Pielke, 1994). Pollutants from the San Francisco Bay Area spill into the San Joaquin Valley through Altamont Pass.

An example of transboundary pollution is the transport of forest fire smoke from Indonesia to six other Asian countries in September 1997. Sulfur dioxide emissions from China are also suspected of causing a portion of acid deposition problems in Japan. Aerosol particles and ozone precursors from Asia travel long distances over the Pacific Ocean to North America (Prospero and Savoie, 1989; Zhang et al., 1993; Jacob et al., 1999; Song and Carmichael, 1999). Hydrocarbons, ozone, and PAN travel long distances across Europe (Derwent and Jenkin, 1991) as do pollutants from Europe to Africa (Kallos et al., 1998). Pollutants also travel between the United States and Canada, between the United States and Mexico, and from North America to Europe.

The long-range transport of pollution from Asia to North America is generally strongest during March and April. During these months, winds are particularly fast, picking up dust from the Gobi Desert, mixing it with air pollution from populated cities in China and Japan, and sending it rapidly across the Pacific, counterclockwise around the Aleutian low-pressure system and clockwise around the Pacific high-pressure system (e.g., Figure 6.7) to the west coast of North America. During one such event in April 1998, dust and pollution were transported across the Pacific in five days and increased PM_{10} concentrations in California by 20 to 50 $\mu g \ m^{-3}$ (Husar et al., 2001).

Similarly, pollution from the east coast of North America travels rapidly and sinusoidally around the highs and lows of the Atlantic Ocean to Europe. In one study, pollution emitted over the east coast of the United States during November 2001 was tracked by aircraft. The pollution was first lifted to the midtroposphere, where it was swept across the Atlantic and then sank over Europe, intercepting the Alps a week later (Huntrieser et al., 2005).

Example 6.3

How long would it take pollution emitted at one point on the globe at 30°N latitude and 5-km altitude to traverse the globe back to its original position if it travelled without changing latitude or altitude with a wind speed of 50 m s^{-1}?

Solution

From Appendix A.1.4, the radius of the Earth is $R_e = 6,371$ km. Adding $z = 5$ km to this gives $R_e + z = 6,376$ km. At $\varphi = 30°N$, the distance from the axis of rotation of the Earth to a point 5 km in the air is approximately $(R_e + z)\cos\varphi = 5,528$ km, giving the circumference of the Earth as $2\pi(R_e + z)\cos\varphi = 34,732$ km. Dividing this distance by a wind speed of 50 m s^{-1} gives a time of 8.04 days to travel around the world at 30°N.

6.6.3. Cloud Cover

Clouds affect pollution in two major ways. First, they reduce the penetration of UV radiation, therefore decreasing rates of photolysis below them. Second, rainout and washout remove pollutants from the air. Thus, rain-forming clouds help cleanse the atmosphere. In some cases, though, the pollutants in raindrops are returned to the air upon evaporation of the drops before they land. Because cloud cover and precipitation are often greater and mixing depths, higher in surface low-pressure systems than they are in surface high-pressure systems, photochemical smog concentrations are usually lower in the boundary layer of low-pressure systems than high-pressure systems.

6.7. Effects of Local Meteorology on Air Pollution

Although large-scale pressure systems control the prevailing meteorology of a region, local factors also affect meteorology and thus air pollution. Some of these factors are discussed briefly.

6.7.1. Ground Temperatures

Ground temperatures affect local meteorology in at least three ways, and meteorology feeds back to air pollution. Figure 6.14, which gives insight into the first mechanism, shows that warm ground surfaces produce high inversion base heights (thick mixing depths), which, in turn, reduce pollution mixing ratios. Conversely, cold ground surfaces produce thin mixing depths and high pollution mixing ratios.

Second, ground temperatures affect pollution by modifying wind speed. Warm surfaces enhance convection, causing surface air to mix with air aloft and vice versa. Because horizontal wind speeds at the ground are zero and those aloft are faster, the vertical mixing of horizontal winds increases wind speeds near the surface and reduces them aloft. Thus, warmer ground surfaces increase near-surface winds. Faster near-surface winds increase dispersion of near-surface pollution but may also increase the resuspension of loose soil dust and other aerosol particles from the ground. Conversely, lower ground temperatures have the opposite effect, slowing down near-surface winds and enhancing near-surface pollution buildup.

Third, changes in ground temperatures change near-surface air temperatures, and air temperatures affect rates of several processes. For example, temperatures influence the rates of biogenic gas emissions from trees, carbon monoxide emissions from vehicles, and gas-to-particle conversion rates. In particular, higher temperatures increase ozone and other pollutants due to chemical and physical feedbacks (Section 12.5.6).

6.7.2. Soil Liquid Water Content

An important parameter that affects ground temperatures, and therefore pollutant concentrations, is soil liquid water content (**soil moisture**). Increases in soil moisture cool the ground, reducing convection, decreasing mixing depths, and slowing near-surface winds. Thinner mixing depths and slower wind speeds enhance pollutant buildup. Conversely, decreases in soil liquid water increase convection, increasing mixing depths and near-surface winds, reducing pollution.

Soil moisture cools the ground in two major ways. First, evaporation of liquid water in soil cools the soil. Therefore, the more liquid water a soil has, the greater the evaporation and cooling of the soil during the day. Second, liquid water in soil increases the average specific heat of a soil-air-water mixture. The wetter the soil, the less the soil can heat up when solar radiation is added to it.

In a study of the effects of soil liquid water on temperatures, winds, and pollution, it was found that increases in soil water of only 4 percent decreased peak near-surface air temperatures by up to 6°C, decreased wind speeds by up to 1.5 m s^{-1}, delayed the times of peak ozone mixing ratio by up to two hours, and increased the magnitude of peak particulate concentrations substantially in Los Angeles over a two-day period (Jacobson, 1999a). Such results imply that rainfall, irrigation, and climate change all affect pollution concentrations.

6.7.3. Urban Heat Island Effect

Land cover affects ground temperatures, which affect pollutant concentrations. Most of the globe is covered with water (71.3 percent) or snow/ice (3.3 percent). The remainder is covered with forests, grassland, cropland, wetland, barren land, tundra, savanna, shrub land, and urban areas. Urban surfaces consist primarily of roads, walkways, rooftops, vegetation cover, and bare soil.

Urban construction material surfaces increase surface temperatures due to the **urban heat island effect**, first recorded in 1807 by English meteorologist **Luke Howard** (1772–1864), who is also known for classifying clouds. He measured temperatures at several sites within and outside London and found that temperatures within the city were consistently warmer than were those outside (Howard, 1833). Urban areas are generally warmer during the day than are vegetated areas around them because urban surfaces replace vegetation, reducing evapotranspiration from it. Evapotranspiration normally keeps the surface cool (because evaporation is a cooling process) (e.g., Oke, 1978, 1988). Urban surfaces also have sufficiently different properties of surface material (e.g., heat capacities, thermal conductivities, albedos, emissivities) to enhance urban warming relative to surrounding vegetated areas. Figure 6.18 shows a satellite-derived image of daytime surface temperatures in downtown Atlanta and its environs. Road surfaces and buildings stand out as being particularly hot.

Increased urban temperatures result in increased mixing depths and faster near-surface winds. Although faster near-surface winds disperse pollutants, higher temperatures increase ozone and other pollutants due to chemical and physical feedbacks (Section 12.5.6). Increased urban temperatures may also be responsible for enhanced thunderstorm activity (Bornstein and Lin, 2000).

6.7.4. Local Winds

Another factor that affects air pollution is the local wind. Winds arise due to pressure gradients. Although large-scale pressure gradients affect winds, local pressure gradients, resulting from uneven ground heating, variable topography, and local turbulence, can modify or override large-scale winds. Important local winds include sea, lake, bay, land, valley, and mountain breezes.

6.7.4.1. Sea, Lake, and Bay Breezes
Sea, lake, and bay breezes form during the day between oceans, lakes, or bays, respectively, and land. Figure 6.19 illustrates a basic sea breeze circulation. During the day, land heats up relative to water because land has a lower specific heat than does water. Rising air over land forces air aloft to diverge horizontally, decreasing surface air pressures (setting up a shallow thermal low-pressure system) over land. As a result of the pressure gradient between land and water, air moves from the water, where the pressure is now relatively high, toward the land. In the case of ocean water meeting land, the movement of near-surface air is the **sea breeze**. Although the ACoF acts on the sea breeze air, the distance traveled by the sea breeze is too short (a few tens of kilometers) for the Coriolis force to turn the air noticeably.

Figure 6.18. Landsat 5 satellite image with buildings superimposed showing daytime surface temperatures in Atlanta, Georgia. Temperatures from hot to cold are represented by white, red, yellow, green, and blue, respectively. Small cool areas in the middle of warm areas are often due to shadows caused by large buildings. NASA-Goddard Space Flight Center, Scientific Visualization Studio; http://svs.gsfc.nasa.gov/imagewall/AAAS/Urban_day_temp.jpg.

Meanwhile, some of the diverging air aloft over land returns toward the water. The convergence of air aloft over water increases surface air pressure over water, prompting a stronger flow of surface air from the water to the land, completing the basic sea breeze circulation cell. At night, land cools to a greater extent than does water, and all the pressures and flow directions in Figure 6.19 reverse themselves, creating a

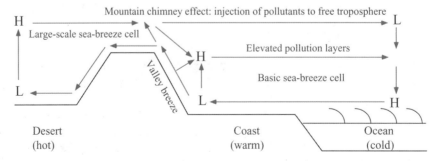

Figure 6.19. Illustration of a large-scale sea breeze circulation cell, basic sea breeze circulation cell, valley breeze, chimney effect, and formation of elevated pollution layers, as described in the text. Pressures shown (L and H) are relative to other pressures in the horizontal at the same altitude.

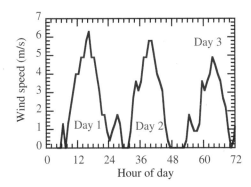

Figure 6.20. Near-surface wind speeds at Hawthorne, California, August 26–28, 1987.

land breeze, a near-surface flow of air from land to water.

Figure 6.19 illustrates that a basic sea breeze circulation cell can be embedded in a large-scale sea breeze cell. The Los Angeles Basin, for example, is bordered on its southwestern side by the Pacific Ocean and on its eastern side by the San Bernardino Mountains. The Mojave Desert lies to the east of the mountains. The desert heats up more than does land near the coast during the day, creating a thermal low over the desert, drawing air in from the coast, and creating the circulation pattern shown.

Figure 6.20 shows the variation of sea and land breeze wind speeds at Hawthorne, California, near the coast in the Los Angeles Basin, over a three-day period. Sea breeze wind speeds peak in the afternoon, when land–ocean temperature differences peak. Similarly, sea breeze winds peak during summer and are minimum during winter. Land breezes are weaker than sea breezes because the land–ocean temperature difference is small at night.

6.7.4.2. Valley and Mountain Breezes

A **valley breeze** is a wind that blows from a valley up a mountain slope and results from the heating of the mountain slope during the day. Heating causes air on the mountain slope to rise, drawing air up from the valley to replace the rising air. Figure 6.19 illustrates that in the case of a mountain near the coast, a valley breeze can become integrated into a large-scale sea breeze cell. The opposite of a valley breeze is a **mountain breeze**, which originates from a mountain slope and travels downward. Mountain breezes typically occur at night, when mountain faces cool rapidly. As a mountain face cools, air above the face also cools and drains downslope.

6.7.4.3. Effects of Sea and Valley Breezes on Pollution

In the Los Angeles Basin, sea and valley breezes are instrumental in transferring primary pollutants, emitted mainly from source regions on the west side of the basin, to receptor regions on the east side of the basin, where they arrive as secondary pollutants. Figure 4.13 shows an example of the daily variation of $NO(g)$, $NO_2(g)$, and $O_3(g)$ at a source and receptor region in Los Angeles. The valley breeze, in particular, moves ozone from Fontana, Riverside, and San Bernardino, in the eastern Los Angeles Basin, up the San Bernardino Mountains to Crestline, historically one of the most polluted locations in the United States in terms of ozone.

6.7.4.4. Chimney Effect and Elevated Pollution Layers

Pollutants in an enclosed basin, such as the Los Angeles Basin, can escape the basin through mountain passes and over mountain ridges. A third mechanism of escape is through the **mountain chimney effect** (Lu and Turco, 1995; Figure 6.19). Through this effect, a mountain slope is heated, causing air containing pollutants to rise, injecting the pollutants from within the mixed layer into the background troposphere.

Instead of dispersing, gases and particles may build up in **elevated pollution layers**. Pollutant concentrations in these layers often exceed those near the ground. Figure 6.21 shows a brilliant sunset through an elevated pollution layer in Los Angeles. Elevated layers form in one of at least four ways.

First, the rising portion of a sea breeze circulation can lift and inject pollutants into an inversion layer. Elevated pollution layers formed by this mechanism have been reported in Tokyo (Wakamatsu et al., 1983), Athens (Lalas et al., 1983), and near Lake Michigan (Lyons and Olsson, 1973; Fitzner et al., 1989). Figure 6.22 shows an example of an elevated pollution layer formed by a sea breeze circulation over Long Beach, California.

Second, some of the air forced up a mountain slope by winds may rise into and spread horizontally in an inversion layer. Of the air that continues up the mountain slope past the inversion, some may circulate back down into the inversion (e.g., Lu and Turco, 1995). Elevated pollution layers formed by this mechanism have been observed adjacent to the San Bernardino and San Gabriel Mountains in Los Angeles (Wakimoto and McElroy, 1986).

Figure 6.21. Brilliant sunset through elevated pollution layer over Los Angeles, California, in May 1972. Gene Daniels, U.S. EPA, May 1972, Still Pictures Branch, U.S. National Archives.

Figure 6.22. Elevated pollution layer formed over Long Beach, California, on July 22, 2000, by a sea breeze circulation. Photo by Mark Z. Jacobson.

Figure 6.23. Elevated layer of smoke trapped in an inversion layer following a greenhouse fire in Menlo Park, California, in June 2001. Photo by Mark Z. Jacobson.

Third, emissions from a smokestack or fire can rise buoyantly into an inversion layer. The inversion limits the height to which the plume can rise and forces the plume to spread horizontally. Figure 6.23 shows an example of a pollution layer formed by this mechanism.

Fourth, **elevated ozone layers** in the boundary layer may form by the destruction of surface ozone. During the afternoon, ozone is diluted uniformly throughout a mixing depth. In the evening, cooling of the ground stabilizes the air near the surface without affecting the stability aloft. In regions of nighttime $NO(g)$ emissions, the $NO(g)$ destroys near-surface ozone. Because nighttime air near the surface is stable, ozone aloft does not mix downward to replenish the lost surface ozone. Figure 6.24 shows an example of an elevated ozone layer formed by this process. The next day, the mixing depth increases, recapturing the elevated ozone and mixing it downward. McElroy and Smith (1992) estimated that daytime downmixing of elevated ozone enhanced surface ozone in certain areas of Los Angeles

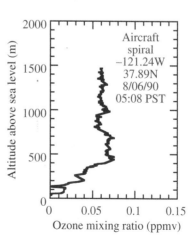

Figure 6.24. Vertical profile of ozone mixing ratio over Stockton, California, on August 6, 1990, at 05:08 PST, showing a nighttime elevated ozone layer that formed by the destruction of surface ozone. Data from the SARMAP field campaign (Solomon and Thuillier, 1995).

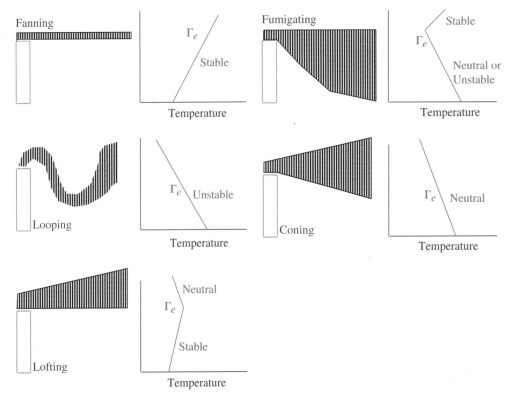

Figure 6.25. Cross sections of plumes under different stability conditions.

by 30 to 40 ppbv above what they would otherwise have been due to photochemistry alone. When few nighttime sources of NO(g) exist, elevated ozone layers do not form by this mechanism.

6.7.5. Plume Dispersion

Meteorology also affects the dispersion of pollution plumes from smokestacks. Figure 6.25 illustrates cross-sections of classic plume configurations that result from different stability conditions. In stable air, pollutants emitted from stacks do not rise or sink vertically; instead, they **fan** out horizontally. Viewed from above, the pollution distribution looks like a giant fan. Fanning does not expose people to pollution immediately downwind of a stack because the fanned pollution does not mix down to the surface. Figure 6.23 shows the side view of a fanning plume originating from a fire, not a stack. The worst meteorological condition for those living immediately downwind of a stack occurs when the atmosphere is neutral or unstable below the stack and stable above. This condition leads to **fumigation**, or the downwashing of pollutants toward the surface. **Looping** occurs when the atmosphere is unstable. Under such a condition, emissions from a stack may alternately rise and sink, depending on the extent of turbulence (Figure 6.26a). **Coning** occurs when the atmosphere is neutrally stratified. Under such a condition, pollutants emitted tend to slowly disperse both upward and downward. When the atmosphere is stable below and neutral above a stack, **lofting** occurs (Figure 6.26b). Lofting results in the least potential exposure to pollutants by people immediately downwind of a stack.

6.8. Summary

This chapter describes forces acting on air, the general circulation of the atmosphere, the generation of large-scale pressure systems, the effects of large-scale pressure systems on air pollution, and the effects of small-scale weather systems on air pollution. Forces examined include the pressure gradient force, apparent Coriolis force, apparent centrifugal force, and friction force. The winds produced by these forces include the geostrophic wind, the gradient wind, and surface winds. On the global scale, the south–north and vertical components of wind flow can be described by three circulation cells – the Hadley, Ferrel, and polar cells – in

(a)   (b)

Figure 6.26. (a) Looping, © Liliya Zakharchenko/Dreamstime.com; and (b) lofting plumes of pollution from smokestacks. © Artem Sapegin/Dreamstime.com.

both the Northern and Southern Hemispheres. Descending and rising air in these cells results in the creation of large-scale semipermanent high- and low-pressure systems over the oceans. Pressure systems over continents (thermal pressure systems) are seasonal and controlled by ground heating and cooling. Large-scale pressure systems affect pollution through their effects on stability and winds. Surface high-pressure systems enhance the stability of the boundary layer and increase pollution buildup. Surface low-pressure systems reduce the stability of the boundary layer and decrease pollution buildup. Soil moisture affects stability, winds, and pollutant concentrations. Sea and valley breezes are important local-scale winds that enhance or clear out pollution in an area. Elevated pollution layers, often observed, are a product of the interaction of winds, stability, topography, and chemistry.

6.9. Problems

6.1. Draw a diagram showing the forces and the gradient wind around an elevated low-pressure center in the Southern Hemisphere.

6.2. Draw a diagram showing the forces and resulting winds around a surface low-pressure center in the Southern Hemisphere.

6.3. Should horizontal winds turn clockwise or counterclockwise with increasing height in the Southern Hemisphere? Demonstrate your conclusion with a force diagram at the surface and aloft.

6.4. Characterize the stability of each of the six layers in Figure 6.27 with one of the stability criteria given in Equation 6.4.

6.5. If the Earth did not rotate, how would you expect the current three-cell model of the general circulation of the atmosphere to change?

6.6. In Figure 6.27, to approximately what altitude will a parcel of pollution rise adiabatically from the surface in unsaturated air if the parcel's initial temperature is 25°C? What if its initial temperature is 30°C?

6.7. In Figure 6.27, to approximately what altitude will a parcel of pollution rise adiabatically from the surface in saturated air if the parcel's initial temperature is 25°C? What if its initial temperature is 30°C? Assume a wet adiabatic lapse rate of 6°C km^{-1}.

6.8. Why do ozone mixing ratios peak in the afternoon, even though mixing depths are usually maximum in the afternoon?

6.9. Explain why ozone mixing ratios are higher on the east side of the Los Angeles Basin, even though ozone

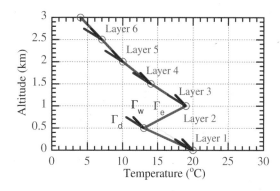

Figure 6.27. Environmental temperature profile with wet and dry adiabatic lapse rate profiles superimposed.

precursors are emitted primarily on the west side of the basin.

6.10. Summarize the meteorological conditions that tend to enhance and reduce, respectively, photochemical smog buildup in an area.

6.11. If emissions did not change with season, why might $CO(g)$ and aerosol particle concentrations be higher in winter than in summer?

6.12. If a diesel engine truck is driving in front of you, and much of the exhaust is entering your car and not rising in the air, how would you define the plume dispersion shape?

6.13. From Figure 6.27, determine the following quantities: (a) inversion base height, (b) inversion top height, (c) inversion base temperature, (d) inversion top temperature, (e) inversion strength, (f) inversion thickness, and (g) mixing depth.

6.14. What wind speed is needed for a pollution plume emitted at 45°N latitude and 3-km altitude to circumnavigate the globe back to its original position in six days if it travels without changing latitude or altitude?

Chapter 7

Effects of Pollution on Visibility, Ultraviolet Radiation, and Colors in the Sky

Visibility, UV radiation intensity, and optical phenomena are affected by gases, aerosol particles, and hydrometeor particles interacting with solar radiation. In clean air, gases and particles affect how far we can see along the horizon and the colors of the sky, clouds, and rainbows. In polluted air, gases and aerosol particles affect visibility, optical phenomena, and UV radiation intensity. In this chapter, visibility, optics, and UV transmission in clean and polluted atmospheres are discussed. An understanding of these phenomena requires a study of the interaction of solar radiation with gases, aerosol particles, and hydrometeor particles in light of several optical processes, including reflection, refraction, diffraction, dispersion, scattering, absorption, and transmission. This chapter describes such interactions and processes.

7.1. Processes Affecting Solar Radiation in the Atmosphere

The solar spectrum is comprised of **UV** (0.01–0.38 μm), **visible** (0.38–0.75 μm), and **solar-IR** (0.75–4.0 μm) wavelengths of light (Section 2.2). The UV portion of the spectrum drives most of the photochemistry of the atmosphere, controls the color of our skin, and causes most of the health problems, including skin cancer and cataracts, associated with solar radiation (Chapter 11). The visible portion of the spectrum provides most of the energy that keeps the Earth warm. It also affects the distance we can see and colors in the

atmosphere. It is no coincidence that the acuity (keenness) of our vision peaks at 0.55 μm, in the green part of the visible spectrum, which is near the wavelength of the sun's peak radiation intensity. Our eyes have evolved to take advantage of the peak intensity in this part of the visible spectrum. The solar-IR portion of the spectrum is important primarily for heating the Earth, but not so much as is the visible portion of the spectrum.

In 1666, **Sir Isaac Newton** (1642–1727; Figure 7.1), an English physicist and mathematician, showed that when visible light passes through a glass prism, each wavelength of light bends to a different degree, resulting in the separation of the total light, which is white, into a variety of colors that he called the **light spectrum**. Although the spectrum is continuous (the eye can distinguish 10 million colors), Newton discretized the spectrum into seven colors: red, orange, yellow, green, blue, indigo, and violet, to correspond to the seven notes on a musical scale. When the colors of the spectrum were recombined, they reproduced white light.

Newton also defined **primary colors** as blue, green, and red. These are referred to today as Newton's primary colors because they differ from artist's primary colors, which are blue, yellow, and red. When pairs of primary colors are added together, they produce new colors. For example, equal amounts of red plus green produce yellow, equal amounts of red plus blue produce magenta, and equal amounts of green plus blue produce cyan. Equal amounts of red, green, and blue produce white. For simplicity, the visible spectrum here

Table 7.1. Wavelengths of absorption in the visible and ultraviolet spectra by atmospheric gases

Gas name	Chemical formula	Absorption wavelengths (μm)
Visible/Near-UV/Far-UV Absorbers		
Ozone	$O_3(g)$	<0.35, 0.45–0.75
Nitrate radical	$NO_3(g)$	<0.67
Nitrogen dioxide	$NO_2(g)$	<0.71
Near-UV/Far-UV Absorbers		
Formaldehyde	$HCHO(g)$	<0.36
Nitric acid	$HNO_3(g)$	<0.33
Far-UV Absorbers		
Molecular oxygen	$O_2(g)$	<0.245
Carbon dioxide	$CO_2(g)$	<0.21
Water	$H_2O(g)$	<0.21
Molecular nitrogen	$N_2(g)$	<0.1

UV, ultraviolet.

Figure 7.1. Sir Isaac Newton (1642–1727). Edgar Fahs Smith Collection, University of Pennsylvania Library.

is divided into wavelengths of Newton's primary colors, **blue** (0.38–0.5 μm), **green** (0.5–0.6 μm), and **red** (0.6–0.75 μm).

When radiation passes through the Earth's atmosphere, it is attenuated or redirected by absorption and scattering by gases, aerosol particles, and hydrometeor particles. These processes are discussed next.

7.1.1. Gas Absorption

Absorption occurs when radiative energy (e.g., from the sun or Earth) enters a substance and is converted to internal energy, changing the temperature of the substance. The substance then reradiates some of the internal energy back to the air at a longer (thermal-IR) wavelength, heating the air. Absorption removes radiation from an incident beam, reducing the amount of radiation received past the point of absorption. Next, absorption by gases is discussed with respect to the solar spectrum.

7.1.1.1. Gas Absorption at Ultraviolet and Visible Wavelengths

Gases selectively absorb radiation in different portions of the electromagnetic spectrum. Table 7.1 lists some gases that absorb visible or UV radiation. Of the gases, all absorb UV, but only a few absorb visible radiation. Gases that absorb visible or UV radiation are often, but not always, photolyzed by this radiation into simpler products.

The gases that affect UV radiation the most are **molecular oxygen** [$O_2(g)$], **molecular nitrogen** [$N_2(g)$], and **ozone** [$O_3(g)$]. Oxygen absorbs wavelengths shorter than 0.245 μm, and nitrogen absorbs wavelengths shorter than 0.1 μm. Oxygen and nitrogen prevent nearly all solar wavelengths shorter than 0.245 μm from reaching the troposphere.

In 1880, M. J. Chappuis found that ozone also absorbed visible radiation between the wavelengths of 0.45 μm and 0.75 μm. The absorption bands in this wavelength region are now called **Chappuis bands**. In 1881, Walter N. Hartley first suggested that ozone was present in the upper atmosphere and hypothesized that the reason that the Earth's surface received little radiation shorter than 0.31 μm was because ozone also absorbs these wavelengths. The absorption bands of ozone below 0.31 μm are now called the **Hartley bands**. In 1916, English physicists Alfred Fowler (1868–1940) and Robert John Strutt (1875–1947; the son of Lord Baron Rayleigh) showed that ozone also weakly absorbs wavelengths between 0.31 μm and 0.35 μm. The bands of absorption in this region are now called the **Huggins bands**. Ozone absorption of UV and solar radiation, followed by reemission of thermal-IR radiation, heats the stratosphere, causing the stratospheric temperature inversion (Figure 3.3). Ozone absorption also protects the surface of the Earth by preventing nearly all UV wavelengths 0.245 to 0.29 μm and most wavelengths 0.29 to 0.32 μm from reaching the troposphere.

Although the gases in Table 7.1 absorb UV radiation, the mixing ratios of all except $O_3(g)$, $O_2(g)$, and $N_2(g)$ are too low to have much effect on UV penetration to the surface. For instance, stratospheric mixing ratios of **water vapor** (<6 ppmv) are much lower than are stratospheric mixing ratios of $O_2(g)$, which absorbs many of the same UV wavelengths as does water vapor. Thus, water vapor has little effect on UV attenuation in the stratosphere. Similarly, stratospheric mixing ratios of **carbon dioxide** (393 ppmv in 2011) are much lower than are those of $O_2(g)$ or $N_2(g)$, both of which absorb the same wavelengths as does carbon dioxide.

The only gas that absorbs visible radiation sufficiently to affect visibility is **nitrogen dioxide** [$NO_2(g)$], but its effect is important only in polluted air, where its mixing ratios are sufficiently high. Mixing ratios of the **nitrate radical** [$NO_3(g)$], which absorbs even further into the visible spectrum than does nitrogen dioxide, are low, except at night or in the early morning. Thus, $NO_3(g)$ does not affect visibility. Although ozone mixing ratios can be high, ozone is a relatively weak absorber of visible light. However, ozone absorption of green light after a volcano can cause beautiful colors in the sky (Section 7.3.5).

7.1.1.2. Gas Absorption Extinction Coefficient

Visibility is affected by all processes that attenuate or enhance radiation. In this subsection, attenuation by gas absorption is briefly discussed.

Figure 7.2 illustrates how radiation passing through a gas is reduced by absorption. Suppose incident radiation of intensity I_0 travels a distance $dx = x - x_0$ through a uniformly mixed absorbing gas q of number concentration N_q (molecules per cubic centimeter of air). As the radiation passes through the gas, gas molecules

intercept it. If the molecules absorb the radiation, the intensity of the radiation diminishes. The ability of a gas molecule to absorb radiation is embodied in the absorption cross section of the gas, $b_{a,g,q}$ (cm^2 per molecule), where the subscripts a, g, and q mean "absorption," "by a gas," and "by gas q," respectively. Wavelength subscripts were omitted. An **absorption cross section** of a gas is an effective cross section that results in radiation reduction by absorption. Its size is on the order of, but usually not equal to, the real cross-sectional area of a gas molecule.

The product of the number concentration and absorption cross section of a gas is called an absorption extinction coefficient. An **extinction coefficient** measures the loss of electromagnetic radiation due to a specific process, per unit distance. Extinction coefficients, symbolized with σ, have units of inverse distance (cm^{-1}, m^{-1}, or km^{-1}) and vary with wavelength. The **absorption extinction coefficient** (cm^{-1}) of gas q is

$$\sigma_{a,g,q} = N_q b_{a,g,q} \qquad (7.1)$$

The gas absorption extinction coefficient due to the sum of all gases ($\sigma_{a,g}$) is the sum of Equation 7.1 over all absorbing gases. The greater the absorption extinction coefficient of a gas in the visible spectrum, the more the gas reduces visibility. Figure 7.3 shows absorption extinction coefficients of both $NO_2(g)$ and $O_3(g)$ at two mixing ratios. It indicates that nitrogen dioxide affects extinction (and therefore visibility) primarily at high mixing ratios and at wavelengths below about 0.5 μm (when $\sigma_{a,g,q} > 0.1$ km^{-1}). In polluted air, such as in Los Angeles, $NO_2(g)$ mixing ratios typically range from 0.01 to 0.1 ppmv and peak near 0.15 ppmv during the morning. A typical value is 0.05 ppmv. $O_3(g)$ has a larger effect on extinction than does $NO_2(g)$ at wavelengths below about 0.32 μm. $O_3(g)$ mixing ratios

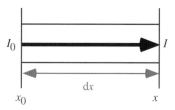

Figure 7.2. Attenuation of incident radiance I_0 due to absorption in a column of gas.

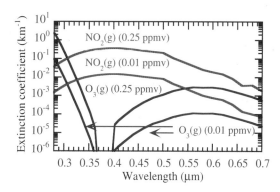

Figure 7.3. Extinction coefficients due to $NO_2(g)$ and $O_3(g)$ absorption when $T = 298$ K and $p_a = 1,013$ hPa.

in polluted air usually peak between 0.05 ppmv and 0.25 ppmv. The reduction in radiation intensity (where I denotes intensity) at a given wavelength with distance through an absorbing gas can now be defined as

$$I = I_0 e^{-\sigma_{a,g,q}(x-x_0)} = I_0 e^{-N_q b_{a,g,q}(x-x_0)} \quad (7.2)$$

Equation 7.2 states that incident radiation I_0 is reduced by a factor $\exp[-N_q b_{a,g,q}(x - x_0)]$ between points x_0 and x by gas absorption. From this equation, it is evident that the absorption cross section of a gas can be determined experimentally by measuring the attenuation of radiation through a homogenous gas column, such as the one shown in Figure 7.2, of known path length $N_q(x - x_0)$ (molec cm^{-2}), and then extracting the cross section from Equation 7.2.

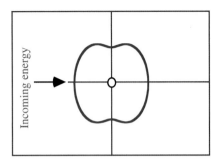

Figure 7.4. Probability distribution of where incident energy is scattered by a gas molecule (located at the center of the diagram).

> **Example 7.1**
> Find the fraction of incident radiation intensity (the transmission) passing through a uniform gas column of length 1 km when the number concentration of the gas is $N_q = 10^{12}$ molec cm^{-3} and the absorption cross section of a gas molecule is $b_{a,g,q} = 10^{-19}$ cm^2 per molecule.
>
> **Solution**
> From Equation 7.1, the extinction coefficient through the gas is $\sigma_{a,g,q} = N_q b_{a,g,q} = 10^{-7}$ cm^{-1}. From Equation 7.2, the transmission is $I/I_0 = \exp(-10^{-7} \times 10^5) = 0.99$. Thus, 99 percent of incident radiation traveling through this column reaches the other side.

7.1.2. Gas Scattering

Gas scattering is the redirection of radiation by a gas molecule without a net transfer of energy to the molecule. When a gas molecule scatters, incident radiation is redirected symmetrically in the forward and backward direction and somewhat off to the side, as shown in Figure 7.4. The figure gives the probability distribution of incident energy scattered by a gas molecule.

The scattering of radiation by gas molecules or aerosol particles much smaller than the wavelength of light is called **Rayleigh scattering**. Because gas molecules are on the order of 0.0005 μm in diameter and a typical wavelength of visible light is 0.5 μm, all gas molecules are Rayleigh scatterers. Because molecular nitrogen [$N_2(g)$], molecular oxygen [$O_2(g)$], argon [$Ar(g)$], and water vapor [$H_2O(g)$] are the most abundant gases in the air, these are the most important Rayleigh scatterers. Rayleigh scattering is named after **Lord Baron Rayleigh**, born John William Strutt (1842–1919; Figure 7.5), who also discovered argon gas with Sir William Ramsay in 1894 (Chapter 1). Rayleigh's theoretical work on gas scattering was

Figure 7.5. Lord Baron Rayleigh (John William Strutt) (1842–1919). Edgar Fahs Smith Collection, University of Pennsylvania Library.

published in 1871 (Rayleigh, 1871), 23 years before his discovery of argon.

7.1.2.1. Gas Scattering Extinction Coefficient

The **gas scattering extinction coefficient** (cm^{-1}) of total air (Rayleigh scattering extinction coefficient), which is analogous to that of the gas absorption extinction coefficient, is defined as

$$\sigma_{s,g} = N_a b_{s,g} \qquad (7.3)$$

where N_a is the number concentration of air molecules (molec cm^{-3}) at a given altitude and $b_{s,g}$ is the scattering cross section (cm^2 $molec^{-1}$) of a typical air molecule. The **scattering cross section** of an air molecule is an effective cross section that results in radiation reduction by scattering and is proportional to the inverse of the fourth power of the wavelength. This means that gas molecules in the air scatter short (blue) wavelengths preferentially over long (red) wavelengths.

7.1.2.2. Colors of the Sky and Sun

The variation of the Rayleigh scattering cross section with wavelength explains why the sun appears white at noon, yellow in the afternoon, and red at sunset. It also explains why the sky is blue. White sunlight that enters the Earth's atmosphere travels a shorter distance before reaching a viewer's eye at noon than at any other time during the day, as illustrated in Figure 7.6. During white light's travel through the atmosphere, blue wavelengths are preferentially scattered out of the direct beam by air molecules, but not enough blue light is scattered for a person looking at the sun to notice that the incident beam has changed its color from white. Thus, a person looking at the sun at noon often sees a **white sun**. The blue light that scatters out of the direct beam is scattered by gas molecules multiple times, and some of it eventually enters the viewer's eye when the viewer looks away from the sun. As such, a viewer looking away from the sun sees a **blue sky**.

Although Lord Rayleigh formalized the theory of scattering by gases and aerosol particles much smaller than the wavelength of light, it may have been **Leonardo da Vinci** (1452–1519) who first suggested that blue colors in the sky were due to interactions of sunlight with atmospheric constituents. He wrote in his notebooks (c. 1500, Chapter VI, Section 300):

> I say that the blueness we see in the atmosphere is not intrinsic color, but is caused by warm vapor evaporated in minute and insensible atoms on which the solar rays fall, rendering them luminous against the infinite

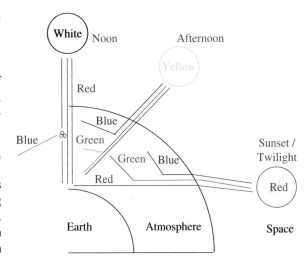

Figure 7.6. Colors of the sun. At noon, the sun appears white because red, green, and some blue light transmit to a viewer's eye. In the afternoon, sunlight traverses a longer path in the atmosphere, removing more blue. At sunset, most green is removed from the line of sight, leaving a red sun. After sunset, the sky appears red due to refraction between space and the atmosphere.

> darkness of the fiery sphere which lies beyond and includes it. . . . If you produce a small quantity of smoke from dry wood and the rays of the sun fall on this smoke and if you place (behind it) a piece of black velvet on which the sun does not fall, you will see that the black stuff will appear of a beautiful blue color. . . . Water violently ejected in a fine spray and in a dark chamber where the sunbeams are admitted produces then blue rays. . . . Thus it follows, as I say, that the atmosphere assumes this azure hue by reason of the particles of moisture which catch the rays of the sun.

What da Vinci saw, however, was not scattering by gas molecules but scattering by small aerosol particles, namely, dry wood smoke particles and small liquid water drops. Because such particles are smaller than the peak wavelength of visible light, they preferentially scatter blue light, as do gas molecules. Small aerosol particles are well known to scatter blue light preferentially. Preferential scattering of blue light by small aerosol particles in the air is responsible for the blue appearance of the **Blue Ridge Mountains** in Virginia and the **Blue Mountains** in Australia. Occasionally, after a forest fire, a **blue moon** appears due to the scattering of blue light by small organic aerosol particles. Similarly, after a heavy rain, the relative humidity is low, and aerosol particles lose much of their water, shrinking

Figure 7.7. Yellow sun in the early evening off the island of Maui in the South Pacific Ocean. Photo by Mark Z. Jacobson.

Figure 7.8. Red horizon after the sun dips below a deck of stratus clouds over the Pacific Ocean, signifying the beginning of twilight. Photo by Mark Z. Jacobson.

enough to scatter preferentially blue light and give the sky a deep blue appearance.

In the afternoon, light takes a longer path through the air than it does at noon; thus, more blue and some green light is scattered out of the direct solar beam in the afternoon than at noon. Although a single gas molecule is less likely to scatter a green than a blue wavelength, the number of gas molecules along a viewer's line of sight is so large in the afternoon that the probability of green light scattering is sizable. Nearly all red and some green are still transmitted to the viewer's eye in the afternoon, causing the sun to appear yellow. In clean air, the sun can remain yellow until just before it reaches the horizon, as shown in Figure 7.7.

When the sun reaches the horizon at sunset, sunlight traverses its longest distance through the atmosphere, and all blue and green and some red wavelengths are scattered out of the sun's direct beam. Only some red light transmits, and a viewer looking at the sun sees a **red sun**. Sunlight can be seen after sunset because sunlight refracts as it enters Earth's atmosphere, as illustrated in Figure 7.6. **Refraction**, discussed in Section 7.1.4.2, is the bending of light as it passes from a medium of one density to a medium of a different density.

During and after sunset, the whole horizon often appears red (Figure 7.8) due to the presence of aerosol particles along the horizon. Aerosol particles scatter blue, green, and red wavelengths of direct sunlight, sending such light in all directions along the horizon. At all points along the horizon, additional particles rescatter some of the scattered blue, green, and red light toward the viewer's eye. As these wavelengths pass through the air, blue and green are scattered preferentially out of the viewer's line of sight by gas molecules, whereas most red wavelengths are transmitted, causing the horizon to appear red. Red horizons are common over the ocean, where sea spray particles are present, and over land, where soil and pollution particles are present.

The time after sunset during which the sky is still illuminated due to refraction is called **twilight**. Twilight also occurs before sunrise. The length of twilight increases with increasing latitude. At midlatitudes in summer, the length of twilight is about one-half hour. At higher latitudes in summer, twilight in the morning may merge with that in the evening, creating twilight that lasts all night, known as a **white night**.

7.1.3. Aerosol and Hydrometeor Particle Absorption

All aerosol and hydrometeor particles absorb thermal- and solar-IR radiation, but only a few absorb visible and near-UV radiation. Next, visible and UV absorption properties of aerosol particles and the effect of aerosol particle absorption on UV radiation and pollution are discussed.

7.1.3.1. Important Absorbers of Visible and Ultraviolet Radiation

The strongest aerosol particle absorber across the solar spectrum (UV, visible, and solar-IR wavelengths) is **black carbon**. It substantially absorbs all wavelengths of light and does not transmit any, thereby appearing black. Black carbon is the main component of diesel soot particles, which also contain organic carbon, some sulfate, and trace metals. Diesel soot particles appear black because they are dominated by black carbon. Black carbon also appears in soot particles from biofuel and biomass burning. Such particles contain less black carbon and more organic carbon, and so generally appear brown, yellow, or gray.

All organic aerosol particle constituents absorb short UV wavelengths, but only some absorb long UV wavelengths and short visible (blue and some green) wavelengths. The strongest long UV and short visible wavelength–absorbing organics include certain **nitrated aromatics**, **polycyclic aromatic hydrocarbons** (PAHs), benzaldehydes, benzoic acids, aromatic polycarboxylic acids, and phenols (Jacobson, 1999b). Several of these compounds may be present in **tar balls** (Figure 7.9), which are amorphous, spherical aerosol particles 30 to 500 nm in diameter that form downwind of biomass or biofuel burning emissions and absorb UV and short visible light wavelengths (Pósfai et al., 2004). Tar balls comprise about 15 percent of biomass burning particles by number 1.6 km from a fire. Although individually they are weaker absorbers than black carbon aggregates (Figure 7.10), their size, relative abundance, and strong absorption cause them to heat biomass burning plumes only slightly less than do black carbon emissions in the plumes (Adachi and Buseck, 2011). Together, all visible-absorbing organics appear brown

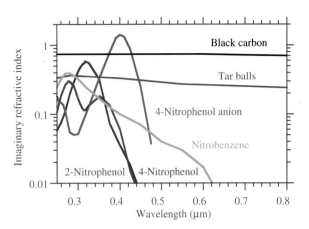

Figure 7.10. Imaginary indices of refraction of black carbon, nitrated aromatics, and tar balls versus wavelength. Black carbon data are from Krekov (1993) and references therein; nitrated aromatic data are from Jacobson (1999b) and references therein; tar balls data are from Alexander et al. (2008).

or yellow and are referred to as **brown carbon**. Biofuel and biomass burning particles often contain both black carbon and brown carbon, although others, such as tar balls, contain only brown carbon.

Another important aerosol particle type that absorbs solar radiation is soil dust. Soil dust particles contain different proportions of $Fe_2O_3(s)$, $Al_2O_3(s)$, $SiO_2(s)$, $CaCO_3(s)$, $MgCO_3(s)$, clays, and other substances. Such particles often appear brown or red because they contain hematite. Hematite is also found in industrial particles. **Hematite** [$Fe_2O_3(s)$] strongly absorbs blue, moderately absorbs green, and weakly absorbs red wavelengths. Because it reflects red and some green, it appears red or reddish-brown in high concentrations. Because of hematite primarily, soil dust particle absorption increases from the visible to the UV spectra (e.g., Gillette et al., 1993; Sokolik et al., 1993).

Soil dust particles also contain **aluminum oxide** [$Al_2O_3(s)$, alumina], which absorbs moderately to weakly across the whole visible spectrum and appears silvery white in pure form. Aluminum oxide is also found in combustion particles.

Silicon dioxide [$SiO_2(s)$, silica], which is a crystalline compound found in quartz, sand, and other minerals present in soil dust, is a weak absorber of visible and UV radiation. Because it appears white or colorless, it does not contribute much to the color of soil dust particles.

Most other components in soil dust and other particles are weak absorbers of visible and UV radiation. For

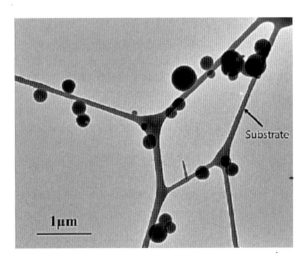

Figure 7.9. Transmission electron microscope image of tar balls on fibers of lacey carbon substrate, collected on March 22, 2006, in Mexico. From Adachi and Buseck (2011).

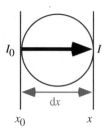

Figure 7.11. Attenuation of incident radiation I_0 due to absorption by a spherical aerosol particle.

example, **sodium chloride [NaCl(s)]**, **ammonium sulfate [(NH$_4$)$_2$SO$_4$(s)]**, and **sulfuric acid [H$_2$SO$_4$(aq)]** are all very weak absorbers of visible and UV radiation.

7.1.3.2. The Imaginary Refractive Index

Figure 7.11 shows a possible path of radiation through a single spherical aerosol particle.

The attenuation, due to absorption, of incident radiation I_0 of wavelength λ as it travels through a single particle is

$$I = I_0 e^{-4\pi\kappa(x-x_0)/\lambda} \qquad (7.4)$$

where κ is the imaginary index of refraction, and $x–x_0$ is the distance through the particle. The **imaginary index of refraction** is a measure of the extent to which a particle absorbs radiation. The term $4\pi\kappa/\lambda$ is an absorption extinction coefficient for a single aerosol particle. Table 7.2 gives imaginary refractive indices for some substances at wavelengths of 0.3, 0.5, and 10 μm. Black carbon has the largest imaginary indices of refraction across the entire spectrum among the substances shown. Tar balls also strongly absorb UV and visible light but

less so than black carbon, whereas nitrobenzene is a strong UV absorber but a weaker visible absorber.

Example 7.2
Find the fraction of incident radiation intensity that transmits through a uniform particle of diameter 0.1 μm at a wavelength of 0.5 μm when the particle is composed of (a) black carbon and (b) water.

Solution
From Table 7.2, the imaginary refractive indices of black carbon and water at $\lambda = 0.5$ μm are $\kappa = 0.74$ and 10^{-9}, respectively. From Equation 7.4, the transmission of light through (a) black carbon is $I/I_0 = \exp(-4\pi \times 0.74 \times 0.1/0.5) = 0.16$ and that through (b) liquid water is 0.999999997. Thus, a 0.1-μm black carbon particle absorbs 84 percent of incident radiation, whereas a 0.1-μm water particle absorbs only 0.0000003 percent of incident visible radiation passing through it. As such, black carbon is a strong absorber of visible light, but liquid water is not.

7.1.3.3. Effects of Aerosol Particle Absorption on Ultraviolet Radiation

Figure 7.10 shows the wavelength-dependent imaginary refractive indices of important absorbing components in aerosol particles, namely, black carbon, nitrated aromatic compounds, and tar balls. It indicates that black carbon absorbs evenly across the UV and visible spectra. Tar balls also absorb across these spectra but with greater preference for UV absorption. Although nitrated aromatics absorb preferentially in the UV spectrum,

Table 7.2. Real and imaginary indices of refraction for substances at $\lambda = 0.3, 0.5,$ and 10 μm

	$\lambda = 0.3$ μm		$\lambda = 0.5$ μm		$\lambda = 10$ μm	
	Real (n)	Imaginary (κ)	Real (n)	Imaginary (κ)	Real (n)	Imaginary (κ)
H$_2$O(aq)[a]	1.35	1.0×10^{-8}	1.34	1.0×10^{-9}	1.22	0.051
Black carbon (s)[b]	1.80	0.74	1.82	0.74	2.40	1.0
Tar balls (s)	1.47[c]	0.38[c]	1.67[c]	0.27[c]	1.77[b]	0.12[b]
Nitrobenzene (aq)	1.64[d]	0.32[d]	1.59[d]	0.03[d]	1.77[b]	0.12[b]
Low-absorbing OM[b]	1.45	0.001	1.45	0.001	1.77	0.12
H$_2$SO$_4$(aq)[b]	1.47	1.0×10^{-8}	1.43	1.0×10^{-8}	1.89	0.46

OM, organic matter.
[a] Hale and Querry (1973); [b] Krekov (1993); [c] Alexander et al. (2008); [d] Foster (1992).

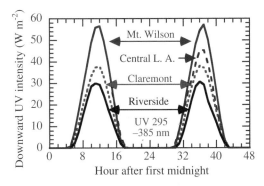

Figure 7.12. Effect of pollution on ultraviolet (UV) radiation. Measurements of downward UV radiation intensity at four sites in Los Angeles, California, on August 27 and 28, 1987. The three nonmountain sites are progressively further inland. The elevations of the four sites are as follows: Mt. Wilson, 1.7 km; Central Los Angeles, 87 m; Claremont, 364 m; and Riverside, 249 m. UV radiation between the mountain and surface was reduced the most at Riverside, where the highest pollution level occurred. Measurements in central Los Angeles were available only on the second day. Data provided by the California Air Resources Board.

they also absorb blue and, in some cases, green wavelengths of visible light. Tar balls are found to contain more nitrogen than other types of organic-containing particles (Adachi and Buseck, 2011). Because the addition of nitrogen to an organic compound extends the compound's absorption to longer wavelength, some of the strong UV and short visible wavelength absorptivity due to tar balls may be due to the presence of nitrogen in these particles.

Absorption of UV radiation by organic compounds in aerosol particles affects the amount of UV radiation reaching the ground in polluted air. For example, measurements in the Los Angeles Basin in 1987, shown in Figure 7.12, indicate that UV radiation of wavelength from 0.295 μm to 0.385 μm was reduced by 22 percent in central Los Angeles (near the coast), 33 percent at Claremont (further inland), and 48 percent at Riverside (much further inland) in comparison with UV radiation at Mount Wilson, 1.7 km above the basin. Measurements also indicate that total solar radiation between 0.285 μm and 2.8 μm was reduced by only 8 to 14 percent between Riverside and Mount Wilson. Some components of photochemical smog must have preferentially reduced UV radiation over total solar radiation.

Whereas some of the preferential UV reductions were due to gas absorption, Rayleigh scattering, and aerosol particle scattering, the sum of these effects could not account for the up to 50 percent of observed decreases in UV radiation at Riverside. Thus, aerosol particle absorption most likely played a role in the reductions. Because black carbon is not a strong preferential absorber of UV radiation (it absorbs UV and visible wavelengths relatively equally), and because soil dust concentrations were relatively low, these UV absorbers could not account for the remainder of preferential UV reductions. Nitrated and aromatic aerosol components, which preferentially absorb UV (Figure 7.10) likely accounted for good portion of the remaining UV absorption (Jacobson, 1999b). Figure 5.15 shows that nitrate concentrations at Riverside, where the largest UV reductions occurred, were high (up to 60 μg m^{-3}); thus, it is likely that organics at Riverside were heavily nitrated, absorbing UV radiation.

7.1.3.4. Effects of Ultraviolet Radiation Reductions on Ozone

The effect of UV radiation loss on ozone cannot be measured, but it can be examined with a model. A modeling study of air pollution in Los Angeles found that decreases in UV radiation due to aerosol particles in smog decreased photolysis rates, decreasing near-surface ozone mixing ratios by an average of 5 to 8 percent in August 1987 (Jacobson, 1998). The study also discovered the following:

- In regions of the boundary layer where aerosol particle UV radiation absorption was strong, the reduction in UV radiation reduced photolysis rates of gases, decreasing ozone.
- In regions of the boundary layer where aerosol particle UV radiation scattering dominated absorption, the increase in scattering increased photolysis rates of gases, increasing ozone.

In a study of relatively nonabsorbing aerosols in Maryland, Dickerson et al. (1997) found that highly sscattering aerosol particles increased ozone, consistent with the second result.

Although reduced UV radiation and ozone may appear to be ironic benefits of smog, the cause of UV reductions is heavy particle loadings. Particles, particularly small ones, cause harmful health effects that far outweigh the benefits of reduced UV radiation or the small level of reduced ozone that they trigger. In addition, although absorbing aerosol particles slightly decrease ozone, they increase the mixing ratios of other pollutant gases.

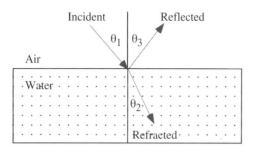

Figure 7.13. Examples of reflection and refraction. During reflection, the angle of incidence (θ_1) equals the angle of reflection (θ_3). During refraction, the angles of incidence and refraction are related by Snell's law. The line perpendicular to the air–water interface is the surface normal.

7.1.4. Aerosol and Hydrometeor Particle Scattering

Particle scattering is the redirection of incident energy by a particle without a loss of energy to the particle. Particle scattering is really the combination of several processes, including reflection, refraction, and diffraction. These processes are discussed next.

7.1.4.1. Reflection

Reflection occurs when radiation bounces off an object at an angle equal to the angle of incidence. No energy is lost during reflection. Figure 7.13 shows an example of reflection. Radiation can reflect off aerosol particles, cloud drops, and other surfaces. The colors of most objects that we see are due to preferential reflection of certain wavelengths by the object. For example, an apple appears red because the apple's skin absorbs blue and green wavelengths and reflects red wavelengths to our eye.

7.1.4.2. Refraction

Refraction occurs when a wave or photon leaves a medium of one density and enters a medium of another density. In such a case, the speed of the wave changes, changing the angle of the incident wave relative to a surface normal, as shown in Figure 7.13. If a wave travels from a medium of one density to a medium of a higher density, it bends (refracts) toward the **surface normal** (the vertical line in Figure 7.13). The angle of refraction is related to the angle of incidence by **Snell's law:**

$$\frac{n_2}{n_1} = \frac{\sin \theta_1}{\sin \theta_2} \qquad (7.5)$$

where n is the real index of refraction (dimensionless), θ is the angle of incidence or refraction, and subscripts 1 and 2 refer to media 1 and 2, respectively. The **real index of refraction** is the ratio of the **speed of light in a vacuum** ($c = 2.99792 \times 10^8$ m s^{-1}) to that in a different medium (c_1, m s^{-1}). Thus,

$$n_1 = \frac{c}{c_1} \qquad (7.6)$$

Because light cannot travel faster than its speed in a vacuum, the real index of refraction of a medium other than a vacuum must exceed unity. The refractive index of air at a wavelength of 0.5 μm is 1.000279. Real refractive indices of some liquids and solids are given in Table 7.2.

Example 7.3

(a) Suppose light at a wavelength of 0.5 μm travels between the atmosphere (medium 1) and liquid water (medium 2) and the angle between the incident light and the surface normal is $\theta_1 = 45°$. By how many degrees is the light bent toward the surface normal when it enters medium 2?

(b) Do the same calculation for light traveling between outer space (medium 1) and the atmosphere (medium 2).

Solution

(a) At a wavelength of 0.5 μm, the real index of refraction of air is $n_1 = 1.000279$ and that of liquid water is $n_2 = 1.335$. From Equation 7.5, the angle between the light and the surface normal in medium 2 is $\theta_2 = 32°$; thus, the light is bent by 13 degrees toward the surface normal when it enters water from the atmosphere.

(b) At a wavelength of 0.5 μm, the real index of refraction of a vacuum is $n_1 = 1.0$. From Equation 7.5, the angle between the light and the surface normal in medium 2 is $\theta_2 = 44.984°$; thus, the light is bent by 0.016 degrees toward the surface normal when it enters the atmosphere from space.

In sum, the angle of refraction between the atmosphere and water is much greater than that between space and the atmosphere.

Because the real index of refraction is wavelength dependent, different wavelengths are refracted by different angles when they pass from one medium to another. For instance, when visible light passes from

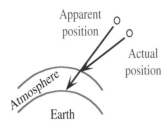

Figure 7.14. Refraction of starlight by the atmosphere makes a star appear to be where it is not.

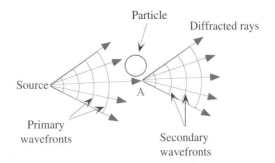

Figure 7.15. Diffraction of light around a spherical particle. Any point along a wavefront may be taken as the source of a new series of secondary waves. Rays emitted from point A appear to cause waves from the original source to bend around the particle.

air to liquid water at an incident angle of $\theta_1 = 45°$, wavelengths of 0.4 and 0.7 μm are bent by 13.11 and 12.90 degrees, respectively. Thus, refraction bends short (blue) wavelengths of visible light more than it bends long (red) wavelengths. Separation of white visible light into individual colors by this selective refraction is called **dispersion** (or **dispersive refraction**). When Sir Isaac Newton separated white light into multiple colors by passing it through a glass prism (Section 7.1), he discovered dispersive refraction.

As shown in Figures 7.6 and 7.8, refraction between space and the atmosphere is responsible for twilight, which is the sunlight seen after the sun sets and before the sun rises. Such refraction also causes stars to appear positioned where they are not, as shown in Figure 7.14. Layers of air at different densities in the Earth's atmosphere cause starlight to refract multiple times, and thus, **flicker**, **twinkle**, or **scintillate**.

7.1.4.3. Diffraction

Diffraction is a process by which the direction of propagation of a wave changes when the wave encounters an obstruction. In terms of visible wavelengths, it is the bending of light as it passes by the edge of an obstruction. In the air, waves diffract as they pass by the surface of an aerosol particle, cloud drop, or raindrop.

Diffraction can be explained in terms of **Huygens's principle**, which states that each point of an advancing wavefront may be considered the source of a new series of secondary waves. If a stone is dropped in a tank of water, waves move out horizontally in all directions, and wavefronts are seen as concentric circles around the stone. If a point source emits waves in three dimensions, wavefronts are concentric spherical surfaces. When a wavefront encounters the edge of an obstacle, waves appear to bend (diffract) around the obstacle because a series of secondary concentric waves is emitted at the edge of the obstacle along the advancing wavefront (e.g., in Figure 7.15) and at each other point along the

wavefront. The diffracted rays in Figure 7.15 appear to cause light to bend around the obstacle.

7.1.4.4. Summary of Particle Scattering

Particle scattering is the combination of the effects of reflection, refraction, and diffraction. When a wave approaches a spherical particle, such as a cloud drop, it can reflect off the particle, diffract around the edge of the particle, or refract into the particle. Once in the particle, the wave can be absorbed, transmitted through the particle and refracted out, or reflected internally one or more times and then refracted out. Figure 7.16 illustrates these processes, except for absorption, which is not a scattering process. The processes that affect particle scattering the most are diffraction and double refraction, identified by rays C and B, respectively.

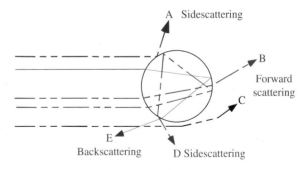

Figure 7.16. Radiative scattering by a sphere. Ray A is reflected; B is refracted twice; C is diffracted; D is refracted, internally reflected twice, and then refracted; and E is refracted, reflected once, and then refracted. Rays A–D scatter in the forward or sideward direction, whereas E scatters in the backward direction.

Figure 7.17. Forward and side scattering of sunlight by a cloud. The thickness of the cloud prevents most sunlight from transmitting through it. Photo by Mark Z. Jacobson.

Figure 7.18. Rainbow over an Alaskan lake in September 1992. Commander John Bortniak, NOAA Corps, available from the National Oceanic and Atmospheric Administration Central Library; www.photolib.noaa.gov/.

Thus, particles scatter light primarily in the forward direction. They also scatter some light to the side and backward. **Backscattered** light results primarily from a single internal reflection (ray E). The light rays seen in Figure 7.17 are the result of light scattering off cloud drops in the forward and sideward directions.

7.1.4.5. Rainbows

A rainbow results from two light-scattering processes, dispersive refraction and reflection, and can be seen only if the sun is at the viewer's back and raindrops are falling in front of the viewer. The seven most prominent colors in a rainbow are red, orange, yellow, green, blue, indigo, and violet. In a **primary rainbow**, red appears on the top and violet appears on the bottom. Figure 7.18 shows an example of a primary rainbow. In a **secondary rainbow**, sometimes seen faintly above a primary rainbow, violet appears on top and red appears on the bottom. For convenience, the discussion of rainbows below considers only red, green, and blue wavelengths.

Figure 7.19 shows how light interacts with raindrops to form a primary rainbow. As a beam of visible light enters a raindrop, all wavelengths bend toward the surface normal due to refraction. Blue light bends the most as a result of dispersive refraction. When light hits the back of the drop, much of it reflects internally. When the reflected light hits the front edge of the drop, it leaves the drop and refracts away from the surface normal. The angles of the blue and red wavelengths that reach a viewer's eye are 40 and 42 degrees, respectively, in relation to the incident beam. Only one wavelength from each raindrop impinges upon a viewer's eye. Thus,

a rainbow appears when individual waves from many raindrops hit the viewer's eye. As seen in Figure 7.19, red appears on the top of a primary rainbow. A secondary rainbow occurs if a second reflection occurs inside each raindrop.

Because winds at midlatitudes originate from the west or southwest (Figure 6.5) and a rainbow appears only when the sun is at a viewer's back, sailors at midlatitudes knew that if they saw a rainbow in the morning, the rainbow was to the west and the winds were driving the storm creating the rainbow toward them. If they saw a rainbow in the evening, the rainbow was to the east and the winds were driving the storm creating the rainbow away from them. These factors led to the following rhyme:

Rainbow in the morning, sailors take warning,
Rainbow at night, sailor's delight.

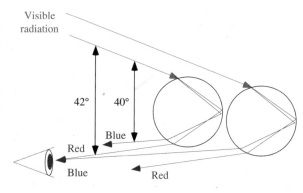

Figure 7.19. Geometry of a primary rainbow.

7.1.5. Particle Scattering and Absorption Extinction Coefficients

The quantification of particle scattering and absorption is more complex than is that of gas scattering or absorption due to the variety of sizes and compositions of aerosol particles. Aerosol particle absorption and scattering extinction coefficients (cm^{-1}) at a given wavelength can be estimated with

$$\sigma_{a,p} = \sum_{i=1}^{N_B} n_i \pi r_i^2 Q_{a,i} \qquad \sigma_{s,p} = \sum_{i=1}^{N_B} n_i \pi r_i^2 Q_{s,i} \qquad (7.7)$$

respectively, where N_B is the number of particle sizes; n_i is the number per cubic centimeter of air (number concentration) of particles of radius r_i (cm); πr_i^2 is the actual cross section of a particle, assuming it is spherical (cm^2 per particle); and $Q_{a,i}$ and $Q_{s,i}$ are single-particle absorption and scattering efficiencies (dimensionless), respectively, for each size i.

A **single-particle scattering efficiency** is the ratio of the effective scattering cross section of a particle to its actual cross section. The scattering efficiency can exceed unity because a portion of the radiation diffracting around a particle can be intercepted and scattered by the particle. Scattering efficiencies above unity account for this additional scattering.

A **single-particle absorption efficiency** is the ratio of the effective absorption cross section of a particle to its actual cross section. The absorption efficiency can exceed unity because a portion of the radiation diffracting around a particle can be intercepted and absorbed by the particle. Absorption efficiencies above unity account for this additional absorption. The larger the imaginary index of refraction of a particle, the greater its absorption efficiency.

Single-particle absorption and scattering efficiencies vary with particle size, radiation wavelength, and refractive indices. Figures 7.19 and 7.20 show Q_a and Q_s for black carbon and liquid water, respectively, at a wavelength of 0.5 μm. They also show the **single-particle forward scattering efficiency**, Q_f, which is the efficiency with which a particle scatters light in the forward direction. The forward scattering efficiency is always less than is the total scattering efficiency. The proximity of Q_f to Q_s in Figures 7.19 and 7.20 indicates that aerosol particles scatter strongly in the forward direction. The difference between Q_s and Q_f equals the scattering efficiency in the backward direction.

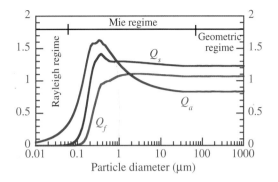

Figure 7.20. Single-particle absorption (Q_a), total scattering (Q_s), and forward scattering (Q_f) efficiencies of black carbon particles of different sizes at λ = 0.50 μm (n = 1.94, κ = 0.66).

When a particle's diameter (D) is much smaller than the wavelength of light (λ) (e.g., when (D/λ < 0.03)), the particle is in the **Rayleigh regime** and is called a **Tyndall absorber or scatterer. John Tyndall** (1820–1893) was an English experimental physicist who demonstrated experimentally that the sky's blue color results from scattering of visible light by gas molecules and that a similar effect occurs with small particles.

When a particle's diameter is near the wavelength of light (0.03 ≤ D/λ < 32), the particle is in the **Mie regime**. The German physicist, **Gustav Mie** (1868–1957), derived equations describing the scattering of radiation by particles in this regime (Mie, 1908). When a particle's diameter is much larger than the wavelength of light (D/λ ≥ 32), the particle is in the **geometric regime**. Figures 7.19 and 7.20 show the diameters corresponding to these regimes for a wavelength of 0.5 μm.

Figure 7.20 shows that visible light absorption efficiencies of black carbon particles peak when the particles are 0.2 to 0.4 μm in diameter. Figure 7.21 shows that water particles 0.3 to 2.0 μm in diameter scatter visible light more efficiently than do smaller or larger particles. Such particles in both figures are in the accumulation mode with respect to particle size and in the Mie regime with respect to the ratio of particle size to the wavelength of light.

Because the accumulation mode contains a relatively high particle number concentration, and because particles in this mode have high scattering efficiencies and, in the presence of black carbon, high absorption efficiencies, *the accumulation mode almost always causes more light reduction than do the nucleation or coarse particle modes* (Waggoner et al., 1981). In many urban

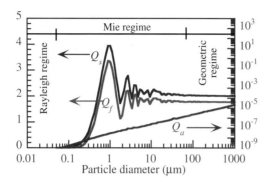

Figure 7.21. Single-particle absorption (Q_a), total scattering (Q_s), and forward scattering (Q_f) efficiencies of liquid water drops of different sizes at $\lambda = 0.50$ μm ($n = 1.335$, $\kappa = 1.0 \times 10^{-9}$).

regions, 20 to 50 percent of the accumulation mode mass is sulfate. *Thus, sulfate is correlated with particle scattering more closely than is any other particulate species, aside from liquid water.*

Figure 7.21 shows that liquid water hardly absorbs visible light until particles are the size of raindrops ($>1,000$ μm in diameter). The absorptivity of raindrops causes the bottoms of precipitating clouds to appear gray or black.

7.2. Visibility

Visibility is a measure of how far we can see through the air. Even in the cleanest air, our ability to see along the Earth's horizon is limited to a few hundred kilometers by background gases and aerosol particles. However, if we look up through the sky at night, we can discern light from stars that are millions of kilometers away. The reason that our visibility is much lower horizontally than vertically is that more gas molecules and aerosol particles lie in front of us in the horizontal than in the vertical.

Several terms describe maximum visibility. Two subjective terms are visual range and prevailing visibility. **Visual range** is the actual distance away from an ideal dark object at which a person can discern the object against the horizon sky. **Prevailing visibility** is the greatest visual range a person can see along 50 percent or more of the horizon circle (360 degrees), but not necessarily in continuous sectors around the circle. It is determined by a person who identifies landmarks known distances away in a full 360-degree circle around an observation point. The greatest visual range observed over 180 degrees or more of the circle (not necessarily in continuous sectors) is the prevailing visibility. Thus, half the area around an observation point may have visibility worse than the prevailing visibility, which is important because most prevailing visibility observations are made at airports. If the visual range in a sector is significantly different from the prevailing visibility, the observer at an airport usually denotes this information in the observation record.

Example 7.4
If the visual range around 25 percent (90 degrees) of the horizon circle is 5 km, that around another 25 percent is 10 km, that around another 25 percent is 15 km, and that around the remainder of the circle is 20 km, what is the prevailing visibility?

Solution
The prevailing visibility is 15 km because the viewer can see 15 km or more around at least 50 percent of the horizon circle, but not necessarily in consecutive sectors around the circle.

A less subjective and, now, a regulatory definition of visibility is the **meteorological range**, which can be explained in terms of the following example. Suppose a perfectly absorbing dark object lies against a white background at a point x_0, as shown in Figure 7.22. Because the object is perfectly absorbing, it reflects

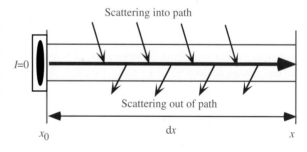

Figure 7.22. Change of radiation intensity along a beam. A radiation beam originating from a dark object has intensity $I = 0$ at point x_0. Over a distance dx, the beam's intensity increases due to scattering of background light into the beam. This added intensity is diminished somewhat by absorption along the beam and scattering out of the beam. At point x, the net intensity of the beam has increased close to that of the background intensity.

and emits no visible radiation; thus, its visible radiation intensity (I) at point x_0 is zero, and it appears black. As a viewer backs away from the object, background white light of intensity I_B scatters into the field of view, increasing the intensity of light in the viewer's line of sight. Although some of the added background light is scattered out of or absorbed along the field of view by gases and aerosol particles, at some distance away from the object, so much background light has entered the path between the viewer and the object that the viewer can barely discern the black object against the background light.

The meteorological range is a function of the **contrast ratio**, defined as

$$C_{ratio} = \frac{I_B - I}{I_B} \qquad (7.8)$$

The contrast ratio gives the difference between the background intensity and the intensity in the viewer's line of sight, all relative to the background intensity. If the contrast ratio is unity, the contrast is clear and an object is perfectly distinguishable from background light, whereas if it is zero, the object cannot be distinguished from background light.

The meteorological range is the distance from an object at which the contrast ratio equals the liminal contrast ratio of 0.02 (2 percent). The **liminal** or **threshold contrast ratio** is the lowest visually perceptible brightness contrast a person can see. It varies from individual to individual. Koschmieder (1924) selected a value of 0.02. Middleton (1952) tested 1,000 people and found a threshold contrast ratio range of between 0.01 and 0.20, with the mode of the sample between 0.02 and 0.03. Campbell and Maffel (1974) found a liminal contrast of 0.003 in laboratory studies of monocular vision. Nevertheless, 0.02 has become an accepted liminal contrast value for meteorological range calculations. In sum, the meteorological range is the distance from an ideal dark object at which the object has a 0.02 liminal contrast ratio against a white background.

The meteorological range can be derived from the equation for the change in object intensity along the path described in Figure 7.22. This equation is

$$\frac{dI}{dx} = \sigma_t(I_B - I) \qquad (7.9)$$

where all wavelength subscripts have been removed, σ_t is the **total extinction coefficient**, $\sigma_t I_B$ accounts for the scattering of background light radiation into the path, and $-\sigma_t I$ accounts for the attenuation of radiation

along the path due to scattering out of the path and absorption along the path. A total extinction coefficient is the sum of extinction coefficients due to scattering and absorption by gases and particles. Thus,

$$\sigma_t = \sigma_{a,g} + \sigma_{s,g} + \sigma_{a,p} + \sigma_{s,p} \qquad (7.10)$$

Integrating Equation 7.9 from $I = 0$ at point $x_0 = 0$ to I at point x with constant σ_t yields the equation for the contrast ratio:

$$C_{ratio} = \frac{I_B - I}{I} = e^{-\sigma_t x} \qquad (7.11)$$

When $C_{ratio} = 0.02$ at a wavelength of 0.55 μm, the resulting distance x is the meteorological range (also called the **Koschmieder equation**).

$$x = \frac{3.912}{\sigma_t} \qquad (7.12)$$

In polluted tropospheric air, the only important gas-phase visible light attenuation processes are Rayleigh scattering and absorption by $NO_2(g)$. Table 7.3 shows the meteorological ranges derived from calculated extinction coefficients, resulting from these two processes in isolation. For $NO_2(g)$, Table 7.3 shows values at two mixing ratios, representing clean and polluted air, respectively. For Rayleigh scattering, one meteorological range is shown because Rayleigh scattering is dominated by molecular nitrogen and oxygen, whose mixing ratios do not change much between clean and polluted air.

At a wavelength of 0.55 μm, the meteorological range due to Rayleigh scattering alone in Table 7.3 is 334 km, indicating that, *in the absence of all pollutants, the furthest one can see along the horizon, assuming a liminal contrast ratio of 2 percent, is near 350 km.* Waggoner et al. (1981) reported a total extinction coefficient

Table 7.3. Meteorological ranges (km) resulting from Rayleigh scattering and $NO_2(g)$ absorption at selected wavelengths (λ)

λ(μm)	Rayleigh scattering	$NO_2(g)$ absorption 0.01 ppmv $NO_2(g)$	0.25 ppmv $NO_2(g)$
0.42	112	296	11.8
0.50	227	641	25.6
0.55	334	1,590	63.6
0.65	664	13,000	520

Figure 7.23. Deer at Golden, Colorado, in front of Denver's brown cloud, December 14, 1998. Photo by David Parsons, available from National Renewable Energy Laboratory, U.S. Department of Energy, www.nrel.gov.

at Bryce Canyon, Utah, corresponding to a meteorological range of less than 400 km, suggesting that visibility was limited by gas scattering.

In Table 7.3, the meteorological range due to $NO_2(g)$ absorption decreases from 1,590 to 63.6 km when the $NO_2(g)$ mixing ratio increases from 0.01 to 0.25 ppmv. Thus, $NO_2(g)$ absorption reduces visibility more than does Rayleigh scattering when the $NO_2(g)$ mixing ratio is high. Results from a project studying **Denver's brown cloud** (Figure 7.23) showed that $NO_2(g)$ accounted for about 7.6 percent of the total reduction in visibility, averaged over all sampling periods, and 37 percent of the total reduction during periods of maximum $NO_2(g)$. Scattering and absorption by aerosol particles caused most remaining extinction (Groblicki et al., 1981). In sum, $NO_2(g)$ attenuates visibility in urban air when its mixing ratios are high.

Although the effects of Rayleigh scattering and $NO_2(g)$ absorption on visibility are nonnegligible in polluted air, they are less important than are aerosol particle scattering and absorption. Scattering by aerosol particles causes between 60 and 95 percent of visibility reductions in smog, whereas absorption by aerosol particles causes between 5 and 40 percent of such reductions (Cass, 1979; Tang et al., 1981; Waggoner et al., 1981).

Table 7.4 shows meteorological ranges derived from extinction coefficient measurements for a polluted and less polluted day in Los Angeles. Particle scattering dominated light extinction on both days. On the less polluted day, gas absorption, particle absorption, and

Table 7.4. Meteorological ranges (km) resulting from gas scattering, gas absorption, particle scattering, particle absorption, and all processes at wavelength of 0.55 μm on polluted and less polluted days in Los Angeles.

Day	Gas scattering	Gas absorption	Particle scattering	Particle absorption	All
Polluted (8/25/83)	366	130	9.6	49.7	7.42
Less polluted (4/7/83)	352	326	151	421	67.1

Meteorological ranges derived from extinction coefficients of Larson et al. (1984).

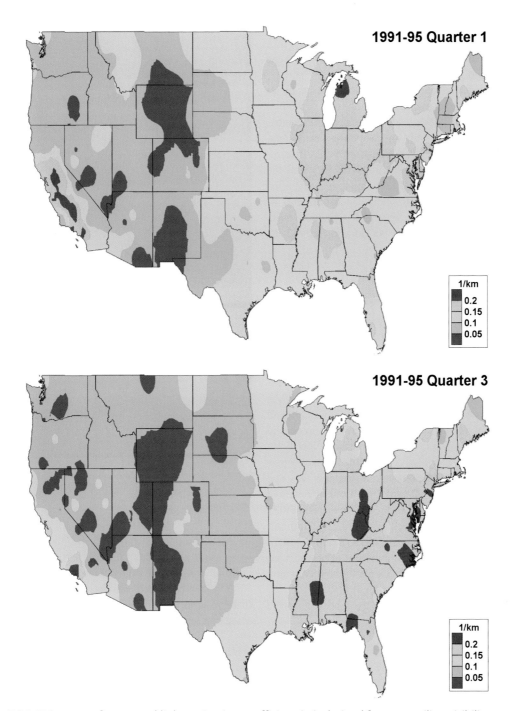

1991-95 Quarter 1

1991-95 Quarter 3

1/km	
▓	0.2
░	0.15
░	0.1
▓	0.05

Figure 7.24. U.S. maps of corrected light extinction coefficient (σ_t), derived from prevailing visibility measurements (V) with the equation $\sigma_t \approx 1.9/V$ for winter (quarter 1) and summer (quarter 3) 1991–1995 (Schichtel et al., 2001). Large extinction coefficients correspond to poor visibility. Data obtained during rain, snow, and fog were eliminated. Corrected extinction coefficients were obtained by applying relative humidity correction factors to adjust all extinction coefficients to a "dry" relative humidity of 60 percent, and then selecting the "dry" extinction coefficient at each monitoring site and for each season corresponding to the 75th percentile of all extinction coefficients at the site and for the season.

gas scattering all had similar small effects. On the polluted day, the most important visibility-reducing processes were particle scattering, particle absorption, gas absorption, and gas scattering, in that order.

Equation 7.12 relates a theoretical quantity, meteorological range, to a measured extinction coefficient. When prevailing visibility, a subjective quantity, is measured simultaneously with an extinction coefficient, they usually do not satisfy Equation 7.12. Instead, the relationship between prevailing visibility and the measured extinction coefficient must be obtained empirically. Griffing (1980) studied measurements of **prevailing visibility** (V, in km) and extinction coefficients (km^{-1}) over a 5-year period and derived the empirical relationship

$$V \approx \frac{1.9}{\sigma_t} \qquad (7.13)$$

Figure 7.24 shows a map of estimated extinction coefficients in the United States between 1991 and 1995 derived from prevailing visibility measurements substituted into Equation 7.13. During winters, visibility was poorest (extinction coefficients were highest) in central and Southern California, Illinois, Indiana, Iowa, Kentucky, and Michigan. During summers, visibility was poorest in Los Angeles and much of the midwestern and southern United States. Winter visibility loss in central California was likely due to a combination of low-lying winter inversions, high relative humidity, and heavy particle loadings. Visibility loss in Los Angeles was due to the presence of smog all year long. Poor visibility in the winter in the Midwest was due to power plant emissions combined with high relative humidity. Summer visibility degradation in the midwestern and southern United States was due to a combination of power plant emissions, organic particles from photochemical smog and vegetation, and high relative humidity.

7.3. Colors in the Atmosphere

Red sunsets and blue skies arise from selective scattering of visible light wavelengths by gas molecules. Rainbows arise from interactions of light with raindrops. Some additional optical phenomena are discussed next.

7.3.1. White Hazes and Clouds

When the relative humidity is high (but less than 100 percent), aerosol particles increase in size by absorbing liquid water. As their diameters approach the wavelength of visible light, the particles enter the Mie regime (Section 7.1.5) and scatter all wavelengths with equal

Figure 7.25. Haze over Los Angeles, May 1972. Gene Daniels, U.S. EPA, May 1972, Still Pictures Branch, U.S. National Archives.

intensity, producing a whitish haze. Hazes resulting from water growth on pollution particles are common in urban areas under sunny skies (Figure 7.25).

Cloud and fog drops, which contain more liquid water than do aerosol particles, appear white because they scatter all wavelengths of visible light with equal intensity. When pollution is heavy and the relative humidity is high, clouds and fogs are difficult to distinguish from hazes, as shown in Figure 7.26.

7.3.2. Reddish and Brown Colors in Smog

Reddish and brown colors in smog, such as those seen in Figure 7.27, are due to three factors. The first is preferential absorption of blue and some green light by $NO_2(g)$, which allows most green and red to be transmitted, giving smog a yellow, brown, or reddish-green color. Brown layers resulting from $NO_2(g)$ can be seen most frequently from 9 to 11 AM in polluted

Figure 7.26. Combined haze and fog over Los Angeles, near the San Gabriel Mountains, May 1972. Gene Daniels, U.S. EPA, May 1972, Still Pictures Branch, U.S. National Archives.

air because these are the hours that $NO_2(g)$ mixing ratios are highest. The second cause of reddish or brown colors is preferential absorption of blue light by nitrated aromatics, PAHs, and tar balls in aerosol particles. Figure 7.10 shows that a variety of nitrated aromatic compounds preferentially absorb not only UV, but also short visible wavelengths. PAHs and tar balls are also absorbers of short visible wavelengths. The blue light absorption and brown transmission by these compounds enhances the brownish color of smog. Third, when soil dust concentrations are high, air can appear red or brown because hematite in soil dust particles

Figure 7.27. Reddish and brown colors in Los Angeles smog on December 19, 2000. Photo by Mark Z. Jacobson.

Figure 7.28. Red sky in late afternoon over transmission lines near Salton Sea, California, May 1972. The smog causing the optical effect originated from Los Angeles. Charles O'Rear, U.S. EPA, May 1972, Still Pictures Branch, U.S. National Archives.

preferentially absorb blue and green light over red light. In the absence of a dust storm and away from desert regions, the concentration of soil dust particles is generally low, and the effect of soil dust on atmospheric optics is limited.

7.3.3. Black Colors in Smog

Black colors in smog (e.g., Figures 5.7 and 5.8) are due solely to absorption of solar radiation by black carbon, a component of soot. Soot is emitted primarily during coal, diesel and jet fuel, biofuel, and biomass burning. Black carbon absorbs all wavelengths of white light, transmitting none, thereby appearing black against the background sky.

7.3.4. Red Skies and Brilliant Horizons in Smog

Whereas the sun itself appears red at sunset and sunrise, and the horizon appears red during sunset and sunrise when aerosol particles are present (Figure 7.8), the sky can also appear red in the afternoon, and red horizons can become more brilliant in the presence of air pollution. As discussed in Section 7.1.2.2, red horizons occur because aerosol particles scatter the sun's white (blue, green, plus red) light, providing a source of white light to a viewer from all points along the horizon. Gas molecules then scatter the blue and green wavelengths out of the viewer's line of sight, allowing only the red to transmit from the horizon to the viewer. When heavy smog is present, particle concentrations near the surface and aloft increase sufficiently to allow this phenomenon to occur in the afternoon, as seen in Figure 7.28. Smog particles similarly enhance the brilliance of the red horizon at sunset, as seen in Figure 6.21.

7.3.5. Purple Glow in the Stratosphere

After a strong volcanic eruption, a purple glow may appear in the stratosphere, as shown in Figure 7.29. This glow results because volcanos emit sulfur dioxide gas, which converts to sulfuric acid–water particles, many of which reach the stratosphere. Such particles scatter light through the stratospheric ozone layer. Ozone weakly absorbs green and some red wavelengths (Figure 7.3), transmitting blue and some red, which combine to form purple. The purple light is scattered back to a viewer's eye by the enhanced stratospheric particle layer, causing the stratosphere to appear purple.

7.4. Summary

In this chapter, processes that affect radiation were discussed with an emphasis on the effects of gases and

Figure 7.29. Purple sunset observed from Palos Verdes, California, following the 1982 El Chichon volcanic eruption. Photo by Jeffrey Lew, UCLA.

aerosol particles on visibility degradation and UV light reduction. Processes affecting radiation through the air include absorption and scattering. Gas absorption of UV radiation is responsible for the stratospheric inversion and most photolysis reactions of gases. Gas absorption of visible light is important only when $NO_2(g)$ mixing ratios are high. Gas (Rayleigh) scattering affects only UV and visible wavelengths. The most important gas scatterers are molecular oxygen and molecular nitrogen. All aerosol particles absorb near- and thermal-IR radiation, but only a few, including black carbon and certain soil components, absorb visible (including UV) radiation. Several organic aerosol components, particularly nitrated and aromatic compounds, preferentially absorb UV radiation, causing UV and ozone reductions in polluted air. All aerosol components scatter radiation. The most important visibility-degrading processes in polluted air are particle scattering, particle absorption, gas absorption, and gas scattering, in that order. In clean air, visibility is limited in the horizontal to a few hundred kilometers by gas scattering.

7.5. Problems

7.1. Calculate the fractional transmission of radiation through a swimming pool 2.0 m deep at a wavelength of 0.5 μm when only absorption of light by liquid water is considered. At what water depth do you expect the incident light to be reduced by 99 percent due to absorption?

7.2. If the angle of incidence, relative to the surface normal, of 0.5-μm-wavelength light entering water from the air is 30 degrees, determine the angle of refraction in the water.

7.3. If 0.5-μm-wavelength light travels from water to air, what angle of incidence (with respect to the surface normal in water) is necessary for the angle of refraction to be 90 degrees? This angle of incidence is referred to as the critical angle. What happens if the angle of incidence in water exceeds the critical angle?

7.4. If gas molecules in the atmosphere preferentially scattered longer wavelengths over shorter wavelengths, what colors would you expect the sky and the sun to be at noon, afternoon, and sunset?

7.5. Explain why nitrogen dioxide gas causes more visibility reduction than does ozone gas.

7.6. Calculate the meteorological range at 0.55 μm resulting from aerosol particle (a) scattering alone; (b) absorption alone; and (c) absorption plus scattering, assuming the atmosphere contains 10^4 particles cm^{-3} of 0.1-μm-diameter spherical particles made of black carbon. Use Figure 7.20 to obtain single-particle absorption and scattering efficiencies, and assume efficiencies at 0.5 μm are similar to those at 0.55 μm.

7.7. If a person's liminal contrast ratio is 0.01 instead of 0.02, what would be the expression for their meteorological range? For a given total extinction coefficient, how much further (in percent) can the person see when their liminal contrast ratio is 0.01 instead of 0.02?

7.8. Assume that the atmosphere contains 10,000 μg m^{-3} of liquid water and the density of liquid water is 1.0 g cm^{-3}. For which size spherical particles is visibility reduced the most at a wavelength of 0.55 μm: (a) 0.5-μm-diameter particles, (b) 10-μm-diameter particles, or (c) 100-μm-diameter particles? Assume single-particle scattering efficiencies in Figure 7.21 can be used for wavelengths of 0.55 μm. Show calculations.

7.9. Explain why soil often appears brown. What is the difference between soil that appears brown and soil that appears red?

7.10. If the visual range around 37 percent of the horizon circle is 8 km, that around 15 percent of the circle is 3 km, and that around the remaining 48 percent of the circle is 12 km, what is the prevailing visibility?

7.11. What color would you see during the day due to the following constituents if they were in the air in high concentrations: (a) soot particles, (b) nitrogen dioxide gas, (c) haze particles containing mostly water, and (d) ozone gas? Explain the process causing the color (e.g., absorption of blue light, scattering of green and red light).

7.6. Visibility Project

Find an elevated location outside where, on a clear day, it is possible to see a long distance over at least 75 percent (270 degrees) along the horizon circle without obstruction by buildings or mountains. Predetermine the distances to a set of landmarks that cover at least 50 percent (180 degrees) of the horizon circle, but not necessarily in continuous sectors around the circle. At two times during a single day (a morning and an afternoon), (a) estimate the prevailing visibility and (b) make a list of the colors you see in the sky along the horizon and where you see them. Discuss your findings in terms of differences between times of day. Comment on how the relative humidity, wind, and large-scale pressure system may have affected results if that information is available. Which of these factors do you believe had the greatest effect on visibility?

Chapter 8

International Regulation of Urban Smog Since the 1940s

Until the 1940s, efforts to control air pollution in the United States were limited to a few municipal ordinances and state laws regulating smoke (Chapter 4). In the UK, regulations were limited to the British Alkali Act of 1863 and some relatively weak public health bills designed to abate smoke. Regulations in other countries were similarly weak or nonexistent. The main reason for the lack of regulation in polluted cities was that the coal, oil, chemical, and auto industries – the ultimate sources of much of the pollution – had political power and used it to resist efforts of government intervention (e.g., Section 4.1). Because the long-term health effects of pollutants on outdoor concentrations were not well known at the time, it was also difficult for public health agencies to recommend the banning of a pollutant, particularly in the face of political pressure from industry and arguments that such a ban would hurt economic growth (e.g., Midgley's defense of tetraethyl lead in Section 3.6.9). From the late 1940s to the mid-1950s, damage due to photochemical smog in many U.S. cities was so significant that the federal government decided to take steps to address the problem. Similarly, deadly London-type smog events in the UK spurred government legislation in the 1950s. Today, nearly all countries have instituted air pollution regulations of some kind or another. Nevertheless, regulations in most countries are still weak or unenforced, resulting in the persistence of devastating air pollution worldwide. This chapter discusses regulation of outdoor pollution and air quality trends in several countries since the 1950s.

International regulations involving transboundary air pollution are discussed in Section 10.7.

8.1. Regulation in the United States

In the 1940s, photochemical smog in Los Angeles became a cause for concern and led to the formation of the Los Angeles Air Pollution Control District in 1947. In 1951, the state of Oregon established an agency to oversee and regulate air pollution. Other states followed, so that by 1960, seventeen statewide air pollution agencies existed. In 1948, a heavy air pollution episode in Donora, Pennsylvania, killed 20 people, and in 1948, 1952, and 1956, air pollution episodes in London killed nearly 5,000. Pollution in Los Angeles reached its peak in severity in the mid-1950s, when ozone levels as high as 0.68 ppmv were recorded. In 1952, Arie Haagen-Smit isolated the mechanism of ozone formation in photochemical smog. The combination of public outcry in the United States about the effects of air pollution and a better understanding of its causes contributed to the first federal legislation concerning air pollution in 1955.

8.1.1. Air Pollution Control Act of 1955

Because of the accelerating health and quality-of-life problems associated with air pollution, President Dwight D. Eisenhower asked the U.S. Congress to consider legislation addressing the issue in 1955. Until this time, state and local governments had received no federal guidance in combating air pollution. On

July 14, 1955, the U.S. Congress passed the first of several major air pollution–related federal regulations, the **Air Pollution Control Act of 1955** (Public Law (PL) 84-159).

The Air Pollution Control Act of 1955 granted $3 million per year for 5 years to the U.S. Public Health Service (PHS), Department of Health, Education, and Welfare, to study air pollution. The act directed the PHS to provide technical assistance to states for combating air pollution, train individuals in air pollution, and perform more research on air pollution control.

In 1959, the act was extended by 4 years, with funding of $5 million per year. On June 6, 1960, it was amended (PL 86-493) to authorize the U.S. Surgeon General to study the health effects of automobile exhaust. It was amended again on October 9, 1962 (PL 87-761). Although the 1955 law raised air pollution issues to a federal level, it did not impose any federal regulations on air pollution and delegated air pollution control and prevention to state and local levels.

8.1.2. California Vehicle Emission Regulations

In the 1950s, new cars typically emitted about 13 grams per mile (g/mi) of total hydrocarbon (THC) gases, 87 g/mi of carbon monoxide [$CO(g)$], and 3.6 g/mi of nitrogen oxides [$NO_x(g)$]. Because of the severity of air pollution resulting from vehicle exhaust, the California state legislature in 1959 created the **Motor Vehicle Pollution Control Board**, which set the first automobile emission standards in the world. The first rule implemented by the board was to reduce crankcase emissions of unburned hydrocarbons, beginning with 1963 model cars sold in California. The crankcase is the housing of the crankshaft, which turns the up-and-down motion of pistons into rotation. At the time, crankcase emissions were responsible for about 20 percent of THC emissions from automobiles. Fuel tank evaporation (9 percent), carburetor evaporation (9 percent), and tailpipe exhaust (62 percent) accounted for the remainder of the THC emissions. Per the new regulations, new cars were required to reroute crankcase THC emissions back to the intake manifold, which supplies the air/fuel mixture to the pistons, to be reburned instead of emitted into the air. The first emission standard did not control emissions of $CO(g)$ or $NO_x(g)$.

8.1.3. Clean Air Act of 1963

By 1963, public awareness and concern about air pollution and its impacts had grown significantly. Aware-

ness was enhanced further by the 1962 publication of *Silent Spring* by **Rachel Carson**, which brought forth scientific research outlining the relationship between aerial spraying of the pesticide dichlorodiphenyltrichloroethane (**DDT**) and bird deaths, as well as other data on air and water pollution effects.

Due primarily to the concern and visual impact of air pollution in cities, the U.S. Congress enacted the **Clean Air Act of 1963** (CAA63, PL 88-206) in December 1963. The purpose of this act was

> to improve, strengthen, and accelerate programs for the prevention and abatement of air pollution.

CAA63 contained provisions giving the federal government authority to reduce interstate air pollution. Such reductions would be obtained by specifying emission standards for stationary pollution sources, including power plants and steel mills, and by encouraging the use of technologies to remove sulfur from coal and oil. CAA63 did not specify controls for automobiles, the most serious source of air pollution. It did, however, initiate a process for reviewing the status of automobile emissions by setting up a technical committee of representatives from the Department of Health, Education, and Welfare, as well as the automobile, control device, and oil industries.

8.1.4. Motor Vehicle Air Pollution Control Act of 1965

Investigations by a technical committee and subsequent hearings in 1964 by the Senate Public Works Subcommittee on Air and Water Pollution led to the first amendment to CAA63, the **Motor Vehicle Air Pollution Control Act of 1965** (PL 89-272, October 20, 1965). With this act, the U.S. government set its first federal automobile emission standards for THCs and $CO(g)$. Table 8.1 lists federal light-duty vehicle emission standards since then. The first standards were applicable to 1968 model cars and were patterned after California state standards developed for 1966 model cars sold in California. The federal standards were intended to reduce emissions to 72 percent of their 1963 values for tailpipe THCs, 56 percent for tailpipe $CO(g)$, and 100 percent for crankcase THCs. Despite the good intentions of the act, more than half of all 1968 and 1969 model cars failed to meet the new emission standards. In 1966, Congress passed a second amendment to the Clean Air Act of 1963 to expand local air pollution control programs.

Table 8.1. U.S. federal and California/Northeast states full-life (10 years/100,000–120,000 miles) emission standards for passenger cars since 1968

Year	THC(g) (g/mi)	NMOG(g) (g/mi)	CO(g) (g/mi)	NO$_x$(g) (g/mi)	HCHO(g) (g/mi)	Pb(s) (g/gal)	PM (g/mi)
Federal Prior to Tier I							
1968–1970	3.2	–	33	–	–	–	–
1971–1972	4.6	–	47	4.0	–	–	–
1972	3.4	–	39	–	–	–	–
1973–1974	3.4	–	39	3.0	–	–	–
1975–1976	1.5	–	15	3.1	–	–	–
1977–1979	1.5	–	15	2.0	–	0.8	–
1980	0.41	–	7.0	2.0	–	0.5	–
1981	0.41	–	3.4	1.0	–	0.5	–
1982–1986	0.41	–	3.4	1.0	–	0.5	0.6
1987–1993 (Tier 0)	0.41	–	3.4	1.0	–	0.5	0.2
Federal Tier I Program (1994–2003)							
1994–2003, gas	0.31	–	4.2	0.6	–	0.5	0.10
1994–2003, diesel	0.31	–	4.2	1.25	–	0.5	0.10
Federal Tier II Program (2004–Present)							
2004–2008, Bin 10a	–	0.156	4.2	0.6	0.018	0.0	0.08
2004 onward, Bin 8a	–	0.125	4.2	0.20	0.018	0.0	0.02
2004 onward, Bin 5		0.09	4.2	0.07	0.018	0.0	0.01
2004 onward, Bin 3	–	0.055	2.1	0.03	0.011	0.0	0.01
2004 onward, Bin 1	–	0.0	0.0	0.0	0.0	0.0	0.0
California and Northeast LEV I Program (2001–2006)							
LEV I, gas	–	0.09	4.2	0.3	0.018	0.0	–
LEV I, diesel	–	0.09	4.2	0.3	0.018	0.0	0.08
ULEV I, gas	–	0.055	2.1	0.3	0.011	0.0	–
ULEV I, diesel	–	0.055	2.1	0.3	0.011	0.0	0.04
California and Northeast LEV II Program (2004–Present)							
LEV II	–	0.09	4.2	0.07	0.018	0.0	0.01
ULEV II	–	0.055	2.1	0.07	0.011	0.0	0.01
SULEV II	–	0.01	1.0	0.02	0.004	0.0	0.01
PZEV	–	0.01	1.0	0.02	0.004	0.0	0.01
ZEV	–	0.0	0.0	0.0	0.0	0.0	0.0

Bin, category of emissions. Eleven bins are defined (four are shown here). Manufacturers can make vehicles fit into any bin as long as a fleet average of Bin 5 emissions is met.

LEV I, California's first "low emission vehicle" program.

LEV II, replaced the LEV I program with tighter emission standards.

LEV, low emission vehicle. LEV I and II vehicles correspond to the overall LEV I or II programs, respectively. LEV II vehicle standards correspond to Tier II, Bin 5, federal standards.

NMOG, nonmethane organic gas.

PZEV, partial zero emission vehicle. PZEV standards are similar to those for SULEV II, but must meet standards for 15 yr/150,000 mi, plus have a fully sealed zero emission fuel system.

SULEV, super ultra low emission vehicle.

THC(g), total hydrocarbon in the gas phase (all hydrocarbons, including methane).

Tier I, U.S. EPA's first emission standard program after 1994.

Tier II, replaced the Tier I program with tighter emission standards, including limits of sulfur in fuel.

TLEV, transitional low emission vehicle.

ULEV, ultra low emission vehicle.

ZEV, zero emission vehicle.

Source: U.S. EPA (2007).

8.1.5. Air Quality Act of 1967

On November 21, 1967, the U.S. Congress passed a third amendment to the Clean Air Act of 1963, the **Air Quality Act of 1967** (PL 90-148). This act divided the United States into several inter- or intrastate **Air Quality Control Regions** (AQCRs). Officials in each AQCR conducted studies to determine the extent of air pollution in the region, collect ambient air quality data, and develop emission inventories. The act also specified that the Department of Health, Education, and Welfare develop and publish **Air Quality Criteria** (AQC) reports, which were science-based reports containing information about the effects, as a function of concentration, of pollutants that damage human health and welfare. The reports also contained suggestions about acceptable levels of pollution. Each state was then required to set and enforce its own air quality standards based on the acceptable levels suggested in the AQC reports. The standards had to be at least as stringent as those suggested in the reports. Each state was required to submit a **State Implementation Plan** (SIP) to the federal government discussing how the state intended to meet its air quality standards. A SIP consisted of a list of regulations that the state would implement to clean a polluted region. If the SIP was not approved by the federal government or if the state did not enforce its regulations, the federal government had the authority to bring suit against the state. If the SIP was approved, the state was delegated federal authority to regulate air pollution in the state.

The act also recommended the publication of air pollution control methods through **control technology documents**, provided funds to states for motor vehicle inspection programs, and allowed California to set its own automobile emission standards. In 1969, the CAA63 was amended again to authorize research on low emission fuels and automobiles.

8.1.6. Clean Air Act Amendments of 1970

In late 1970, President Richard Nixon combined several existing federal air pollution programs to form the **U.S. Environmental Protection Agency** (U.S. EPA), whose purpose was to enforce federal air pollution regulations. On December 31, 1970, Congress passed the **Clean Air Act Amendments of 1970** (CAAA70, PL 91-604). The purpose of CAAA70 was

> to amend the Clean Air Act to provide for a more effective program to improve the quality of the Nation's air,

and it resulted in the transfer of administrative duties for air pollution regulation from the Department of Health, Education, and Welfare to the U.S. EPA. CAAA70 specified that the U.S. EPA design **National Ambient Air Quality Standards (NAAQS) for criteria air pollutants**, so called because their permissible levels were based on health-based guidelines, or criteria, obtained from AQC reports.

NAAQS were divided into **primary standards**, designed to protect the public health (in particular, that of people most susceptible to respiratory problems, such as asthmatics, the elderly, and infants), and **secondary standards**, designed to protect the public welfare (visibility, buildings, statues, crops, vegetation, water, animals, transportation, other economic assets, and personal comfort and well-being). The U.S. EPA was required to set primary standards based on health considerations alone, not on the cost of or technology available for attaining the standard. Regions in which primary standards for criteria pollutants were met were called **attainment areas**, and those in which primary standards were not met were called **nonattainment areas**.

The six original criteria pollutants specified by CAAA70 were $CO(g)$, $NO_2(g)$, $SO_2(g)$, total suspended particulates (TSPs), THCs, and photochemical oxidants. Particulate lead [Pb(s)] was added to the list in 1976, ozone [$O_3(g)$] replaced photochemical oxidants in 1979, THCs were removed from the list in 1983, and TSPs were changed to include only particulates with diameter ≤ 10 μm and called PM_{10}, and a $PM_{2.5}$ standard was added in 1997. PM_{10} and $PM_{2.5}$ are, more precisely, the concentration of aerosol particles that pass through a size-selective inlet with a 50-percent efficiency cutoff at 10- and 2.5-μm aerodynamic diameter, respectively. Table 8.2 lists criteria pollutants for which NAAQS primary and, in most cases, secondary standards have been set. Secondary standards are the same as primary standards for most criteria air pollutants. One exception is $CO(g)$, for which no secondary standard has been set because its major impact is on human health. A second exception is $SO_2(g)$, for which a separate secondary standard has been set. Table 8.2 also shows California state standards. For all pollutants, California state standards are stricter than are NAAQS.

CAAA70 further specified that the U.S. EPA design **New Source Performance Standards** (NSPS), which were emission limits for new stationary sources of pollution. Each state was required to inspect new stationary sources and certify that pollution controls indeed worked and would remain working for the

Table 8.2. California ambient air quality standards (CAAQS) and U.S. federal primary and secondary national ambient air quality standards (NAAQS), as of 2011

Pollutant	California standard (CAAQS)	Federal primary standard (NAAQS)	Federal secondary standard (NAAQS)
Ozone [$O_3(g)$][a]			
1-hour average	90 ppbv[b]	120 ppbv[b]	Same as primary
8-hour average	70 ppbv[c]	75 ppbv[c]	Same as primary
Carbon Monoxide [$CO(g)$][a]			
8-hour average	9 ppmv	9 ppmv	None
1-hour average	20 ppmv	35 ppmv	None
Nitrogen Dioxide [$NO_2(g)$][a]			
Annual arithmetic average	30 ppbv	53 ppbv	Same as primary
1-hour average	180 ppbv	100 ppbv	None
Sulfur Dioxide [$SO_2(g)$][a]			
Annual average	None	0.03 ppmv	None
24 hours	0.04 ppmv	0.14 ppmv	None
3 hours	None	–	0.5 ppmv
1 hour	0.25 ppmv	0.075 ppmv	None
Particulate Matter $\leq$10 μm in Diameter (PM_{10})[a]			
Annual arithmetic average	20 μg m^{-3}	None	None
24-hour average	50 μg m^{-3}	150 μg m^{-3d}	Same as primary
Particulate Matter $\leq$2.5 μm in Diameter ($PM_{2.5}$)[a]			
Annual arithmetic average	12 μg m^{-3}	15 μg m^{-3e}	Same as primary
24-hour average	None	35 μg m^{-3f}	Same as primary
Lead [$Pb(s)$][a]			
30-day average	1.5 μg m^{-3}	None	None
Rolling 3-month average	None	0.15 μg m^{-3}	Same as primary
Quarterly average	None	1.5 μg m^{-3}	Same as primary
Particulate Sulfates			
24-hour average	25 μg m^{-3}	None	None
Hydrogen Sulfide [$H_2S(g)$]			
1-hour average	0.03 ppmv	None	None
Vinyl Chloride			
24-hour average	0.01 ppmv	None	None

[a] Criteria air pollutants.
[b] Standard exceeded if the daily maximum 1-hour average concentration exceeds given value more than once per year, averaged over 3 consecutive years. The California 1-hour standard applies in California. The federal standard applies only to areas designated as nonattainment areas prior to the 1997 Clean Air Act revision.
[c] Standard exceeded if the 3-year average of the fourth highest daily maximum 8-hour average ozone mixing ratio exceeds the given value.
[d] Standard exceeded if the 99th percentile of the distribution of the 24-hour concentrations for a period of 1 year, averaged over 3 years, exceeds 150 μg m^{-3} at each monitor within an area.
[e] Standard exceeded if the 3-year average of the annual arithmetic mean of 24-hour measured concentrations exceeds 15 μg m^{-3}.
[f] Standard exceeded if the 98th percentile of the distribution of 24-hour concentrations measured for 1 year, averaged over 3 years, exceeds 35 μg m^{-3} at each monitor within an area.

lifetime of the source. CAAA70 also required **National Emission Standards for Hazardous Air Pollutants (NESHAPS)**. Hazardous pollutants were defined as pollutants

> to which no ambient air standard is applicable and that . . . causes, or contributes to air pollution which may be anticipated to result in an increase in mortality or an increase in serious irreversible, or incapacitating reversible illness.

In 1973, the list of hazardous pollutants included only asbestos, beryllium, and mercury. By 1984, the list had been expanded to include benzene, arsenic, coke oven emissions, vinyl chloride, and radionuclides. CAAA70 also specified that

> . . . each state shall have the primary responsibility for assuring air quality within the entire geographic area comprising each state by submitting an implementation plan for such state which shall specify the manner in which national primary and secondary ambient air quality standards will be achieved and maintained within each air quality control region in each state.

Thus, the use of SIPs, which originated with the Air Quality Control Act of 1967, continued under CAAA70. CAAA70 required that SIPs address primary and secondary standards. Through a SIP, each state was required to set ambient air quality standards at least as stringent as federal standards, evaluate air quality in each AQCR within the state, and establish methods and timetables for improving air quality in each AQCR to meet state standards. The SIP was required to address approval procedures for new pollution sources and methods of reducing pollution from existing sources. Once submitted, a SIP required U.S. EPA approval; otherwise, the U.S. EPA had the power to take control of the state's air pollution program.

CAAA70 further required that the U.S. EPA develop aircraft emission standards; expand the number of AQCRs; and establish tough fines and criminal penalties for violations of SIPs, emission standards, and performance standards. It also permitted citizen's suits

> against any person, including the United States, alleged to be in violation of emission standards or an order issued by the administrator.

The national aircraft emission standards set by CAAA70 were enacted after California set a state standard in 1969.

CAAA70 required that new automobiles emit 90 percent less THCs and CO(g) in 1975 than in 1970 and 90 percent less $NO_x(g)$ in 1976 than in 1971. Thus,

Congress, and not the U.S. EPA, set the initial automobile emission standards, but the U.S. EPA was authorized to extend deadlines for autoemission reductions. The U.S. EPA extended the deadline in 1975 for all reductions by 1 year in 1973, by a second year in 1974, and by a third year for THCs and CO(g).

CAAA70 also gave the U.S. EPA authority to set and revise

> . . . standards applicable to the emission of any air pollutant from any class or classes of new motor vehicles or new motor vehicle engines, which in his judgment cause, or contribute to, air pollution which may reasonably be anticipated to endanger public health or welfare. [Section 202(a)(1)]

Thus, the U.S. EPA could regulate additional pollutants from automobile exhaust found to affect health or welfare and could revise existing emission standards from motor vehicles. *This clause would serve as the basis for regulating greenhouse gas emissions from vehicles* almost 40 years later, as discussed in Section 8.1.12.

8.1.7. Catalytic Converters

In direct response to automobile emission regulations instigated by CAAA70, the **catalytic converter**, an important automobile emission control technology, was developed by 1975. Automobile engines produce incompletely combusted $NO_x(g)$, CO(g), and THCs, as well as completely combusted $CO_2(g)$ and $H_2O(g)$. Catalytic converters convert $NO_x(g)$ to $N_2(g)$, CO(g) to $CO_2(g)$, and unreacted THCs to $CO_2(g)$ and $H_2O(g)$. Since 1975, three major types of catalytic converters have been developed: (1) single-bed converters, (2) dual-bed converters, and (3) single-bed three-way converters.

A catalyst is a substance that causes or accelerates a chemical reaction without affecting the substance. The catalysts in catalytic converters are generally the noble metals **platinum** (Pt), **palladium** (Pd), **rhodium** (Rh), **ruthenium** (Ru), **gold** (Au), or a combination of these. These catalysts are applied as a coating over porous alumina spherical pellets, ceramic honeycombs, or metallic honeycombs within the converter to increase the surface area contacting the exhaust. Exhaust gases travel through the converter with a residence time of about 50 milliseconds. The use of noble metal catalysts requires the simultaneous use of unleaded fuel because lead deactivates the catalyst through chemical reaction. Thus, *the catalytic converters forced the emission reduction of another pollutant, lead.*

The first converters were single-bed converters that used Pt and Pd in the ratio 2:1 as the catalyst. In this converter, called an **oxidation converter**, CO(g) and unreacted THCs were burned (oxidized) over the catalyst at temperatures of 250°C to 600°C to produce $CO_2(g)$ and $H_2O(g)$. Single-bed converters did not control $NO_x(g)$.

A second bed (**reduction converter**) was developed to reduce $NO_x(g)$ to $N_2(g)$ and $O_2(g)$. The catalysts in this bed are typically Rh, Ru, Pt, and/or Pd. When NO(g) or $NO_2(g)$ impinge upon the metal catalyst, the catalyst briefly bonds to the nitrogen atom and shears off the oxygen atom(s), allowing the oxygen to form $O_2(g)$. When a second nitrogen atom bonds to the catalyst adjacent to the first, the two combine to form $N_2(g)$, which is released back to the exhaust stream.

The three-way catalyst, developed in 1979, allowed for the simultaneous oxidation of unreacted THCs and CO(g) and reduction of $NO_x(g)$ in a single bed. The use of this catalyst requires a specific input air-to-fuel ratio of 14.8:1 to 14.9:1 and a temperature range of 350°C to 600°C for it to remain effective in converting all three groups of compounds. At high air-to-fuel ratios, CO(g) and THCs are converted efficiently, but $NO_x(g)$ is not. At low ratios, the reverse is true. At temperatures less than 350°C, conversion efficiency falls off fast. At 250°C, it is near zero. The catalysts in the three-way structures are usually platinum and rhodium at a ratio of Pt:Rh = 5:1.

8.1.8. Corporate Average Fuel Economy Standards

In response to the 1973/1974 oil embargo, which reduced the supply and increased the cost of oil-derived fuels, the U.S. Congress passed the Energy Policy Conservation Act of 1975. This act gave power to the Administrator of the National Highway Transportation and Safety Administration to set **Corporate Average Fuel Economy (CAFE)** standards for cars and light trucks (trucks and vans lighter than 3,900 kg and sport utility vehicles lighter than 4,500 kg). Under this law, automobile manufacturers pay fines if the average fuel economy of the fleet of vehicles they sell is less than the CAFE standard set. The U.S. EPA measures vehicle fuel economy to determine compliance under the act.

In 1978, the CAFE standard was first set for passenger vehicles at 18 miles per gallon (mpg). It fluctuated but gradually increased to 27.5 mpg by 1990, where it stayed through 2010. In 2011, the standard was increased to 30.2 mpg. The first light truck standard was 17.2 mpg, set in 1979. It gradually increased to 20.7 mpg, where it stayed until 2004 before gradually increasing again to 24.1 mpg in 2011.

CAFE standards do not directly limit the emissions of smog-forming pollutants, but they do indirectly reduce such emissions. CAFE standards set a lower limit for average vehicle mileage, reducing fuel use. Because carbon dioxide and smog-forming pollutant emissions are proportional to fuel use, the standards indirectly reduce emissions of such pollutants.

8.1.9. Clean Air Act Amendments of 1977

On August 7, 1977, Congress passed the **Clean Air Act Amendments of 1977** (CAAA77, PL 95-95), which extended the date for mandated automobile emission reductions of THCs and CO(g) to 1980. The $NO_x(g)$ emission standard was relaxed from 0.4 g/mi to 1.0 g/mi, and the deadline for compliance extended to 1981. A 0.8 grams per gallon (g/gal) standard was also introduced for lead. The lead standard was tightened in 1980 to 0.5 g/gal. In 1980, the U.S. EPA also set limits on diesel fuel particulate emissions and required CO(g) emissions from heavy-duty trucks to be reduced by 90 percent by 1984.

In 1977, most U.S. states had nonattainment areas where at least one NAAQS had not been achieved. CAAA77 required states that had at least one nonattainment area to describe, in a revised SIP, how they would achieve attainment by December 31, 1982.

CAAA77 also formalized a permitting program, initiated by the U.S. EPA in 1974, to **prevent significant deterioration (PSD)** of air quality in regions that were already in attainment of NAAQS. Under the program, Class I, II, and III regions were designated. Class I regions included pristine areas, such as national and international parks and national wilderness areas, where no new sources of pollution were allowed. Class II regions comprised areas where moderate changes in air quality were allowed, but where stringent regulations were desired. Class III regions involved areas where major growth and industrialization were allowed as long as pollutant levels did not exceed NAAQS. Before a new pollution source can be built or an existing source can be modified to increase pollution in a PSD region that allows growth, a PSD permit must be obtained. To obtain a permit, the polluter proposing the new source or change must ensure that the **best available control technology (BACT)** will be installed and the resulting pollution will not lead to a violation of an NAAQS. A BACT is a pollution control technology that results in the removal of the greatest amount of emissions from a particular industry or process.

CAAA77 also mandated that computer modeling be performed to check whether each proposed new source of pollution could result in an exceedance of emission limits or in a violation of an NAAQS. CAAA77 contained the first regulations in which the U.S. government attempted to control the emissions of CFCs, precursors to the destruction of stratospheric ozone.

8.1.10. Clean Air Act Amendments of 1990

The overall financial **benefit-to-cost ratio of the Clean Air Act Amendments** between 1970 and 1990 has been estimated as 4:1 (U.S. EPA, 2010). The benefits have been in the form of reduced damage to human health, agricultural crops, timber yields, aquatic ecosystems, and coastal estuaries due to reduced particulate matter, ozone, acid deposition, nitrogen deposition, ultraviolet radiation exposure, and stratospheric ozone loss.

Despite air quality and financial benefits of the CAAA70 regulations, problems related to urban air pollution, air toxics, acid deposition, and stratospheric ozone reduction persisted through the 1980s. For example, in 1990, 96 U.S. cities were still in violation of the NAAQS for ozone, 41 were in violation of the NAAQS for carbon monoxide, and 70 were in violation of the NAAQS for PM_{10}. Only seven air toxics had been regulated with NESHAPS between 1970 and 1990, although many more had been identified.

In response to these continued problems, the U.S. Congress passed the **Clean Air Act Amendments of 1990** (CAAA90, PL 101-549) on November 15, 1990. This was described as an act

> to amend the Clean Air Act to provide for attainment and maintenance of health protective national ambient air quality standards, and for other purposes.

CAAA90 was similar to CAAA70 with respect to NAAQS, NSPS, and PSD standards. However, some major changes were enacted. In particular, nonattainment areas for $O_3(g)$, $CO(g)$, and PM_{10} were divided into six classifications, depending on the severity of nonattainment, and each district in which nonattainment occurred was given a different deadline for reaching attainment. For ozone, only Los Angeles was designated as "extreme" and given until 2010 to reach attainment of the NAAQS. Baltimore and New York City were designated as "severe" and given until 2007. Chicago, Houston, Milwaukee, Muskegon, Philadelphia, and San Diego were also designated as "severe,"

although slightly less so, and given until 2005 to reach attainment.

For attainment to be achieved, all new pollution sources in nonattainment areas, regardless of size, were required to obtain their **lowest achievable emissions rate (LAER)**, which is the lowest emissions rate achieved for a specific pollutant by a similar source in any region. LAERs were required to be less stringent than NSPSs for the source. To achieve LAERs, states or ACQRs were required to adopt **reasonably achievable control technologies (RACTs)** for all existing major emission sources. RACTs are control technologies that are reasonably available and technologically and economically feasible. They are usually applied to existing sources in nonattainment areas.

Per CAAA90, state or local air quality districts overseeing a nonattainment area were required to develop emission inventories for ROGs, $NO_x(g)$, and $CO(g)$. Emissions from mobile, stationary point, area, and biogenic sources were to be included in the inventories. The act also mandated that computer modeling be carried out with current and projected future inventories to demonstrate that attainment could be obtained under proposed reductions in emissions.

CAAA90 also created a list of 189 **hazardous air pollutants (HAPs)** from hundreds of source categories. Under CAAA90, the U.S. EPA was required to develop emission standards for each source category under a timetable. For each new or existing source anticipated to emit more than 9.1 tonnes per year of 1 HAP, or 22.7 tonnes per year of a combination of HAPs, the U.S. EPA was required to establish a **maximum achievable control technology (MACT)** to reduce hazardous pollution from the source. In selecting MACTs, the U.S. EPA was permitted to consider cost, non–air quality health and environmental impacts, and energy requirements. Because the use of a MACT does not necessarily mean that hazardous pollutant concentrations will be reduced to a safe level, the U.S. EPA was also required to consult with the Surgeon General to evaluate the risk resulting from the implementation of each MACT.

The control of toxic air pollutants under CAAA90 differed from the control of criteria air pollutants. In the former case, the U.S. EPA was required to develop a program to control toxic emissions; in the latter, states were required to develop programs to control criteria air pollutant emissions and their ambient concentrations.

CAAA90 also tightened emission standards for automobiles and trucks, required additional reductions in emissions of acid deposition precursors, established a federal permitting program for point sources of pollution (previously, states were responsible for permits), and mandated reductions in the emission of CFCs to combat stratospheric ozone reduction. With respect to acid deposition, CAAA90 required a 10-million-ton reduction in sulfur dioxide emissions from 1980 levels and a 2-million-ton reduction in nitrogen oxide emissions from 1980 levels. With respect to CFCs, CAAA90 required a complete phase-out of CFCs, halons (synthetic bromine-containing compounds), and carbon tetrachloride by 2000 and methyl chloroform by 2002. CAAA90 also required the U.S. EPA to publish a list of safe and unsafe substitutes for these compounds and a list of the ozone depletion potential, atmospheric lifetimes, and global warming potential of all regulated substances suspected of damaging stratospheric ozone. CAAA90 banned the production of nonessential products releasing ozone-depleting chemicals. Such products included aerosol spray cans releasing CFCs and certain noninsulating foam.

8.1.11. Clean Air Act Revision of 1997

In 1997, the U.S. Congress passed the **Clean Air Act Revision of 1997** (CAAR97), by which it modified the NAAQSs for ozone and particulate matter (Table 8.2). The revised federal ozone standard was based on an 8-hour average rather than a 1-hour average. The new averaging time was intended to protect those who spend a long time each day working or playing outdoors, the people most vulnerable to the effects of ozone.

The particulate standard was modified to include a new $PM_{2.5}$ standard because studies have shown that aerosol particles ≤ 2.5 μm in diameter have more effect on respiratory illness, premature death, and visibility than do larger aerosol particles (Section 5.6). The purpose of the new $PM_{2.5}$ standard was to reduce risks associated with disease and early death associated with $PM_{2.5}$.

8.1.12. California Waiver

By 2007, neither the United States nor any state had set an emission standard for carbon dioxide or other long-lived gases that cause global warming. Emissions of some global warming agents, including black carbon and carbon monoxide, were regulated as health-damaging air pollutants. Black carbon, for example, was indirectly regulated through automobile and stationary source particulate matter emission standards. Carbon monoxide was regulated directly under emission standards. Outdoor concentrations of $PM_{2.5}$ (which contains black carbon), carbon monoxide, and ozone, another

global warming agent, were also regulated through NAAQS.

Nevertheless, CAAA70 permitted the U.S. EPA to regulate motor vehicle emissions of pollutants that "cause, or contribute to, air pollution which may reasonably be anticipated to endanger public health or welfare" [Section 202(a)(1)], implying that emissions of $CO_2(g)$ and other long-lived global warming agents could be regulated if the U.S. EPA were convinced that such agents endangered public health or welfare. The issue of whether the U.S. EPA could consider emissions of such chemicals was settled by the U.S. Supreme Court in April 2007. It ruled in *Massachusetts v. Environmental Protection Agency* (549 U.S. 497) that carbon dioxide and other greenhouse gases are air pollutants covered under CAAA70; thus, the U.S. EPA had the authority to *consider* regulating these gases. More specifically, the U.S. EPA must determine whether

emissions of greenhouse gases from new motor vehicles cause or contribute to air pollution which may reasonably be anticipated to endanger public health or welfare.

The basis for this ruling was the fact that vehicles enhanced 30,000 premature mortalities per year in the United States, and the Court determined that the U.S. EPA had the authority to determine whether greenhouse gases contributed to such mortality.

In the meantime, on September 27, 2006, California passed the **Global Warming Solutions Act of 2006** (AB32), which mandated that the state prepare a plan to reduce greenhouse gas emissions in the state by 2020. On February 21, 2007, the state subsequently requested the U.S. EPA to grant a **waiver of Clean Air Act preemption** to allow the state to set carbon dioxide emission standards for passenger vehicles, light-duty trucks, and medium-duty passenger vehicles sold in the state. The CAAA70 gives California authority to set stricter motor vehicle emission standards than the federal government, but only if the state submits a waiver request to the U.S. EPA to do so and if the U.S. EPA approves the waiver. The waiver can be granted as long as California can demonstrate that the standard is at least as tough as the federal standard, that the state needs the standard "to meet compelling and extraordinary conditions," and that the standard complies with the Clean Air Act Amendments of 1970. Historically, waivers have been granted regularly to California because the state has had the most severe air pollution in the United States.

Following the U.S. Supreme Court ruling permitting the U.S. EPA to consider long-lived greenhouse gas emission regulations, the U.S. EPA considered California's waiver request. However, the U.S. EPA Administrator denied the waiver request on March 6, 2008, arguing that California's request did not meet "compelling and extraordinary conditions." Specifically, he based his decision on two beliefs, namely, that because carbon dioxide is a well-mixed gas

1. It does not affect the health of people in California any more or less than it affects the health of people in other states.
2. Local California emissions do not affect the state's air pollution any more than do carbon dioxide emissions from outside California affect the state's air pollution.

Subsequently, when the U.S. EPA Administrator was replaced, California repetitioned the agency to reconsider the waiver issue. By that time, additional research had been performed showing that

1. Carbon dioxide emissions increase temperatures and water vapor, and both increase ozone where the ozone is already high, but not where it is low. Because California had six of the ten most polluted cities in the United States, carbon dioxide affected the health of people in California disproportionately compared with the rest of the United States (Jacobson, 2008a, 2008b).
2. Local carbon dioxide domes that form over cities increase temperatures and water vapor locally, both of which can increase local air pollution (Jacobson, 2010a).

The agency accounted for this research and other research demonstrating the relationship between higher temperatures and wildfire pollution, heat stress, coastal land loss, and other factors, and, on June 30, 2009, granted California a **waiver of Clean Air Act preemption** to allow the state to set carbon dioxide emission standards for 2009 model-year passenger vehicles, light-duty trucks, and medium-duty passenger vehicles sold in the state. The U.S. EPA published its decision on December 15, 2009, in an **endangerment finding**, stating in part,

> The Administrator finds that the current and projected concentrations of the six key well-mixed greenhouse gases – carbon dioxide (CO_2), methane (CH_4), nitrous oxide (N_2O), hydrofluorocarbons (HFCs), perfluorocarbons (PFCs), and sulfur hexafluoride (SF_6) – in the atmosphere threaten the public health and welfare of current and future generations. (U.S. EPA, 2009a)

It also published a cause and contribute finding, stating that

> [t]he Administrator finds that the combined emissions of these well-mixed greenhouse gases from new motor vehicles and new motor vehicle engines contribute to the greenhouse gas pollution which threatens public health and welfare. (U.S. EPA, 2009a)

The waiver allowed California to set the first regulation of carbon dioxide emissions in the United States. Under CAAA70, other states are permitted to set regulations as stringent as California's standards, so the granting of the waiver to California allowed other states to set carbon dioxide and other greenhouse gas emission regulations as well.

8.1.13. Regulation of U.S. Interstate and Transboundary Air Pollution

Long-range transport of air pollution is a problem that affects most countries (Section 6.6.2.4). International efforts to control transboundary air pollution are discussed with respect to acid deposition in Section 10.7. Here, control of long-range transport in the United States, and between the United States and its neighbors, is examined.

Interstate transport of air pollution in the United States is recognized by the government through Section 110 of CAAA70, which requires that SIPs be submitted to the U.S. EPA by each state and contain provisions to address problems of emissions transported to downwind states. Section 126 of CAAA70 allows downwind states to file petitions with the U.S. EPA to take action to reduce emissions in upwind states when such emissions make it difficult for the downwind state to meet federal air quality standards.

In 1994, the U.S. EPA recognized the difficulty for some states to meet federal standards merely by reducing emissions in their own state. As a result, the **Ozone Transport Assessment Group (OTAG)** – a partnership between the U.S. EPA, the Environmental Council of the States (ECOS) (thirty-seven easternmost states and the District of Columbia), and several industry and environmental groups – was established in 1995 (and concluded in 1997) to develop a mechanism to reduce ozone buildup due to interstate ozone transport, particularly in the northeastern United States. In 1997, during the period of OTAG, eight northeastern states filed petitions under Section 126 of CAAA70 requesting the U.S. EPA to take action against twenty-two upwind states and the District of Columbia for emitting excess oxides of nitrogen [NOx(g)]. Such emissions were suspected of exacerbating ozone in the petitioning states. In September 1998, the U.S. EPA implemented a rule (known as the **Nitrogen Oxide SIP Call**) that required these twenty-two states and the District of Columbia to submit SIPs that addressed the regional transport of ozone. The rule required reductions of NO$_x$(g), but not nonmethane organic gases (NMOGs), in these states by 2003.

In 2005, the U.S. EPA established the **Clean Air Interstate Rule (CAIR)**, which mandated reductions in power plant SO$_2$(g) emissions by 70 percent and NO$_x$(g) emissions by 60 percent relative to 2003 levels in twenty-eight eastern states and the District of Columbia. The emission reductions would be obtained by a **cap-and-trade** system by which overall emission limits and pollution allowances were set each year, and industries could trade among each other for the pollution allowances. Enforcement of the regulation was based on monitoring of emissions and penalties for exceeding emission allowances. On July 6, 2010, the U.S. EPA proposed a modification to CAIR, expanding it to thirty-one states and reducing power plant SO$_2$(g) emissions by 71 percent and NO$_x$(g) emissions by 52 percent relative to 2005 levels by 2014.

Transboundary pollution between Canada and the United States is formally recognized through the Canada-U.S. Air Quality Agreement. Article V of the agreement states that Canada and the United States must notify each other of any proposed projects within 100 km of the border that would be likely to emit more than 82 tonnes/yr of SO$_2$(g), NO$_x$(g), CO(g), TSPs, or ROGs. Between 1998 and 2010, the United States notified Canada of sixty-nine new facilities.

On a smaller scale, the United States and Mexico implemented a cooperative study, the Big Bend Regional Preliminary Visibility Study, to examine causes of visibility reduction in Big Bend National Park. In 1998, this preliminary study concluded that sources in both the United States and Mexico degraded visibility in the park, depending on the wind conditions. In September 2002, a joint U.S.-Mexico program called **Border 2012** was proposed with the goal of reducing air pollution and other environmental problems within 100 km on each side of the border in ten U.S. and Mexican border states. This program is administered by the U.S. EPA and the Mexican Ministry of the Environment and Natural Resources and provides funding for projects in local communities to study and reduce air pollution.

8.1.14. Smog Alerts

Because ozone is a criteria pollutant and its mixing ratios exceed federal and state standards more than do those of any other pollutant and because high levels of ozone are often good indicators of the severity of photochemical smog problems, many cities in the United States and worldwide issue smog alerts when ozone levels reach certain plateaus. Smog alert levels exist for pollutants aside from ozone as well. Table 8.3 identifies the mixing ratios of ozone, carbon monoxide, and nitrogen dioxide required to trigger a California standard violation; a federal standard violation; a California health advisory; and a California Stage 1, 2, and 3 smog alert.

Smog alerts in Los Angeles have been in place since the 1950s. Ozone levels required for a Stage 1, 2, or 3 alert are based on the relative health risk associated

Table 8.3. Mixing ratios required before the given 1-hour standard is exceeded

Health standard level	Ozone 1-hour average mixing ratio (ppbv)	Carbon monoxide 1-hour average mixing ratio (ppmv)	Nitrogen dioxide 1-hour average mixing ratio (ppbv)
California standard	90	20	180
National Ambient Air Quality Standards (NAAQS)	120[a]	35	100
Health advisory	150[b]	–	–
Stage 1 smog alert[b]	200	40	600
Stage 2 smog alert[b]	350	75	1,200
Stage 3 smog alert[b]	500	100	1,600

[a] Prior to 1997.
[b] Applies to California.

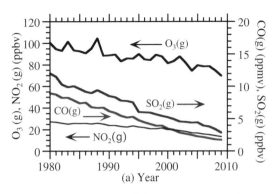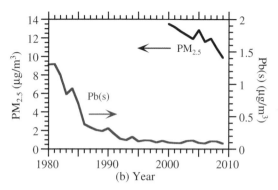

Figure 8.1. Air quality trends for (a) gases and (b) particle components or types in the United States, 1980–2009. From U.S. EPA (2011c). Ozone data are based on annual fourth maximum 8-hour averages at 255 sites; carbon monoxide data are based on annual second maximum 8-hour averages at 114 sites; nitrogen dioxide data are based on annual arithmetic averages at 81 sites; sulfur dioxide data are based on annual arithmetic averages at 134 sites; $PM_{2.5}$ data are based on seasonally weighted annual averages at 724 sites; lead data are based on annual maximum 3-month averages at 20 sites.

with each level. When an ozone smog alert occurs, individuals are advised to limit outdoor activity to morning or early evening hours because ozone levels peak in the midafternoon. Outdoor activities in schools are usually eliminated during a smog alert. **Health advisories** for ozone were initiated in California in 1990 by the California Air Resources Board because evidence indicated that ozone levels above the federal standard and below a Stage 1 smog alert may affect healthy, exercising adults. When a health advisory occurs, all individuals, including children and healthy adults, are advised to limit prolonged and vigorous outdoor exercise. Individuals with heart or lung disease are advised to avoid outdoor activity until the advisory is no longer in effect.

8.1.15. U.S. Air Quality Trends from the 1970s to Present

Between 2006 and 2008, 175 million people (58 percent of the U.S. population) lived in U.S. cities with unhealthful air (American Lung Association, 2010). In terms of ozone, eight of the ten most polluted cities were in California, with Los Angeles being the most polluted. In terms of annual particulate matter pollution, five of the ten most polluted cities were in California, with most in the Central Valley. Historically, the regions of the United States in which air quality problems have been most severe are in central and Southern California, the Boston through Washington (BoWash) corridor, the Milwaukee through Chicago Great Lakes Region, Phoenix, El Paso, Dallas-Ft. Worth, Houston,

Baton Rouge, and Atlanta. These locations shift over time, however.

Although U.S. pollution levels are still severe, they have improved since the 1950s. Much of the improvement can be attributed to regulations arising at the district, state, and federal levels. Figure 8.1 shows ambient levels of several gas and particle pollutants, averaged over U.S. air monitoring stations, from 1980 to 2009. It also indicates a **marked decrease in lead concentrations**, due primarily to the phase-out of leaded gasoline, and more gradual but noticeable decreases in the other pollutants. $NO_2(g)$ and $O_3(g)$ decreases have lagged behind $CO(g)$ and $SO_2(g)$ decreases. $PM_{2.5}$ measurements commenced only in 2000.

The Los Angeles Air Pollution Control District, formed in 1947 and now part of the SCAQMD, has been at the forefront in initiating regulations that were ultimately adopted at the federal level. Despite the fact that Los Angeles is still the most polluted city in the United States in terms of ozone (Table 8.4), improvements in air quality resulting from regulation there have been significant. Figure 8.2a shows the maximum ozone levels and the number of days per year that California state standards, federal standards, health advisory levels, Stage 1 alerts, and Stage 2 alerts were exceeded between 1973 and 2009 in the Los Angeles Basin. Between 1973 and 2009, the highest ozone level recorded in the Los Angeles Basin was 0.51 ppmv (1974). The California state ozone standard, which was exceeded 219 days in 1976, was exceeded only 102 days in 2009. Between 1976 and 2009, NAAQS exceedances were reduced from 176 to 15 days per year, and Stage 1 and 2 alerts were

Table 8.4. Ranking of most polluted U.S. cities in terms of ozone and year-round particulate matter pollution, 2006–2008

Ozone	Particulate matter pollution
1. Los Angeles-Long Beach-Riverside, CA	1. Phoenix-Mesa-Scottsdale, AZ
2. Bakersfield, CA	2. Bakersfield, CA
3. Visalia-Porterville, CA	3. Los Angeles-Long Beach-Riverside, CA
4. Fresno-Madera, CA	3. Visalia-Porterville, CA
5. Sacramento-Arden-Arcade-Yuba City, CA-NV	5. Pittsburgh-New Castle, PA
6. Hanford-Corcoran, CA	6. Fresno-Madera, CA
7. Houston-Baytown-Huntsville, TX	7. Birmingham-Hoover-Cullman, AL
8. San Diego-Carlsbad-San Marcos, CA	8. Hanford-Corcoran, CA
9. San Luis Obispo-Paso Robles, CA	9. Cincinnati-Middletown-Wilmington, OH-KY-IN
10. Charlotte-Gastonia-Salisbury, NC-SC	10. Charleston, WV; Detroit-Warren-Flint, MI; Weirton-Steubenville, WV-OH

Source: American Lung Association (2010).

eliminated. Stage 3 alerts, frequent in the 1950s, when peak ozone mixing ratios reached 0.68 ppmv, have not occurred in Los Angeles since before 1976. The number of days of $PM_{2.5}$ standard exceedances has similarly decreased in Los Angeles, although maximum $PM_{2.5}$ levels have not decreased to the same degree as the number of standard exceedances (Figure 8.2b). Occasional spikes in the maximum PM_{10} concentration are likely due to severe Santa Ana wind dust events.

Figure 8.3 compares the trends in the maximum hourly ozone and annual $PM_{2.5}$ among four air districts in California. Ozone improvements are greatest in the South Coast district (home to Los Angeles), but improvements in other districts, including the clean

North Central Coast district, have also occurred. Maximum particulate matter pollution increased in the San Joaquin Valley between 2004 and 2009, but it decreased or stayed constant in the other districts.

Part of the reason for the ozone reductions over time in California has been the decrease in ozone-producing organic gases (Figure 8.4a). Concentrations of atmospheric metals, primarily from industrial sources and also from vehicles (in the case of lead primarily), have also decreased over time (Figure 8.4b).

The measured improvement in air quality in the United States due to air pollution regulations correlates with an improvement in the economy during the same period. For example, between 1980 and 2009,

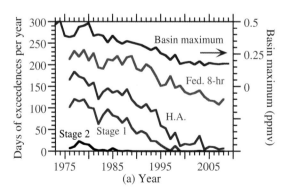
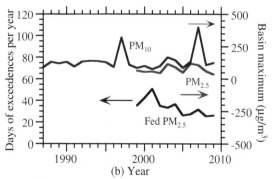

Figure 8.2. (a) Days per year that the ozone mixing ratio in the Los Angeles Basin exceeded the California state standard (State), the U.S. National Ambient Air Quality Standard (NAAQS) (Fed.), the California health advisory (H.A.) level, the Stage 1 smog alert level, and the Stage 2 smog alert level over time. Also shown is the maximum ozone mixing ratio each year in the basin. (b) Same as (a), but for days of exceedances of the federal $PM_{2.5}$ standard and the maximum $PM_{2.5}$ and PM_{10} levels in the basin. From California Air Resources Board (2011); South Coast Air Quality Management District (SCAQMD) (2011).

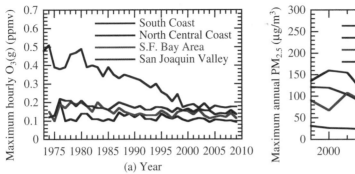 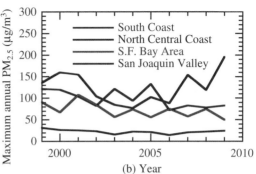

Figure 8.3. Trends in (a) maximum hourly ozone and (b) maximum annual PM₂.₅ in four California Air Quality Control Regions. From California Air Resources Board (2011).

the U.S. population increased from 226.5 to 307.0 million (+35.5 percent), the gross domestic product (GDP) increased from $2.77 to $14.26 trillion (+415 percent), and the per capita GDP increased from $12,200 to $46,500 (+280 percent). Simultaneously, maximum annual ozone levels decreased from 100.5 ppbv in 1980 to 69 ppbv in 2009 (−31 percent) (Figure 8.1). These data suggest that *air pollution regulations did not correlate with damage to the overall economy during this period.* Instead, air pollution regulation led to inventions and overall job growth in new or expanded industries. Areas of invention included air pollution control, engine, renewable energy, and improved fuel technologies. New or expanded industries included the renewable energy, pollution control device, pollution measurement device, pollution remediation, pollution software, and pollution modeling industries. Regulations also resulted in the employment of public and educational sector workers in the areas of pollution/climate regulation, policy, science, and engineering, and led to the expansion of the supercomputer industry to satiate

the demand for researchers devoted to studying and mitigating air pollution and climate problems.

8.1.16. Visibility Regulations and Trends

In the United States, California first set a **visibility standard** in 1959 and modified it in 1969. The 1969 standard required that the prevailing visibility outside Lake Tahoe exceed 10 miles (16.09 km) when the relative humidity was less than 70 percent. In Lake Tahoe, the minimum allowable visibility was set to 30 miles (48.3 km). Measurements were made by a person looking for landmarks a known distance away. The furthest landmark that could be seen along 180 degrees or more of the horizon circle, not necessarily in continuous sectors around the circle, defined the prevailing visibility.

Because prevailing visibility is a subjective measure of visibility, the California Air Resources Board changed the California visibility standard in 1991 to one based on the use of the meteorological range. The revised standard required that the meteorological range

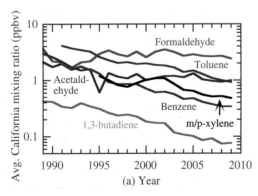 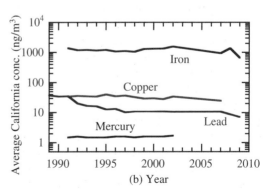

Figure 8.4. Trends in average California (a) mixing ratios of selected organic gases and (b) concentrations of selected metals in the air. From California Air Resources Board (2011).

at a wavelength of 0.55 μm not be less than 10 miles (16.09 km) outside Lake Tahoe and 30 miles (48.3 km) in Lake Tahoe when the relative humidity was less than 70 percent. In 1990 and 1994, visibility in the eastern Los Angeles Basin was less than the state standard on 50 percent or more of the days of each year. Such visibility degradation was due primarily to aerosol particle buildup in the eastern basin.

At the federal level, prevailing visibility data have been collected at 280 monitoring stations at airports in the United States since 1960. These data have been used primarily by air traffic controllers. From an analysis of the data, the U.S. EPA Office of Air and Radiation found that visibility deteriorated in the eastern United States between 1970 and 1980, but improved slightly between 1980 and 1990. Schichtel et al. (2001) similarly found that, between 1980 and 1995, visibility improved by about 10 percent in the eastern United States and in hazy parts of California. According to both studies, *visibility improvements in the eastern United States coincided with decreases in sulfur dioxide emissions*. Figure 7.24 compares summer versus winter visibility in the United States for 1991 to 1995, from the second study.

In July 1999, the U.S. EPA set **regional haze rules** designed to improve visibility in 156 Class I pristine areas (Section 8.1.9), primarily large national parks, wilderness areas, memorial parks, and international parks in existence as of August 1977, the month that CAAA77 was passed. The regulations required states to make short- and long-term efforts to reduce emissions affecting haze in these areas down to natural levels within 60 years. Part of these efforts required the use of a **best available retrofit technology (BART)** for industrial facilities that emit visibility-reducing air pollutants. State plans were required to be submitted to the U.S. EPA for approval. In 2006, the U.S. EPA finalized a rule that allowed states to use emission trading to satisfy the visibility improvement requirements. In 2009, the U.S. EPA found that thirty-seven states had missed deadlines for submitting regional haze rule plans.

Figure 8.5 shows the change in visibility in national parks and wilderness areas in the western and eastern United States between 1992 and 2008. Visibility increased in both regions during this period not only on good visibility days, but also on midrange and poor visibility days. Visibility was uniformly lower in the east than in the west because the relative humidity is generally higher in the east, so aerosol particles have greater liquid water contents and larger sizes in the east than in the west. The largest contributor to visibility reduction in both the east and west is sulfate, which causes more

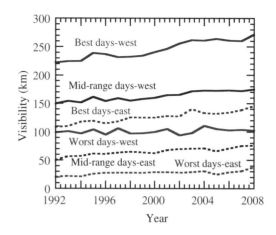

Figure 8.5. Visibility (visual range) in thirty national parks and wilderness areas in the western United States and eleven in the eastern United States, 1992–2008, on best, midrange, and worst visibility days. From Debell et al. (2006); IMPROVE (n.d.).

than 60 percent of the visibility loss in the east and 25 to 65 percent of the loss in the west. Organic matter, nitrates, black carbon, and soil dust cause most of the remaining visibility loss in both regions.

8.2. Pollution Trends and Regulations outside the United States

The United States began to consider urban air pollution reduction formally beginning in 1955, and the UK enacted legislation in 1956 (Section 8.2.2). However, national air pollution regulation and/or enforcement in most other countries of the world lagged until the 1970s through the 1990s. In the 1950s, Los Angeles, California, was the most polluted city in the world in terms of ozone, and London may have been the most polluted in terms of particulate matter. In 2011, Linfen, Beijing, Calcutta, Mexico City, New Delhi, Cairo, São Paulo, Shanghai, Jakarta, Bangkok, and Tehran were among the most polluted.

Worldwide, 2.5 to 3 million people die each year prematurely, and tens of millions of others suffer illness from air pollution. Of the fatalities, 1.6 million deaths/yr are from indoor air pollution (World Health Organization (WHO), 2005), with the rest being from outdoor air pollution (WHO, 2002; Jacobson, 2010b). More than half the air pollution health problems worldwide are in developing countries. In many of the world's cities, annual averaged outdoor PM_{10} concentrations exceed 70 μg m^{-3}, much higher than the recommended levels of 20 μg m^{-3}. In this section, air quality trends

Table 8.5. European union full-life (10 yr/100,000–120,000 mi) emission standards for passenger cars in g/km (g/mi)

Year	THC(g)	NMHC(g)	CO(g)	NO$_x$(g)	THC + NO$_x$(g)	PM
Diesel						
Euro 1 (1992)	–	–	2.72 (4.4)	–	0.97 (1.6)	0.14 (0.23)
Euro 2 (1996)	–	–	1.0 (1.6)	–	0.7 (1.1)	0.08 (0.13)
Euro 3 (2000)	–	–	0.64 (1.0)	0.5 (0.8)	0.56 (0.9)	0.05 (0.08)
Euro 4 (2005)	–	–	0.5 (0.8)	0.25 (0.4)	0.30 (0.48)	0.025 (0.04)
Euro 5 (2009)	–	–	0.5 (0.8)	0.18 (0.29)	0.23 (0.37)	0.005 (0.008)
Euro 6 (2014)	–	–	0.5 (0.8)	0.08 (0.13)	0.17 (0.27)	0.0025 (0.004)
Gasoline						
Euro 1 (1992)	–	–	2.72 (4.4)	–	0.97 (1.6)	–
Euro 2 (1996)	–	–	2.2 (3.5)	–	0.5 (0.8)	–
Euro 3 (2000)	0.2 (0.32)	–	2.3 (3.7)	0.15 (0.24)	–	–
Euro 4 (2005)	0.1 (0.16)	–	1.0 (1.6)	0.08 (0.13)	–	–
Euro 5 (2009)	0.1 (0.16)	0.068 (0.11)	1.0 (1.6)	0.06 (0.10)	–	0.005[a] (0.008)
Euro 6 (2014)	0.1 (0.16)	0.068 (0.11)	1.0 (1.6)	0.06 (0.10)	–	0.005[a] (0.008)

NMHC, nonmethane hydrocarbon; THC(g), total hydrocarbon in the gas phase (all hydrocarbons, including methane).
[a] Applies to direct injection engines only.
Source: European Commission (2011).

and regulations in a region and several countries are discussed.

8.2.1. European Union

The European Community was established in 1957 (originally with six member countries) under the Treaty of Rome to integrate the countries of Europe economically and politically. It gradually grew and, in 1993, changed its name to the **European Union** (EU). Its greatest enlargement occurred in 2005, when it expanded from fifteen to twenty-five countries with the addition of several Eastern European nations. In 2011, it consisted of twenty-seven member countries: Austria, Belgium, Bulgaria, Cyprus, Czech Republic, Denmark, Estonia, Finland, France, Germany, Greece, Hungary, Ireland, Italy, Latvia, Lithuania, Luxembourg, Malta, the Netherlands, Poland, Portugal, Romania, Slovakia, Slovenia, Spain, Sweden, and the United Kingdom.

In 1970, the EU began enacting air pollution regulations in the form of directives, regulations, and decisions. Most laws are **directives**, which are binding regulations on all member nations that take into account particular conditions in each nation. For example, a directive for emission reductions from a power plant may allow each member nation to reduce emissions by a different level. About 10 percent of EU laws are **regulations**, which are laws applied uniformly to all

member nations. An example of a regulation is the appointment of a qualified person to inspect the use of a dangerous chemical. **Decisions**, which are the least common form of environmental law, are requirements that may be directed at specific member nations or may be a modification of a regulation.

European Union emission standards for new passenger vehicles (Table 8.5), light-duty trucks, heavy-duty trucks and buses, and nonroad vehicles are set by directives and are designed to be increasingly stringent over time. Standards are set for THCs, nonmethane hydrocarbons (NMHCs), CO(g), NO$_x$(g), and particulate matter. The standards for gasoline vehicles roughly parallel those of the United States (Table 8.1). However, about 50 percent of EU passenger cars run on diesel fuel, whereas about 2 percent of U.S. passenger cars run on diesel. Diesel vehicles emit more nitrogen oxides and particulate matter than do gasoline vehicles per unit distance, and this is reflected by the fact that the EU diesel standards for nitrogen oxides and particulate matter are less stringent than are EU gasoline standards. Diesel vehicles emit less CO(g) and THCs than do gasoline vehicles, and this is also reflected in diesel versus gasoline regulations in Table 8.5. Because of the high penetration of diesel in the EU, EU emissions of NO$_x$(g) and particulate matter from passenger vehicles per unit distance driven exceed those in the United States, whereas THC(g) emissions per unit driven are

Table 8.6. European union ambient air quality standards, 2011

Pollutant	Standard
Ozone [$O_3(g)$]	
8-Hour average	60 ppbv (120 μg m^{-3})
Carbon Monoxide [$CO(g)$]	
8-Hour average	9 ppmv (10 mg m^{-3})
Nitrogen Dioxide [$NO_2(g)$]	
Annual average	21 ppbv (40 μg m^{-3})
1-Hour average	105 ppbv (200 μg m^{-3})
Sulfur Dioxide [$SO_2(g)$]	
24 Hours	0.05 ppmv (125 μg m^{-3})
1 Hour	0.13 ppmv (350 μg m^{-3})
PM_{10}	
Annual average	40 μg m^{-3}
24-Hour average	50 μg m^{-3}
$PM_{2.5}$	
Annual average	25 μg m^{-3}
Lead [$Pb(s)$]	
Annual average	0.5 μg m^{-3}

Source: European Commission (2011).

higher in the United States. As of 2011, the EU did not have a standard limiting $CO_2(g)$ emissions from vehicles, although proposals for such a standard had been floated.

Table 8.6 summarizes **ambient air quality standards in the EU**. European Union ozone standards in 2011 were more stringent than in the United States (60 ppbv vs. 75 ppbv), primarily because photochemical ozone production in Europe is, on average, slower than in the United States due to the fact that most of Europe is at higher latitudes, and UV radiation penetration to the surface is weaker than in the United States. Because ozone forms less readily in Europe, its buildup is not so significant of a problem as in many U.S. cities, so a standard for it can be more stringent in the EU without invoking opposition. In the United States, industry groups have continuously opposed reducing ozone standards. Carbon monoxide and the 1-hour nitrogen dioxide standards are similar in the EU and the United States. The 1-hour sulfur dioxide standard and the annual $PM_{2.5}$ standard are more stringent in the United States. The weaker particulate matter standard in the EU occurs because of the greater difficulty in meeting a stringent $PM_{2.5}$ standard in the EU due to the deep penetration of diesel passenger vehicles, which are high particle emitters. The high $PM_{2.5}$ concentration throughout the EU reduces the average life expectancy by about 8.6 months (WHO, 2008).

8.2.2. United Kingdom

In January 1956, soon after the Air Pollution Control Act was enacted in the United States, the **Clean Air Act** became law in the UK. The UK act was a response to the devastating smog event in London that killed 4,000 people in 1952. It controlled both household and industrial emissions of pollution for the first time in the UK, but it dealt only with smoke, particularly black and dark smoke. It did not deal with sulfur dioxide, although presumably reductions in smoke would also reduce sulfur dioxide. The act resulted in 90 percent of London being controlled by smoke regulations. It mandated smokeless zones in London and the relocation of many power plants to rural areas.

The next major piece of air pollution legislation in the United Kingdom was the **Clean Air Act of 1968**, which required that industries burning fossil fuels (solid, liquid, or gas) construct tall chimneys to prevent their emissions from depositing to the ground locally.

The UK joined the EU in 1972. Since then, EU directives on the controls of pollution have been followed in the UK. Some of these directives were formalized in the UK's **Environmental Protection Act of 1990**, which permitted local governments to control emissions from small industrial processes and establish a statutory system for local air quality management. The national government still controlled emissions from large sources under the 1956 and 1968 UK Clean Air Acts.

In 1992, UK vehicle emission standards were strengthened in accordance with an EU directive. In 1993, catalytic converters were required in all new gasoline-powered vehicles, again in accordance with an EU directive. In 1995, the UK passed the **Environment Act**, which outlined the need for scientific studies of the effects of air pollution and the need for new health-based air quality standards by 2005. The act required the publication of a plan to combat air pollution. In 1997, a plan called the United Kingdom National Air Quality Strategy was published. Under this strategy, ambient limits for ozone, nitrogen dioxide, sulfur dioxide, carbon monoxide, benzene, 1-,3-butadiene, lead, and particulate matter were set in accordance with EU standards.

Prior to the 1980s, most energy in the UK was obtained from coal, a fuel that caused much of the smoke problems experienced in London and other big cities. Between 1980 and 1998, coal consumption decreased by 57 percent, improving air quality. Although there are fewer vehicles in the UK than in many other EU countries, vehicles in the UK are driven more miles than in other EU countries, resulting in vehicles being one of the largest sources of pollution in the UK.

In 2007, the UK passed the **Air Quality Standards Regulations 2007**, which was subsequently superseded by the **Air Quality Standards Regulations 2010**. These laws set ambient air quality standards, allowing different standards to be set for different zones of the UK. For zones in which a standard is exceeded, the laws require a plan to be submitted for the zone to meet the standard. The laws also require air pollution monitoring. The 2010 law also incorporates EU directives specifying ambient limits on toxic pollutants and ozone precursors.

Regulations have improved air quality in the UK to levels not seen since the onset of the Industrial Revolution (European Environment Agency, 2010a). However, UK air pollution is still estimated to reduce the average life span by 6 months and kill 50,000 people per year prematurely. Between 1992 and 2009, average annual ambient PM_{10} concentrations decreased by about 44 percent, and $NO_2(g)$ mixing ratios decreased by about one-third across the country at urban locations. Ozone levels, in contrast, increased by about 4 percent. Emissions of most primary pollutants decreased significantly during this period (European Environment Agency, 2010a).

8.2.3. France

Because it lacked domestic fossil fuel energy sources and desired independence from reliance on imported fuel for electricity, France developed a large nuclear power industry starting in the 1970s, resulting in lower than European-averaged air pollution emissions from the electric power sector. Instead, vehicles are the major source of air pollution in France. The country had a population of about 65 million people and 32 million vehicles in 2011.

Air pollution monitoring in Paris began in 1956, when the Laboratoire d'Hygiène installed an outdoor surveillance network. In 1972, a network that measured pollution from automobiles near roads was introduced, and in 1973, a twenty-five–station monitoring network that measured pollution near electric utilities was funded. In 1979, the French Ministry of the Environment created a government agency responsible for measuring pollution and assessing its impacts in the Paris

area. The agency combined existing with new monitoring stations to produce a network of seventy-five stations. Pollutants measured include $CO(g)$, $NO_2(g)$, $SO_2(g)$, $O_3(g)$, lead, and total suspended particulates.

In 1985, France instituted a tax on $SO_2(g)$ emissions. The tax was subsequently extended to include $NO_x(g)$, VOCs, and $HCl(g)$. Since December 1990, the French Environment and Energy Management Agency, a public agency with private partnerships, has overseen air pollution regulation in France. In 1994, the agency first established a nationwide system of alerting the public to high levels of pollution. In 1996, France passed the **Law on Air and the Rational Use of Energy**, which set ambient air quality standards, requirements for monitoring pollution, emission standards, and requirements for regions to submit 5-year air quality plans in order to meet air pollution regulations. The law also required ambient and emission standards to be in compliance with EU standards.

Regulation of air pollution in France due to EU directives resulted in improved air quality for several pollutants between 2000 and 2009. The greatest improvements were for $SO_2(g)$ (a 65-percent reduction in ambient levels) and $NO_2(g)$ (a 14-percent reduction in ambient levels). However, ozone increased slightly and PM_{10} decreased only slightly during this period (European Environment Agency, 2010b).

8.2.4. Germany

From the late 1800s to the 1980s, the largest source of energy in Germany was coal. Much of the coal in Germany was high in sulfur and originated from the **Ruhr region**. In the 1920s, particulate soot, sulfur dioxide, and chemical pollution became so severe and industry so powerful that around 1930, the school in Solingen in the Ruhr region had to shut down for 18 months to accommodate industrial emissions (McNeill, 2000). The leveling of many factories in the Ruhr region during World War II reduced pollution temporarily. In the 1950s, industry rebounded, increasing pollution levels to new heights. In the 1960s, a law in West Germany required new smokestacks to be taller than before, reducing pollution locally, but increasing it downwind.

In East Germany after World War II, the primary coal burned was lignite (brown) coal, which is relatively dirty with high sulfur levels. The region between Dresden (Germany), Prague (Czech Republic), and Krakow (Poland) is rich in lignite coal and is called the **sulfur triangle**. As in the Ruhr region, steel, coal, cement,

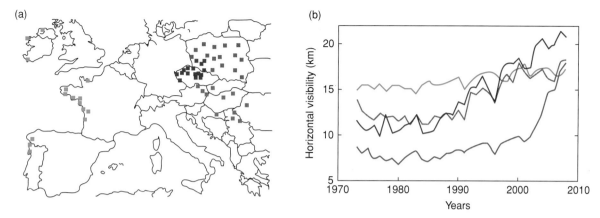

Figure 8.6. Horizontal annual mean visibility from 1973 to 2008 for stations in coastal western Europe (green) and the sulfur triangle (red, black, and blue). From Stjern et al. (2011).

glass, ceramic, and iron factories flourished in the sulfur triangle following World War II. The pollution in the sulfur triangle may have killed up to 7,000 people per year in the mid-1970s (McNeill, 2000) and caused up to 50 percent forest dieback between 1972 and 1989 (Ardo et al., 1997). After the reunification of Germany in 1990, most lignite mines and associated power plants in former East Germany were shut down, and the remaining power plants were retrofitted with pollution control equipment. Nevertheless, coal is still Germany's only domestic source of nonrenewable energy, and Germany burns oil and coal at one of the highest rates in the EU.

Figure 8.6 shows visibility changes between 1983 and 2008 in coastal western Europe and in three regions of the sulfur triangle. Pollution reduction during this period increased horizontal visibility by about 15 km in the sulfur triangle, compared with less than 2.5 km in coastal western Europe (Stjern et al., 2011). Most visibility improvements in the sulfur triangle occurred following the 1990 reunification of Germany and the resulting elimination of the lignite-burning power plants.

In the 1990s and early 2000s, Germany had a strong environmental political party, the Greens. As a result, Germany had tougher environmental legislation than many other EU nations. To promote clean air, the federal government of Germany instituted **eco taxes** on gasoline and electricity. The first tax went into effect on April 1, 1999. Revenues from the tax were used to fund renewable energy programs.

Germany's air pollution control strategy is designed to meet EU ambient and emission standards. To meet these standards, the federal government uses an air pollution regulation tool referred to as the **Technical Instructions on Air Quality Control** (TA Luft). The TA Luft was first instituted in West Germany in 1964 and revised in 1974, 1983, 1988, and 2002. The TA Luft regulates emissions of dust, sulfur dioxide, nitrogen oxides, fluorine, arsenic, lead, cadmium, nickel, mercury, thallium, ammonia, and particulate matter, among other chemicals. It also sets procedures for licensing new industrial facilities and monitoring pollution. In 1974, West Germany enacted the **Federal Air Pollution Control Act**, which was subsequently amended several times, including in 1985 and 2002. This act set broad goals for reducing air pollution in Germany using specific regulations set under TA Luft. Another air pollution control tool, the **Federal Immission Control Act**, was passed in 1990 and amended in 2000. The purpose of the act was to control emissions from the installation and operation of facilities; the production and testing of fuels, products, and vehicles; and the construction of public roads and railways.

8.2.5. Russia

Russia had a population of 142 million people in 2011. Emissions from 30 million vehicles, old factories, and uncontrolled burning of waste and wood have caused significant environmental degradation and health impacts. Cherepovets ("city of skulls"), Norilisk, and Karabash, Russia, for example, are among the most polluted cities in the world. Moscow (Figure 8.7a) and St. Petersburg (Figure 8.7b) similarly suffer from power plant, industrial facility, and vehicular pollution. Air pollution particles from **wildfires in Russia** during August 2010 increased the all-cause death rate

Figure 8.7. (a) Power station pollution against a backdrop of a brown layer of smog in the center of Moscow, Russia, on a frosty January 18, 2010. © Vladimir Zhuravlev/Dreamstime.com. (b) Pollution in St. Petersburg, Russia, on January 16, 2011. © Alexander Lebedev/Dreamstime.com. (c) Woman with a face mask embracing a man in front of devastating wildfire pollution obscuring the Kremlin in Moscow on August 7, 2010. © Sergey Gordeev/Dreamstime.com. (d) Toxic industrial pollution in Zaporozhye, Ukraine. © Kateryna Moskalenko/Dreamstime.com.

from 360–380 to 700 deaths per day in Moscow alone (Figure 8.7c).

Russia regulates air pollution through national laws, but few are enforced. Economic problems have resulted in environmental protection taking a low priority. In 1999, the national environmental body, the Russian State Committee on Environmental Protection, called about 15 percent of Russian territory (the combined area of England, France, Germany, Sweden, and Finland) "ecologically unfavorable." Most Russians live in these areas. Under pressure by the powerful oil industry, the environmental protection body was abolished and its duties transferred to the Ministry of Natural Resources on May 17, 2000 (Sinitsyna, 2007). The new ministry made little effort to design or enforce regulations. The loss of national regulatory strength has resulted in all Russian cities, including Moscow, experiencing continued severe air pollution since then.

New vehicle emission standards in Russia are regulated according to EU emission standards (Table 8.5), with on-road light-duty emissions following Euro 1 standards in 1999, Euro 2 standards in 2006, Euro 3 standards in 2008, Euro 4 standards in 2010, and Euro 5 standards in 2014. Standards are also set for heavy-duty vehicle and nonroad engine emissions, as well as for fuel quality.

About 44 percent of Russians may live in areas with heavy air pollution, which may be responsible for 17 and 10 percent of disease in children and adults, respectively (Eurasian Development Bank, 2009). In December 2010, Russian President Dmitry Medvedev instructed the Natural Resources and Ecology Minister to suggest amendments to Russia's environmental law that required the use of new technologies to protect the environment. Companies that used low-polluting technologies would face lower taxes, whereas those who did not would be prosecuted.

Air pollution in Russia is not new. Historically, the Soviet Union suffered from emissions resulting from industrial and military development that had little regard for the environment. Such pollution was compounded by the 1986 **Chernobyl nuclear disaster** in Ukraine, a Soviet Republic at the time. The disaster contributed to the radiation contamination of a large populated region, not only in Ukraine and Russia, but also in most of the rest of eastern and central Europe. Today, air pollution in Ukraine and other former Soviet Republics, now called the **Commonwealth of Independent States**, is still severe (e.g., Figure 8.7d).

8.2.6. Israel

Israel imports almost all its raw energy resources. About 75 percent of its electricity comes from coal, and about 85 percent of all coal consumed in the Middle East is consumed in Israel. Most of the rest is consumed in Iran (Energy Information Administration, 2011a). Vehicle traffic is a major source of pollution in Tel Aviv, Jerusalem, and Haifa. In 2009, Israeli roads supported 2.5 million vehicles, an increase from 70,000 in 1960. Diesel vehicles emitted about 80 percent of the nitrogen oxide and particulate matter emissions from all vehicles in Israel.

Air pollution in Israel was first controlled under the **Abatement of Nuisances Law of 1961**, which authorized the Minister of the Environment to define and regulate "unreasonable" air pollution. The law also allowed the minister to issue decrees against specific emitters, require factories to comply with emission standards, and monitor air pollution. The first ambient standards under this law were set in 1971 and revised in 1992.

In 1994, the Ministry of the Environment used its power to design a program to control smog by reducing vehicle emissions and setting up a national air quality monitoring network. Under the program, all new cars were required to have catalytic converters, lead was gradually phased out of cars, and vehicle emission standards were required to conform with those set by the EU. Industrial emissions of sulfur dioxide, nitrogen oxides, VOCs, heavy metals, particulate matter, hazardous inorganic particulate matter, and other substances were also controlled.

In 2007, a **National Plan for the Reduction of Vehicular Pollution** was approved by the Minister of the Environment. This plan required more stringent emission standards and steps for reducing vehicle emissions, including a plan to pay to remove old vehicles from the road, a plan to have inspectors order owners of polluting vehicles not to operate their vehicle, and incentives to increase the use of public transportation.

In January 2008, a new **Clean Air Law** was passed in Israel to plan, regulate, and monitor air quality. The law came into force on January 1, 2011. Aside from setting ambient and emission standards, the law also established economic incentives for reducing pollution, and it expanded powers given to the Ministry of Environmental Protection to implement and enforce the regulations. Furthermore, it required industries to obtain emission permits and a monitoring network in which all stations operated under uniform standards. Ambient

Figure 8.8. Downtown Cairo, Egypt, February 15, 2009. © Jakezc/Dreamstime.com.

standards under the law were set for $CO(g)$, $NO_x(g)$, $SO_2(g)$, $O_3(g)$, and PM_{10}.

8.2.7. Egypt

Air pollution in Egypt is regulated under Law No. 4 of 1994. This law set maximum permissible emission limits for gases and particles from industries, vehicles, and open burning. It also established ambient air quality limits for $CO(g)$, $SO_2(g)$, $NO_2(g)$, $O_3(g)$, lead, total suspended particles, black smoke, and PM_{10}.

Egypt's most severe air pollution is in its capital, Cairo (Figure 8.8), which houses one-fourth of the country's population. Cairo is one of the most polluted cities in the world. This is due in part to the large number of vehicles, many of which are old, with poor control devices. About 4.5 million vehicles operate in Cairo, of which half are passenger vehicles, about 20 percent are trucks and trailers, 16 percent are motorcycles, and 8 percent are taxis. Air pollution in Cairo is also exacerbated by unregistered copper smelters and the open burning of trash and waste. The WHO indicates that concentrations of air pollutants, mainly particulate matter, are often up to 100 times above safe limits in Cairo.

8.2.8. Iran

Iran has severe air pollution problems, primarily due to its vehicular traffic and coal combustion. Gasoline is the most common motor vehicle fuel used in Iran because it is inexpensive, and oil is a domestic product of Iran. Vehicles are estimated to cause 75 to 80 percent of air pollution in its capital city, Tehran. Iran has more than 12 million cars and 7 million motorcycles. More than 30 percent of motor vehicles use roads in or near Tehran. The road infrastructure in Tehran was not designed for the number of vehicles used in the city. Iran's vehicle fleet is aged and thus highly polluting. Pollution in Tehran is also exacerbated by the fact that the city is bounded by mountains to the north, which slow the winds, particularly when a large-scale subsidence inversion is present. Tehran's air pollution is among the worst in the world.

Tehran's strategy for reducing air pollution has been through command and control and improved public transportation. The city, for example, designated a required green space of 10 square meters per person in 1993. It also converted many diesel buses to natural gas buses, implemented vehicle inspection rules, encouraged the use of electric bicycles, and expanded its subway and bus fleet.

Nevertheless, smog events in Tehran are so severe that they have frequently forced closures of schools and government offices, including twice in late 2010. Respiratory illness in Tehran, in particular, has increased substantially with the growth of the city and the slow pace of air pollution abatement.

8.2.9. India

India is the second most populous country in the world after China. The rapid population growth in India, from

(a)

(b)

Figure 8.9. (a) Motorcycle emitting smoke on a street in Calcutta, India, on April 24, 2007; © Samrat35/Dreamstime.com. (b) New Delhi winter morning through air pollution haze. © Digitalfestival/Dreamstime.com.

300 million in 1947 to 1.2 billion in 2011, has strained resources and increased pollution (Figure 8.9). In 2007, about 40 percent of India's energy was from coal burning, and 27 percent was from the burning of **solid biofuels** (e.g., wood, grass, agricultural waste, dung). Solid biofuels are burned on a large scale within individual homes for heating and cooking. Although India had only 13 million cars in 2008, the number is increasing rapidly. In addition, about 10 percent of households own a motorcycle. The combination of coal, solid biofuel, and transportation fuel combustion has resulted in some of the world's most intense air pollution, not only in major cities such as New Delhi and Calcutta, but also on a continental scale. Such emissions have contributed, along with pollution from China, to the **Asian Brown Cloud**, a phenomenon that can be seen by satellite (Figure 8.10; Ramanathan et al., 2001).

From the 1880s until recently, Calcutta, near the coal fields of West Bengal, was the most polluted city in India. Calcutta built iron, steel, glass, jute, chemical, and paper industries powered by coal combustion. By the 1920s, smokestacks from 2,500 coal-fired steam boilers covered Calcutta's landscape (McNeill, 2000). Early smoke control was facilitated by the willingness of the authoritarian colonial government of India to clamp down on smoke emissions. Regulatory controls in Calcutta reduced smoke from the 1910s to the 1950s, at which time smoke emissions started to increase again. In the 1980s, an increase in the number of vehicles exacerbated pollution problems. Today, New Delhi and Calcutta are the most polluted cities in India.

In 1976, India passed an amendment to its constitution allowing the government to intervene to protect public health, forests, and wildlife. This amendment was ineffective because another clause in the constitution stated that the amendment was not enforceable in court.

In 1981, India passed the **Air (Prevention and Control of Pollution) Act**, a first effort to control and abate air pollution. In 1984, an accidental release of methyl isocyanate from a pesticide manufacturing plant in **Bhopal**, India, killed 4,000 people and injured more than 200,000 others. As a result, **The Environment (Protection) Act of 1986** was passed, giving the Ministry of Environment and Forests responsibility over controlling emissions from all sources. A series of laws and rules were subsequently enacted that enforced the siting, transportation, emissions, and storage of hazardous waste. The first emission standards for vehicles in India were for idling vehicles and became effective in 1989. Motor vehicle emission standards were set for gasoline vehicles in 1991 and diesel vehicles in 1992. By 2000, India adopted Euro 1 passenger vehicle emission standards (Table 8.5), which were tightened to Euro 4 standards by 2010. In 2000, India set rules on the production and consumption of ozone-depleting substances.

8.2.10. China

China is the world's most populous country, with more than 1.3 billion inhabitants in 2011. Two major

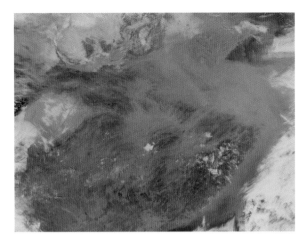

Figure 8.10. Satellite image of the Asian Brown Cloud over central eastern China. The brown in the cloud is due to a mix of absorbing aerosol particles and nitrogen dioxide gas from vehicles, factories, and power plants, as well as from wood burning for home heating and cooking. Courtesy NASA/GeoEye, Inc. NASA/GeoEye, Inc., http://eoimages.gsfc.nasa.gov/ve/12878/china_shanghai.tif.

pollutants common in China are sulfur dioxide and aerosol particles, both emitted during coal burning. China, the world's largest producer and consumer of coal, has the third largest coal reserves (after the United States and Russia). Large sources of pollution in China are industrial boilers and furnaces and residential burning of coal and solid biofuels for home heating and cooking. Northern China also experiences severe **dust storms** originating from the **Gobi Desert**. In the spring of 2000, the storms were the most severe in 50 years, darkening the skies over northern China and large cities, such as Beijing, more than twelve times. The intensity and frequency of the dust storms were enhanced by soil erosion and deforestation, coupled with drought. Pollution from China's cities, along with dust from the Gobi Desert, are sources that affect local residents and are also transported by the westerly winds to Japan, South Korea, and across the Pacific to North America. The pollution in China is so widespread and thick that it can be seen in satellite images (Figure 8.10), comprising part of the Asian Brown Cloud.

In 1995, China amended its **Air Pollution Control Law of 1987**, calling for the phase-out of leaded gasoline by 2000. On July 1, 2000, sulfur and alkene contents of gasoline in Beijing, Shanghai, and Guangzhou were limited to 0.08 and 35 percent, respectively. These limits applied to the rest of China in 2003.

On September 1, 2000, China amended its Air Pollution Control Law again. The new law aimed to reduce sulfur dioxide emissions from 16.9 million tonnes in 1999 to 9.1 million tonnes by 2010. It also required the phase-out of dirty coal and provided incentives for the use of low-sulfur, low-ash coal in medium and large cities. New and expanded power plants in these cities were required to install equipment that reduced sulfur dioxide and aerosol particle emissions. Under the previous law, power plant emitters were fined only if their emissions exceeded a certain standard. Under the new law, emitters were fined on the basis of their total emissions. In provinces where pollution standards were not met, provincial governments could require all factories in a given zone of the province to obtain an emission permit.

Under the 2000 law, cities were given incentives to replace coal heating stoves in households with centralized heating. All new coal-burning heating boilers were banned from areas in which centralized heating was available. Restaurants in large- and medium-size cities were required to convert from coal to natural gas, liquefied petroleum gas, or electricity. Household cooking stoves were required to be converted from coal to gas, electricity, or coal briquettes. Construction companies were required to reduce resuspended dust emissions. Cities were given incentives to pave or grow plants over bare soil and reduce road dust.

Under the 2000 law, only emissions from industrial sources and power plants were controlled; under the new law, emissions from vehicles, ships, domestic heating and cooking, and construction were also controlled. Passenger vehicle emission standards for year 2000 vehicles were set to Euro 1 standards (Table 8.5). Existing vehicles were required to be retrofitted to meet the standards and pass annual emission checks and random inspections. In 2004, emission standards were tightened to meet Euro 2 standards. In 2007, they were tightened further to meet Euro 3 standards and, in 2010, to meet Euro 4 standards.

In 1998, China had seven of the ten most polluted cities in the world. The government, itself, found that two-thirds of 338 cities in which the air was monitored were polluted. Despite significant efforts to control emissions through regulation, 16 of the world's 20 most polluted cities in 2006 were in China (Worldwatch Institute, 2006) (e.g., Figure 8.11). The most polluted city among these may be **Linfen, China**, a city of 3 million in the midst of China's coal mining region. The air is so polluted on many days that laundry turns black before drying. Birth defects are nearly thirty times the

(a) (b)

(c) (d)

Figure 8.11. (a) Smog in Beijing, China, August 2007. Photo by Andrew Chang. (b) Montage of pollution during 5 consecutive days in Beijing, China during the city government's pollution control testing for the Olympics, August 3–7, 2007. Photo by Andrew Chang and Kimberly Ong. (c) Pollution under a subsidence inversion layer, as seen from the top of Mount Taishan, Shandong Province, China, November 2007. Photo by Andrew Chang. (d) Air pollution in Guangzhou, China. Photo by HSC/Dreamstime.com.

world average. Living there is equivalent to smoking three packs of cigarettes a day.

Overall, in China, about 750,000 people die each year prematurely, 350,000 to 400,000 due to outdoor pollution and the rest due to indoor pollution (McGregor, 2007). However, China is not the only country in Southeast Asia with poor air quality. Other countries, such as Taiwan (Figure 8.12a) and Thailand (Figure 8.12b), suffer as well.

8.2.11. Japan

Some early pollution in Japan was due to mining. From the seventeenth century to 1925, the Besshi copper mine and smelter on Shikoku Island produced pollution that damaged crops, enraging farmers. In 1925, a high smokestack and desulfurization equipment reduced pollution from the smelter. The implementation of a tall stack at the Hitachi copper mine in 1905 similarly reduced local complaints about pollution.

In the early 1900s, the expansion of industry increased urban emissions in Japan. In the 1910s, pollution in Osaka, a city developed around coal combustion, was as severe as that in other industrial cities of the world (McNeill, 2000). In 1932, Osaka passed a law requiring industry to increase combustion efficiency and reduce smoke emissions. Instead, the lack of inspectors and an increase in the number of chemical

(a)   (b)

Figure 8.12. Air pollution in (a) Taipei, Taiwan, on July 12, 2009. © Christopher Rawlins/Dreamstime.com and (b) Bangkok, Thailand, on March 11, 2010. © Anthony Aneese Totah, Jr./Dreamstime.com.

and metallurgy plants exacerbated pollution. During World War II, much of Osaka was leveled, decreasing pollution temporarily. However, following the war, industry rebounded, and the population and number of vehicles increased. By the early 1970s, Osaka, Kobe, and Kyoto had merged into a large metropolitan region with heavy smoke and vehicle pollution. Tokyo, which had also grown since the late 1800s, similarly experienced pollution problems, mostly from vehicles.

In 1960, the city of Ube, in southern Japan, passed regulations controlling soot and smoke. The success of the regulations catalyzed a national law in 1962 that regulated some soot and smoke emissions throughout Japan. The regulations did not solve problems immediately, however, because pollution in major cities continued to intensify.

In 1968, Japan passed a national **Air Pollution Control Law**, amended most recently in 1996. The intent of the original law was to establish an air quality monitoring system and emission standards for industrial sources, on-road vehicles, and building construction and demolition in order to "protect public health and preserve the living environment." Ultimately, the law was managed by Japan's Environment Agency, formed in 1970, after a severe smog episode in Tokyo. The main pollutants designated for control under the law were sulfur oxides, soot aerosol particles, cadmium, chlorine, hydrogen fluoride, and lead from stationary sources; soil dust from construction and building demolition; and carbon monoxide, hydrocarbons, and lead from on-road motor vehicles. In 1982, hydrocarbons from stationary sources were also regulated under the law,

and, in 1987, a program to phase out lead in gasoline was implemented.

In 1992, the Environment Agency (which evolved into the Ministry of the Environment in 2001) enacted an additional requirement to control $NO_x(g)$ from on-road diesel passenger cars and all trucks, vans, buses, and special-purpose vehicles in 196 metropolitan areas. This law was expanded to include particulate matter emissions in 2001, and the result was the **Automotive $NO_x(g)$ and PM Law**. $NO_x(g)$ and PM emission standards under this law were tightened in 2005 and 2009 to the point that they were similar to U.S. 2010 and Euro 5 emission standards.

Construction machine emissions [CO(g), THCs, $NO_x(g)$, PM] in Japan were controlled first in 1996, with modifications in 2003. Nonroad vehicle emissions of the same pollutants were controlled first in 2003.

Although the highest pollution levels in Japan today are much lower than they were in the 1950s to 1970s, increases in population since the 1980s have spread pollution over a larger portion of the country. Health problems related to air pollution have also been exacerbated by a large number of diesel vehicles and transport of urban air pollution and Gobi Desert dust from China.

8.2.12. Canada

Canadian air is relatively clean compared with that in many other countries. Nevertheless, pollution is a problem in many Canadian cities, and acid deposition affects Canadian lakes and forests.

Figure 8.13. Cathedral, National Palace, and Zocalo (main square) through the smog of Mexico City. © Alexandre Fagundes/Dreamstime.com.

Vehicles are the largest source of air pollution in Canada. The country has about 560 vehicles for every 1,000 people, the seventh highest penetration in the world after the United States, Luxembourg, Iceland, Australia, Puerto Rico, and Italy.

Canada has been concerned with air pollution since at least the late 1960s. In 1969, it initiated an air pollution monitoring network called the National Air Pollution Surveillance (NAPS) Network. This network contained about 150 air quality monitoring stations, located mostly in urban areas. In 1970, Canada passed a **Clean Air Act** to regulate emissions of lead, asbestos, mercury, and vinyl chloride.

In 1999, Canada enacted the **Canadian Environmental Protection Act (CEPA)**, under which the government obtained new powers to control emissions, primarily of toxic substances. PM_{10} was declared toxic, allowing it to be regulated under the act. The act reduced the allowable level of sulfur in gasoline and diesel, reduced the allowable level of benzene in gasoline, and doubled funding for outdoor air pollution monitoring.

Canadian on-road vehicle emission standards were first set under the **Motor Vehicle Safety Act of 1971**. The standards became aligned closely with those in the United States starting in 1988. Under CEPA, the on-road regulations were tightened according to U.S. standards in 2003, and nonroad emissions were first regulated based on U.S. standards in 1999.

8.2.13. Mexico

In the 1940s, Mexico City's air was relatively clean, resulting in visibility of about 100 km. From 1940 to 2011, the population of Mexico City and its surroundings grew from 1.8 million to about 20 million, making it the most populated metropolitan area in the world. Mexico has 209 vehicles per 1,000 people, suggesting that about 4.1 million vehicles operate in the Mexico City metropolitan area. Due to vehicle and factory emissions, as well as emissions from cooking, heating, and electricity generation, Mexico City's air is among the most polluted in the world (Figure 8.13).

Like Los Angeles, Mexico City sits in a basin surrounded by mountains, and pollution is frequently trapped beneath the Pacific high-pressure system. Unlike Los Angeles, Mexico City is not bounded by an ocean on one side. The low-latitude, relatively thin stratospheric ozone layer above it (because the ozone layer becomes thinner with decreasing latitude – Chapter 11) and high elevation also enhance the UV radiation penetration to Mexico City, increasing ozone formation there.

Air pollution regulations in Mexico City were first enacted in 1971, with a **Decree to Prevent and Control Environmental Pollution**, under which monitors were required to measure outdoor pollution. A full set of monitors was implemented only by 1986, at which time ozone was found to be the main pollutant.

Ambient standards were then set for CO(g), NO$_2$(g), SO$_2$(g), O$_3$(g), lead, and total suspended particulate matter. The government also provided incentives to relocate industry out of Mexico City (1978), introduced low-leaded gasoline into the fleet (1986), and began to require vehicle inspections (1989).

In 1988, Mexico passed its first emission standards for light- and heavy-duty vehicles. These standards applied to 1993 model cars. The light-duty standards were made more stringent, in line with U.S. Tier I standards in 2001. Since 2004, such standards have followed a mix of U.S. and EU standards. In 2006, new emission standards for heavy-duty trucks and buses were implemented based on U.S. 2004 and Euro 4 standards for 2008 model vehicles (DieselNet, 2011).

Ozone levels in Mexico City exceeded national standards during 324 days in 1995. That year, Mexico initiated a 5-year National Environmental Program (1996–2000) designed to clean up air pollution in Mexico City. The government also passed a tax incentive program for the purchase of pollution control equipment. As a result of these measures and the phase-out of older cars between 2000 and 2006, ozone mixing ratios decreased by about 24 percent. However, mixing ratios for half the hours of 2006 were still above 83 ppbv, which compares with half the hours exceeding 108 ppbv in 2000 (Sanchez and Ayala, 2008).

8.2.14. Brazil

Most of Brazil is covered by the Amazon rainforest, which makes up 30 percent of the world's remaining tropical forests. São Paulo, Brazil's largest city, is the second most populated metropolitan area in the Americas after Mexico City. Air pollution in São Paulo results from a large number of high-emitting automobiles, poor road systems, low fuel prices, and local burning of sugarcane fields. Electric power production is not a major source of air pollution in Brazil because about 95 percent of the country's electricity is obtained from hydroelectric power. Instead, a large source of air pollution in Brazil is the burning of parts of the Amazon rainforest to clear land for agriculture and grazing (**biomass burning**) and the annual burning of many agricultural fields.

In 1975, Brazil started an **alcohol fuel program**, the Brazilian National Alcohol Program, to reduce Brazil's reliance on imported fuel following the worldwide spike in oil prices in 1973. The program ensured that all gasoline sold in Brazil contained 22 percent anhydrous ethanol and that the price of the ethanol-gasoline blend would remain similar to that of gasoline alone. Ethanol for the project was produced primarily from sugarcane. Because the market price of an ethanol-gasoline blend was higher than that of pure gasoline, the program required government subsidies. When gasoline prices fell and sugarcane prices increased in the late 1980s, ethanol prices rose sharply, and the program disintegrated. In fact, by 1997, only 1 percent of new cars sold in Brazil used ethanol. In 1999, the Brazilian government revived the alcohol fuel program by encouraging the replacement of taxis and government vehicles with new vehicles that ran on 100-percent ethanol. Although the use of ethanol fuel instead of gasoline reduces the emission of aromatic hydrocarbons in smog, it increases the production of acetaldehyde (Section 4.3.7), generally leading to greater ozone and PAN formation.

A damaging by-product of the **sugarcane ethanol** program in Brazil is the annual burning of sugarcane fields to facilitate manual harvesting of the sugarcane. For example, Brazil's largest sugarcane-producing state, São Paulo, has more than 100 ethanol mills and sugarcane plantations. Annual burning of the sugarcane in the state occurs over more than 2 million acres, creating extended plumes of smoke (Figure 8.14) that choke rural communities and contribute to pollution over cities. The use of mechanized harvesting can eliminate the need for cane burning, and many ethanol mills in the state of São Paulo have agreed to convert to mechanization by 2017. However, burning continues, and other states have not followed suit.

In 1993, Brazil adopted its first vehicle emission standards based on EU standards. It strengthened the standards in 2002, effective from 2006 to 2009, based on Euro 3 and 4 light-duty vehicle standards, and again in 2009, effective from 2013 to 2015, based on Euro 5 standards. Brazil does not permit diesel engines in passenger cars, so the country did not originally have a diesel vehicle emission standard. However, because Brazil's standards are used by adjacent South American countries that do allow diesel passenger cars, Brazil set diesel emission standards in 2002. Brazil's vehicle emission regulations cover heavy-duty vehicles as well (DieselNet, 2011).

8.2.15. Chile

The air in Santiago, Chile, ranks among the most polluted in the world, especially in terms of particulate matter. The pollution is due to vehicles, industrial emissions, unpaved road dust, and dust from eroded land. The city, nestled in a valley, frequently experiences a

Figure 8.14. Sugarcane field burn, April 12, 2009. © Sean van Tonder/Dreamstime.com.

large-scale subsidence inversion. As such, winds are often light and rainfall is scarce, exacerbating pollution problems.

Pollution regulations in Santiago were first enacted in 1987. That year, a law was passed allowing only cars with specified last digits on their license plates to operate within the city limits on a given workday. Starting in 1989, inspections of carbon monoxide, particulate matter, and hydrocarbon emissions from vehicles were mandated.

In 1992, the Chilean government outlined a long-term plan to combat air pollution. The two main control strategies were emission regulations and economic incentives. This plan culminated in the **Basic Environmental Law**, passed in 1994, which set up a framework for environmental management. Just before that, in 1993, a law was passed requiring catalytic converters in all new passenger vehicles. The same year, emissions from stationary industrial sources were regulated and open-burning wood stoves were banned unless they contained control devices to reduce particle emissions. In 1995, a measure requiring stationary sources to shut down when particle concentrations rose above a critical level was passed. In 1997, laws controlling emissions from nonindustrial stationary sources and residential heating boilers were passed. Another method of reducing pollution in Santiago was to eliminate old buses. Between 1980 and 1990, the number of buses in the city increased from 9,500 to 13,000, but by 1997, the number had decreased back down to 9,000 (Jorquera et al., 2000).

Chile's first emission standards for light-duty passenger vehicles and trucks were passed in 1992. Medium-duty vehicle standards were set in 1995. Heavy-duty vehicle standards were first set in 1994. More stringent standards were set for the Santiago metropolitan region than for the rest of the country. All standards were updated in 2005 and 2006. Dual emission standards exist in Chile: new engines can meet either U.S. or EU standards.

Pollution abatement in Santiago has been modestly effective. Ozone levels decreased between 1989 and 1995, but they did not change from 1995 to 1998. Between 1989 and 1998, particulate matter levels decreased by 1.5 to 7 percent per year. Monthly average $PM_{2.5}$ levels still ranged from 100 to 150 $\mu g\ m^{-3}$ (Jorquera et al., 2000), and black carbon and organic matter levels in Santiago were still more than seven times those in Los Angeles (Didyk et al., 2000) However, since 2000, particulate matter pollution has doubled (Morales, 2010).

8.2.16. South Africa

Ninety percent of South Africa's electricity is obtained from coal. Coal is also burned in boilers and stoves in factories and hospitals. Millions of people in townships and villages without electricity burn coal and wood for home heating and cooking. Coal burning produces sulfur dioxide and soot aerosol particles. Coal emissions from homes and industry, combined with other pollutants from automobile exhaust, frequently create dark skies over Johannesburg. Cape Town, a coastal city, is cleaner than Johannesburg. However, it still has pollution problems, particularly from nitrogen dioxide formed following vehicle emissions of $NO(g)$.

In 1965, South Africa initiated the **Atmospheric Pollution Prevention Act**. This law resulted in the regulation of fuel burning emissions from industries, hotels, dairies, and dry cleaners. It also declared specified residential areas where open burning normally occurred as smokefree zones and required smoke and $SO_2(g)$ monitoring. The law, however, was largely ineffective because it ignored numerous sources that, in aggregate, degraded air quality.

Following the end of apartheid, the 1996 Constitution of South Africa guaranteed everyone the right (a) "to an environment that is not harmful to their health or well-being" and (b) "to have the environment protected for the benefit of present and future generations through reasonable legislative and other measures." In October 1997, a white paper on environmental management called for South Africa's Department of Environment Affairs and Tourism to regulate and combat air pollution, particularly from coal and fuel burning, vehicle exhaust, mining, industrial activity, and incineration. Although emission regulations followed, they were not enforced strictly.

To address pollution more formally, South Africa enacted the **National Environment Management: Air Quality Act** in 2004. The purpose of the act was to protect human health and well-being, prevent air pollution and ecological degradation, and secure ecologically sustainable development while promoting justifiable economic and social development. The act gave the Minister of the Environment the authority and duty to monitor and regulate ambient pollution concentrations and emissions from point, nonpoint, and mobile sources. It also allowed provinces to set standards equal to or stricter than national standards. The act provided for an ambient PM_{10} standard, but not a $PM_{2.5}$ standard. Sources of $PM_{2.5}$ in South Africa include not only industrial and vehicle combustion, but also desert dust from the **Kalahari Desert** and other arid regions of the country. Particulate matter air pollution in some regions of South Africa is still among the greatest worldwide. In an effort to chip away at additional sources of pollution, South Africa passed a **National Air Quality Act** that came into effect April 1, 2010. The law requires regulation of emissions from the mineral, metallurgical, chemical processing, and waste disposal industries.

8.2.17. Australia

Air pollution problems have historically been less severe in Australia than in many other countries, particularly because Australian cities are less populated and have fewer sources of pollution than do more polluted cities. Australia is also surrounded by oceans, so it does not receive much transboundary air pollution. The sulfur contents of coal and oil in Australia are lower than are those in other countries. Coal-fired power plants are also located away from urban areas. As a result, sulfur dioxide mixing ratios in Australian cities are relatively low. Sydney and Melbourne occasionally have bad air pollution days, particularly in summer and autumn. Street-level pollution is also significant due to the omnipresence of automobiles, trucks, and buses.

Vehicles are the largest source of air pollution in Australia. The first light-duty vehicle emission standards were set in the early 1970s. In 1976, **Australian Design Rule** (ADR) 30/00 required all new diesel vehicles to meet smoke opacity standards. Catalytic converters were required for new passenger vehicles starting in 1986. The first diesel emission standards became effective under an ADR in 1995. In 2002, more stringent ADRs regulated emissions from both gasoline and diesel vehicles. These standards began to converge with U.S. and EU emission standards.

Emission standards in Australia are generally set by states and territories, not by the national government. For example, the South Australian government passed the Environmental Protection (Air Quality) Policy in 1994, which set standards for emissions of dark smoke, lead, antimony, arsenic, cadmium, mercury, nitrogen oxides, nitric acid, sulfur dioxide, hydrogen sulfide, fluorine, hydrofluoric acid, chlorine, and carbon monoxide from stationary sources. The government similarly set rules for the open burning of waste in 1994 and for vehicle fuel composition in 2002.

The first national ambient air quality standards in Australia were passed as part of the **National Environment Protection Measure (NEPM) for Ambient Air Quality** in 1998. Pollutants covered included $CO(g)$, $NO_2(g)$, $O_3(g)$, $SO_2(g)$, lead, and PM_{10}. In 2001, an NEPM was passed to monitor and repair dirty diesel vehicles for their $NO_x(g)$ and PM emissions. This NEPM was updated in 2009. In 2004, an NEPM for Air Toxics was passed to monitor five toxics:

benzene, formaldehyde, benzo(a)pyrene, toluene, and xylenes.

8.3. Summary

In this chapter, regulation of and trends in urban air pollution are discussed with respect to several countries. Until the 1950s, air pollution in the United States was not addressed at the national level. Increased concern over photochemical smog and fatal London-type smog events encouraged the U.S. Congress to implement the Air Pollution Control Act of 1955 and the United Kingdom to implement the U.K. Clean Air Act of 1956. The U.S. act provided funds to study air pollution and provided for technical assistance to states, many of which had already enacted air pollution legislation, but it did not regulate air pollution at the federal level. The U.S. Congress first regulated stationary sources of pollution through the Clean Air Act of 1963. This act led to the Motor Vehicle Air Pollution Control Act of 1965, which was the first federal legislation controlling emission from motor vehicles. The Air Quality Act of 1967 divided the country into Air Quality Control Regions, specified the development of Air Quality Criteria for certain pollutants, and required states to submit State Implementation Plans for combating air pollution. In landmark legislation, Congress passed the Clean Air Act Amendments of 1970. This act required the newly created U.S. EPA to develop National Ambient Air Quality Standards for certain pollutants identified as being unhealthful and specified that the U.S. EPA set performance standards for new sources of air pollution. Regulations under CAAA70 led to the 1975 invention of the catalytic converter, the most important device to date for limiting emissions from vehicles. The Clean Air Act Amendments of 1990 led to the control of many hazardous air pollutants not addressed in previous legislation. The Clean Air Act Revision of 1997 led to revised ozone and particulate matter standards. The California Waiver decision by the U.S. EPA in 2009 paved the way for the first law worldwide permitting control of $CO_2(g)$ from motor vehicles. State and federal air pollution legislation have resulted in a remarkable improvement in urban air quality in the United States. Strong legislation by the EU and several industrialized countries has also led to improved air quality. Although legislation has been enacted in many developing countries, legislation in several of those countries is relatively new, has not been enforced, or has not been sufficiently strong. As such, air pollution in many cities of the world has increased.

8.4. Problems

8.1. Suppose you drive a gasoline engine car 15,000 miles per year.

(a) How many kilograms per year of nitrogen oxides and carbon monoxide would your car have emitted in 1971 versus today if your car exactly met federal emission standards in both years? Use California and Northeast Low Emission Vehicle (LEV) II Standards for today's emissions.

(b) If each of 10 million cars in your city were driven 15,000 miles per year, how many tonnes per day of carbon monoxide would have been emitted in 1971 versus today?

8.2. Explain how the three-way catalyst works.

8.3. The British Alkali Act of 1863 resulted in new control technologies for converting hydrochloric acid from soda ash factory chimneys to reusable forms of chlorine, and the Clean Air Act Amendments resulted in the development of the catalytic converter. If air pollution control regulation leads to improved technologies, why is tough air pollution legislation generally so difficult to pass?

8.4. Identify any 1-hour California state or U.S. federal standards, health advisories, or smog alert levels exceeded by $NO_2(g)$ or $O_3(g)$ in Figures 4.13a and 4.13b.

8.5. What are two possible explanations for the improvements in Los Angeles air pollution between 1975 and today, as shown in Figure 8.2?

8.6. Today, what criteria pollutants exceed United States federal standards the most? Which exceed the standards the least?

8.7. Although more people in the United States are exposed to levels of $CO(g)$ than PM_{10} above the federal standard, exposure to PM_{10} is of greater concern in terms of health effects. Why?

8.8. Match each city with an air pollution–related characteristic of the city.

(a) Cairo	(1) Heavy reliance on alcohol fuel burning
(b) Santiago	(2) Subject to severe desert dust pollution events
(c) Mexico City	(3) Energy sector relies heavily on nuclear power
(d) Los Angeles	(4) Severe pollution from open burning of waste and copper smelters

(e) Beijing

(f) New Delhi

(g) Paris

(5) Experienced a devastating smog event in 1952

(6) Heavy particulate pollution, particularly from soil erosion

(7) Environmental protection body abolished in 2000

(h) São Paulo

(i) Moscow

(j) London

(8) High ozone due to high-elevation basin surrounded by mountains

(9) Ozone levels exceeded 0.6 ppmv in the 1950s

(10) Biofuels for heating and cooking have contributed to the Asian Brown Cloud

Chapter 9

Indoor Air Pollution

People spend most of their time indoors, so the composition and quality of indoor air has a significant impact on human health. Because people's time is often divided between home and work, it is important to examine air quality in both residences and workplaces. Sources of indoor air pollution include outdoor air that infiltrates indoors and indoor emissions. Outdoor air contains the constituents of smog, but some of these constituents dissipate quickly indoors because of the lack of UV radiation to regenerate them indoors. Major indoor sources of pollution include stoves, heaters, carpets, fireplaces, tobacco smoke, motor vehicle exhaust from garages, building materials, and insulation. In developing countries, major sources of indoor air pollution include the products of solid biofuel and coal combustion for home heating and cooking. Such pollution is responsible for significant premature mortality worldwide. Whereas indoor air pollution is often regulated in workplaces, it is not regulated in residences. In this chapter, characteristics, sources, and regulation of indoor air pollution are discussed.

9.1. Pollutants in Indoor Air and Their Sources

Throughout the world, people spend most of their time indoors. One study found that, in the United States, about 89 percent of time was spent indoors, 6 percent was spent in vehicles, and 5 percent was spent outdoors (Robinson et al., 1991). A study of less developed countries showed that people in urban areas spent about 79 percent of their time indoors, and those in rural areas spent about 65 percent of their time indoors (Smith, 1993). Because people breathe indoor air more than outdoor air, an examination of indoor air is warranted. Nearly 1.6 million people worldwide die each year prematurely from indoor air pollution (not including smoking), making it one of the leading causes of death worldwide (World Health Organization (WHO), 2005). Most of these deaths are due to indoor burning of solid biofuels and coal for home heating and cooking with inefficient cook stoves and heaters.

Table 9.1 identifies major pollutants in indoor air and their primary sources. Many of the pollutant gases in indoor air are also found in outdoor air. Outdoor pollutants enter indoor air by infiltration, natural ventilation, and forced ventilation. **Infiltration** is natural air exchange through cracks and leaks, such as through door and window frames, chimneys, exhaust vents, ducts, plumbing passages, and electrical outlets. **Natural ventilation** is air exchange resulting from the opening or closing of windows or doors to enhance the circulation of air. **Forced ventilation** is the air exchange resulting from the use of whole house fans or blowers (Masters, 1998). Next, the pollutants in Table 9.1 are discussed briefly.

9.1.1. Carbon Dioxide

Carbon dioxide [$CO_2(g)$], which is present in background air, is also produced indoors from breathing and the burning of wood, coal, oil, and gas. It does not pose

Table 9.1. Important indoor air pollutants and their emission sources

Pollutant	Emission sources
Gases	
Carbon dioxide	Metabolic activity, combustion, garage exhaust, tobacco smoke
Carbon monoxide	Boilers, gas or kerosene heaters, gas or wood stoves, fireplaces, tobacco smoke, garage exhaust, outdoor air
Nitrogen dioxide	Outdoor air, garage exhaust, gas or kerosene heaters, gas or wood stoves, tobacco smoke
Ozone	Outdoor air, photocopy machines, electrostatic air cleaners
Sulfur dioxide	Outdoor air, kerosene space heaters, gas stoves, coal appliances
Formaldehyde	Particleboard, insulation, furnishings, paneling, plywood, carpets, ceiling tile, tobacco smoke
Volatile organic compounds	Adhesives, solvents, building materials, combustion appliances, paints, varnishes, tobacco smoke, room deodorizers, cooking, carpets, furniture, draperies
Radon	Diffusion from soil
Aerosol Particles	
Allergens	House dust, domestic animals, insects, pollens
Asbestos	Fire-retardant materials, insulation
Fungal spores	Soil, plants, foodstuffs, internal surfaces
Bacteria, viruses	People, animals, plants, air conditioners
Polycyclic aromatic hydrocarbons	Fuel combustion, tobacco smoke
Other	Resuspension, tobacco smoke, wood stoves, fireplaces, outdoor air

Sources: Spengler and Sexton (1983); Nagda et al. (1987).

a health problem until its mixing ratio reaches 15,000 ppmv, much higher than its background mixing ratio of 393 ppmv in 2011. Mixing ratios and health effects of carbon dioxide are discussed in Section 3.6.2.

9.1.2. Carbon Monoxide

Carbon monoxide [CO(g)], produced outdoors by automobile and other fossil fuel and biofuel combustion sources, is emitted indoors by boilers, heaters, stoves, fireplaces, cigarettes, and cars in garages. In the absence of indoor sources, mixing ratios of CO(g) indoors are usually less than are those outdoors (Jones, 1999). In the presence of indoor sources, indoor mixing ratios of CO(g) can reach a factor of 4 or more times those outdoors. Health effects of carbon monoxide are discussed in Section 3.6.3.

9.1.3. Nitrogen Dioxide

Nitrogen dioxide [$NO_2(g)$] is produced chemically from the oxidation of nitric oxide [NO(g)] and emitted in small quantities indoors. Sources of NO(g)

and $NO_2(g)$ indoors include in-garage cars, kerosene and gas space heaters, wood stoves, gas stoves, and cigarettes. In the absence of indoor sources, indoor mixing ratios of $NO_2(g)$ are similar to those outdoors (Jones, 1999). High mixing ratios of $NO_2(g)$ occur over short periods during the operation of $NO_2(g)$-producing appliances. Because UV sunlight does not penetrate indoors, the photolysis of $NO_2(g)$ and subsequent production of ozone is not a concern in indoor air. The health effects of $NO_2(g)$ are discussed in Section 3.6.8.

9.1.4. Ozone

Ozone [$O_3(g)$], produced photochemically in urban air outdoors following photolysis of nitrogen dioxide, is rarely produced indoors because UV sunlight, required for its production, is usually unavailable indoors. The major indoor source of ozone is outdoor air. Photocopy machines and electrostatic air cleaners, which emit UV radiation, can also produce ozone indoors. In residences and most workplaces, however, ozone mixing ratios indoors are almost always less than they are outdoors. The indoor-to-outdoor ratio of ozone ranges from 0.1:1 to 1:1, with typical values of 0.3:1 to 0.5:1 (Finlayson-Pitts and Pitts, 1999). Ozone is lost indoors by reaction with wall, floor, and ceiling surfaces; reaction with indoor gases; and deposition to floors. Health effects of ozone are discussed in Section 3.6.5.

9.1.5. Sulfur Dioxide

Sulfur dioxide [$SO_2(g)$], emitted outdoors during gasoline, diesel, and coal combustion, is emitted indoors during combustion of kerosene for space heaters or wood for heating and cooking. In the absence of indoor sources, indoor $SO_2(g)$ mixing ratios are typically 10 to 60 percent those of outdoor air (Jones, 1999). Once indoors, $SO_2(g)$ does not chemically degrade quickly in the gas phase because the hydroxyl radical [$OH(g)$] required to initiate its breakdown is not produced indoors. Because the *e*-folding lifetime of the hydroxyl radical is about 1 second, $OH(g)$ brought indoors from the outside disappears quickly. Losses of $SO_2(g)$ include deposition to wall and floor surfaces, dissolution into liquid water (e.g., in bathtubs and sinks), and dissolution into aerosol particles containing liquid water. The health effects of sulfur dioxide are discussed in Section 3.6.6.

9.1.6. Formaldehyde

Formaldehyde [$HCHO(g)$], produced during fossil fuel and solid and liquid biofuel burning and chemical reaction outdoors, is emitted from particleboard, insulation, furnishings, paneling, plywood, carpets, ceiling tile, tobacco smoke, and combustion indoors. Formaldehyde mixing ratios indoors are usually greater than are those outdoors. In outdoor air, formaldehyde breaks down by photolysis and reactions with $HO_2(g)$ and $OH(g)$. UV sunlight and $OH(g)$ are not present indoors, but $HO_2(g)$ sometimes is, and it is the most likely indoor chemical breakdown source of formaldehyde. Formaldehyde is also removed by deposition to the ground and reaction with wall, floor, and ceiling surfaces. Health effects of formaldehyde are discussed in Section 4.2.6.

9.1.7. Radon

Radon (Rn) is a radioactive but chemically unreactive, colorless, tasteless, and odorless gas that forms naturally in soils. Its decay products are believed to be carcinogenic and have been measured in high concentrations near uranium mines and in houses, particu-

larly in their basements, overlying soils with uranium-rich rocks (Nazaroff and Nero, 1988). Radon plays no role in acid deposition, stratospheric ozone, or outdoor air pollution problems. Because of its indoor effects, radon is considered a hazardous pollutant in the United States under the Clean Air Act Amendments of 1990 (CAAA90).

9.1.7.1. Sources and Sinks.

The ultimate source of radon gas is the radioactive decay of solid mineral **uranium-238** (^{238}U), where the 238 refers to the isotope, or number of protons plus neutrons in the nucleus of a uranium atom. Of all uranium on Earth, 99.2745 percent is ^{238}U, 0.72 percent is ^{235}U, and 0.0055 percent is ^{234}U. ^{238}U has a half-life of 4.5 billion years.

Radon formation from uranium involves a long sequence of radioactive decay processes. During radioactive decay of an element, the element spontaneously emits radiation in the form of an alpha (α) particle, beta (β) particle, or gamma (γ) ray. An **alpha particle** is the nucleus of a helium atom, which is made of two neutrons and two protons (Figure 1.1). It is the least penetrating form of radiation and can be stopped by a thick piece of paper. Alpha particles are not dangerous unless the emitting substance is inhaled or ingested. A **beta particle** is a high-velocity electron. Beta particles penetrate deeper than do alpha particles, but less than do other forms of radiation, such as gamma rays. A **gamma ray** is a highly energized, deeply penetrating photon emitted from the nucleus of an atom not only during nuclear fusion (e.g., in the sun's core), but also sometimes during radioactive decay of an element.

The French physicist **Antoine Henri Becquerel** (1871–1937; Figure 9.1) discovered radioactive decay on March 1, 1896. To obtain his discovery, Becquerel placed a uranium-containing mineral on top of a photographic plate wrapped by thin, black paper. After letting the experiment sit in a drawer for a few days, he developed the plate and found that it had become fogged by emissions (Figure 9.2) that he traced to the uranium in the mineral. He referred to the emissions as **metallic phosphorescence**. What he had discovered was the emission of some type of particle due to radioactive decay. He repeated the experiment by placing coins under the paper and found that their outlines were traced by the emissions. Two years later, the New Zealand–born, British physicist **Ernest Rutherford** (1871–1937; Figure 9.3) found that uranium emitted two types of particles, which he named alpha and

Figure 9.1. Antoine Henri Becquerel (1871–1937). Edgar Fahs Smith Collection, University of Pennsylvania Library.

Figure 9.2. Ernest Rutherford (1871–1937). American Institute of Physics Emilio Segrè Visual Archives, William G. Myers Collection.

beta particles. Rutherford later discovered the gamma ray as well.

Equation 9.1 summarizes the radioactive decay pathway of ^{238}U to ^{206}Pb. Numbers shown are half-lives of each decay process.

^{234}Pa decays further to uranium-234 (^{234}U), then to thorium-230 (^{230}Th), then to radium-226 (^{226}Ra), and then to radon-222 (^{222}Rn).

Whereas radon precursors are bound in minerals (Lyman, 1997), ^{222}Rn is a gas and can escape through

$$
\begin{array}{ccccccccccccc}
4.5 \times 10^9 \text{ yr} & & 24 \text{ d} & & 1.2 \text{ min} & & 2.5 \times 10^5 \text{ yr} & & 8 \times 10^4 \text{ yr} & & 1620 \text{ yr} & & 3.8 \text{ d} \\
^{238}U & \longrightarrow & ^{234}Th & \longrightarrow & ^{234}Pa & \longrightarrow & ^{234}U & \longrightarrow & ^{230}Th & \longrightarrow & ^{226}Ra & \longrightarrow & ^{222}Rn & \longrightarrow & ^{218}Po \\
\alpha & & \beta & & \beta & & \alpha & & \alpha & & \alpha & & \alpha \\
\end{array}
$$

$$
\begin{array}{ccccccccccccc}
3 \text{ min} & & 27 \text{ min} & & 30 \text{ min} & & 0.00016 \text{ s} & & 22 \text{ yr} & & 5 \text{ d} & & 138 \text{ d} \\
\longrightarrow & ^{214}Pb & \longrightarrow & ^{214}Bi & \longrightarrow & ^{214}Po & \longrightarrow & ^{210}Pb & \longrightarrow & ^{210}Bi & \longrightarrow & ^{210}Po & \longrightarrow & ^{206}Pb \\
\alpha & & \beta & & \beta & & \alpha & & \beta & & \beta & & \alpha \\
\end{array}
$$

(9.1)

When it decays to produce radon, ^{238}U first releases an alpha particle, producing thorium-234 (^{234}Th), which decays to protactinium-234 (^{234}Pa), releasing a beta particle. ^{234}Pa has the same number of protons and neutrons in its nucleus as does ^{234}Th, but ^{234}Pa has one less electron than does ^{234}Th, giving ^{234}Pa a positive charge.

soil and unsealed floors into houses, where its mixing ratio builds up in the absence of ventilation. ^{222}Rn has a half-life of 3.8 days. It decays to polonium-218 (^{218}Po), which has a half-life of 3 minutes and decays to lead-214 (^{214}Pb). ^{218}Po and ^{214}Pb, referred to as **radon progeny**, are electrically charged and can be inhaled or attach to

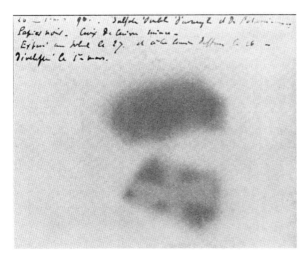

Figure 9.3. First evidence of radioactivity, along with Becquerel's notes. Edgar Fahs Smith Collection, University of Pennsylvania Library.

particles that are inhaled (Cohen, 1998). In the lungs or in ambient air, ^{214}Pb decays to bismuth-214 (^{214}Bi), which decays to polonium-214 (^{214}Po). ^{214}Po decays almost immediately to lead-210 (^{210}Pb), which has a lifetime of 22 years and usually settles to the ground if it has not been inhaled. It decays to bismuth-210 (^{210}Bi), then to polonium-210 (^{210}Po), and then to the stable isotope, lead-206 (^{206}Pb), which does not decay further.

9.1.7.2. Concentrations

Outdoor radon concentrations are generally low and do not pose a human health risk. However, because of the lack of ventilation in many houses, indoor concentrations can become thousands of times larger than outdoor concentrations (Wanner, 1993). Indoor concentrations depend on the abundance of radon in soil and the porosity of floors. Nero et al. (1986) found mean indoor concentrations of ^{222}Rn in 552 homes as 56 Bq (becquerels) m^{-3}, where a **Becquerel unit** is the number of disintegrations of atomic nuclei per second to another isotope or element. A variety of other studies found indoor concentrations of 46 to 116 Bq m^{-3} (Jones, 1999). Marcinowski et al. (1994) estimated that about 6 percent of U.S. homes contained radon concentrations in excess of U.S. EPA levels considered safe (148 Bq m^{-3}). Radon levels in homes can be reduced by installing check valves in drains, sealing basement walls and floors, and installing fans in crawl spaces to speed mixing of outside air with radon-laden air under the house.

9.1.7.3. Health Effects

^{222}Rn, a gas, is not itself harmful, but its progeny, ^{218}Po and ^{214}Pb, which enter the lungs directly or on the surfaces of aerosol particles, are believed to be highly carcinogenic (Polpong and Bovornkitti, 1998). Any activity increasing the inhalation of aerosol particles enhances the risk of inhaling radon progeny; thus, the combination of radon and cigarette smoking increases lung cancer risks above the normal risks associated with smoking (Hampson et al., 1998). Whereas several studies argue that a link exists between high radon levels and enhanced cancer rates (Henshaw et al., 1990; Lagarde et al., 1997), other studies argue that no such link exists (Etherington et al., 1996). Nevertheless, radon progeny were the second largest cause of lung cancer in the United States after smoking in 2007, causing about 21,000 lung cancer deaths (U.S. EPA, 2009c). Of these, about 18,000 occurred in smokers and 3,000 in nonsmokers, indicating that radon exacerbates the health effects of smoking.

9.1.8. Volatile Organic Compounds

Volatile organic compounds (VOCs) are organic compounds that have relatively low boiling points (50°C–260°C). Because of the their low boiling points, VOCs often evaporate from materials containing them. Sources of VOCs indoors include adhesives, solvents, building materials, combustion appliances, paints, varnishes, tobacco smoke, room deodorizers, cooking, carpets, furniture, and draperies. More than 350 VOCs with mixing ratios greater than 1 ppbv have been measured in indoor air (Brooks et al., 1991). Carpets alone emit at least 99 different VOCs (Sollinger et al., 1994). Some common VOCs in indoor air include propane, butane, pentane, hexane, n-decane, benzene, toluene, xylene, styrene, acetone, methylethylketone, and limonene, among many others. VOC mixing ratios indoors can easily exceed those outdoors by a factor of 5 (Wallace, 1991). Many VOCs are hazardous. The health effects of some VOCs were described in Table 3.15.

9.1.9. Allergens

Allergens are particles such as pollens, foods, or microorganisms that cause an allergy, which is an abnormally high sensitivity to a substance. Symptoms of an allergy include sneezing, itching, and skin rashes. Indoor sources of allergens are dust mites, cats, dogs, rodents, cockroaches, and fungi. Pollens originate

Figure 9.4. Three-dimensional rendered close-up of dust mite. © Sebastian Kaulitzki/Dreamstime.com.

mostly from outdoor trees, plants, grasses, and weeds. Omnipresent indoor allergens are **dust mite feces**. Dust mites (Figure 9.4) grow to adult size of about 250 to 350 μm from an egg in about 25 days (D'Amado et al., 1994). Their feces range in size from 10 to 40 μm. A gram of dust may contain up to 100,000 feces. The allergen in dust mite feces is a protein in the intestinal enzyme coating of the feces (Jones, 1999). Allergens in airborne dust mite feces may exacerbate symptoms in 85 percent of asthmatics (Platts-Mills and Carter, 1997). In cockroaches, the sources of allergens are body parts and feces. Cat allergens are present in the saliva, dander, and skin of the cat. Dog allergens are found in the saliva and dander of the dog. Allergens from cats, dogs, and cockroaches can trigger rapid asthmatic responses.

9.1.10. Coal Dust

A major worldwide health problem directly linked to aerosol particles is coal workers' pneumoconiosis, the medical name coined by British doctors studying the disease in 1942. This disease was first referred to as "miner's asthma" in 1822. The name was changed to **black lung disease** in 1831 by medical workers who examined the blackened lungs of deceased miners. Coal workers develop black lung disease over many years of exposure to **coal dust**. The dust first builds up in air sacs in the lungs then scars the sacs, making breathing difficult. Symptoms of the disease include chronic cough and shortness of breath.

Black lung disease is the greatest occupational hazard worldwide. Between the 1950s and 2005 in China alone, 140,000 miners died from the disease, and another 440,000 contracted it (People's Daily Online, 2005). In the United States, black lung disease killed 75,000 coal miners between 1968 and 2005 (National Institute for Occupational Safety and Health, 2008). The death rate averaged about 2,500 per year from 1968 to 1986, then declined linearly to 650 per year in 2005. The decline was due primarily to better ventilation and respiratory protection for coal miners, as well as a decrease in the number of miners. However, in 2005 and 2006, 9 percent of miners with 25 years or more experience had contracted the disease, an increase over the 4 percent rate in the late 1990s. Measures to improve conditions for coal miners in the United States started with the **Coal Mine Health and Safety Act of 1969**. Between 1969 and 2010, this act also resulted in payout of $45 billion to coal miners disabled due to black lung disease or their survivors through the **Federal Black Lung Benefits Program**. Each year, more than 300,000 recipients typically receive about $4,800 annually from the program.

9.1.11. Asbestos

Asbestos is a class of natural impure hydrated silicate minerals that can be separated into flexible fibers (e.g., Figure 9.5). It is mined in open pits by blasting rock until asbestos is exposed and removed by machines. The raw asbestos mineral is waxy, greasy, and soft, but

Figure 9.5. Unmilled rock sample of mineral asbestos. © Pancaketom/Dreamstime.com.

contains brittle fibers that easily fracture. Asbestos is chemically inert, does not conduct heat or electricity, and is fire resistant. As a result, it is has been widely used as an insulator and fire retardant.

Inhabitants of east Finland used asbestos as far back as 4,500 years ago to strengthen earthenware pots and cooking utensils. The Egyptians were aware of asbestos' fire-resistant properties and used asbestos in burial cloth. Around 300 BC, the Greek author Theophrastus described asbestos in detail in his book, *On Stones*. The name asbestos derives from *asbestinon* ("unquenchable"), a term used in the Roman Pliny the Elder's manuscript, *Natural History*. The French emperor **Charlemagne** (742–814) amused his guests at dinner parties by wiping his mouth with a napkin made of asbestos, throwing the napkin into the fireplace, and then retrieving the unburned napkin to their surprise.

The adverse health effects of asbestos were suggested as early as 1898, when the Chief Inspector of Factories in the UK stated that asbestos had "easily demonstrated" health risks. A London doctor testified that the 1899 death of a young asbestos worker was due in part to asbestos dust. Asbestos-related illnesses were reported among UK textile workers in the 1920s and U.S. ship workers during World War II. Nevertheless, asbestos use grew throughout the construction industry as an electrical and thermal insulator, particularly in pipe and boiler insulation, cement board, floor and ceiling tiles, paint, and wallpaper (Maroni et al., 1995; Jones, 1999). Asbestos has also been used in sheetrock tape, mud and texture coats, plasters and stuccos, tars, felts, shingles, putty, caulk, brake pads, lamp wicks, stage curtains, electric ovens, fire blankets and doors, firefighter clothing, and fireproofing. More than 3,000 products contained asbestos at one time or another. The early lack of concern about asbestos health effects is evidenced by the use of asbestos as a filter in Kent cigarettes between 1952 and 1956.

Between 1920 and 2006, the five leading asbestos-using countries worldwide were the United States, the UK, Germany, Japan, and France, with the United States consuming 42.6 percent of all asbestos among 56 countries (Park et al., 2011). The United States first banned the production of asbestos for new fireproofing and insulation in 1973 and for decorative use in 1978, although existing stock could be used until 1986. Asbestos is also defined as a hazardous air pollutant under CAAA90. However, asbestos can still be used in the construction of, for example, cement asbestos pipes in the United States. Although most new asbestos production is banned in the United States and all new production is banned in the EU, Australia, Hong Kong, Japan, and New Zealand, much of the asbestos installed in buildings before the bans occurred still exists. Furthermore, many countries, particularly developing countries, still allow construction with asbestos, particularly for roofing and side walls in homes, factories, and schools. In 2006, the five leading asbestos-mining countries were Russia, China, Kazakhstan, Canada, and Brazil.

People who were most likely to be exposed to asbestos in the past were miners and insulation manufacturers and installers. From 1913 to 1990, a mine in Libby, Montana, became the world's largest producer of vermiculite, supplying 80 percent of the world's supply. Vermiculite is a clay mineral intertwined with asbestos. Upon heating, vermiculite expands to featherweight pieces used in wall insulation, wallboard, nursery products, and wood products. Unknown to the miners, the vermiculite was contaminated with asbestos. Between the early 1980s and 2000, 192 people from the town of 2,700 died, and 375 others were diagnosed with asbestos-related lung problems. Some miners inadvertently exposed their families to asbestos when they brought their clothes, carrying asbestos-laden dust, into their homes. Today, the Libby, Montana mine, owned from 1963 to 1990 by the W. R. Grace Company, is an U.S. EPA Superfund site.

9.1.11.1. Sources and Sinks

Once insulation containing asbestos has been installed, the asbestos is not expected to cause damage to humans unless the insulation is disturbed. At that time, fibers can be scattered into the air, where they can remain for minutes to days until they deposit to the ground or are inhaled. Thus, the only source of asbestos in indoor air is turbulent uplift, and the only sink is deposition.

9.1.11.2. Concentrations

Indoor concentrations of asbestos vary from building to building. Lee et al. (1992) found an average of 0.02 structures per cubic centimeters of air in 315 public, commercial, residential, school, and university buildings in the United States. The concentration of fibers longer than 5 μm was only 0.00013 structures per cubic centimeter.

9.1.11.3. Health Effects

The primary health effects of asbestos exposure are lung cancer, mesothelioma, and asbestosis. In 2007, about

4,800 people in the United States died of asbestos-induced lung cancer (U.S. EPA, 2009b). **Mesothelioma** is a cancer of the mesothelial membrane lining the lungs and chest cavity. Between 1994 and 2008, about 174,300 people in fifty-six countries contracted mesothelioma, causing more than 92,100 deaths (Park et al., 2011). **Asbestosis** is a slow, debilitating disease of the lungs, whereby bodily produced acids scar the lungs as they try to dissolve asbestos fibers in lung tissue. Significant scarring causes crackling of the lungs and inhibits oxygen from transferring from the airway to the blood, making it difficult to breathe. The most dangerous asbestos fibers are those longer than 5 to 10 μm but with a diameter of 0.01 to 1 μm. Exposure to 0.0004 structures per cubic centimeter of air is expected to cause a lifetime cancer risk of 160 cases of mesothelioma per 1 million people (Turco, 1997). People with lung cancer attributable to asbestos typically have more than 100 fibers per gram of lung tissue (Wright et al., 2008). The time between first exposure to asbestos and the appearance of tumors is estimated to be 20 to 50 years (Jones, 1999). Cigarette smoking and exposure to asbestos are believed to amplify the rates of lung cancer in comparison with the rates of cancer associated with just smoking or just exposure to asbestos. Short-term acute exposure to asbestos can lead to skin irritation and itching (Spengler and Sexton, 1983). Today, miners and those who remove asbestos from buildings are the people most likely to be exposed to asbestos in concentrations high enough to cause health problems.

9.1.12. Fungal Spores, Bacteria, Viruses, and Pollen

Fungal spores, bacteria, viruses, and pollen are common indoor air contaminants. They can infiltrate indoors from outdoor air or grow indoors. **Fungal spores** are reproductive or resting organisms released by fungi and algae growing on leaf surfaces, soils, animals, and foodstuffs. They are generally 2 to 3.5 μm in diameter. Fungi grow well on damp surfaces. Common fungi in buildings include *Penicillium*, *Cladosporium*, and *Aspergillus* (Jones, 1999).

Bacteria and viruses live in water, soil, plants, animals, and foodstuffs, and they can be picked up readily by wind or turbulence. Common bacteria include *Bacillus*, *Staphylococcus*, and *Micrococcus*. Bacteria are generally 0.5 to 10 μm in diameter.

Pollens are large granules containing male genetic material released from flowers and blown by the wind to other flowers for fertilization. Emitted pollen grains are generally large, ranging from 10 to 125 μm in diameter. A typical corn plant emits 14 to 50 million pollen grains/yr, all within a short period (Miller, 1985). Pollen grains are released more during spring than in any other season (Pasken and Pietrowicz, 2005).

Some diseases associated with fungi, bacteria, and viruses include rhinitis (a respiratory illness), asthma, humidifier fever, extrinsic allergic alveolitis, and atopic dermatitis. Most human illnesses due to viral and bacterial infection are due to human-to-human transmission of microorganisms rather than to building-to-human transmission (Ayars, 1997).

9.1.13. Environmental Tobacco Smoke

When a person smokes a cigarette, some of the smoke is inhaled and swallowed, some is inhaled and exhaled (**mainstream smoke**), and the rest is emitted from the burning cigarette between puffs (**sidestream smoke**) (Figure 9.6). The mainstream plus sidestream smoke is called **environmental tobacco smoke** (ETS), a mixture of more than 4,000 aerosol particle components and gases, at least 50 of which are known carcinogens. However, because of the different conditions resulting in mainstream versus sidestream smoke, the relative gas and particle composition of the two types of smoke differ somewhat, with sidestream smoke possibly being more dangerous (Schick and Glantz, 2005). Also called **second-hand smoke**, ETS builds up in enclosed spaces, increasing danger to others in the vicinity. Even in well-ventilated indoor areas, particle and gas concentrations associated with ETS increase. Although the cumulative effect of ETS on outdoor air pollution is relatively small compared with the effects of other sources of pollution, such as automobiles, ETS concentrations can build up outdoors in the vicinity of smokers.

Outdoor ETS emission rates are not regulated in the United States, although regulations prohibiting smoking in many public and private indoor facilities exist at the local, state, and federal levels. Many chemical constituents of ETS are classified as hazardous air pollutants under CAAA90, and ETS itself has been classified by the U.S. EPA as a carcinogen. However, CAAA90 controls only sources emitting more than 9.1 tonnes of a hazardous substance per year, and individual cigarettes emit less than 1 g of all pollutants combined. The product of the number of cigarettes smoked per year in the United States multiplied by the emission rate per cigarette is much larger than

(a)   (b)

Figure 9.6. (a) Mainstream and (b) sidestream smoke. © Robert Kneschke/Dreamtime.com.

9.1 tonnes per year, but this statistic is not recognized by CAAA90.

9.1.13.1. Sources

Table 9.2 shows main- and sidestream emissions for some components emitted from cigarettes. The actual emission rate of a cigarette depends on the type of tobacco, the density of its packing, the type of wrapping paper, and the puffing rate of the smoker (Hines et al., 1993).

Table 9.2 indicates that emission rates of sidestream smoke are greater than are those of mainstream smoke

Table 9.2. Main- and sidestream emission rates per cigarette

Substance	Mainstream smoke (μg/cigarette)	Sidestream smoke (μg/cigarette)
Carbon dioxide	10,000–80,000	81,000–640,000
Carbon monoxide	500–26,000	1,200–65,000
Nitrogen oxides	16–600	80–3,500
Ammonia	10–130	400–9,500
Hydrogen cyanide	280–550	48–203
Formaldehyde	20–90	1,000–4,600
Acrolein	10–140	100–1,700
N-nitrosodimethylamine	0.004–0.18	0.04–149
Nicotine	60–2,300	160–7,600
Total particulates	100–40,000	130–76,000
Phenol	20–150	52–390
Catechol	40–280	28–196
Naphthalene	2.8	45
Benzo(a)pyrene	0.008–0.04	0.02–0.14
Aniline	0.10–1.20	3–36
2-Naphthylamine	0.004–0.027	0.02–1.1
4-Aminobiphenyl	0.002–0.005	0.06–0.16
N'-nitrosonornicotine	0.2–3.7	0.02–18

Sources: Rando et al. (1997); Jones (1999).

Table 9.3. Comparison of emissions from cigarettes and mobile sources in the United States

Substance	(a) Average cigarette emissions[a] (g/cigarette)	(b) Average vehicle emissions[b] (g/mi)	(c) Number of cigarettes resulting in same emissions as 1 mile of driving[c]	(d) Estimated cigarette emissions in United States[d] (tonnes/day)	(e) Estimated mobile source direct emissions in United States[e] (tonnes/day)
Carbon monoxide	0.0464	4.2	90.5	61	193,000
Nitrogen oxides	0.0021	0.07	33.3	2.7	40,600
Particles	0.058	0.01	0.17	76	12,200

[a] Sum of average main- and sidestream emissions in Table 9.2.
[b] 2004 U.S. EPA Tier II, Bin 5, light-duty vehicle emission standards.
[c] Column (b) divided by column (a).
[d] Assumes that the population of the United States in 2011 is 311,000,000 and 21 percent of the population smokes 20 cigarettes per day.
[e] Estimated 2008 emissions from U.S. EPA (2011a).

for many pollutants. Thus, a person standing a short distance from a cigarette is often exposed to more pollution than is the smoker (Schlitt and Knöppel, 1989).

Table 9.3 compares emissions from a cigarette with those from a vehicle. Driving a vehicle that meets 2004 U.S. EPA Tier II emission standards for 1 mile results in particle emissions equivalent to emissions from about one-fifth of one cigarette. Emissions of CO(g) and NO_x(g) from driving 1 mile are equivalent to those from smoking 91 and 33 cigarettes, respectively. The emissions per day of these chemicals, summed over all cigarettes smoked in the United States each day, is much less than those from all vehicle miles driven each day, but ETS is often emitted in enclosed spaces, where its concentrations build up.

9.1.13.2. Concentrations

ETS contributes to the buildup of gas and particle concentrations indoors. Spengler et al. (1981) found that one pack of cigarettes per day contributes to about 20 $\mu g\, m^{-3}$ of particle concentration indoors over a 24-hour period. During the time a cigarette is actually smoked, particle concentrations increase to 500 to 1,000 $\mu g\, m^{-3}$ near the smoker. Leaderer et al. (1990) found that particle concentrations in homes with a cigarette smoker were up to three times those in homes without a smoker.

9.1.13.3. Health Effects

Approximately 1.3 billion people smoked worldwide in 2010. In 2009, about 5 million smokers died from direct smoking-related illnesses. This number is expected to increase to 8 million/yr by 2030, with most occurring in middle- and low-income countries. *The total number of premature deaths due to smoking during the twenty-first century is projected to be about 1 billion* (WHO, 2009b).

In 1986, 70 percent of all children in the United States lived in households in which at least one parent smoked (Weiss, 1986). Between 1978 and 2010, though, the percent of the population that smoked regularly in the United States decreased from 34 to 21 percent. In the United States, direct smoking causes about 250,000 deaths/yr, 160,000 of which are from lung cancer. Most of the remaining direct smoking-related deaths are due to cancer of the pancreas, esophagus, urinary tract, and stomach (U.S. EPA, 1993).

Worldwide, about 600,000 people die prematurely each year from ETS. This represents about 1 percent of the all-cause mortality. About 62.8 percent of ETS deaths are from heart disease, 27.4 percent from lower respiratory infections, 6.1 percent from asthma, and the rest from lung cancer. About 47 percent of the deaths are in adult women, 26 percent in adult men, and 28 percent in children (Oberg et al., 2010). In the United States, ETS causes 150,000 to 300,000 lower respiratory tract infections, such as bronchitis and pneumonia, in infants and children younger than 18 months each year. About 7,500 to 15,000 of these illnesses result in hospitalization. ETS also causes 3,000 U.S. lung cancer deaths annually (U.S. EPA, 1993).

Health effects studies suggest that short-term exposure to ETS results in eye, nose, and throat irritation for most individuals, and allergic skin reactions for some

(Maroni et al., 1995). ETS also elevates symptoms for people who have asthma and may induce asthma in some children (Jones, 1999). ETS exposure has also been linked to lower respiratory tract illness (Somerville et al., 1988). Children exposed to ETS are hospitalized more often than children who are not exposed (Harlap and Davies, 1974). Janerich et al. (1990) found that ETS may cause up to 17 percent of lung cancers among nonsmokers. Rando et al. (1997) found a 30-percent increase in the cancer risk to women whose husbands smoked, and Ryan et al. (1992) linked brain cancer tumors to ETS. Long-term exposure also results in cardiovascular disease and pulmonary malfunction, contributing to further mortality (e.g., Centers for Disease Control and Prevention, 2005; U.S. Department of Health and Human Services, 2006). There is no low threshold to the health effects of ETS (U.S. Department of Health and Human Services, 2006).

9.1.14. Indoor Solid Biofuel and Coal Burning

One of the leading causes of death worldwide is the indoor burning of solid biofuels and coal for home heating and cooking (e.g., Figure 9.7). Such burning is carried out by large segments of the population in many developing countries. **Solid biofuels** used for home heating and cooking are generally wood, grass, agricultural waste, and dung. The use of solid biofuels is often favored over the use of kerosene or other fossil fuels due to the lower price or better availability of biofuels, particularly in places such as sub-Saharan Africa.

Indoor fuel burning kills about 1.6 million people prematurely each year, with 56 percent of the deaths

Figure 9.7. Soot-blackened iron pot cooking over an open fire in a rural kitchen, August 24, 2009. © Henrischmit/Dreamstime.com.

occurring in children younger than 5 years (WHO, 2005). Indoor burning may also put half the world's population at risk of air pollution–related disease or mortality (WHO, 2008). Indoor PM levels during burning are ten to fifty times higher than recommended safe values. Women and children are disproportionately affected because they spend more time near the fires than do adult men. Whereas indoor air pollution mortality causes about 2.7 percent of all deaths worldwide annually, it is responsible for about 3.7 percent of such mortality in developing countries.

Most deaths from indoor fuel burning are due to pneumonia, chronic respiratory disease, and lung cancer. Indoor smoke significantly increases the risk of pneumonia primarily among those 5 years old and younger. Exposure to smoke doubles the risk of pneumonia; thus, it triggers more than 900,000 of the world's 2 million annual deaths from pneumonia (WHO, 2005). Indoor smoke also strongly increases the risk of chronic respiratory disease and lung cancer, primarily among adults older than 30 years. In addition, women exposed to indoor smoke have three times the chance of contracting **chronic obstructive pulmonary disease (COPD)**, such as chronic bronchitis, than do women who cook with cleaner fuels. Smoke exposure to men doubles their risk of COPD. Indoor air pollution may cause about 700,000 COPD-related deaths, whereas indoor smoke may cause about 15,000 lung cancer deaths per year. The link between biofuel smoke and cataracts and tuberculosis is more uncertain.

The ultimate solution to the problems associated with indoor fuel burning is to replace wood- and coal-burning stoves and heaters with electric heaters and/or to ventilate smoke exhaust better. This is not a technological barrier to overcome, but an economic and social barrier, because many societies are resistant to change.

9.2. Sick Building Syndrome

In some workplaces, employees experience an unusually high rate of headaches, nausea, nasal and chest congestion, eye and throat problems, fatigue, fever, muscle pain, dizziness, and dry skin. These symptoms, present during working hours, often improve after a person leaves work. The situation described is called **sick building syndrome (SBS)**. The cause of SBS is not certain, but it may be due to certain VOCs, inadequate building ventilation systems, or molds. SBS may also be caused by enhanced stress levels and heavy workloads, or a combination of psychological and chemical factors (Jones, 1999).

SBS may alternatively be caused by exposure to many pollutants in low doses simultaneously. This condition is referred to as **multiple chemical sensitivity (MCS)**. People who are exposed over a long period of time to low levels of many chemicals due to poor ventilation in an office building can have reactions that people without MCS usually tolerate. Statistically, one-third of people working in a sealed building may be sensitive to one or more common chemicals. Triggers for MCS include tobacco smoke, perfume, traffic exhaust, nail polish remover, newspaper ink, hair spray, paint thinner, paint, insecticide, artificial colors and sweeteners, carpeting, adhesive tape, flame retardant, felt-tip pens, and chlorine (MedicineNet, n.d.).

9.3. Personal Clouds

Concentrations of aerosol particles and gases measured in the vicinity of an individual who is indoors are often greater than are concentrations measured from a stationary indoor monitor away from the individual. The relatively high concentration of pollution measured near an individual is called a **personal cloud** (e.g., Rodes et al., 1991; McBride et al., 1999; Wallace, 2000). A personal cloud may arise when a person's movement stirs up gases and particles on clothes and nearby surfaces, increasing pollutant concentrations. People also release thermal-IR radiation, which rises and thus stirs and lifts pollutants.

Personal cloud concentrations of $PM_{2.5}$ and PM_{10} were found in one study to be 1.4 and 1.6 times, respectively, those of background indoor concentrations when the person was doing normal activity, such as walking or sitting (Ferro et al., 2004). Personal cloud concentrations increased to six and seventeen times, respectively, those of background values when activities such as dusting, folding clothes and blankets, and making a bed were performed. Dusting resulted in peak PM_5 concentrations of 300 μg m^{-3} for 3 minutes, whereas two people walking resulted in peak concentrations of 130 μg m^{-3} for 3 minutes. For comparison, concentrations with no one in the house were 11 μg m^{-3}. In sum, personal cloud concentrations are higher than background concentrations, which are determined by stationary monitors. Thus, personal clouds cause additional health impacts not considered when stationary monitors are used to measure pollutant concentrations.

9.4. Regulation of Indoor Air Pollution

Worldwide, many countries regulate indoor air pollutant concentrations in workplaces, public buildings, and schools; however, few enforceable regulations exist in residences. In addition, indoor and some outdoor smoking regulations have been enacted in more than 100 countries (WHO, 2009a). In this section, indoor regulation of air pollution is briefly discussed, with a focus on U.S. regulations as an example.

In the United States, NAAQS control outdoor air pollutants only. No regulations control air pollution concentrations in indoor residences. In 1994, California was the first state to ban smoking in indoor workplaces. However, as of 2011, smoking was still legal in California Indian casinos, resulting in high concentrations of pollution particles in these venues (Jiang et al., 2011).

Aside from smoking regulations, regulatory standards for individual pollutant concentrations in indoor workplaces are set by the **Occupational Safety and Health Administration (OSHA)**, which obtains recommendations for standards from another government agency, the **National Institute for Occupational Safety and Health (NIOSH)**, and an independent professional society, the **American Conference of Governmental Industrial Hygienists (ACGIH)**. OSHA and NIOSH were created by the Occupational Safety and Health Act in 1970. OSHA, an agency within the U.S. Department of Labor, is responsible for setting and enforcing workplace standards, whereas NIOSH, an agency within the U.S. Department of Health and Human Services, is responsible for researching workplace health issues. ACGIH's primary mission is to promulgate workplace safety standards. Workplace regulations apply to public buildings and schools, which are also considered workplaces.

NIOSH recommends **permissible exposure limits (PELs)**, **short-term exposure limits (STELs)**, and **ceiling concentrations**. A PEL is the maximum allowable concentration of a pollutant in an indoor workplace over an 8-hour period during a day. A STEL is the maximum allowable concentration of a pollutant over a 15-minute period. A ceiling concentration is a concentration that may never be exceeded. ACGIH sets 8-hour **time-weighted average threshold limit values (TWA-TLVs)**, which are similar to PELs. For most pollutants, PELs and TWA-TLVs are the same. When differences occur, they are small.

Indoor standards exist for more than 150 compounds. Table 9.4 compares indoor standards with outdoor standards for four compounds. For ozone and carbon monoxide, the 8-hour PEL/TWA-TLV standards are less stringent than are the outdoor standards. The reason is that outdoor standards are designed to protect the entire population, particularly infants and people afflicted with disease or illness. Indoor standards are designed to protect workers, who are assumed to be

Table 9.4. Comparison of indoor workplace standards with outdoor federal and California State standards for selected gases

Gas	Indoor 8-hr PEL and TWA-TLV[a] (ppmv)	Indoor 15-min STEL[a] (ppmv)	Indoor ceiling[a] (ppmv)	Outdoor NAAQS (ppmv)	Outdoor California standard (ppmv)
Carbon monoxide	35	–	200	9.0 (8-hr)	9 (8-hr)
Nitrogen dioxide	–	1	–	0.053 (annual)	0.18 (1-hr)
Ozone	0.1	0.3	–	0.075 (8-hr)	0.07 (8-hr)
Sulfur dioxide	2	5	–	0.14 (24-hr)	0.04 (24-hr)

NAAQS, National Ambient Air Quality Standards; PEL, permissible exposure limit; STEL, short-term exposure limit; TWA-TLV, time-weighted average threshold limit value.
[a] National Institute for Occupational Safety and Health (2010).

healthier than is the average person. Table 9.4 also indicates that the 15-minute STEL for nitrogen dioxide is five times higher (less stringent) than is the 1-hour outdoor California standard. Stringent outdoor standards for nitrogen dioxide are set because it is a precursor to photochemical smog. Because UV sunlight does not penetrate indoors, nitrogen dioxide does not produce ozone indoors, and indoor regulations of nitrogen dioxide as a smog precursor are not necessary. Indoor standards for nitrogen dioxide are based solely on health concerns.

9.5. Summary

People spend most of their time indoors; thus, the quality of indoor air has a significant impact on human health risk. Indoor air contains many of the same pollutants as outdoor air, but pollution concentrations in indoor and outdoor air usually differ. Although indoor mixing ratios of ozone and sulfur dioxide are usually less than are those outdoors, indoor mixing ratios of formaldehyde are usually greater than are those outdoors. Indoor mixing ratios of carbon monoxide and nitrogen dioxide are generally the same as or less than are those outdoors, unless appliances or other indoor combustion sources are turned on. Indoor concentrations of radon, asbestos, and ETS, when present, are usually greater than are those outdoors, giving rise to potentially serious health problems for people exposed to these pollutants indoors. Indoor air also contains VOCs, allergens, fungi, bacteria, and viruses. A major source of human mortality and illness worldwide is the indoor burning of solid biofuels and coal for home heating and cooking. This problem could be mitigated with current technologies. In the United States, indoor air is regulated only in the workplace and in public buildings; residential air is not regulated.

9.6. Problems

9.1. Why are ozone mixing ratios almost always lower indoors than outdoors?

9.2. Why are workplace standards for pollutant concentrations generally less stringent than standards for outdoor air?

9.3. What would be the volume mixing ratio (ppmv) of carbon monoxide and the mass concentration ($\mu g \ m^{-3}$) of particles if ten cigarettes were smoked in a $5 \times 10 \times 3$-m room? How do these values compare with the U.S. federal primary 1-hour standard for CO(g) and 24-hour average standard for PM (Table 8.2)? Based on the results, which pollutant do you believe is more of a cause for concern with respect to indoor air quality? Assume that the dry air partial pressure is 1,013 hPa and the temperature is 298 K, and use cigarette emission rates from Table 9.2.

9.4. Why is radium less of a concern than radon?

9.5. Why is removing asbestos from buildings often more dangerous than leaving it alone?

9.6. Why does the gas-phase chemical decay of organic compounds generally take longer indoors than outdoors?

9.7. Identify three methods that could be used to reduce indoor pollution or its exposure due to the indoor burning of solid biofuels and coal for home heating and cooking.

9.8. Identify five indoor activities that could increase your personal cloud concentration of air pollution.

Chapter 10

Acid Deposition

Acid deposition occurs when an acid – primarily sulfuric acid, nitric acid, or hydrochloric acid – is emitted into or produced in the air and deposits to soils, lakes, grass, forests, or buildings. Acid deposition can be dry or wet. **Dry acid deposition** is the direct deposition of acid gases to surfaces. **Wet acid deposition** is the deposition to the surface of acids dissolved in rainwater (**acid rain**), fog water (**acid fog**), or liquid aerosol particles (**acid haze**). On the Earth's surface, acids have a variety of environmental impacts, including damage to microorganisms, fish, forests, agriculture, and structures. When breathed in, acids in high concentrations are harmful to humans and animals. Acid deposition problems have occurred since coal was first combusted and increased during the Industrial Revolution in the eighteenth century. The problems became more severe with the growth of the alkali industry in nineteenth-century France and the UK. In this chapter, the history, science, and regulatory control of acid deposition problems are discussed.

10.1. Historical Aspects of Acid Deposition

Acid deposition is caused by the emission or atmospheric formation of gas- and aqueous-phase **sulfuric acid** (H_2SO_4), **nitric acid** (HNO_3), or **hydrochloric acid** (HCl). Historically, coal was the first and largest source of anthropogenically produced atmospheric acids. Coal combustion emits gas-phase sulfur dioxide [$SO_2(g)$], hydrochloric acid [HCl(g)], and nitrogen oxides [$NO_x(g)$]. $SO_2(g)$ oxidizes to gas- and aqueous-

phase sulfuric acid, and $NO_x(g)$ oxidize to nitric acid. Humans have combusted coal for thousands of years. In the 1200s, sea coal was brought to London and used in lime kilns and forges (Section 4.1). It was later burned in furnaces to produce glass and bricks, in breweries to produce beer and ale, and in homes to provide heat. Beginning with the eighteenth-century **Industrial Revolution**, coal combustion provided energy for the steam engine.

During the late eighteenth century, a second major source of atmospheric acids emerged. In France and England, the demand for soaps, detergents, cleansers, glass, paper, bleaches, and dyes was increasing rapidly. **Soap** was produced by combining an alkali, such as **potash** [**potassium carbonate**, $K_2CO_3(s)$] or **soda ash** [**sodium carbonate**, $Na_2CO_3(s)$], with animal fat. The source of potash was the ashes of wood fires (pot ashes) from which the potash was extracted using hot water. Because England and France were largely deforested, wood for potash was imported from North America, Scandinavia, or Russia, increasing the cost of potassium carbonate. Sodium carbonate was also expensive. It could be extracted from coastal barilla plants in Spain and the Canary Islands, mined from mineral deposits in Egypt's Lakes of Natron (Section 1.2.1.6), or extracted from sea kelp washing up on the beaches of Scotland and Ireland. However, these sources were either far (Mediterranean and Egypt) or labor intensive (sea kelp). For example, more than 100,000 seasonal workers were required to collect and process sea kelp along Scotland's shores in order to provide sodium carbonate for England, thus driving up its cost.

Sensing the urgency for a better source of alkali for France, the French Academy of Sciences, in 1783, offered a prize to the person who could develop the most efficient and economic method of producing soda ash from **common salt** (sodium chloride), which was readily available. **Nicolas Leblanc** (1742–1806), encouraged by the competition, began experimenting in 1784 to find a new process. In 1789, he devised a two-step set of reactions to produce soda ash. In the first step, he dissolved common salt into a sulfuric acid solution at high temperature (800°C–900°C) in an iron pan to produce dissolved sodium sulfate (**Glauber's salt**) and hydrochloric acid gas by

$$2NaCl(s) + H_2SO_4(aq) \xrightarrow{\text{High temperature}} Na_2SO_4(aq) + 2HCl(g)$$

Sodium chloride Sulfuric acid Sodium sulfate Hydrochloric acid

$$(10.1)$$

In the second step, he heated sodium sulfate together with charcoal and chalk in a kiln to form sodium carbonate by

$$Na_2SO_4(aq) + 2C + CaCO_3(s) \xrightarrow{\text{High temperature}} Na_2CO_3(aq)$$

Sodium sulfate Carbon from charcoal Calcium carbonate from chalk Sodium carbonate

$$+ CaS(s) + 2CO_2(g) \quad (10.2)$$

Calcium sulfide Carbon dioxide

By-products of the second reaction, when complete, included **calcium sulfide** [CaS(s)], an odorous, yellow-to-light gray powder, and carbon dioxide gas. In practice, the reaction was incomplete and produced gas-phase sulfuric acid and soot as well. The sodium carbonate residue was separated from calcium sulfide by adding water, which preferentially dissolved the sodium carbonate. The resulting solution was then dried, producing sodium carbonate crystals. Finally, sodium carbonate was combined with animal fat to produce soap.

A necessary ingredient for the production of sodium carbonate was aqueous sulfuric acid (Reaction 10.1). One source of sulfur was combustion of elemental sulfur (S) powder with saltpeter [KNO$_3$(s)], and then dissolving the resulting H$_2$SO$_4$(g) in water, as was done by Libavius in 1585. This process was inefficient, releasing volumes of gas-phase sulfuric acid and nitric oxide [NO(g)] (which converts to nitrogen dioxide and then to nitric acid in air). Another source was the burning of iron pyrites [FeS$_2$(s)] and arsenopyrites [FeAsS$_2$(s)]. Burning of the latter resulted in the release of arsenic vapor, which was emitted to the air along with the other pollutants.

In sum, although the production of sodium carbonate from common salt and sulfuric acid allowed the **alkali industry** to become self-sufficient and escape reliance on the import of natural sodium carbonate, it resulted in the release of HCl(g), H$_2$SO$_4$(g), HNO$_3$(g), and soot, causing widespread acid deposition and air pollution in France and the UK (Figure 10.1a). The Leblanc process also produced large amounts of impure solid alkali waste containing calcium sulfide. The waste was piled near each factory and commonly referred to as **galligu**, a name ascribed specifically to waste from the Leblanc process. When rainwater fell on the waste, some of the

(a)

(b)

Figure 10.1. (a) Alkali factory emissions in the industrial town of Widnes, Cheshire County, England, in the early 1800s. Courtesy Halton Borough Council, UK; www.ceb.cam.ac.uk/exemplarch2002/mcp21/leblanc.jpg. (b) Piles of waste galligu from an alkali factory. Courtesy Halton Borough Council, UK; www2.halton.gov.uk/images/main/conlanddittonalps.

calcium sulfide dissolved, producing dissolved **hydrogen sulfide** [$H_2S(aq)$], which evaporated as a harmful, odorous, colorless gas. Much of the rest of the calcium sulfide oxidized over time to form calcium sulfate [$CaSO_4–2H_2O(s)$, gypsum]. Piles of galligu from soda ash factories can still be seen today (Figure 10.1b).

Leblanc patented the soda ash technique and started an alkali factory at St. Denis, a town near Paris, in 1791. The factory produced soda ash daily, and his product was in high demand because barilla supplies from Spain were scarce due to a war between France and Spain. However, in 1793, Leblanc lost his patent and factory to the state, which nationalized patents and factories during the French Revolution. In 1801, Napoleon returned Leblanc's St. Denis plant to him, but the plant was in poor condition. The government also refused to award the prize money for the French Academy of Sciences competition he had won. Unable to recover from his business losses or compete with new alkali factories, Leblanc committed suicide on January 6, 1806.

Nevertheless, Leblanc's soda ash process dominated the alkali industry until the 1880s. In 1861, Belgian chemist **Ernst Solvay** (1838–1922) developed a more efficient process of producing soda ash by raining sodium chloride and ammonia, both dissolved in water, down a tower (**Solvay tower**) over an up-current of carbon dioxide gas, produced by heating calcium carbonate. The technique required no sulfuric acid or potassium nitrate (saltpeter). The only solid waste product was calcium chloride, which was used as road salt. The main disadvantage was the cost of the tower. Only by the 1880s did the Solvay process overtake the Leblanc process in economic efficiency and popularity. However, Leblanc's idea did not disappear quickly. The last Leblanc process soda ash factory shut down only in the early 1920s.

In 1863, an estimated 1.6 million tonnes of raw material were burned to form soda ash, producing only 0.25 million tonnes of useful products. Most of the remaining 1.35 million tonnes was emitted as $HCl(g)$ or other gases and produced as solid waste (Brock, 1992). Environmental damage due to the alkali industry was severe. $HCl(g)$ and other pollutants rained down onto agricultural property and cities in France and the UK, causing property and health damage. In the UK, St. Helens, Newcastle, and Glasgow countrysides were decimated (Brimblecombe, 1987).

Early complaints against alkali factories were in the form of civil litigation. For example, in 1838, a Liverpool landowner filed a complaint against an alkali factory, charging that it destroyed his crops and interfered with his hunting. Ultimately, $HCl(g)$ from alkali manufacturers was regulated in France and the UK. In France, regulation took the form of planning laws that controlled the location of alkali factories. In the UK, the **British Alkali Act of 1863** required alkali manufacturers to reduce 95 percent of their $HCl(g)$ emissions. The act also called for the appointment of an alkali inspector to watch over the industry.

Impetus for the British Alkali Act came in the early 1860s, when **William Gossage** invented a technique to wash $HCl(g)$ from waste gases before it was released from chimneys. Gossage built his own soda ash factory in Worcestershire in 1830. Spurred by his neighbors' complaints about the hydrochloric acid emitted from his factory, he worked to mitigate the problem. His solution was to convert a windmill into a tower (later called a **Gossage tower**), fill the tower with brushwood, and spray water down the top of the tower as smoke rose from the bottom. Gossage had invented the first air pollution **scrubber**. The water dissolved most of the hydrochloric acid (just as rain does) and drained it into a nearby waterway. Whereas the Gossage tower reduced $HCl(g)$, the same hydrochloric acid was now dissolved in wastewater from the tower, where it destroyed nearly all marine life and corroded all structures in its path downstream.

Nevertheless, because the technique was so simple and inexpensive and because the consequences of not implementing the technique were so severe, the British Alkali Act easily passed shortly after Gossage's invention. Subsequent to the act, Walter Weldon (in 1866) and Hugh Deacon (in 1868) developed processes for converting $HCl(g)$ to chlorine that could be used for bleaching powder. These inventions allowed chlorine, which otherwise would have been wasted, to be recycled.

Even though the British Alkali Act was successful at reducing $HCl(g)$, the alkali industry continued to emit sulfuric acid, nitric acid, soot, and other pollutants in abundance. Because the act did not control emissions from factories other than those producing soda ash, pollution problems in the UK worsened. The British Alkali Act was modified in 1881 to regulate other sections of the chemical industry, but the number of factories and the volume of emissions from them had grown so much that the new law had little effect. In 1899, one writer described St. Helens as

> . . . a sordid ugly town. The sky is a low-hanging roof of smeary smoke. The atmosphere is a blend of railway tunnel, hospital ward, gas works and open sewer.

Figure 10.2. Robert Angus Smith (1817–1884). Reproduced courtesy of the Library and Information Centre, Royal Society of Chemistry.

The features of the place are chimneys, furnaces, steam jets, smoke clouds and coal mines. (Blatchford, 1899, p. 15)

The first British Alkali Act inspector in the UK was Scottish chemist **Robert Angus Smith** (1817–1884; Figure 10.2). He was charged with ensuring that the industry reduced HCl(g) emissions by 95 percent. Smith was also a field experimentalist. In 1872, he published *Air and Rain: The Beginnings of a Chemical Climatology*, in which he discussed results of the first monitoring network for air pollution in Great Britain. As part of the analysis, he recorded the gas-phase mixing ratios of molecular oxygen and carbon dioxide, and measured the composition of chloride, sulfate, nitrate, and ammonium in rainwater in the British Isles. In his book, Smith introduced the term **acid rain** to describe the high sulfate concentrations in rain near coal-burning facilities.

Throughout the twentieth century, acid deposition problems continued to plague cities in the UK and municipalities in other countries where coal and chemical burning occurred. The fatal pollution episodes in London, discussed in Chapter 4, included contributions from acidic compounds in smoke.

Acid deposition became an issue of international interest in the 1950s and 1960s, when a relationship was found between sulfur emissions in continental Europe and acidification of Scandinavian lakes. These studies led to a 1972 United Nations Conference on the Human Environment in Stockholm that called for an international effort to reduce acidification. Since then, numerous studies on acid deposition have been carried out, and, as a result of these studies, national governments have intervened to control acid deposition problems. Such efforts are discussed at the end of this chapter. First, the basic science of acid deposition is discussed.

10.2. Causes of Acidity

In Chapter 5, pH was defined as

$$pH = -\log_{10}[H^+] \qquad (10.3)$$

where $[H^+]$ is the **molarity** (mol L^{-1}) of H^+ in a solution containing a solvent and one or more solutes. The pH scale, shown in Figure 10.3 for a limited range, varies from less than 0 (a lot of H^+ and very acidic) to greater than 14 (very little H^+ and very basic or alkaline). Neutral pH, which is the pH of distilled water, is 7.0. At this pH, the molarity of H^+ is 10^{-7} mol L^{-1}. A pH of 4 means that the molarity of H^+ is 10^{-4} mol L^{-1}. Thus, a pH of 4 is 1,000 times more acidic (contains 1,000 times more H^+ ions) than is water at a pH of 7. A pH of 2 corresponds to an H^+ molarity 100,000 times that of distilled water.

The quantity of H^+ is related to that of OH^- by the equilibrium relationship,

$$\underset{\substack{\text{Liquid}\\\text{water}}}{H_2O(aq)} \rightleftharpoons \underset{\substack{\text{Hydrogen}\\\text{ion}}}{H^+} + \underset{\substack{\text{Hydroxide}\\\text{ion}}}{OH^-} \qquad (10.4)$$

The equilibrium constant for this relationship is approximately 10^{-14} mol^2 L^{-2}, meaning that the product of $[H^+]$ and $[OH^-]$ must always equal approximately 10^{-14} mol^2 L^{-2} for water to be in equilibrium with H^+ and OH^-.

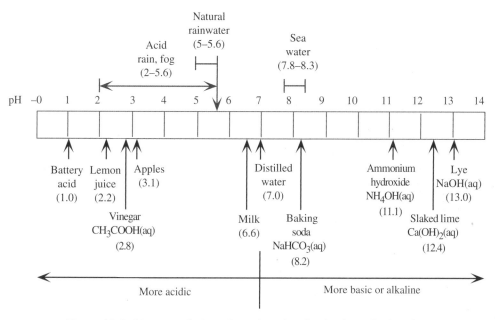

Figure 10.3. Diagram of pH scale and pH levels of selected solutions.

10.2.1. Carbonic Acid

Water can be acidified in one of several ways. When gas-phase carbon dioxide dissolves in water, it reacts rapidly with a water molecule to form aqueous **carbonic acid** [H_2CO_3(aq)], a weak acid, which partly dissociates by the reversible reactions

$$\underset{\substack{\text{Dissolved} \\ \text{carbon dioxide}}}{CO_2(aq)} + \underset{\substack{\text{Liquid} \\ \text{water}}}{H_2O(aq)} \rightleftharpoons \underset{\substack{\text{Dissolved} \\ \text{carbonic acid}}}{H_2CO_3(aq)}$$

$$\rightleftharpoons \underset{\substack{\text{Hydrogen} \\ \text{ion}}}{H^+} + \underset{\substack{\text{Bicarbonate} \\ \text{ion}}}{HCO_3^-}$$

$$\rightleftharpoons \underset{\substack{\text{Hydrogen} \\ \text{ion}}}{2H^+} + \underset{\substack{\text{Carbonate} \\ \text{ion}}}{CO_3^{2-}} \quad (10.5)$$

At a pH of distilled water (7), only the first dissociation (to the bicarbonate ion) occurs. The added H^+ (proton) decreases the pH of the solution, increasing its acidity. In the background air, the mixing ratio of CO_2(g) in 2011 was about 393 ppmv. A fraction of this CO_2(g) always dissolves in rainwater. Thus, rainwater, even in the cleanest environment on Earth, is naturally acidic due to the presence of background carbonic acid in it. The *pH of rainwater affected by only carbonic acid is about 5.6*, indicating that its hydrogen ion molarity is twenty-five times that of distilled water.

10.2.2. Sulfuric Acid

When sulfuric acid gas dissolves in or condenses onto raindrops, the resulting aqueous-phase **sulfuric acid** [H_2SO_4(aq)], a strong acid, dissociates by

$$\underset{\substack{\text{Sulfuric} \\ \text{acid gas}}}{H_2SO_4(g)} \rightarrow \underset{\substack{\text{Dissolved} \\ \text{sulfuric acid}}}{H_2SO_4(aq)} \rightleftharpoons \underset{\substack{\text{Hydrogen} \\ \text{ion}}}{H^+} + \underset{\substack{\text{Bisulfate} \\ \text{ion}}}{HSO_4^-}$$

$$\rightleftharpoons \underset{\substack{\text{Hydrogen} \\ \text{ion}}}{2H^+} + \underset{\substack{\text{Sulfate} \\ \text{ion}}}{SO_4^{2-}} \quad (10.6)$$

At pH levels greater than $+2$, complete dissociation to the sulfate ion is favored, adding two protons to solution. The increase in $[H^+]$ decreases the pH further, increasing the acidity of rainwater. Sulfur dioxide can also dissolve in rainwater and produce the sulfate and hydrogen ions chemically, acidifying the water. This process is discussed in Section 10.3.2.

10.2.3. Nitric Acid

When nitric acid gas dissolves in raindrops, it forms aqueous nitric acid [$HNO_3(aq)$], a strong acid that dissociates almost completely by

$$HNO_3(g) \rightleftharpoons HNO_3(aq) \rightleftharpoons H^+ + NO_3^- \quad (10.7)$$

$$\text{Nitric} \quad \text{Dissolved} \quad \text{Hydrogen} \quad \text{Nitrate}$$
$$\text{acid gas} \quad \text{nitric acid} \quad \text{ion} \quad \text{ion}$$

adding one proton to solution. As with sulfuric acid, nitric acid decreases the pH of rainwater below that of rainwater affected by only carbonic acid.

10.2.4. Hydrochloric Acid

When gas-phase hydrochloric acid dissolves in raindrops, it forms aqueous **hydrochloric acid** [$HCl(aq)$], a strong acid that dissociates almost completely by

$$HCl(g) \rightleftharpoons HCl(aq) \rightleftharpoons H^+ + Cl^- \quad (10.8)$$

$$\text{Hydrochloric} \quad \text{Dissolved} \quad \text{Hydrogen} \quad \text{Chloride}$$
$$\text{acid gas} \quad \text{hydrochloric acid} \quad \text{ion} \quad \text{ion}$$

adding one proton to solution. Hydrochloric acid also decreases the pH of rainwater below that of rainwater affected by only carbonic acid.

10.2.5. Natural and Anthropogenic Sources of Acids

Some of the enhanced acidity of rainwater from sulfuric, nitric, and hydrochloric acids is natural. Volcanos, for example, emit $SO_2(g)$, a source of sulfuric acid, and $HCl(g)$. Phytoplankton over the oceans emit dimethylsulfide [$DMS(g)$], which oxidizes to $SO_2(g)$. The main natural source of $HNO_3(g)$ is gas-phase oxidation of $NO_2(g)$ originating from lightning-produced $NO(g)$. The addition of natural acids to rainwater already containing carbonic acid results in typical natural rainwater pHs of between 5.0 and 5.6, as shown in Figure 10.3.

Acid deposition occurs when anthropogenically produced acids are deposited to soils, lakes, plants, tree leaves, or buildings as gases or within aerosol particles, fog drops, or raindrops. The two most important anthropogenically produced acids today are sulfuric and nitric acids, although hydrochloric acid can be important in some areas. In the United States, about 70 percent of $SO_2(g)$ and more than 85 percent of $NO_x(g)$ emissions are anthropogenic in origin. Thus, the excess acidification of rain due to sulfuric and nitric acids is a result of primarily anthropogenic rather than natural acids. In the eastern United States, about 60 to 70 percent of excess acidity of rainwater has been due to sulfuric acid, whereas 30 to 40 percent has been due to nitric acid (Glass et al., 1979). Thus, sulfuric acid has been the predominant acid of concern.

In polluted cities where fog is present, such as in Los Angeles or London, nitric acid fog is a problem. In locations where $HCl(g)$ is emitted anthropogenically, such as near wood burning or industrial processing, $HCl(aq)$ affects the acidity of rainwater. Today, however, $HCl(aq)$ contributes to less than 5 percent of total rainwater acidity by mass. Other acids that are occasionally important in rainwater include formic acid [$HCOOH(aq)$, produced from formaldehyde] and acetic acid [$CH_3COOH(aq)$, produced from acetaldehyde].

10.2.6. Acidity of Rainwater and Fog Water

Rainwater with a pH less than that of natural rainwater is acid rain. The *pH of acid rain varies between 2 and 5*, although typical values are near 4 and extreme values of less than 2 have been observed (Likens, 1976; Marsh, 1978; Graves, 1980; Graedel and Weschler, 1981). A pH of 4 corresponds to an H^+ molarity 40 times that of natural rainwater, whereas a pH of 2 corresponds to an H^+ molarity 4,000 times that of natural rainwater.

In Los Angeles, where fogs are common and nitric acid mixing ratios are high, fog water pHs are typically 2.2 to 4.0 (Waldman et al., 1982; Munger et al., 1983), but levels as low as 1.7 have been recorded (Jacob, 1985). Nitrate ion molarities in those studies were about 2.5 times those of sulfate ions. A fog with a pH below 5 is an **acid fog**.

10.3. Sulfuric Acid Deposition

The most abundant acid in the air is usually sulfuric acid, whose source is sulfur dioxide gas, emitted anthropogenically from coal-fired power plants, metal smelter operations, and other sources (Section 3.6.6).

Power plants usually emit $SO_2(g)$ from high stacks so that the pollutant is not easily deposited to the ground nearby. The higher the stack, the further the wind carries the gas before it is removed from the air. The wind transports $SO_2(g)$ long distances, sometimes hundreds to thousands of kilometers. Thus, acid deposition is often a regional and **long-range transport** problem. When acids or acid precursors are transported across political boundaries, they create **transboundary air pollution** problems, prevalent between the United States and

Table 10.1. Names and formulae of S(IV) and S(VI) species

S(IV) Family		S(VI) Family	
Chemical name	Chemical formula	Chemical name	Chemical formula
Sulfur dioxide	$SO_2(g,aq)$		
Sulfurous acid	$H_2SO_3(aq)$	Sulfuric acid	$H_2SO_4(g,aq)$
Bisulfite ion	HSO_3^-	Bisulfate ion	HSO_4^-
Sulfite ion	SO_3^{2-}	Sulfate ion	SO_4^{2-}

Canada; among western, northern, and eastern European countries; and among Asian countries.

Sulfur dioxide and sulfuric acid are but two of several sulfur-containing species in the air. Table 10.1 lists some additional species. The species are conveniently divided into two families, the **S(IV) and S(VI) families**, in which the IV and the VI represent the oxidation states (+4 and +6, respectively) of the members of the respective families. Thus, S(VI) members are more oxidized than are S(IV) members. Because sulfur dioxide is in the S(IV) family and sulfuric acid is in the S(VI) family, the oxidation of gas-phase sulfur dioxide to aqueous-phase sulfuric acid represents a conversion from the S(IV) family to the S(VI) family. This conversion occurs along two pathways, described next.

10.3.1. Gas-Phase Oxidation of S(IV)

The first conversion mechanism of S(IV) to S(VI) involves (1) gas-phase oxidation of $SO_2(g)$ to $H_2SO_4(g)$, (2) condensation of $H_2SO_4(g)$ and water vapor onto aerosol particles or cloud drops to produce an $H_2SO_4(aq)$-$H_2O(aq)$ solution, and (3) dissociation of $H_2SO_4(aq)$ to SO_4^{2-} in the solution. The gas-phase chemical conversion process (Step 1) is

$$\underset{\substack{\text{Sulfur} \\ \text{dioxide}}}{SO_2(g)} \xrightarrow{+\dot{O}H(g),M} \underset{\substack{\text{Bisulfite}}}{H\dot{S}O_3(g)} \xrightarrow[H\dot{O}_2(g)]{+O_2(g)} \underset{\substack{\text{Sulfur} \\ \text{trioxide}}}{SO_3(g)} \xrightarrow{+H_2O(g)} \underset{\substack{\text{Sulfuric} \\ \text{acid}}}{H_2SO_4(g)}$$

(10.9)

Because sulfuric acid has a low SVP (Section 5.3.2.1), nearly all $H_2SO_4(g)$ produced by Reaction 10.9 condenses onto aerosol particle or cloud drop surfaces (Step 2). At typical pHs of aerosol particles and cloud drops, nearly all condensed $H_2SO_4(aq)$ dissociates to SO_4^{2-} by Reaction 10.6 (Step 3). The dissociation releases two protons, decreasing pH and increasing acidity.

Whereas this is the dominant mechanism by which S(IV) produces S(VI) in aerosol particles, particularly when the relative humidity is below 70 percent, a second mechanism more rapidly produces S(VI) from S(IV) in cloud drops and raindrops.

10.3.2. Aqueous-Phase Oxidation of S(IV)

The second conversion process of S(IV) to S(VI) involves (1) dissolution of $SO_2(g)$ into liquid water drops to produce $SO_2(aq)$, (2) in-drop conversion of $SO_2(aq)$ to $H_2SO_3(aq)$ and dissociation of $H_2SO_3(aq)$ to HSO_3^- and SO_3^{2-}, and (3) in-drop oxidation of HSO_3^- and SO_3^{2-} to SO_4^{2-}. The dissolution process (Step 1) is represented by the reversible reaction

$$\underset{\substack{\text{Sulfur} \\ \text{dioxide} \\ \text{gas}}}{SO_2(g)} \rightleftharpoons \underset{\substack{\text{Dissolved} \\ \text{sulfur} \\ \text{dioxide}}}{SO_2(aq)} \qquad (10.10)$$

because $SO_2(aq)$ can evaporate as well. The formation and dissociation of **sulfurous acid** [$H_2SO_3(aq)$] (Step 2) occurs by

$$\underset{\substack{\text{Dissolved} \\ \text{sulfur} \\ \text{dioxide}}}{SO_2(aq)} + \underset{\substack{\text{Liquid} \\ \text{water}}}{H_2O(aq)} \rightleftharpoons \underset{\substack{\text{Sulfurous} \\ \text{acid}}}{H_2SO_3(aq)} \rightleftharpoons \underset{\substack{\text{Hydrogen} \\ \text{ion}}}{H^+} + \underset{\substack{\text{Bisulfite} \\ \text{ion}}}{HSO_3^-}$$

$$\rightleftharpoons \underset{\substack{\text{Hydrogen} \\ \text{ion}}}{2H^+} + \underset{\substack{\text{Sulfite} \\ \text{ion}}}{SO_3^{2-}} \qquad (10.11)$$

Step 3 involves the irreversible conversion of the S(IV) family (primarily HSO_3^- and SO_3^{2-}) to the S(VI) family (primarily SO_4^{2-}). At pH levels of 6 or less, the most important reaction converting S(IV) to S(VI) is

$$\underset{\substack{\text{Bisulfite} \\ \text{ion}}}{HSO_3^-} + \underset{\substack{\text{Dissolved} \\ \text{hydrogen} \\ \text{peroxide}}}{H_2O_2(aq)} + H^+ \rightarrow \underset{\substack{\text{Sulfate} \\ \text{ion}}}{SO_4^{2-}} + H_2O(aq) + 2H^+$$

(10.12)

This reaction is written in terms of HSO_3^- and SO_4^{2-} because at pHs of 2 to 6, most S(IV) exists as HSO_3^- and most S(VI) exists as SO_4^{2-}.

At pH levels greater than 6, which occur only in cloud drops that contain basic substances, such as ammonium or sodium, the most important reaction converting S(IV) to S(VI) is

$$\underset{\substack{\text{Sulfite} \\ \text{ion}}}{SO_3^{2-}} + \underset{\substack{\text{Dissolved} \\ \text{ozone}}}{O_3(aq)} \rightarrow \underset{\substack{\text{Sulfate} \\ \text{ion}}}{SO_4^{2-}} + \underset{\substack{\text{Dissolved} \\ \text{oxygen}}}{O_2(aq)} \quad (10.13)$$

This reaction is written in terms of SO_3^{2-} and SO_4^{2-} because the $HSO_3^- + O_3(aq)$ reaction is relatively slow, and at pH levels greater than 6, most S(VI) exists as SO_4^{2-}.

When $SO_2(g)$ dissolves in a drop to form $H_2SO_3(aq)$, the $H_2SO_3(aq)$ reacts to form SO_4^{2-}, forcing more $SO_2(g)$ to be drawn into the drop to replace the lost $H_2SO_3(aq)$. The more $SO_2(g)$ that dissolves and reacts, the more SO_4^{2-} that forms. In cloud drops, dissolution and aqueous reaction can convert 60 percent of $SO_2(g)$ molecules to SO_4^{2-} molecules within 20 minutes (Liang and Jacobson, 1999).

10.4. Nitric Acid Deposition

Nitric acid deposition occurs in and downwind of urban areas and is enhanced by the presence of clouds or fog. The origin of nitric acid is usually nitric oxide, emitted from vehicles and power plants. In the air, $NO(g)$ is oxidized to $NO_2(g)$, some of which is also directly emitted. $NO_2(g)$ is oxidized to nitric acid by

$$\underset{\substack{\text{Hydroxyl} \\ \text{radical}}}{\dot{O}H(g)} + \underset{\substack{\text{Nitrogen} \\ \text{dioxide}}}{\dot{N}O_2(g)} \overset{M}{\longrightarrow} \underset{\substack{\text{Nitric} \\ \text{acid}}}{HNO_3(g)} \qquad (10.14)$$

Gas-phase nitric acid dissolves into aerosol particles or fog drops to form $HNO_3(aq)$, which dissociates to a proton $[H^+]$ and the nitrate ion $[NO_3^-]$ by Reaction 10.7. Thus, the addition of nitric acid to cloud water decreases the pH and increases the acidity of the water. Gas-phase nitric acid also deposits directly to the ground, where it can cause environmental damage.

10.5. Effects of Acid Deposition

The most severe pollution episode in the twentieth century involving sulfuric acid–containing fog was probably that in London in 1952, discussed in Section 4.1.6.1. During that episode, coal burning combined with a heavy fog resulted in more than 4,000 excess deaths. Although other pollutants were also responsible, the acidified fog contributed to the disaster.

Acid deposition affects lakes, rivers, forests, agriculture, and building materials. The regions of the world that have been affected most by acid deposition include provinces of eastern Canada, the northeastern United States (particularly the Adirondack Mountain region), southern Scandinavia, middle and eastern Europe, India, South Korea, Russia, China, Japan, Thailand, and South Africa.

10.5.1. Effects on Lakes and Streams

Acids reduce the pH of lakes and streams. Because fish and microorganisms can survive only in particular pH ranges, the change in a lake's pH kills off many varieties of fish (including trout and salmon), invertebrates, and microorganisms. Most aquatic insects, algae, and plankton live only above pH levels of 5. The reduction in lake pH below 5 kills off these organisms and fish eggs, causing starvation at higher levels of the food chain. Low pHs (less than 5.5) in lakes have also been associated with reproductive failures and mutations in fish and amphibians.

Lake acidification has particularly been a problem in Scandinavian countries. Most damage occurred during the 1950s and 1960s, when the average pH of Swedish lakes fell by 1 pH unit. By the end of the 1970s, about 25,000 of Sweden's 90,000 lakes were so acidified that only acid-resistant plants and animals could survive. Of the acidified lakes, about 8,000 were naturally acidic, suggesting that 17,000 had been acidified anthropogenically. Today, many lakes in Sweden and in other countries have been restored.

Figure 10.4 shows rainwater acidity in the United States in 1994 and 2009. In both years, most acid rain occurred in Ohio, West Virginia, Pennsylvania, New York, Indiana, Michigan, Maryland, and parts of Florida. However, *minimum and average rainwater acidity decreased (pH increased) markedly between 1994 and 2009* due to reductions in sulfur emissions during this period (Section 10.7.2).

The effects of acid rain on lakes are most pronounced after the first snowmelt of a season. Because acids accumulate in snow, runoff from melted snow can send a shock wave of acid to a lake. In some cases, the acidity of meltwater is ten times greater than is that of the original snowfall.

10.5.2. Effects on Biomass

Acids damage plant and tree leaves and roots. When sulfuric acid deposits onto a leaf or needle, it forms a liquid film of low pH that erodes the cuticle wax, leading to the drying out (desiccation) of, and injury to, the leaf or needle. When acid gases, aerosol particles, or raindrops enter forest groundwater, they damage plants at their roots in two ways. First, sulfuric and nitric acid solutions dissolve and carry away important mineral nutrients, including calcium, magnesium, potassium, and sodium. Second, in acidic solutions, hydrogen ions $[H^+]$ react with aluminum- and iron-containing minerals, such as **aluminum hydroxide** $[Al(OH)_3(s)]$ and **iron hydroxide** $[Fe(OH)_3(s)]$, releasing Al^{3+} and Fe^{3+}, respectively. At high enough concentrations, these metal ions are toxic to root systems (Tomlinson, 1983).

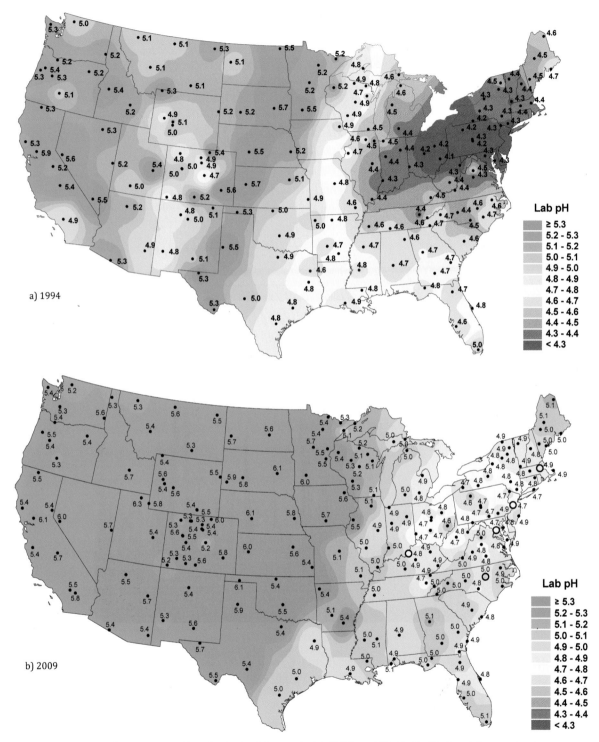

Figure 10.4. Map of rainwater pH in the United States in (a) 1994 and (b) 2009. National Atmospheric Deposition Program/National Trends Network, 2011; http://nadp.sws.uiuc.edu.

(a)

(b)

Figure 10.5. (a) Acidified forest, Oberwiesenthal, Germany, near the border with the Czech Republic, in 1991. The trees are of the *Picea* family. Photo by Stefan Rosengren, available from Johnner. (b) Acidified forest in the Erzgebirge Mountains, north of the town of Most, Czech Republic, taken in 1987. Photo by Owen Bricker, USGS.

Due to acid deposition, whole forests have been decimated. In Poland and the Czech Republic, 60 to 80 percent of trees died in the 1980s and 1990s (Stockholm Environment Institute, 1998). Figure 10.5a shows a forest near the border between Germany and the Czech Republic in which all the lower foliage died. Figure 10.5b shows a forest near Most, Czech Republic, that was decimated by acid deposition and air pollution. Forest damage is also evident in central Europe, the United States, Canada, China, Japan, and many other countries.

Acid deposition destroys crops in the same way that it destroys forests.

Not all damage to forests and crops is a result of acid deposition. Ozone also reacts with leaves, increasing plant and tree stress and making plants and trees more susceptible to disease, infestation, and death. Soot particles smother leaves, increasing plant and tree stress. PAN discolors the leaves of plants and trees.

10.5.3. Effects on Buildings and Sculptures

Acid deposition erodes materials. In particular, acids erode sandstone, limestone, marble, copper, bronze, and brass. Of note are buildings and sculptures of historical and archeological interest, such as the Parthenon in Greece, that have decayed, partly as a result of acid deposition and partly as a result of other pollutants in the air. Ozone, for example, reduces the detail in statues. Figure 10.6 shows an example of statue erosion, due both to acid deposition and other types of air pollution.

10.6. Natural and Artificial Neutralization of Lakes and Soils

One way to reduce the effect of acid deposition on lakes is to add a **neutralizing agent** (often called a **buffer**) to the lake. A neutralizing agent increases the pH of an acidified lake toward that of natural rainwater. Certain natural chemicals in soil and aerosol particles also act as neutralizing agents, preventing some soils and lakes from acidifying.

10.6.1. Ammonium Hydroxide

One anthropogenic neutralizing agent is **ammonium hydroxide** [$NH_4OH(aq)$], obtained by dissolving ammonia gas [$NH_3(g)$] in water. The net effect of adding $NH_4OH(aq)$ to acidified water is

$$\underset{\substack{\text{Ammonium} \\ \text{hydroxide}}}{NH_4OH(aq)} + \underset{\substack{\text{Hydrogen} \\ \text{ion}}}{H^+} \rightleftharpoons \underset{\substack{\text{Ammonium} \\ \text{ion}}}{NH_4^+} + \underset{\substack{\text{Liquid} \\ \text{water}}}{H_2O(aq)}$$

$$(10.15)$$

which reduces the H^+ molarity, reducing acidity and increasing pH.

10.6.2. Sodium and Calcium Hydroxide

When lye [$NaOH(aq)$, **sodium hydroxide**], the common component of "Drano," or **slaked lime** [$Ca(OH)_2(aq)$, **calcium hydroxide**] is added to an acidified lake, it reacts with H^+ to form water, decreasing

Figure 10.6. Sandstone figure over the portal of a castle, built in 1702, in Westphalia, Germany, photographed in 1908 (left) and in 1968 (right). The erosion of the figure is due to a combination of acid deposition and air pollution produced from the industrialized Ruhr region of Germany. Photo courtesy of Herr Schmidt-Thomsen, available from www.ultranet.com/~jkimball/BiologyPages/A/AcidRain.html#westphalia.

Figure 10.7. Liming of a lake in Sweden by helicopter. Photo by Tero Niemi, available from Johnner.

acidity. The net effect of adding slaked lime to acidified water is

$$Ca(OH)_2(aq) + 2H^+ \rightleftharpoons Ca^{2+} + 2H_2O(aq)$$

Calcium hydroxide | Hydrogen ion | Calcium ion | Liquid water

(10.16)

Lime is commonly added to lakes in large amounts (Figure 10.7) to reduce the effects of acid deposition. Sweden, which had the largest liming program in the world, added 200,000 tonnes of fine-ground limestone to lakes and watercourses each year in the 1990s and early 2000s. Since the 1970s, more than 7,000 of Sweden's 17,000 anthropogenically acidified lakes have been limed. Because lime is consumed 2 to 3 years after its application, acidified lakes need to be relimed regularly. Other countries that have had large liming programs include Norway, Finland, and Canada. Although liming of forest soil has also been tried, it is not so cost effective as lake liming (Bostedt et al., 2010).

10.6.3. Calcium Carbonate

Some soils that contain the minerals calcite or aragonite [$CaCO_3(s)$, **calcium carbonate**] have a natural ability to neutralize acids. When acid rain falls onto calcite-containing soils, the H^+ is removed by

$$CaCO_3(s) + 2H^+ \rightleftharpoons Ca^{2+} + CO_2(g) + H_2O(aq)$$

Calcium carbonate | Hydrogen ion | Calcium ion | Carbon dioxide gas | Liquid water

(10.17)

The same result occurs when soil dust particles that contain $CaCO_3(s)$ collide with acidified raindrops. The erosion of farmland and desert borders in many locations worldwide has enhanced the quantity of soil dust in the air, inadvertently increasing the calcium carbonate content of rainwater, decreasing rainwater acidity, and increasing rainwater pH in nearby regions.

Unfortunately, the same process described by Reaction 10.17 that decreases soil acidity is partly responsible for the erosion of great statues and buildings made of or containing **marble** or **limestone**. These materials, both of which contain calcite, erode when they become coated with acidified water. Coating can occur in at least two ways. The first is when acidified raindrops or aerosol particles deposit directly onto a marble or limestone surface. The second is when a gas dissolves and forms an acid in dew or rainwater that has recently coated a surface. For example, when $SO_2(g)$ dissolves in water, it oxidizes to sulfuric acid (Section 10.3.2).

When water containing sulfuric acid coats a calcite surface, the hydrogen ion dissociated from the acid dissolves the calcite by Reaction 10.17, and the sulfate ion reacts with the dissociated calcium ion to form the mineral **gypsum** by

$$Ca^{2+} + SO_4^{2-} + 2H_2O(aq) \rightleftharpoons CaSO_4 - 2H_2O(s)$$

Calcium ion | Sulfate ion | Liquid water | Calcium sulfate dihydrate (gypsum)

(10.18)

(a) 1930   (b) 1934

Figure 10.8. Soiling of the limestone exterior of the Cathedral of Learning at the University of Pittsburgh between (a) 1930 and (b) 1934 (Davidson et al., 2000). The building was constructed between 1929 and 1937. Sulfate and soot from coal smoke caused erosion and darkening of the building after only 4 years. Photo courtesy University Archives, University of Pittsburgh.

The net result is the formation of a clear-to-white gypsum crust over the marble or limestone. Bombardment by rain over time removes some of the brittle gypsum crust. Because the crust now contains part of the statue or building material (the calcium), its removal creates tiny crevices, or pits, causing erosion (Davidson et al., 2000). Because the gypsum crust and the crevices roughen the surface of a statue or building, other pollutants, such as soot, more readily bond to the surface, darkening it, as illustrated in Figure 10.8. It shows photographs of the **Cathedral of Learning** at the University of Pittsburgh, taken in 1930, soon after the start of its construction, and in 1934. During a 4-year period, sulfate from coal smoke emitted by steel mills and locomotives (Section 4.1.6.3) roughened the limestone exterior of the building, and soot from the same smoke bonded with the roughened exterior, darkening it.

Gypsum forms not only on buildings, but also in soils and aerosol particles. When rainwater containing the sulfate ion falls on soil containing calcite, the calcite dissociates by Reaction 10.17. When the soil dries, the calcium ion reacts with the sulfate ion by Reaction 10.18, producing gypsum. The same process occurs when soil dust particles containing calcite collide with acidified raindrops. Deposition of rain containing sulfuric acid over soils containing calcite and deposition of particles already containing gypsum have, over time, produced worldwide **deposits of gypsum soil**.

10.6.4. Sodium Chloride

Some acidic soils near nonpolluted coastal areas are naturally neutralized by cations originating from sea spray (e.g., Na^+, Ca^{2+}, Mg^{2+}, K^+) that have deposited onto soils over the millennia. The pH of natural seawater ranges from 7.8 to 8.3, and that of uncontaminated large sea spray drops is similar, indicating that little H^+ exists in such drops. The deposition of sea spray drops to coastal soils, and the subsequent desiccation of these drops, produces common salt [$NaCl(s)$]. When sulfate-containing water enters $NaCl(s)$-containing soils, $NaCl(s)$ dissolves and dissociates, H^+ combines with the chloride ion [Cl^-] to form $HCl(aq)$, which evaporates to the gas phase. The net process is

$$NaCl(s) + H^+ \rightleftharpoons Na^+ + HCl(g) \qquad (10.19)$$

Sodium   Hydrogen   Sodium   Hydrochloric
chloride   ion   ion   acid

which reduces H^+ and the acidity of soil water and increases pH.

10.6.5. Ammonia

Ammonia gas [$NH_3(g)$] is considered an anthropogenic pollutant, but it also neutralizes raindrop and soil water acidity. Figure 5.14 gave worldwide emission sources of ammonia gas. Once in the air, ammonia gas dissolves in water. The dissolved gas then reacts with the hydrogen ion to form the ammonium ion by

$$\underset{\substack{\text{Dissolved}\\\text{ammonia}}}{NH_3(aq)} + \underset{\substack{\text{Hydrogen}\\\text{ion}}}{H^+} \rightleftharpoons \underset{\substack{\text{Ammonium}\\\text{ion}}}{NH_4^+} \qquad (10.20)$$

The loss of H^+ that results from this reaction increases pH, reducing acidity. In many cases, the pH of rainwater containing the ammonium ion exceeds 6. Aerosol particles and raindrops containing the ammonium ion deposit to soils and lakes, providing these surfaces with a neutralizing agent. Soils downwind of high ammonia gas emissions tend to have a better neutralizing capacity against acid deposition than do soils far from ammonia sources, if all other conditions are the same.

10.7. Recent Regulatory Control of Acid Deposition

The first major effort to control acid deposition was the British Alkali Act of 1863, which mandated large reductions in hydrochloric acid gas emissions by soda ash manufacturers. In more recent years, the U.S. Clean Air Act Amendments of 1970 led to lower emissions of acid deposition precursors, namely $SO_2(g)$ and $NO_2(g)$. In 1977, the United States initiated the National Atmospheric Deposition Program (NADP), whose purpose was to monitor trends of acidity in precipitation. In 1980, the U.S. Congress passed the **Acid Precipitation Act**, which funded a program, the National Acid Precipitation Assessment Program. Under the program, the network of monitoring stations under NADP was enlarged to produce a National Trends Network.

The Clean Air Act Amendments of 1990 mandated a 10-million-ton reduction in sulfur dioxide [$SO_2(g)$] emissions from 1980 levels and a 2-million-ton reduction in nitrogen oxide [$NO_x(g)$] emissions from 1980 levels by 2010. To implement these reductions, the U.S. EPA established an emission trading system, whereby emitters could trade among themselves for limited rights to release $SO_2(g)$. Power plants were also required to install emission monitoring systems. In January 2000, the U.S. EPA issued a new rule requiring U.S. refiners to cut the sulfur content of gasoline to one-tenth its value by 2006.

Meanwhile, several studies in the 1970s concluded that winds were transporting acid deposition precursors over long distances and political boundaries. Such studies culminated in the 1979 **Geneva Convention on Long-Range Transboundary Air Pollution**. The convention, originally signed by thirty-four governments and the European Community, was the first agreement to deal with an international air pollution problem. Since then, fifty-one countries have signed on to the treaty, and it has been amended eight times. As part of a 1985 amendment (the Sulfur Protocol), member countries were required to reduce their emissions or transboundary fluxes of sulfur by 30 percent below 1980 levels by 1993. A 1988 amendment (the Nitrogen Oxide Protocol) required countries to reduce their emissions or transboundary fluxes of nitrogen oxide to their 1987 levels by December 1994. Because the first Sulfur Protocol did not address forest loss in central Europe sufficiently, a second Sulfur Protocol was signed in 1994 that was intended to result in a 60 percent reduction in sulfur emissions compared with 1980 values by 2010.

10.7.1. Methods of Controlling Emissions

Several mechanisms are available to control emissions of acid deposition precursors. These include the mandatory use of low-sulfur coal instead of high-sulfur coal and the use of emission control technologies. The quantity of $SO_2(g)$ emitted during coal combustion depends on the **sulfur content of the coal**. In the United States, about 40 percent of coal in 2009 was mined in Wyoming, and 32 percent was mined in the Appalachian Mountains (Energy Information Administration, 2011c). Coal from the Appalachian Mountains has a high sulfur content. The cost of transporting Appalachian coal to power plants, most of which are in the midwestern and eastern United States, is lower than is the cost of transporting low-sulfur coal from Wyoming or other western states to these plants. As such, coal burners prefer to use high-sulfur coal. However, CAAA90 requirements to reduce $SO_2(g)$ emissions caused an increase in Wyoming's share of U.S. coal from 31 to 40 percent between 2000 and 2009 and a decrease in Appalachia's share from 39 percent to 32 percent.

The use of low-sulfur coal is one mechanism to reduce emission of $SO_2(g)$ during coal burning. Another is to remove a certain fraction of sulfur from high-sulfur coal before burning it. A technology available for reducing $SO_2(g)$ emission from a stack is the **scrubber**, first developed by William Gossage to reduce HCl(g) emission. A modern-day scrubber technique is called **flue gas desulfurization**. With this technique, a fan first extracts a hot, gaseous exhaust stream containing $SO_2(g)$ into a reaction tower. A mixture of aerosolized water and dissolved limestone [$CaCO_3(aq)$] or slaked lime [$Ca(OH)_2(aq)$] is then sprayed into the stream. Sulfur dioxide can dissolve in the spray drops, producing calcium sulfate [$CaSO_4(s)$], which is readily removed as a solid. The remaining exhaust and pollutants are vented to the air.

10.7.2. Effects of Regulation

Between the 1970s and 2011, reductions in sulfur dioxide emissions reduced the extent of acid deposition in most of North America and Europe, but increases in such emissions exacerbated acid deposition in parts of Asia.

In the United States, rainwater acidity decreased and pH increased between the 1990s and late 2000s, as illustrated in Figure 10.4. For example, between 1994 and 1996, more than 40 percent of the United States experienced acid rain in at least 50 percent of all rainwater samples. One percent of all samples had a pH < 5. Most acidity occurred in the northeastern United States, particularly Ohio, Pennsylvania, and nearby areas. Between 2002 and 2004, the area over which at least 50 percent of rainwater samples were acidic dropped to 35 percent (Lehmann et al., 2007). Although the pH increase occurred simultaneously with a decrease in the sulfate ion in rainwater, it also coincided with an increase in the ammonium ion. For example, between 1985 and 2004, the sulfate ion content of U.S. rainwater decreased by 46 percent and the ammonium ion content increased by 29 percent (Lehmann et al., 2007). Ammonia is a neutralizing agent that helps increase the pH (Section 10.6.5).

Reductions in sulfur dioxide emissions in Canada have similarly reduced the acidity of some Canadian lakes and forests. For example, in the late 1960s, the Sudbury, Ontario nickel smelting plant was the largest individual source of $SO_2(g)$ in North America, emitting 4,500 tonnes of $SO_2(g)$ per day, devastating nearby lakes and forests. In 1972, a large stack was built to disperse the pollutants further downwind (Section 6.6.2.4), and in the 1990s, pollution scrubbing equipment was added to reduce emissions to less than 450 tonnes of $SO_2(g)$ per day. As a result, nearby lakes and forests partially regenerated, although significant pollution still occurred. Reductions in the acidity of lakes in Quebec, Atlantic Canada, and other areas of Ontario were less dramatic.

In Europe, reductions in sulfur dioxide emissions have similarly reduced the acidity of rainwater. During the last three decades, many lakes in Sweden have been restored, but many more are still damaged by acid deposition. Acid deposition problems in the sulfur triangle (Section 8.2.4) are still severe, although reductions in emissions of acid gases and other pollutants resulted in visibility improvements during the 2000s (Figure 8.6), suggesting that acid deposition problems have also become less severe over time.

Despite success in some parts of the world, acid rain became more pronounced in China during the 2000s than in prior decades, primarily due to the substantial growth in coal combustion there. For example, in 2010, 258 cities experienced acid rain, primarily from sulfuric and nitric acid (Press Trust of India, 2011). In many of these cities, every raindrop measured was acidic. Despite having some of the highest concentrations of sulfate and nitrate ions in rainwater in the world, the *pH of rainwater in Beijing is less acidic than in many other cities*, often between 5 and 6.8 (Xu and Han, 2009). The reason is that frequent dust storms from the Gobi Desert pass through Beijing. Dust contains calcium carbonate, which is a neutralizing agent (Section 10.6.3). Thus, although rainwater in Beijing contains sulfate and nitrate, it also contains calcium, which moderates the pH of rainwater. *The pH of rainwater in many other cities in China is frequently between 4 and 5, as is the pH of rainwater in Tokyo, Istanbul, Mexico City, and Itatiaia, Brazil* (Xu and Han, 2009).

10.8. Summary

In this chapter, the history, science, effects, and control of acid deposition are discussed. Acidity is determined by pH, which ranges from less than 0 (very acidic) to more than 14 (very basic). The pH of distilled water is 7, of natural rainwater is 5 to 5.6, and of acid rain or fog is less than 5. Acid deposition occurs when sulfuric, nitric, or hydrochloric acid is emitted into or forms chemically in the air and is subsequently deposited as a gas or liquid to the ground, where it harms microorganisms, fish, forests, agriculture, and structures. In high concentrations in the air, acids can also harm humans. Severe acid deposition problems arose from increased coal combustion in the UK during the Industrial Revolution and from the growth of the alkali industry in France and the UK during the 1800s. Today, sulfuric acid is usually the most abundant acid in rainwater. Sulfuric acid is produced by gas- and aqueous-phase oxidation of sulfur dioxide. The latter process is most efficient when cloud drops are present. In polluted coastal air, nitric acid fog is often a problem. A method of ameliorating the effects of acid deposition on lakes is to add a neutralizing agent, such as slaked lime. In the United States, Canada, and western Europe, government intervention in the form of regulations limiting the emissions of acid deposition precursors has resulted in reductions in the acidity of rainwater. Acid deposition problems in eastern Europe are still severe but improving; those in China have grown due to a large increase in coal combustion.

10.9. Problems

10.1. Identify the atmospheric acids produced by Leblanc's soda ash process.

10.2. In terms of acid deposition precursors, what were the advantages of the Solvay versus the Leblanc soda ash process?

10.3. Although Leblanc's process produced $HCl(g)$, which caused widespread acid deposition problems in the early 1800s, $HCl(g)$ was no longer the most dangerous by-product of this process by the late 1800s. Why?

10.4. Describe the two important conversion pathways for $S(IV)$ to $S(VI)$. Which pathway is more important when aerosol particles are present? Why?

10.5. What are the most important aqueous-phase oxidants of $S(IV)$?

10.6. Why are nitric and hydrochloric acid depositions less of a problem in most parts of the world than sulfuric acid deposition?

10.7. How do neutralizing agents reduce the acidity of a lake?

10.8. Suppose rainwater containing the sulfate ion enters a soil containing magnesium carbonate $[MgCO_3(s)]$. Would the magnesium carbonate act as a neutralizing agent or enhance the acidity of the water? Show the pertinent chemical process.

10.9. Identify three products that you use or activities that you do that result in the emission of acids into the atmosphere.

10.10. Identify three ways that acids or acid precursors can be controlled through legislative action.

Chapter 11

Global Stratospheric Ozone Reduction

The stratospheric ozone layer began to form soon after the Great Oxygenation Event, whereby oxygen started to accumulate in the atmosphere due to photosynthesis about 2.45 b.y.a. The ozone layer did not develop fully until after 400 m.y.a., when green plants evolved and molecular oxygen mixing ratios began to approach their present levels. Absorption of UV radiation by ozone is responsible for the temperature inversion that defines the present-day stratosphere. This absorption is critical for preventing UV radiation from reaching the surface of the Earth, where it can harm life. The anthropogenic emission of long-lived chlorine- and bromine-containing compounds into the air since the 1930s, as well as the slow transfer of these compounds to the stratosphere, have caused a nontrivial reduction in the global stratospheric ozone layer since the 1970s. In addition, during September, October, and November of each year since 1979, such emissions have caused up to a 70 percent destruction of the ozone layer over the Antarctic. Lesser reductions have occurred over the Arctic in March, April, and May of each year. Recent international cooperation has helped replace emissions of most ozone-depleting gases with less destructive ones, reducing ozone loss. However, global warming, which warms the surface but cools the stratosphere, has delayed the recovery of the ozone layer. In this chapter, the natural stratospheric ozone layer, global ozone reduction, and Antarctic/Arctic ozone destruction and regeneration are discussed.

11.1. Structure of the Present-Day Ozone Layer

Christian Schönbein discovered the ozone molecule in 1839 (Section 1.2.3.4). However, the ozone layer was not identified until more than 70 years later. In 1879, the French physicist **Alfred Cornu** (1841–1902) measured a sharp reduction in short wavelengths of UV light, relative to longer wavelengths, reaching the Earth's surface, suggesting that an atmospheric constituent might be absorbing the short wavelengths. In 1881, **Walter N. Hartley** explained the absorption of short UV wavelengths as due to ozone high in the atmosphere. Finally, in 1913, French physicists **Charles Fabry** (1867–1945) and **Henri Buisson** (1873–1944) discovered and began to quantify the thickness of the stratospheric ozone layer.

About 90 percent of all ozone molecules in the atmosphere reside in the stratosphere; most of the remaining ozone molecules reside in the troposphere. Whereas ozone molecules near the surface harm humans, animals, plants, trees, and structures, the same ozone molecules, whether in the stratosphere or in polluted air, shield the Earth from harmful UV radiation.

A common measure of the quantity of ozone in the air is the ozone column abundance, which is the number of ozone molecules above a square centimeter of the ground, summed to the top of the atmosphere. When this number is divided by 2.7×10^{16}, the result is the

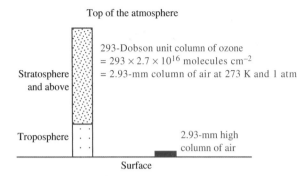

Top of the atmosphere

293-Dobson unit column of ozone
$= 293 \times 2.7 \times 10^{16}$ molecules cm^{-2}
$= 2.93$-mm column of air at 273 K and 1 atm

Stratosphere and above

Troposphere

2.93-mm high column of air

Surface

Figure 11.1. Example of globally averaged column abundance of ozone. The number of ozone molecules per unit area of surface in a 293-DU column of ozone is equivalent to the number of air molecules in a 2.93-mm-high column near the surface. (The figure is not to scale.)

column abundance in **Dobson units** (DUs). Thus, 1 DU equals 2.7×10^{16} molecules of ozone per square centimeter of surface. The DU is named after **Gordon M. B. Dobson** (1889–1976), a researcher at Oxford University, who, in the 1920s, built the first instrument, now called a Dobson meter, to measure total ozone column abundance from the ground. In 2010, the globally averaged column abundance of ozone from 90°S to 90°N was 291.5 DUs. This column abundance contains the

same number of molecules as a column of air 2.92 mm high at 1 atm of pressure and 273 K (near-surface conditions). Figure 11.1 illustrates ozone column abundance.

Figure 11.2 shows the variation in the ozone column abundance with latitude (zonally averaged, or averaged over all longitudes) and day of the year in 2010. It exhibits the following features:

- A year-round equatorial ozone minimum due to upward motion of ozone-poor air from the troposphere that displaces ozone-rich air horizontally to higher latitudes. The column abundance over the Equator is typically 250 to 290 DUs all year.
- A Northern Hemisphere (NH) spring (March–May) maximum, ranging from 350 to 460 DUs, near the North Pole. The maximum is due to the northward transport of stratospheric ozone from the Equator. As ozone converges at the pole, it descends, increasing the ozone column abundance.
- A Southern Hemisphere (SH) spring (September–November) subpolar (60°S–65°S) maximum, usually ranging from 350 to 440 DUs. The maximum is due to the southward transport of ozone from the Equator. As the ozone moves south, much of it is forced to descend in front of the **polar vortex**, a polar front jet stream wind system that travels around the Antarctic continent in the upper troposphere and stratosphere.

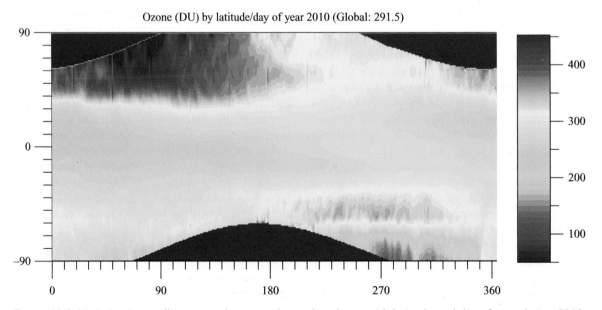

Ozone (DU) by latitude/day of year 2010 (Global: 291.5)

Figure 11.2. Variation in zonally averaged ozone column abundance with latitude and day of year during 2010. Blank regions near the poles indicate locations where data were not available. Data for the figure were obtained from the satellite-based Total Ozone Mapping Spectrometer (TOMS) and made available by NASA Goddard Space Flight Center, Greenbelt, Maryland.

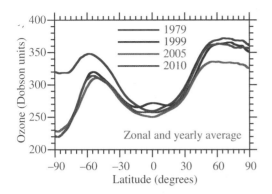

Figure 11.3. Variation of yearly and zonally averaged ozone column abundance with latitude during 1979, 1999, 2005, and 2010. Data were obtained from the satellite-based Total Ozone Mapping Spectrometer (TOMS) and made available by NASA Goddard Space Flight Center, Greenbelt, Maryland.

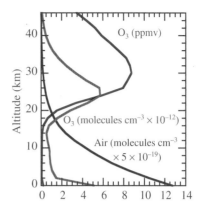

Figure 11.4. Example vertical variation in ozone mixing ratio, ozone number concentration, and air number concentration with altitude. In this example, the ozone mixing ratio at the surface is 0.20 ppmv, the level of a Stage 1 smog alert in the United States.

The polar vortex prevents outside stratospheric air from entering the Antarctic region and Antarctic air from escaping.

- A SH spring minimum of less than 150 DUs over the South Pole due to chemical reactions of chlorine and bromine radicals with ozone. This minimum is the **Antarctic ozone hole**, which is defined as the area of the Antarctic over which the ozone column abundance decreases below 220 DUs. The minimum column abundance at a specific location (not zonally averaged) within the ozone hole can drop to 80 DUs.

Figure 11.3 shows the variation with latitude of the zonally and yearly averaged ozone column abundance in 1979, 1999, 2005, and 2010. The ozone layer was thin near the Equator for all 4 years. In 1999, the ozone column abundance between 15°S and 15°N increased compared with 1979. Such increases, relative to 1979 ozone, occurred in about one-third of the years between 1979 and 2010.

The yearly averaged ozone column abundance 60°S to 90°S is always greater than over the Equator; however, since 1979, ozone 60°S to 90°S has declined due to the seasonal Antarctic ozone hole (Section 11.4). The decline held steady from 1999 to 2010.

The yearly averaged ozone column abundance 60°N to 90°N is also greater than over the Equator. For most years since 1979, the column abundance 60°N to 90°N has been lower than in 1979. Two exceptions were during 1999 and 2010, when the column abundance was nearly the same as in 1979. When a reduction over the Arctic occurs, it is called an **Arctic ozone dent** (Section 11.4).

Figure 11.4 shows a typical variation of ozone mixing ratio, ozone number concentration, and total air number concentration with altitude. The ozone number concentration (molecules of ozone per cubic centimeter of air) in the stratosphere generally peaks at 25 to 32 km in altitude. The ozone mixing ratio (number concentration of ozone divided by that of dry air) peaks at a higher altitude than does the number concentration. The peak ozone number concentration in the stratosphere is close to that in polluted urban air. The **peak ozone mixing ratio in the stratosphere** (near 10 ppmv), however, is much higher than is that in very polluted urban air (0.2–0.35 ppmv) or free-tropospheric air (20–40 ppbv).

11.2. Relationship between the Ozone Layer and Ultraviolet Radiation

The ozone layer prevents damaging UV wavelengths from reaching the Earth's surface. The UV portion of the solar spectrum is divided into far- and near-UV wavelengths (Figure 2.5). Near-UV wavelengths are further divided into UV-A, -B, and -C wavelengths (Section 2.2). Gases, particularly ozone and oxygen, and aerosol particles absorb most UV radiation before it reaches the Earth's surface. Decreases in stratospheric ozone increase the transmission of UV to the surface. Enhancements in UV at the surface damage life. In this subsection, processes affecting UV radiation are summarized.

Figure 11.5 shows the intensity of downward UV and visible radiation at the top of the atmosphere (TOA) and at the ground. Of the incident solar radiation at the TOA, only wavelengths longer than 0.29 μm penetrate to the

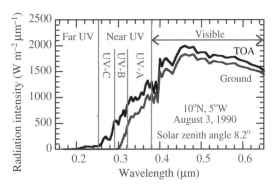

Figure 11.5. Downward solar radiation less than 0.65 μm in wavelength at the top of the atmosphere and at the ground at a location near the Equator in early August. The solar zenith angle is the angle of the sun relative to a line perpendicular to the Earth's surface.

ground. Thus, the air filters out all far-UV and UV-C wavelengths before they reach the surface. Of the UV that reaches the ground, about 9 percent is UV-B, and the rest (91 percent) is UV-A. Of the total solar radiation reaching the surface, about 5.2 percent is UV-A/-B, and the rest (94.8 percent) is visible/solar IR.

Table 11.1 identifies the major absorbing components responsible for reducing near- and far-UV radiation between the TOA and the ground. Molecular nitrogen [$N_2(g)$] absorbs far-UV wavelengths shorter than 100 nm (0.1 μm) in the thermosphere and mesosphere, and molecular oxygen [$O_2(g)$] absorbs wavelengths shorter than 250 nm in the thermosphere, mesosphere, and stratosphere.

Ozone absorbs wavelengths shorter than 345 nm (strongly below 310 nm and weakly from 310 to 345 nm). Ozone also absorbs weakly from 450 to 750 nm.

Stratospheric ozone allows little UV-C, some UV-B, and most UV-A radiation to reach the troposphere. Ozone in the background troposphere absorbs some of the UV-B and -A not absorbed in the stratosphere. In polluted air, additional absorbers of UV-B radiation include nitrated aromatic gases and aerosol particle components, such as black carbon, nitrated aromatics, PAHs, tar balls, and soil-dust components (Section 7.1.3.1).

The major UV-A–absorbing gas is nitrogen dioxide [$NO_2(g)$]. Its mixing ratio in clean air is too low to affect UV-A reaching the surface. In polluted air, the $NO_2(g)$ mixing ratio is high only in the morning, when UV-A intensity is low. Other UV-A absorbers in polluted air include the same aerosol particle components that absorb UV-B radiation.

Gas and aerosol particle absorption is not the only mechanism that reduces incident downward UV radiation. Gas and aerosol particle backscattering, as well as ground and cloud reflection, return some incident UV radiation to space.

11.3. Chemistry of the Natural Ozone Layer

The chemistry of the natural stratospheric ozone layer involves reactions among primarily oxygen-containing compounds; however, the shape of the vertical profile of the layer is affected by the chemistry of natural and anthropogenic nitrogen- and hydrogen-containing compounds as well. Next, the chemistry of the natural ozone layer is discussed.

11.3.1. The Chapman Cycle

The photochemistry of the natural stratosphere is similar to that of the background troposphere, except that

Table 11.1. Major absorbers of ultraviolet radiation in the atmosphere

Name of spectrum	Wavelengths (nm)	Dominant absorbers	Location of absorption
Far-UV	10–100	$N_2(g)$	Thermosphere, mesosphere
	10–250	$O_2(g)$	Thermosphere, mesosphere, stratosphere
Near-UV			
UV-C	250–290	$O_3(g)$	Stratosphere
UV-B	290–345	$O_3(g)$	Stratosphere, troposphere
		Some particle components	Polluted troposphere
UV-A	320–380	$NO_2(g)$	Polluted troposphere
		Some particle components	Polluted troposphere
UV, ultraviolet.			

stratospheric ozone is produced after photolysis of molecular oxygen, whereas tropospheric ozone is produced after photolysis of nitrogen dioxide. Next, reactions that naturally produce and destroy stratospheric ozone are described.

In the stratosphere, far-UV wavelengths shorter than 245 nm break down molecular oxygen by

$$\underset{\substack{\text{Molecular}\\\text{oxygen}}}{O_2(g)} + h\nu \rightarrow \underset{\substack{\text{Excited}\\\text{atomic}\\\text{oxygen}}}{\cdot\dot{O}(^1D)(g)} + \underset{\substack{\text{Atomic}\\\text{oxygen}}}{\cdot\dot{O}(g)} \quad \lambda < 175\,\text{nm}$$

(11.1)

$$\underset{\substack{\text{Molecular}\\\text{oxygen}}}{O_2(g)} + h\nu \rightarrow \underset{\substack{\text{Atomic}\\\text{oxygen}}}{\cdot\dot{O}(g)} + \cdot\dot{O}(g) \quad 175 < \lambda < 245\,\text{nm}$$

(11.2)

The first reaction is important only at the top of the stratosphere because wavelengths shorter than 175 nm do not penetrate lower. Neither reaction is important in the troposphere. Excited atomic oxygen from Reaction 11.1 rapidly converts to the ground state by

$$\underset{\substack{\text{Excited}\\\text{atomic}\\\text{oxygen}}}{\cdot\dot{O}(^1D)(g)} \overset{M}{\rightarrow} \underset{\substack{\text{Atomic}\\\text{oxygen}}}{\cdot\dot{O}(g)}$$

(11.3)

Ozone then forms by

$$\underset{\substack{\text{Atomic}\\\text{oxygen}}}{\cdot\dot{O}(g)} + \underset{\substack{\text{Molecular}\\\text{oxygen}}}{O_2(g)} \overset{M}{\rightarrow} \underset{\text{Ozone}}{O_3(g)}$$

(11.4)

This reaction also occurs in the troposphere, where the O(g) originates from $NO_2(g)$ photolysis rather than from $O_2(g)$ photolysis. Ozone is destroyed naturally in the stratosphere and troposphere by

$$\underset{\text{Ozone}}{O_3(g)} + h\nu \rightarrow \underset{\substack{\text{Molecular}\\\text{oxygen}}}{O_2(g)} + \underset{\substack{\text{Excited}\\\text{atomic}\\\text{oxygen}}}{\cdot\dot{O}(^1D)(g)} \quad \lambda < 310\,\text{nm}$$

(11.5)

$$\underset{\text{Ozone}}{O_3(g)} + h\nu \rightarrow \underset{\substack{\text{Molecular}\\\text{oxygen}}}{O_2(g)} + \underset{\substack{\text{Atomic}\\\text{oxygen}}}{\cdot\dot{O}(g)} \quad \lambda > 310\,\text{nm}$$

(11.6)

Stratospheric ozone is also destroyed by

$$\underset{\substack{\text{Atomic}\\\text{oxygen}}}{\cdot\dot{O}(g)} + \underset{\text{Ozone}}{O_3(g)} \rightarrow \underset{\substack{\text{Molecular}\\\text{oxygen}}}{2O_2(g)}$$

(11.7)

In 1930, the English physicist **Sydney Chapman** (1888–1970; Figure 11.6) suggested that UV photolysis of molecular oxygen produced ozone in the

Figure 11.6. Sydney Chapman (1888–1970). American Institute of Physics, Emilio Segrè Visual Archives, Physics Today collection.

stratosphere. He further postulated that Reactions 11.2, 11.4, 11.6, 11.7, and

$$\underset{\substack{\text{Atomic}\\\text{oxygen}}}{\cdot\dot{O}(g)} + \cdot\dot{O}(g) \overset{M}{\rightarrow} \underset{\substack{\text{Molecular}\\\text{oxygen}}}{O_2(g)}$$

(11.8)

controlled the natural formation and destruction of ozone in the stratosphere (Chapman, 1930). These reactions comprise what is known today as the **Chapman cycle** and describe the process fairly well. Some Chapman reactions are more important than are others. Reactions 11.2, 11.4, and 11.6 affect ozone the most. The non-Chapman reaction, Reaction 11.5, is also important.

Some of the Chapman cycle reactions can be used to explain why the altitudes of peak ozone concentration and mixing ratio occur where they do, as illustrated in Figure 11.7. Oxygen density, like air density, decreases exponentially with increasing altitude. UV radiation intensity, conversely, decreases with decreasing altitude. Peak ozone densities occur where sufficient radiation encounters sufficient oxygen density, which is near 25 to 32 km. At higher altitudes, oxygen density is too low for its photolysis by Reactions 11.1

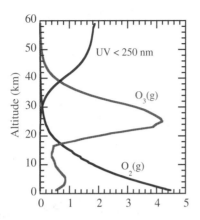

Figure 11.7. Variation with altitude of ultraviolet radiation <250 nm in wavelength (photons/cm^2/s × 10^{-14}), O$_2$(g) (molec./cm^3 × 10^{-18}), and O$_3$(g) (molec./cm^3 × 10^{-12}) on a clear January 1 day at 10°S and 20°E.

and 11.2 to produce high ozone densities, whereas, at lower altitudes, UV radiation is not intense enough for oxygen photolysis to produce high ozone densities.

11.3.2. Effects of Nitrogen on the Natural Ozone Layer

Oxides of nitrogen [NO(g) and NO$_2$(g)] naturally destroy ozone, primarily in the upper stratosphere, helping shape the vertical profile of the ozone layer. In the troposphere, the major sources of nitric oxide are surface anthropogenic and natural emissions and lightning. The major source of NO(g) in the stratosphere is transport from the troposphere and the breakdown of nitrous oxide [N$_2$O(g)] (laughing gas), a colorless gas emitted during denitrification by anaerobic bacteria in soils (Section 2.3.6). It is also emitted by bacteria in fertilizers, sewage, and the oceans, as well as during biofuel and biomass burning, automobile and aircraft combustion, nylon manufacturing, and aerosol spray cans. In the troposphere, N$_2$O(g) is lost by transport to the stratosphere, deposition to the surface, and chemical reaction. Because its loss rate from the troposphere is slow, nitrous oxide is long lived and well diluted in the troposphere, with an average mixing ratio of about 0.33 ppmv in 2011. The mixing ratio of N$_2$O(g) is relatively constant up to about 15 to 20 km, but decreases above that as a result of photolysis. Throughout the atmosphere, N$_2$O(g) produces nitric oxide by

$$\underset{\substack{\text{Nitrous}\\\text{oxide}}}{N_2O(g)} + \underset{\substack{\text{Excited}\\\text{atomic}\\\text{oxygen}}}{\cdot\dot{O}(^1D)(g)} \rightarrow \underset{\substack{\text{Nitric oxide}}}{2\dot{N}O(g)} \qquad (11.9)$$

Nitric oxide naturally reduces ozone in the upper stratosphere by

$$\underset{\substack{\text{Nitric}\\\text{oxide}}}{\dot{N}O(g)} + \underset{\substack{\text{Ozone}}}{O_3(g)} \rightarrow \underset{\substack{\text{Nitrogen}\\\text{dioxide}}}{\dot{N}O_2(g)} + \underset{\substack{\text{Molecular}\\\text{oxygen}}}{O_2(g)} \qquad (11.10)$$

$$\underset{\substack{\text{Nitrogen}\\\text{dioxide}}}{\dot{N}O_2(g)} + \underset{\substack{\text{Atomic}\\\text{oxygen}}}{\cdot\dot{O}(g)} \rightarrow \underset{\substack{\text{Nitric}\\\text{oxide}}}{\dot{N}O(g)} + \underset{\substack{\text{Molecular}}}{O_2(g)} \qquad (11.11)$$

$$\underset{\substack{\text{Atomic}\\\text{oxygen}}}{\cdot\dot{O}(g)} + \underset{\substack{\text{Ozone}}}{O_3(g)} \rightarrow \underset{\substack{\text{Molecular}\\\text{oxygen}}}{2O_2(g)} \qquad (11.12)$$
$$\text{(net process)}$$

The result of this sequence is that one molecule of ozone is destroyed, but neither NO(g) nor NO$_2$(g) is lost. This sequence is called a catalytic ozone destruction cycle because the species causing the O$_3$(g) loss, NO(g), is recycled. This particular cycle is the NO$_x$(g) catalytic ozone destruction cycle, where NO$_x$(g) = NO(g) + NO$_2$(g) and NO(g) is the catalyst. The number of times the cycle is executed before NO$_x$(g) is removed from the cycle by reaction with another gas is the chain length. In the upper stratosphere, the chain length of this cycle is about 10^5 (Lary, 1997). Thus, 10^5 molecules of O$_3$(g) are destroyed before one NO$_x$(g) molecule is removed from the cycle. In the lower stratosphere, the chain length decreases to near 10. When NO$_x$(g) is removed from this cycle, its major loss processes are the formation of nitric acid and peroxynitric acid by the reactions

$$\underset{\substack{\text{Nitrogen}\\\text{dioxide}}}{\dot{N}O_2(g)} + \underset{\substack{\text{Hydroxyl}\\\text{radical}}}{\dot{O}H(g)} \overset{M}{\rightarrow} \underset{\substack{\text{Nitric}\\\text{acid}}}{HNO_3(g)} \qquad (11.13)$$

$$\underset{\substack{\text{Hydroperoxy}\\\text{radical}}}{H\dot{O}_2(g)} + \underset{\substack{\text{Nitrogen}\\\text{dioxide}}}{\dot{N}O_2(g)} \overset{M}{\rightarrow} \underset{\substack{\text{Peroxynitric}\\\text{acid}}}{HO_2NO_2(g)} \qquad (11.14)$$

Nitric acid and peroxynitric acid photolyze back to the reactants that formed them, but such processes are slow. Peroxynitric acid also decomposes thermally, but thermal decomposition is slow in the stratosphere because temperatures are low there.

The natural NO$_x$(g) catalytic cycle erodes the ozone layer above ozone's peak altitude shown in Figure 11.4. Most of this cycle is natural because most NO$_x$(g) in the stratosphere is from natural sources. An increasing anthropogenic source of stratospheric NO$_x$(g) is aircraft emission. *Worldwide, about 24 percent of aircraft emissions occur in the lower stratosphere*, and almost all stratospheric emissions occur between 30°N and 90°N (Whitt et al., 2011). However, such emissions occur simultaneously with emissions of organic gases, carbon monoxide, and $NO_x(g)$, which together slightly increase ozone in the lower stratosphere via reaction mechanisms such as those in Sections 4.2.4 and 4.2.5.

In the 1970s and 1980s, a concern arose about a proposal to introduce a fleet of **supersonic transport** (SST) aircraft into the middle stratosphere. In the middle and upper stratosphere, the $NO_x(g)$ catalytic destruction cycle of ozone is strong; thus, $NO_x(g)$ has more potential to destroy ozone than it does in the lower stratosphere, even in the presence of organic gases and carbon monoxide. However, the plan to introduce a fleet of SSTs never materialized.

11.3.3. Effects of Hydrogen on the Natural Ozone Layer

Hydrogen-containing compounds, particularly the hydroxyl radical [OH(g)] and the hydroperoxy radical [$HO_2(g)$], are responsible for shaping the ozone profile in the lower stratosphere. The hydroxyl radical is produced in the stratosphere by one of several reactions:

$$\cdot\dot{O}(^1D)(g) + \begin{cases} H_2O(g) \\ \text{Water} \\ \text{vapor} \\ \\ CH_4(g) \\ \text{Methane} \\ \\ H_2(g) \\ \text{Molecular} \\ \text{hydrogen} \end{cases} \rightarrow \dot{O}H(g) + \begin{cases} \dot{O}H(g) \\ \text{Hydroxyl} \\ \text{radical} \\ \\ \dot{C}H_3(g) \\ \text{Methyl} \\ \text{radical} \\ \\ \dot{H} \\ \text{Atomic} \\ \text{Hydrogen} \end{cases} \quad (11.15)$$

Excited atomic oxygen

The hydroxyl radical participates in an **$HO_x(g)$ catalytic ozone destruction cycle**, where $HO_x(g) = OH(g) + HO_2(g)$. $HO_x(g)$ catalytic cycles are important in the lower stratosphere. The most effective $HO_x(g)$ cycle, which has a chain length in the lower stratosphere of 1 to 40 (Lary, 1997), is

$$\dot{O}H(g) + O_3(g) \rightarrow H\dot{O}_2(g) + O_2(g) \quad (11.16)$$
Hydroxyl radical   Ozone   Hydroperoxy radical   Molecular oxygen

$$H\dot{O}_2(g) + O_3(g) \rightarrow \dot{O}H(g) + 2O_2(g) \quad (11.17)$$
Hydroperoxy radical   Ozone   Hydroxyl radical   Molecular oxygen

$$2O_3(g) \rightarrow 3O_2(g) \quad (11.18)$$
Ozone   Molecular oxygen   (net process)

$HO_x(g)$ species can be removed temporarily from the catalytic cycle by Reactions 11.13, 11.14, and 11.19:

$$H\dot{O}_2(g) + \dot{O}H(g) \rightarrow H_2O(g) + O_2(g) \quad (11.19)$$
Hydroperoxy radical   Hydroxyl radical   Water vapor   Molecular oxygen

This mechanism is particularly efficient at removing $HO_x(g)$ from the cycle because it removes two $HO_x(g)$ molecules at a time.

11.3.4. Effects of Carbon on the Natural Ozone Layer

Carbon monoxide and methane produce ozone by the chemical reaction mechanisms shown in Sections 4.2.4 and 4.2.5, respectively. The contributions of CO(g) and $CH_4(g)$ to ozone production in the stratosphere are relatively small, but they increase when $NO_x(g)$ simultaneously increases, such as in the case of aircraft emissions in the lower stratosphere (Section 11.3.2).

An important *by-product of methane oxidation in the stratosphere is water vapor*, produced by

$$CH_4(g) + \dot{O}H(g) \rightarrow \dot{C}H_3(g) + H_2O(g) \quad (11.20)$$
Methane   Hydroxyl radical   Methyl radical   Water vapor

Because water vapor mixing ratios in the stratosphere are low and transport of water vapor from the troposphere to stratosphere is slow, this reaction is a relatively important source of water vapor in the stratosphere. An anthropogenic source of water vapor into the stratosphere is aircraft exhaust.

11.4. Recent Changes to the Ozone Layer

Changes in stratospheric ozone since the 1970s can be categorized as global stratospheric changes, Antarctic stratospheric changes, and Arctic stratospheric changes.

11.4.1. Global Stratospheric Changes

Between 1979 and 2011, the global stratospheric ozone column abundance decreased by approximately 5 percent (Figure 11.8). Unusual decreases in global ozone were detected following the El Chichón (Mexico) volcanic eruption in April 1982, and the Mount Pinatubo (Philippines) eruption in June 1991 (Figure 11.9). These eruptions injected aerosol particles into the stratosphere. On the surfaces of these particles, chemical reactions involving chlorine occurred that contributed to ozone loss. Over time, the concentration of these particles decreased, and the global ozone layer partially

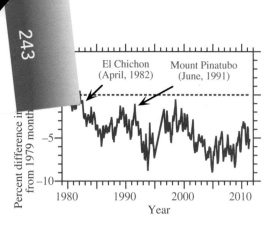

Figure 11.8. Percentage change in the monthly averaged global (90°S to 90°N) ozone column abundance between a given month through May 2011 and the same month in 1979. No data were available from December 1994 to July 1996. Data were obtained from the satellite-based Total Ozone Mapping Spectrometer (TOMS) and made available by NASA Goddard Space Flight Center, Greenbelt, Maryland.

recovered. Because volcanic particles were responsible for only temporary ozone losses, the net ozone loss over the globe from 1979 to 2011 was still about 5 percent.

Figure 11.9. Mount Pinatubo eruption, June 12, 1991. Three days later, a larger eruption, the second largest in the twentieth century, occurred. Photo by Dave Harlow, USGS; http://pubs.usgs.gov/fs/1997/fs113-97/.

11.4.2. Antarctic Stratospheric Changes

Between 1950 and 1980, no measurements from three ground-based stations in Antarctica showed ozone levels less than 220 DUs, the threshold defining Antarctic ozone depletion. Every SH spring (September–November) since 1980, measurements of stratospheric ozone have shown a depletion. Farman et al. (1985) first reported depletions of more than 30 percent relative to pre-1980 measurements. Since then, measurements over the Antarctic have indicated depletions of up to 70 percent of the column ozone for a week in early October. The largest average depletion for the month of October from 60°S to 90°S since 1979 was 42.4 percent and occurred in 2006 (Figure 11.10a). Most ozone depletion occurs between 12 and 20 km in altitude. The large reduction in stratospheric ozone over the Antarctic in the SH spring each year is called the **Antarctic ozone hole**. The areal extent of the ozone hole is commonly greater than the size of North America.

Figure 11.10b shows the zonally and October-averaged ozone column abundance versus latitude for 1979, 1999, 2005, and 2010. In 1999, the October average over the South Pole was 131 DUs, which compares with 286 DUs in 1979, 141 DUs in 2005, and 150 DUs in 2010. October ozone levels over 45°S (the latitude of southern New Zealand, Chile, and Argentina) were all 7 to 8 percent lower in 1999, 2005, and 2010 than in 1979. Ozone levels over 30°S (the latitude of Australia, South Africa, Chile, Argentina, and southern Brazil) were all about 3 percent lower during 1999, 2005, and 2010 than during 1979. Temporary losses of ozone over these countries have caused concern due the effects of the resulting enhanced UV-B radiation on health (Section 11.9).

11.4.3. Arctic Stratospheric Changes

Since 1979, the stratospheric ozone layer over the North Pole has declined during the NH late winter and spring (March–May). This reduction is called the **Arctic ozone dent**. Figure 11.10a indicates that the Arctic ozone dent in March has consistently been less severe than has the corresponding springtime Antarctic ozone hole during October. Figures 11.10a and 11.11, however, show that ozone levels over the Arctic in March 2005 and 2011 were about 15 and 24 percent, respectively, lower than were those in March 1979. Ozone levels over the Arctic in March 2010 were nearly the same as those in March 1979. As discussed in Section 11.8.3, the ozone dent occurs when temperatures in the Arctic

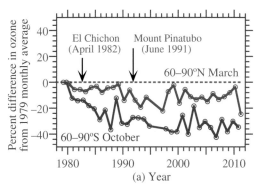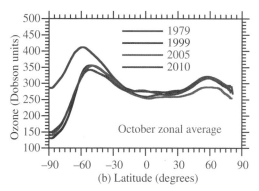

Figure 11.10. (a) Percentage difference in March-averaged 60°N to 90°N and October-averaged 60°S to 90°S ozone column abundances between the given year through March 2011 and 1979. (b) Variation with latitude of October monthly and zonally averaged column abundances of ozone during 1979, 1999, 2005, and 2010. No data were available from December 1994 to July 1996. Data were obtained from the satellite-based Total Ozone Mapping Spectrometer (TOMS) and made available by NASA Goddard Space Flight Center, Greenbelt, Maryland.

stratosphere drop sufficiently for certain ice particles to form and chemical reactions involving anthropogenic chlorine and bromine to occur on their surfaces. The products of such reactions eventually destroy ozone. When the dent does not appear, it is because Arctic stratospheric temperatures are too warm for the ice particles to form.

11.4.4. Effects of Ozone and Air Pollution Changes on Ultraviolet Radiation

Ozone reductions increase UV radiation penetration to the surface. Ozone losses between the 1970s and

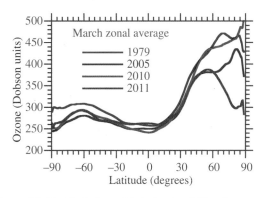

Figure 11.11. Variation with latitude of March monthly and zonally averaged column abundance of ozone during 1979, 2005, 2010, and 2011. Data were obtained from the satellite-based Total Ozone Mapping Spectrometer (TOMS) and made available by NASA Goddard Space Flight Center, Greenbelt, Maryland.

1998 increased ground UV-B radiation during 1998 by about 7 percent in NH midlatitudes in winter and spring, 4 percent in NH midlatitudes in summer and fall, 6 percent in SH midlatitudes during the entire year, 130 percent in the Antarctic in the SH spring, and 22 percent in the Arctic in the NH spring (Madronich et al., 1998). Localized measurements of UV at Lauder, New Zealand (45°S), similarly showed surface UV-B radiation levels 12 percent higher during the SH summer of 1998/1999 than during the early 1990s (McKenzie et al., 1999).

Surface measurements suggest, though, that in the SH, such as in Lauder, New Zealand, and at the South Pole, UV reaching the surface declined slightly between 1999 and 2006, suggesting a slight recovery in stratospheric ozone (McKenzie et al., 2007). However, in the NH, UV levels generally increased during this period, not due to a loss of stratospheric ozone, but due to a reduction in **UV-absorbing particulate matter** air pollution in the NH.

11.5. Effects of Chlorine on Global Ozone Reduction

Ozone reductions since the late 1970s correlate with increases in chlorine and bromine in the stratosphere. Molina and Rowland (1974) first recognized that anthropogenic chlorine compounds could destroy stratospheric ozone. Since then, scientists have strengthened the links among global ozone reduction, Antarctic ozone depletion, and the presence of chlorine- and bromine-containing compounds in the stratosphere.

Table 11.2. Mixing ratios in 2010, overall lifetimes from emission to removal, and 100-year global warming potentials of selected chlorocarbons, bromocarbons, and fluorocarbons

Chemical formula	Trade name	Chemical name	Tropospheric mixing ratio (pptv)	Estimated overall atmospheric lifetime (yr)	100-year global warming potential (GWP)
CHLOROCARBONS AND CHLORINE COMPOUNDS					
Chlorofluorocarbons (CFCs)					
$CFCl_3(g)$	CFC-11	Trichlorofluoromethane	241	45	4,750
$CF_2Cl_2(g)$	CFC-12	Dichlorodifluoromethane	531	100	10,900
$CFCl_2CF_2Cl(g)$	CFC-113	1-Fluorodichloro,2-difluorochloroethane	77	85	6,130
$CF_2ClCF_2Cl(g)$	CFC-114	1-Difluorochloro,2-difluorochloroethane	17	300	10,000
$CF_2ClCF_3(g)$	CFC-115	1-Difluorochloro,2-trifluoroethane	8	1,700	7,370
Hydrochlorofluorocarbons (HCFCs)					
$CF_2ClH(g)$	HCFC-22	Chlorodifluoromethane	251	11.8	1,810
$CH_3CFCl_2(g)$	HCFC-141b	2-Fluorodichloroethane	18	9.3	725
$CH_3CF_2Cl(g)$	HCFC-142b	2-Difluorochloroethane	21	17.9	2,310
Other Chlorocarbons					
$CCl_4(g)$		Carbon tetrachloride	88	26	1,400
$CH_3CCl_3(g)$		Methyl chloroform	7.3	4.8	146
$CH_3Cl(g)$		Methyl chloride	580	1.3	13
Other Chlorinated Compounds					
$HCl(g)$		Hydrochloric acid	10–1,000	<1	–
BROMOCARBONS					
Halons					
$CF_3Br(g)$	H-1301	Trifluorobromomethane	2.9	65	7,140
$CF_2ClBr(g)$	H-1211	Difluorochlorobromomethane	4.1	16	1,890
CF_2BrCF_2Br	H-2402	1-Difluorobromo,2-difluorobromoethane	0.7	20	1,640
Other Bromocarbons					
$CH_3Br(g)$		Methyl bromide	8	0.7	5
FLUOROCARBONS AND FLUORINE COMPOUNDS					
Hydrofluorocarbons (HFCs)					
$CHF_3(g)$	HFC-23	Trifluoromethane	22	270	14,800
$CHF_2CF_3(g)$	HFC-125	Pentafluoroethane	3.7	29	3,500
$CH_2FCF_3(g)$	HFC-134a	1-Fluoro,2-trifluoroethane	35	13.6	1,430
$CH_3CHF_2(g)$	HFC-152a	1,1-Difluoroethane	4	1.4	124
Perfluorocarbons (PFCs)					
$CF_4(g)$	PFC-14	Tetrafluoromethane	74	50,000	7,390
$C_2F_6(g)$	PFC-116	Perfluoroethane	3.5	10,000	12,200
Other Fluorinated Compounds					
$SF_6(g)$		Sulfur hexafluoride	7.2	3,200	22,800

Sources: International Panel on Climate Change (IPCC) (2007, Tables 2.1 and 2.14); Mauna Loa Data Center (n.d.).

11.5.1. Chlorofluorocarbons and Related Compounds

The compounds that play the most important role in reducing stratospheric ozone are **chlorofluorocarbons** (**CFCs**). Table 11.2 identifies important CFCs, which are gases formed synthetically such as by replacing one or more chlorine atoms with fluorine atoms in **carbon tetrachloride** [$CCl_4(g)$]. For example, **CFC-12** [$CF_2Cl_2(g)$, dichlorodifluoromethane] can be formed by replacing two chlorine atoms in carbon tetrachloride with two fluorine atoms. CFCs can also be obtained by replacing all hydrogen atoms in methane [$CH_4(g)$] or ethane [$C_2H_6(g)$] with chlorine and/or fluorine atoms.

11.5.1.1. Invention of Chlorofluorocarbons

The first CFCs were synthesized during the 1890s by the Belgian scientist, Frédéric Swartz (1866–1940). He developed a method to produce $CFCl_3(g)$ and $CF_2Cl_2(g)$ by replacing one or two chlorine atoms with a fluorine atom in $CCl_4(g)$. However, these compounds were not used for practical purposes for several decades.

On a Saturday in 1928, a representative of General Motors' Frigidaire division asked **Thomas Midgley** to find a nontoxic, nonflammable substitute for an existing refrigerant, ammonia, a flammable and toxic gas. The same afternoon, Midgley and his assistants, Albert L. Henne (1901–1967) and Robert R. McNary (1903–1988), working at the Thomas and Hochwalt Laboratory, 127 North Ludlow Street, Dayton, Ohio, hypothesized that $CF_2Cl_2(g)$ had the necessary qualities to replace ammonia. They subsequently developed a more efficient method of synthesizing it than did Swartz.

CFC-12 and subsequent CFCs were inexpensive, nontoxic, nonflammable, nonexplosive, insoluble, and chemically unreactive under tropospheric conditions. As a result, they became popular. Midgley demonstrated the nontoxic and nonflammable properties of CFC-12 to the American Chemical Society in April 1930, by inhaling CFC-12 and then blowing it over a candle flame, extinguishing the flame. The use of CFC-12 for refrigeration was not disclosed previously because the Frigidaire department of General Motors needed time to file patents on the family of CFC compounds that would be used for refrigeration (Bhatti, 1999). Midgley is the same scientist who invented tetraethyl lead (ethyl) as an additive to gasoline (Section 3.6.9). Some argue that Midgley's chemical products led to the two greatest environmental disasters of the twentieth century.

In 1931, **CFC-12** was produced by the DuPont chemical manufacturer under the trade name **Freon**, a name chosen by Midgley and his assistants. Its first use was in small ice cream cabinets. In 1934, it was used in refrigerators and whole-room coolers. Soon after, it was used in household and automotive air conditioning systems. In 1932, **CFC-11** [$CFCl_3(g)$, trichlorofluoromethane] was first produced. Its first use was in large air conditioning units. CFCs became airborne only when coolants leaked or were drained.

During World War II, the U.S. government funded research to develop a portable device that could be used for military personnel to spray insecticide to kill malaria-containing bugs. In 1943, two researchers in the U.S. Department of Agriculture responding to the call, **Lyle Goodhue** and **William Sullivan**, developed a method of propelling the insecticide as an aerosol through the nozzle of a small can by liquefying CFC-11 and -12. CFCs flowed out of the spray can's nozzle, evaporating in the process and carrying with them a mist containing other ingredients.

Spray cans were subsequently used as **propellants** not only of insecticide, but also of hair spray, paint, deodorant, disinfectant, and polish. In 1949, **Robert Abplanalp** invented a method to discharge liquids from a spray can under high pressure of a CFC gas. This led to the widespread use of spray cans ejecting liquid foams, powders, and creams together with the emission of the CFC gas. When CFC use in spray cans was eventually banned, the replacement compound was commonly a mixture of volatile organic compounds selected among propane, n-butane, isobutene, dimethyl ether, and methyl ethyl ether, which are all more flammable than CFCs.

CFC-11 and -12 have also been used as **blowing agents** in foam production. A blowing agent is a volatile compound added in small quantities as a liquid to a chemical mixture containing polyurethane. As the blowing agent evaporates, it causes the polyurethane to expand, producing foam bubbles containing the gas. **Foam** is used in insulation, disposable cups and cartons, and fire extinguishers. CFCs are released to the air during foam production; however, CFCs in the air spaces of foam are usually confined and not an important source of atmospheric CFCs.

Figure 11.12 shows the reported sales of CFC-11 and -12 between 1976 and 2003. In 1976, almost 58 percent of CFC-11 and -12 were sold for use as propellants in spray cans. Secondary use of CFC-11 was as a blowing agent and of CFC-12 was as a refrigerant. The large decreases in the sales of CFC-11 and -12 in Figure 11.12 are due to regulation (Section 11.10). Black market sales, though, continue to this day.

Table 11.2 lists additional CFCs. Of note, CFC-113 was first produced in 1934 and used in air conditioning units. It was subsequently used as a solvent in the microelectronics industry and in the dry cleaning industry, as well as a spray can propellant and a blowing agent in foam production.

11.5.1.2. Other Chlorine Compounds

CFCs are a subset of **chlorocarbons**, which are compounds containing carbon and chlorine.

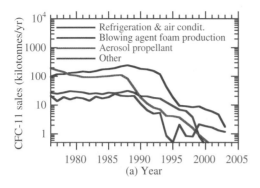
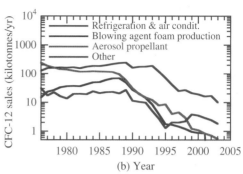

Figure 11.12. Reported sales of CFC-11 and -12 between 1976 and 2003. Percentages are of the total for the year. CFC, chlorofluorocarbon. From Alternative Fluorocarbons Environmental Acceptability Study (AFEAS) (n.d.).

Hydrochlorofluorocarbons (HCFCs) are another subset of chlorocarbons. HCFCs are similar to CFCs, except that HCFCs have at least one hydrogen atom. The hydrogen atom allows HCFCs to be broken down in the troposphere by reaction with OH(g), which does not readily break down CFCs. Because HCFCs react with OH(g) more readily than do CFCs, a smaller percentage of emitted HCFCs than CFCs reaches the stratosphere. Nevertheless, because HCFCs contain chlorine and some HCFCs reach the stratosphere, HCFCs are still a danger to stratospheric ozone. HCFC-22, first produced in 1943, is the most abundant HCFC in the air today. HCFC-22 has been used as a refrigerant, spray can propellant, and blowing agent in foam production.

Other chlorocarbons include **carbon tetrachloride** [$CCl_4(g)$], **methyl chloroform** [$CH_3CCl_3(g)$], and **methyl chloride** [$CH_3Cl(g)$]. Carbon tetrachloride is used as an intermediate in the production of CFCs and HCFCs, and as a solvent and grain fumigant. Methyl chloroform is used as a degreasing agent, a dry cleaning solvent, and an industrial solvent. Methyl chloride is produced synthetically only in small quantities for use in the production of silicones and tetramethyl lead intermediates. Most methyl chloride in the air is produced biogenically in the oceans.

Another chlorine-containing gas in the troposphere is **hydrochloric acid** [HCl(g)]. HCl(g) has larger natural than anthropogenic sources. A natural source includes evaporation of the chloride ion from sea spray and volcanic emissions. Although some anthropogenic emissions of HCl(g) are from waste incineration, about 98 percent are from coal combustion (Saxena et al., 1993).

11.5.1.3. Bromine Compounds
Although chlorine-containing compounds are more abundant than are bromine-containing compounds, the latter compounds are more efficient, molecule for molecule, at destroying ozone. The primary source of stratospheric bromine is **methyl bromide** [$CH_3Br(g)$], which is produced biogenically in the oceans and emitted as a soil fumigant. Other sources of bromine are a group of synthetically produced compounds termed **halons**, which are used in fire extinguishers and as fumigants. The most common halons are **H-1301** [$CF_3Br(g)$], **H-1211** [$CF_2ClBr(g)$], and **H-2402** [$CF_2BrCF_2Br(g)$]. Methyl bromide and halons are **bromocarbons** because they contain both bromine and carbon.

11.5.1.4. Fluorine Compounds
Compounds that contain hydrogen, fluorine, and carbon but not chlorine or bromine are **hydrofluorocarbons** (HFCs). HFCs were produced in abundance only recently as a replacement for CFCs and HCFCs. The most abundantly emitted HFC to date has been **HFC-134a** [$CH_2FCF_3(g)$]. Related to HFCs are **perfluorocarbons** (PFCs), such as perfluoroethane [$C_2F_6(g)$], and **sulfur hexafluoride** [$SF_6(g)$]. Because the fluorine in HFCs and PFCs has little effect on ozone, production of HFCs and PFCs may increase in the future.

Unfortunately, because they strongly absorb thermal-IR radiation, HFCs and PFCs enhance global warming. This is illustrated in Table 11.2, which shows the **100-year global warming potential (GWP)** of these and other chemicals. The GWP gives the 100-year integrated change in global atmospheric heating due to a compound in the air relative to that of carbon dioxide (Section 12.6.4). Table 11.2 indicates that CFCs, halons, HFCs, and PFCs all have high GWPs, indicating that they all contribute to atmospheric warming and will for more than 100 years after their emission.

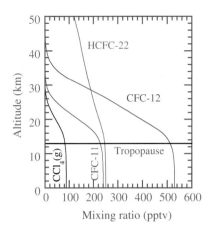

Figure 11.13. Variation of CFC-11 and -12, HCFC-22, and $CCl_4(g)$ with altitude at 30°N latitude. CFC, chlorofluorocarbon; HCFC, hydrochlorofluorocarbon. Smoothed and scaled from Jackman et al. (1996) to 2011 near-surface mixing ratios.

11.5.2. Lifetimes and Mixing Ratios of Chlorinated Compounds

Once emitted, CFCs take about 1 year to mix up to the tropopause. Because they are chemically unreactive and cannot be broken down by solar wavelengths that reach the troposphere, CFCs are not removed chemically from the troposphere. Instead, they become well mixed in the troposphere (Figure 11.13) and slowly penetrate to the stratosphere. Today, the tropospheric mixing ratios of CFC-11 and -12, the two most abundant CFCs, are about 240 and 530 pptv, respectively (Table 11.2).

11.5.2.1. Lifetimes of Chlorofluorocarbons

Because the stratosphere is one large temperature inversion, vertical transport of ozone through it is slow. About 10 megatonnes (Mt) of chlorine in the form of CFCs reside in the troposphere, and the transfer rate of CFC-chlorine from the troposphere to the middle stratosphere is about 0.1 Mt/yr. In this simplified scenario, the average time required for the transfer of a CFC molecule from the troposphere to the middle stratosphere is about 100 years.

CFCs are broken down in the stratosphere only when they are exposed to far-UV radiation (wavelengths of 0.01–0.25 μm), and this exposure occurs at an altitude of 12 to 20 km and higher. At such altitudes, far-UV wavelengths photolyze CFC-11 and -12 by

$$CFCl_3(g) + h\nu \rightarrow \dot{C}FCl_2(g) + \dot{C}l(g) \quad \lambda < 250\,nm$$
CFC-11   Dichlorofluoro-   Atomic
           methyl radical   chlorine   (11.21)

$$CF_2Cl_2(g) + h\nu \rightarrow \dot{C}F_2Cl(g) + \dot{C}l(g) \quad \lambda < 230\,nm$$
CFC-12   Chlorofluoro-   Atomic
           methyl radical   chlorine   (11.22)

At 25 km, *e*-folding lifetimes of CFC-11 and -12 against photolysis under maximum sunlight conditions are on the order of 23 and 251 days, respectively. Average lifetimes are on the order of two to three times these values.

In sum, the limiting factor in CFC decomposition in the stratosphere is not transport from the surface to the tropopause or photochemical breakdown in the stratosphere, but transport from the tropopause to the middle stratosphere. Table 11.2 indicates that the overall lifetimes of CFC-11 and -12 between release at the surface and destruction in the middle stratosphere are about 45 and 100 years, respectively. The lifetime of CFC-12 is longer than that of CFC-11, partly because the former compound must climb to a higher altitude in the stratosphere before breaking apart than must the latter. Because of their long overall lifetimes, some CFCs emitted since the 1930s are still present in the stratosphere. Those emitted today are likely to remain in the air until the second half of the twenty-first century.

11.5.2.2. Lifetimes of Nonchlorofluorocarbons

Overall lifetimes of non-CFC chlorinated compounds are often shorter than are those of CFCs. The lifetimes of $CCl_4(g)$, HCFC-22(g), $CH_3CCl_3(g)$, $CH_3Cl(g)$, and HCl(g) between emission and chemical destruction are about 25 years, 12 years, 7 years, 1.3 years, and less than 0.1 year, respectively (Table 11.2). Non-CFCs generally have shorter lifetimes than do CFCs because the former react faster with OH(g) and are often more water soluble than are the latter. The benefit of a shorter lifetime for a chlorine-containing compound is that, if its breakdown occurs in the troposphere, the chlorine released can be converted to HCl(g), which is highly soluble and can be removed readily by its dissolution in rainwater. Because the stratosphere does not contain clouds, except ice-containing clouds that form seasonally over the poles, HCl(g) cannot be rained out of the stratosphere. Some non-CFCs, such as HCFC-22, photolyze slower than do CFCs, so once HCFC-22 reaches the middle stratosphere, its concentration builds up there to a greater extent than do concentrations of several CFCs, as seen in Figure 11.13.

Of non-CFC chlorine compounds, $CH_3Cl(g)$ and HCl(g) have the largest natural sources. The tropospheric *e*-folding chemical lifetime of $CH_3Cl(g)$ against reaction by OH(g) is about 1.5 years, whereas that of

HCl(g) against reaction by OH(g) is about 15 to 30 days. HCl(g) is also soluble in water and is absorbed by clouds. Volcanos, which emit water vapor and hydrochloric acid, produce clouds and rain that remove HCl(g), preventing most of it from reaching the stratosphere (Lazrus et al., 1979; Pinto et al., 1989; Tabazadeh and Turco, 1993). The fact that the two major natural sources of chlorine, $CH_3Cl(g)$ and HCl(g), have short chemical lifetimes against destruction by OH(g), and the fact that HCl(g) is soluble in water, whereas CFCs have long chemical lifetimes and are insoluble, support the contention that CFCs and not naturally emitted chlorine compounds are responsible for most ozone destruction in the stratosphere.

11.5.2.3. Emissions of Chlorine Compounds to the Stratosphere

Table 11.3 summarizes the relative emissions of anthropogenic and natural chlorine-containing compounds into the stratosphere from one analysis. About 82 percent of chlorine entering the stratosphere originated from anthropogenic sources. Of the remainder, about 15 percent was methyl chloride, emitted almost exclusively by biogenic sources in the oceans, and 3 percent was hydrochloric acid, emitted by volcanos, evaporated from sea spray, and otherwise produced naturally. The relatively large anthropogenic versus natural source of chlorine into the stratosphere supports the

Table 11.3. Relative emissions of selected chlorine compounds into stratosphere

Trade name or chemical name	Chemical formula	Contribution to stratospheric emissions (percent)
Anthropogenic Sources		
CFC-12	$CF_2Cl_2(g)$	28
CFC-11	$CFCl_3(g)$	23
Carbon tetrachloride	$CCl_4(g)$	12
Methyl chloroform	$CH_3CCl_3(g)$	10
CFC-113		6
	$CFCl_2CF_2Cl(g)$	
HCFC-22	$CF_2ClH(g)$	3
Natural Sources		
Methyl chloride	$CH_3Cl(g)$	15
Hydrochloric acid	HCl(g)	3
TOTAL		100

CFC, chlorofluorocarbon; HCFC, hydrochlorofluorocarbon.
Source: World Meteorological Organization (1995).

contention that stratospheric ozone reductions result primarily from anthropogenic chlorine, not natural chlorine.

11.5.3. Catalytic Ozone Destruction by Chlorine

Once released from their parent compounds in the stratosphere, chlorine atoms from CFCs and non-CFCs react along one of several pathways. Chlorine reacts in the **chlorine catalytic ozone destruction cycle**,

$$\dot{C}l(g) + O_3(g) \rightarrow Cl\dot{O}(g) + O_2(g) \quad (11.23)$$

Atomic chlorine   Ozone   Chlorine monoxide   Molecular oxygen

$$Cl\dot{O}(g) + \cdot\dot{O}(g) \rightarrow \dot{C}l(g) + O_2(g) \quad (11.24)$$

Chlorine monoxide   Atomic oxygen   Atomic chlorine   Molecular oxygen

$$\cdot\dot{O}(g) + O_3(g) \rightarrow 2O_2(g) \quad (11.25)$$

Atomic oxygen   Ozone   Molecular oxygen

(net process)

At midlatitudes, the chain length of this cycle increases from about 10 in the lower stratosphere to about 1,000 in the middle and upper stratosphere (Lary, 1997).

The primary removal mechanisms of **active chlorine** [Cl(g)+ClO(g)] from the catalytic cycle are reactions that produce the **chlorine reservoirs**, HCl(g) and **chlorine nitrate** [$ClONO_2(g)$]. Chlorine reservoirs are called such because they temporarily store active chlorine, preventing it from destroying ozone. Conversion of Cl(g) to HCl(g) occurs by

$$\dot{C}l(g) + \left\{ \begin{array}{l} CH_4(g) \\ \text{Methane} \\ \\ H\dot{O}_2(g) \\ \text{Hydroperoxy radical} \\ \\ H_2(g) \\ \text{Molecular hydrogen} \\ \\ H_2O_2(g) \\ \text{Hydrogen peroxide} \end{array} \right\} \rightarrow HCl(g) + \left\{ \begin{array}{l} \dot{C}H_3(g) \\ \text{Methyl radical} \\ \\ O_2(g) \\ \text{Molecular oxygen} \\ \\ \dot{H}(g) \\ \text{Atomic hydrogen} \\ \\ H\dot{O}_2(g) \\ \text{Hydroperoxy radical} \end{array} \right. \quad (11.26)$$

Atomic chlorine   Hydrochloric acid

Conversion of ClO(g) to $ClONO_2(g)$ occurs by

$$Cl\dot{O}(g) + \dot{N}O_2(g) \xrightarrow{M} ClONO_2(g) \quad (11.27)$$

Chlorine monoxide   Nitrogen dioxide   Chlorine nitrate

At any time, about 1 percent of the non-CFC chlorine in the stratosphere is in the form of active chlorine, whereas most of the rest is in the form of a chlorine reservoir. Because CFCs release their chlorine by photolysis in the middle and upper stratosphere, HCl(g) mixing ratios should also peak in the middle and upper stratosphere. Indeed, observations confirm this supposition.

The HCl(g) reservoir leaks back to atomic chlorine by photolysis, reaction with OH(g), and reaction with O(g), all of which are slow processes. The e-folding lifetime of HCl(g) against photolysis, for example, is about 1.5 years at 25 km. HCl(g) also diffuses back to the troposphere, where it can be absorbed by clouds. The $ClONO_2$(g) reservoir leaks back to atomic chlorine by photolysis with an e-folding lifetime of about 4.5 hours at 25 km.

11.6. Effects of Bromine on Global Ozone Reduction

Like chlorine, bromine reduces stratospheric ozone. The primary source of stratospheric bromine is **methyl bromide** [CH_3Br(g)], which is produced biogenically in the oceans and emitted as a soil fumigant. Other sources of bromine are halons, defined in Section 11.5.1. The tropospheric mixing ratios of the most common halons, CF_2ClBr(g) (H-1211) and CF_3Br(g) (H-1301), were about 4 and 3 pptv, respectively, in 2010, less than 1 percent of the mixing ratios of CFC-12 (Table 11.2). Nevertheless, the efficiency of ozone destruction by the bromine catalytic cycle is greater than is that by the chlorine catalytic cycle.

Methyl bromide and halons are photolyzed in the stratosphere above 20 km to produce atomic bromine. Photolysis of methyl bromide occurs by

$$CH_3Br(g) + h\nu \rightarrow \dot{C}H_3(g) + \dot{Br}(g) \quad \lambda < 260\,nm$$

Methyl bromide   Methyl radical   Atomic bromine   (11.28)

The e-folding lifetime of CH_3Br(g) against loss by this reaction is about 10 days at 25 km.

Once atomic bromine forms in the stratosphere, it reacts in the **bromine catalytic ozone destruction cycle**,

$$\dot{Br}(g) + O_3(g) \rightarrow Br\dot{O}(g) + O_2(g) \quad (11.29)$$

Atomic bromine   Ozone   Bromine monoxide   Molecular oxygen

$$Br\dot{O}(g) + \dot{O}(g) \rightarrow \dot{Br}(g) + O_2(g) \quad (11.30)$$

Bromine monoxide   Atomic oxygen   Atomic bromine   Molecular oxygen

$$\dot{O}(g) + O_3(g) \rightarrow 2O_2(g) \quad (11.31)$$

Atomic oxygen   Ozone   Molecular oxygen

(net process)

The chain length of this cycle increases from about 100 at 20 km to about 10^4 at 40 to 50 km (Lary, 1997). The chain length of the bromine catalytic cycle is longer than is that of the chlorine catalytic cycle because Br(g) is removed more slowly from the bromine cycle by reactions with CH_4(g) and H_2(g) than Cl(g) is removed from the chlorine cycle by reactions with the same chemicals.

When atomic bromine is removed from its catalytic cycle, it forms hydrobromic acid [HBr(g)] by

$$\dot{Br}(g) + \begin{cases} H\dot{O}_2(g) \\ \text{Hydroperoxy radical} \\ \\ H_2O_2(g) \\ \text{Hydrogen peroxide} \end{cases} \rightarrow HBr(g) + \begin{cases} O_2(g) \\ \text{Molecular oxygen} \\ \\ H\dot{O}_2(g) \\ \text{Hydroperoxy radical} \end{cases} \quad (11.32)$$

Atomic bromine   Hydrobromic acid

When BrO(g) is removed, it forms **bromine nitrate** [$BrONO_2$(g)] by

$$Br\dot{O}(g) + \dot{N}O_2(g) \xrightarrow{M} BrONO_2(g) \quad (11.33)$$

Bromine monoxide   Nitrogen dioxide   Bromine nitrate

The HBr(g) reservoir leaks slowly back to atomic bromine by reacting with OH(g). The $BrONO_2$(g) reservoir quickly leaks back to atomic bromine by photolysis. The e-folding lifetime of $BrONO_2$(g) against photolysis is about 10 minutes at 25 km.

11.7. Regeneration Rates of Stratospheric Ozone

The presence of chlorine and bromine has caused levels of ozone in the stratosphere to decrease. If chlorine and bromine could be removed easily from the stratosphere, the stratospheric ozone layer could regenerate quickly. However, natural removal of chlorine and bromine is slow because the overall lifetimes of several chlorocarbons and bromocarbons against chemical removal are on the order of 50 to 100 years.

Suppose, however, that all ozone in the stratosphere were destroyed, all ozone-destroying compounds were removed, but all oxygen remained. How long would the ozone layer take to regenerate? An estimate can be obtained from Figure 11.14, which shows results from two computer simulations of the global atmosphere in which all ozone in the present-day atmosphere was initially removed, but oxygen was not. In the first simulation, ozone regeneration was simulated in the absence of

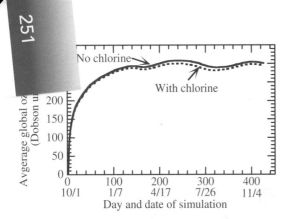

Figure 11.14. Change in globally averaged ozone column abundance during two global model simulations in which chlorine (at 1989 levels) was present and absent, respectively. In both cases, ozone was removed initially from the model atmosphere on October 1, 1988. Bromine was not included in either simulation. From Jacobson (2005a).

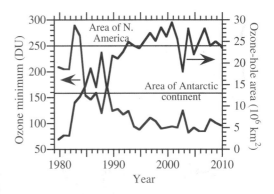

Figure 11.15. Minimum ozone column abundances and areal extent of the ozone hole over the Antarctic region between 1979 and 2010. For comparison, the area of the Antarctic continent is about 13 million km^2 and the area of North America is about 24 million km^2. Data from NASA Goddard Space Flight Center.

chlorine and bromine. In the second, ozone regeneration was simulated in the presence of 1989 levels of chlorine, but in the absence of bromine. *In both simulations, the globally averaged column abundance of ozone regenerated to relatively steady values in less than 1 year.* Regeneration during the simulation in which chlorine was initially present was about 2 to 3 percent less than that during the no-chlorine case, consistent with the estimated global reduction in ozone of 2 to 3 percent between the 1970s and 1989 due to chlorine-containing compounds.

11.8. Antarctic Ozone Hole

The **Antarctic ozone hole** is defined as the area of the globe over which the ozone column abundance decreases below 220 DUs. Such decreases historically have occurred only over the Antarctic during the SH spring (September–November). Between 1981 and 2000, the Antarctic ozone hole area increased from near zero to 29.4 million km^2 in 2000 (Figure 11.15), an area larger than North America. Between 2006 and 2010, the area of the hole gradually declined slightly from 28 to 22 million km^2. Between 1993 and 2010, the minimum ozone anywhere over the Antarctic dropped below 100 DUs twelve times. The lowest minimum was 82 DUs in 2003 (Figure 11.15). Most Antarctic ozone depletion occurs between 14 and 18 km in altitude. Beginning in 1992, springtime ozone decreases were observed up to 24 km and down to 12 to 14 km.

Figure 11.16 shows the extent of the Antarctic ozone hole on October 3, 2010. Although the minimum was within the polar vortex (Section 11.1), an ozone maximum was observed just outside the vortex. The proximity of the maximum to the minimum is one factor that allows the ozone hole to replenish itself from October to December, when the polar vortex breaks down, allowing air outside the Antarctic to penetrate into the Antarctic in the stratosphere.

Whereas the Antarctic ozone hole appears every year during September through November (most of the SH spring), a smaller **Arctic ozone dent** (reduction in ozone to 240–260 DUs) appears from March to May, or most of the NH spring. The greatest ozone dent to date occurred in 2011, resulting in a March ozone column abundance reduction 60°N to 90°N of 24.5 percent compared with March 1979 (Figure 11.11). The second greatest dent occurred in 1997, when March ozone decreased by 22 percent relative to 1979 (Figure 11.10a). During the 1997 event, the minimum ozone value on April 1 was still 247 DUs. This minimum was not low enough to qualify as an ozone hole (less than 220 DUs). In general, the dent is present during some years (e.g., 2000, 2005, 2011) and absent during others (1999, 2010) (Figure 11.11), with dents corresponding to springs with lower Arctic stratospheric temperatures.

The hole and dent are caused by a complex set of factors. One factor linking global ozone reductions to polar ozone depletion is the presence of chlorine and bromine in the stratosphere. Such compounds can cause ozone depletion over the Antarctic or reduction over the

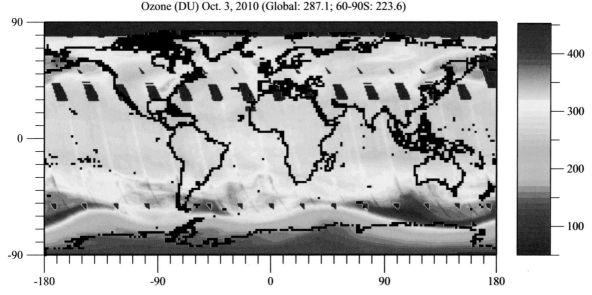

Figure 11.16. Ozone column abundance (in DU) on October 3, 2010. The ozone hole appears over the Antarctic and an ozone maximum appears just outside the Antarctic. No data were available over the North Pole (dark blue). Data were obtained from Total Ozone Mapping Spectrometer (TOMS) satellite and were made available by NASA Goddard Space Flight Center, Greenbelt, Maryland.

Arctic if temperatures are low enough for certain types of ice crystals to form. If such crystals form, reactions occur on their surface releasing compounds that ultimately destroy ozone. These processes are discussed next.

11.8.1. Polar Stratospheric Cloud Formation

The ozone hole over the Antarctic appears in part because the Antarctic winter (June–September) is very cold. Temperatures are low because much of the polar region is exposed to 24 hours of darkness each day during the winter, and a wind system, the **polar vortex**, circles the Antarctic. The vortex is a polar front jet stream wind system that flows around the Antarctic continent, trapping cold air within the Antarctic polar region and preventing warm, ozone-rich air from outside this region into it from penetrating into the Antarctic stratosphere.

Because temperatures are low in the Antarctic and sometimes Arctic stratosphere, optically thin clouds, called **polar stratospheric clouds (PSCs)** form there (Figure 11.17). These clouds have few particles per unit volume of air in comparison with tropospheric clouds. Two major types of clouds form. When temperatures drop to below about 195 K, nitric acid and water vapor grow on small sulfuric acid-water aerosol particles (Toon et al., 1986). Initially, nitric acid and water molecules were believed to deposit to the ice phase in the ratio, 1:3. Such ice crystals have the composition $HNO_3 \cdot 3H_2O$ and are called **nitric acid trihydrate** (NAT) crystals. Subsequently, stratospheric particles were found to contain a variety of phases. Some contain **nitric acid dihydrate** (NAD) (Worsnop et al., 1993), and others contain supercooled liquid water (i.e., liquid water present at temperatures below the freezing point of water) and sulfuric and nitric acids (Tabazadeh et al., 1994). Together, nitrate-containing cloud particles that form at temperatures below about 195 K in the winter polar stratosphere are called **Type I polar stratospheric clouds** (PSCs). Typical diameters of Type I PSCs are 1 μm, varying from 0.01 to 3 μm, and typical number concentrations are ≤1 particles cm^{-3}. As such, these cloud particles are much smaller than are typical tropospheric liquid cloud drops, which are 10 to 20 μm in diameter, and cirrus cloud ice crystals, which are 30 to 80 μm in diameter.

When temperatures drop further, to below the frost point of water (187 K for typical polar stratospheric conditions), water vapor rapidly deposits on an existing aerosol particle's nucleus, creating a more typical ice crystal. Except for its nucleus, the resulting crystals contain almost pure water ice. These crystals are called **Type II polar stratospheric clouds** (PSCs).

Figure 11.17. Polar stratospheric clouds, photographed in the spring of 2000 in the Arctic. NASA.

Typical diameters of Type II PSCs are 20 μm, varying from 1 to 100 μm. Typical number concentrations are ≤0.1 particles cm^{-3}. As such, Type II PCSs are larger in size but fewer in number than are Type I PSCs. The larger size of the Type II PSC suggests it falls from the stratosphere faster than does a Type I PSC. Because Type II PSCs are more difficult to form (because of their lower temperature threshold), and because they fall faster, the number concentration of Type II PSCs is much less than is that of Type I PSCs. One study estimated that about 90 percent of PSCs are Type I, and 10 percent are Type II, when both can form (Turco et al., 1989).

11.8.2. Polar Stratospheric Cloud Surface Reactions

Once PSC particles form in the dark polar winter stratosphere, when temperatures are extremely low, chemical reactions take place on their surfaces. Such reactions are called **heterogeneous reactions** and occur after at least one gas has diffused to and adsorbed to a particle surface. **Adsorption** is a process by which a gas collides with and bonds to a surface. The primary heterogeneous reactions that occur on Type I and II PSC surfaces are

$$\underset{\substack{\text{Chlorine} \\ \text{nitrate}}}{ClONO_2(g)} + \underset{\text{Water-ice}}{H_2O(s)} \rightarrow \underset{\substack{\text{Hypochlorous} \\ \text{acid}}}{HOCl(g)} + \underset{\substack{\text{Adsorbed} \\ \text{nitric} \\ \text{acid}}}{HNO_3(a)}$$

(11.34)

$$\underset{\substack{\text{Chlorine} \\ \text{nitrate}}}{ClONO_2(g)} + \underset{\substack{\text{Adsorbed} \\ \text{hydrochloric} \\ \text{acid}}}{HCl(a)} \rightarrow \underset{\substack{\text{Molecular} \\ \text{chlorine}}}{Cl_2(g)} + \underset{\substack{\text{Adsorbed} \\ \text{nitric} \\ \text{acid}}}{HNO_3(a)}$$

(11.35)

$$\underset{\substack{\text{Dinitrogen} \\ \text{pentoxide}}}{N_2O_5(g)} + \underset{\text{Water-ice}}{H_2O(s)} \rightarrow \underset{\substack{\text{Adsorbed} \\ \text{nitric} \\ \text{acid}}}{2HNO_3(a)}$$

(11.36)

$$\underset{\substack{\text{Dinitrogen} \\ \text{pentoxide}}}{N_2O_5(g)} + \underset{\substack{\text{Adsorbed} \\ \text{hydrochloric} \\ \text{acid}}}{HCl(a)} \rightarrow \underset{\substack{\text{Chlorine} \\ \text{nitrite}}}{ClNO_2(g)} + \underset{\substack{\text{Adsorbed} \\ \text{nitric} \\ \text{acid}}}{HNO_3(a)}$$

(11.37)

$$\underset{\substack{\text{Hypochlorous} \\ \text{acid}}}{HOCl(g)} + \underset{\substack{\text{Adsorbed} \\ \text{hydrochloric} \\ \text{acid}}}{HCl(a)} \rightarrow \underset{\substack{\text{Molecular} \\ \text{chlorine}}}{Cl_2(g)} + \underset{\text{Water-ice}}{H_2O(s)}$$

(11.38)

In these reactions, (g) denotes a gas, $H_2O(s)$ denotes a water-ice surface, HCl(a) denotes HCl adsorbed to either a Type I or II PSC surface, and $HNO_3(a)$ denotes HNO_3 adsorbed to a Type I or II PSC surface. Additional reactions occur for bromine.

Laboratory studies show that HCl(g) readily diffuses to and adsorbs to the surfaces of Type I and II PSCs. When $ClONO_2(g)$, $N_2O_5(g)$, or HOCl(g) impinges upon the surface of a Type I or II PSC, it can react with $H_2O(s)$ or HCl(a) already on the surface. The products of these reactions are adsorbed species, some of which stay adsorbed, whereas others desorb to the vapor phase. Those that desorb to the vapor phase go on ultimately to destroy ozone nearby.

The net result of the heterogeneous reactions is to convert relatively inactive forms of chlorine in the stratosphere, such as $HCl(g)$ and $ClONO_2(g)$, to photochemically active forms, such as $Cl_2(g)$, $HOCl(g)$, and $ClNO_2(g)$. This conversion process is called **chlorine activation**. The most important heterogeneous reaction is Reaction 11.35 (McElroy et al., 1986; Solomon et al., 1986), which generates $Cl_2(g)$.

Reaction 11.36 does not activate chlorine. Its only effect is to remove nitric acid from the gas phase. When nitric acid adsorbs to a Type II PSC particle, which is larger than a Type I PSC particle, the nitric acid can sediment out along with the PSC particle to lower regions of the stratosphere. This removal process is called **stratospheric denitrification**. Denitrification is important because it removes nitrogen that might otherwise reform Type I PSCs or tie up active chlorine as $ClONO_2(g)$.

11.8.3. Springtime Polar Chemistry

Chlorine activation occurs in the stratosphere during the polar winter, a time during which the sun does not rise over the polar region and temperatures are extremely low. When the sun does rise during early spring, Cl-containing gases created by PSC reactions during the winter photolyze by

$$Cl_2(g) + h\nu \rightarrow 2\dot{C}l(g) \qquad \lambda < 450\,nm \quad (11.39)$$
Molecular chlorine   Atomic chlorine

$$HOCl(g) + h\nu \rightarrow \dot{C}l(g) + \dot{O}H(g) \qquad \lambda < 375\ nm$$
Hypochlorous acid   Atomic chlorine   Hydroxyl radical   (11.40)

$$ClNO_2(g) + h\nu \rightarrow \dot{C}l(g) + \dot{N}O_2(g) \qquad \lambda < 370\ nm$$
Chlorine nitrite   Atomic chlorine   Nitrogen dioxide   (11.41)

Once Cl is released, it immediately attacks ozone. The catalytic cycle that destroys ozone in the springtime polar stratosphere differs from that in Reactions 11.23 to 11.25, which reduces ozone on a global scale. A polar stratosphere catalytic ozone destruction cycle is

$$2x\ (\dot{C}l(g) + O_3(g) \rightarrow Cl\dot{O}(g) + O_2(g)) \quad (11.42)$$
Atomic chlorine   Ozone   Chlorine monoxide   Molecular oxygen

$$Cl\dot{O}(g) + Cl\dot{O}(g) \xrightarrow{M} Cl_2O_2(g) \quad (11.43)$$
Chlorine monoxide   Dichlorine dioxide

$$Cl_2O_2(g) + h\nu \rightarrow ClO\dot{O}(g) + \dot{C}l(g) \qquad \lambda < 360\ nm$$
Dichlorine dioxide   Chlorine peroxy radical   Atomic chlorine   (11.44)

$$ClO\dot{O}(g) + \xrightarrow{M} \dot{C}l(g) + O_2(g) \quad (11.45)$$
Chlorine peroxy radical   Atomic chlorine   Molecular oxygen

$$2O_3(g) \rightarrow 3O_2(g) \quad (11.46)$$
Ozone   Molecular oxygen

This mechanism is called the **dimer mechanism** (Molina and Molina, 1986). It is important in the springtime polar stratosphere because, at that time and location, the $ClO(g)$ required for Reaction 11.43 has a high enough concentration for the reaction to proceed rapidly. More specifically, the rate of Reaction 11.43 depends on the square of the concentration of $ClO(g)$. So, the rate increases superlinearly with increasing $ClO(g)$. Once the $ClO(g)$ concentration passes a threshold, the rate of Reaction 11.43 becomes high enough for the entire dimer mechanism to proceed rapidly. If the concentration falls below the threshold, the mechanism is less important than is the normal chlorine catalytic cycle in Reactions 11.23 to 11.25, which is not sufficient for creating the Antarctic ozone hole. A second polar catalytic ozone destruction cycle is

$$\dot{C}l(g) + O_3(g) \rightarrow Cl\dot{O}(g) + O_2(g) \quad (11.47)$$
Atomic chlorine   Ozone   Chlorine monoxide   Molecular oxygen

$$\dot{B}r(g) + O_3(g) \rightarrow Br\dot{O}(g) + O_2(g) \quad (11.48)$$
Atomic bromine   Ozone   Bromine monoxide   Molecular oxygen

$$Br\dot{O}(g) + Cl\dot{O}(g) \rightarrow \dot{B}r(g) + \dot{C}l(g) + O_2(g) \quad (11.49)$$
Bromine monoxide   Chlorine monoxide   Atomic bromine   Atomic chlorine   Molecular oxygen

$$2O_3(g) \rightarrow 3O_2(g) \quad (11.50)$$
Ozone   Molecular oxygen

(McElroy et al., 1986), which is important in the polar lower stratosphere.

In sum, chlorine activation and springtime photochemical reactions convert chlorine from reservoir forms, such as $HCl(g)$ and $ClONO_2(g)$, to active forms, such as $Cl(g)$ and $ClO(g)$, as shown in Figure 11.18. The active forms of chlorine destroy ozone in catalytic cycles.

Every November, the Antarctic warms up sufficiently for the polar vortex to break down and for PSCs to melt and sublimate. Ozone from outside the polar region then

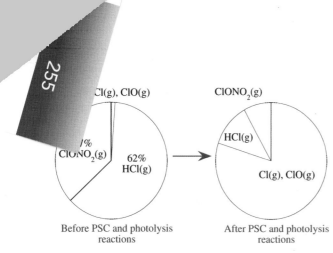

Before PSC and photolysis reactions   After PSC and photolysis reactions

Figure 11.18. Pie chart showing conversion of chlorine reservoirs to active chlorine. During chlorine activation on polar stratospheric cloud surfaces, HCl(g) and $ClONO_2$(g) are converted to potentially active forms of chlorine that are broken down by sunlight in springtime to form Cl(g). Cl(g) forms ClO(g), both of which react catalytically to destroy ozone.

advects into the region. Ozone also regenerates chemically, as in Figure 11.14, and chlorine reservoirs of $ClONO_2$(g) and HCl(g) reestablish themselves. Thus, the Antarctic ozone hole is an annual, regional phenomenon that is controlled primarily by the temperature of the polar stratosphere and the presence of chlorine and bromine.

The ozone dent over the Arctic is not nearly so large or regular as the ozone hole over the Antarctic because stratospheric temperatures over the Arctic do not drop so low as they do over the Antarctic. Because temperatures are higher in the Arctic stratosphere, PSC formation and subsequent chlorine activation over the Arctic are less widespread than over the Antarctic.

Part of the reason that stratospheric temperatures are higher over the Arctic than the Antarctic is that the Arctic surface is an ocean covered with a thin layer of sea ice, whereas the Antarctic is a mountainous continent covered with thick ice sheets. Because the ocean temperature below the Arctic sea ice cannot drop below the freezing point of water, and because energy from the ocean is conducted through the thin sea ice to its surface, sea ice surface temperatures do not decrease so much as do ice sheet surface temperatures over the Antarctic. Because thermal-IR energy fluxes are proportional to temperature to the fourth power (Equation 2.2), more energy is radiated from the Arctic surface to the stratosphere than from the Antarctic surface to the stratosphere.

The other reason Arctic stratospheric temperatures are higher than Antarctic stratospheric temperatures is that the polar vortex around the Arctic is weaker than around the Antarctic. The Arctic vortex is weaker because the Earth at 60°N is covered by land and water, whereas 60°S is covered by water. The rougher land surface slows down surface winds, which in turn slow down winds aloft at 60°N, weakening the vortex there.

11.9. Effects of Enhanced UV-B Radiation on Life and Ecosystems

In the absence of the stratospheric ozone layer, most UV-C radiation incident at the TOA would penetrate to the surface of the Earth, destroying bacteria, protozoa, algae, fungi, plants, and animals in a short time. Fortunately, the ozone layer exists and absorbs almost all UV-C radiation. Ozone also absorbs most UV-B radiation, but some of this radiation penetrates to the surface. UV-B radiation affects human and animal health, terrestrial ecosystems, aquatic ecosystems, biogeochemical cycles, air quality, and materials (United Nations Environmental Program, 1998).

11.9.1. Effects on Humans

Increases in UV-B radiation affect the skin, eyes, and immune system of humans. The layers of skin affected by UV-B are the epidermis (the outer, nonvascular, protective layer of skin that covers the dermis) and the stratum corneum (the top, horny layer of the epidermis, made mainly of peeling or dead cells). The layers of the eye affected by UV-B are the cornea, the iris, and the lens. In the immune system, the Langerhans cells, which migrate through the epidermis, are most susceptible to damage.

11.9.1.1. Effects on Skin

The severity of effects of UV-B radiation on skin depends on skin pigmentation. During the evolution of *Homo sapiens sapiens* (the variant of *Homo sapiens* to which every human belongs), humans who lived near equatorial Africa developed a dark pigment, **melanin**, in their skin to protect their skin against harmful UV radiation.

As humans migrated to higher latitudes, where sunlight was less intense, dark pigmentation prevented what little sunlight was available from catalyzing essential chemical reactions in the skin to produce vitamin D. **Vitamin D** is important because it inhibits the softening of bones (rickets), bone fracturing, and bow-leggedness in humans. Thus, the skin of people who migrated

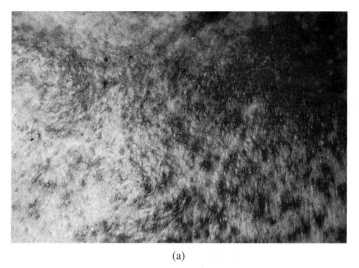
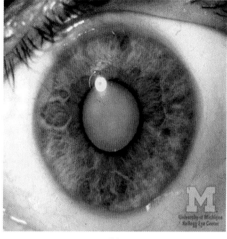

(a)   (b)

Figure 11.19. Some effects of ultraviolet radiation: (a) basal-cell carcinoma, © Michael Ballard/ Dreamstime.com and (b) cataract, University of Michigan Kellogg Eye Center.

poleward became lighter through natural selection (Leakey and Lewin, 1977).

As populations moved across Asia into North America and down toward equatorial South America, the production of melanin again became an advantage. The fact that equatorial South Americans have slightly lighter skin than do equatorial Africans suggests that the former population has not been exposed to the intense equatorial UV radiation for nearly so long as the latter population (Leakey and Lewin, 1977). In recent generations, segments of light-skinned populations whose ancestors inhabited higher latitudes, such as 50°N to 60°N in northern Europe, have migrated to lower latitudes, such as 15°S to 35°S in South Africa, Australia, and New Zealand. Such migration has increased the susceptibility of these populations to the many effects of UV-B radiation on skin. Enhancements of UV-B radiation due to reduced ozone further increase the susceptibility of light-skinned populations to UV-B–related skin problems.

UV-B effects on human skin include sunburn (erythema), photo-aging of the skin, and skin cancer. Symptoms of **sunburn** include reddening of the skin and, in severe cases, blisters. Susceptibility to sunburn depends on skin type. People with the most sensitive skin obtain a moderate to severe sunburn in less than an hour (Longstreth et al., 1998). People who are most resistant to sunburn usually have dark-pigmented skin and become more deeply pigmented with additional exposure to UV-B.

Photoaging is the accelerated aging of the skin due to long-term exposure to sunlight, particularly UV-B radiation. Symptoms include loss of skin elasticity, wrinkles, altered pigmentation, and a decrease in collagen, a fibrous protein in connective tissue.

Skin cancer is the most common cancer among light-pigmented (skinned) humans. Three types of skin cancers occur most frequently: **basal-cell carcinoma** (BCC), **squamous-cell carcinoma** (SCC), and **cutaneous melanoma** (CM). Of all skin cancers, about 79, 19, and 2 percent are BCC, SCC, and CM, respectively.

Basal-cell carcinoma is a tumor that develops in basal cells, which reside deep in the skin. As the tumor evolves, it protrudes through the skin, growing to a large mass that can scab over. BCC rarely spreads and can be removed by surgery or radiation treatment, so it is rarely fatal. Most BCCs occur in the head or neck. They appear as shiny, pearly nodules, red patches, thickened skin, or scarred tissue (Figure 11.19a).

Squamous-cell carcinoma is a tumor that develops in squamous cells, which reside on the outside of the skin. SCC tumors appear as red marks and can spread, but they are readily removed by surgery or radiation treatment and are rarely fatal.

Cutaneous melanoma is a dark pigmented and often malignant tumor arising from a **melanocyte**, which is a cell that produces the pigment melanin in the skin. CM tumors spread quickly and grow into dark, protruding masses that can appear anywhere on the skin. CM is fatal in about one-third of the cases. In some locations, such as northern Europe, CM is as common as is SCC.

Susceptibility to skin cancer depends on a combination of skin pigmentation and exposure. For people with sensitive skin, it is not necessary to be exposed to UV-B over a lifetime for a person to develop skin cancer. Skin cancer rates usually increase from high latitudes (from the poles) to lower latitudes (to the Equator).

11.9.1.2. Effects on Eyes

The **cornea**, which covers the iris and the lens, is the tissue most susceptible to UV-B damage to the eye. Little UV-B radiation penetrates past the lens to the vitreous humor or the retina, the tissues behind the lens. The most common eye problem associated with UV-B exposure is **photokeratitis** or **snow blindness**, an inflammation or reddening of the eyeball. Other symptoms include a feeling of severe pain, tearing, avoidance of light, and twitching (Longstreth et al., 1998). These symptoms are prevalent not only among skiers, but also among people who spend time at the beach or other outdoor locations with highly reflective surfaces.

From a public cost perspective, the most expensive eye-related disease associated with UV-B radiation is **cataract**, a degenerative loss in the transparency, thus cloudiness, of the lens (Figure 11.19b) that frequently leads to blindness unless the damaged lens is removed. Worldwide, cataract is the leading cause of blindness.

More severe, but less widespread, eye-related diseases are squamous-cell carcinoma, which affects the cornea, and **ocular melanoma**, which affects the iris and related tissues.

11.9.1.3. Effects on the Immune System

Human skin contains numerous cells to fight infection that are produced by the immune system. Enhanced UV-B radiation has been linked to suppression of these cells, reducing resistance to certain tumors and infections. Suppressed immune responses to UV-B have been reported for herpes, tuberculosis, leprosy, trichinella, candidiasis, leishmaniasis, listeriosis, and Lyme disease (Longstreth et al., 1998).

11.9.2. Effects on Microorganisms, Animals, and Plants

Increases in UV-B radiation affect microorganisms, animals, and plants in a variety of ways. Phytoplankton, which live on the surfaces of the oceans, are susceptible to slowed growth, reproductive problems, and changes in photosynthetic energy-harvesting enzymes and pigment contents (Hader et al., 1998). Because these microorganisms are near the bottom of the food chain, their deaths affect higher organisms. Algae and sea grasses, which cannot avoid exposure to the sun, are also susceptible to UV-B damage.

Animals are susceptible to several of the same UV-B hazards as humans. SCCs have been found in cats, dogs, sheep, goats, horses, and cattle, usually on unprotected skin, such as eyelids, noses, ears, and tails (Hargis, 1981; Teifke and Lohr, 1996; Mendez et al., 1997). Cataracts and skin lesions have been found in fish (Mayer, 1992).

UV damage to crops varies with species and the crop's ability to adapt. UV-B affects the DNA in some crops and the rate of photosynthesis in others. For many crops, UV-B changes lifecycle timing, plant form, and production of plant chemicals (Caldwell et al., 1998). For others, it makes them more susceptible to disease and attack by insects. Crop yields of some plants are affected by enhanced UV-B, but those of others are not.

11.9.3. Effects on the Global Carbon and Nitrogen Cycles

Changes in UV-B radiation affect the global carbon and nitrogen cycles. UV-B damages phytoplankton, reducing their consumption of $CO_2(g)$ by photosynthesis. UV-B also enhances photodegradation (breakdown by light) of dead plant material, causing the release of more $CO_2(g)$ back to the air. UV-B enhances the release of carbon monoxide gas [$CO(g)$] from charred vegetation (Zepp et al., 1998). It also affects the rate of nitrogen fixation by cyanobacteria (Hader et al., 1998).

11.9.4. Effects on Tropospheric Ozone

Increases in UV-B radiation increase photolysis rates of gases such as ozone, nitrogen dioxide, formaldehyde, hydrogen peroxide, acetaldehyde, and acetone. Increases in photolysis rates of nitrogen dioxide, formaldehyde, and acetaldehyde increase rates of free-tropospheric ozone formation (Tang et al., 1998).

Whereas reductions in stratospheric ozone increase UV-B penetration to the background troposphere, increases in aerosol concentrations in urban air can either decrease or increase UV-B penetration to the surface. Absorption of UV-B by aerosol particles reduces UV-B penetration within and below the particles, depressing ozone formation in polluted air (Section 7.1.3.3). Scattering by aerosol particles increase UV-B, causing the opposite effect.

11.10. Regulation of Chlorofluorocarbons

The effects of CFCs on ozone were first hypothesized by **Mario Molina** and **Sherwood Rowland** in June 1974. As early as December 1974, legislation was introduced in the U.S. Congress to study the problem further and give the U.S. EPA authority to regulate CFCs. This bill died before any action was taken. In 1975, Congress established a committee that ultimately recommended that spray cans, the primary emitters of CFCs, be labeled in such a way as to identify whether they contained CFCs or an alternative compound. In 1976, the U.S. National Academy of Sciences released a report suggesting that CFC emissions were large enough to cause a long-term 6 to 7.5 percent decrease in global stratospheric ozone, potentially increasing surface UV-B radiation by 12 to 15 percent. On the basis of this report, the U.S. Food and Drug Administration, the U.S. EPA, and the Consumer Product Safety Commission issued a joint decision suggesting the phase-out of CFCs from spray cans. In October 1978, the manufacture and sale of CFCs for spray cans was banned in the United States. At the time, the United States was responsible for half the global production of CFCs used in spray cans.

Although overall CFC emissions decreased as a result of the ban of CFCs in spray cans, tropospheric mixing ratios of CFCs continued to increase due to the fact that emission of CFCs from other sources still occurred. Throughout the late 1970s and much of the 1980s, the use of CFCs in refrigeration and foam production and as a solvent increased. To limit damage due to CFCs, rules were contemplated for reducing their emission from all sources. In 1980, the U.S. EPA proposed preventing emission growth of CFCs from refrigeration, but the proposed regulations were thwarted by the new presidential administration.

On the international front, the Vienna Convention for the Protection of the Ozone Layer was convened in March 1985 to discuss CFCs. The result of this convention was an agreement, signed initially by twenty countries, stating that signatory countries had an obligation to reduce CFC emissions and to study further the effect of CFCs on ozone. In September 1987, an international agreement, the **Montreal Protocol**, was signed initially by twenty-seven countries, limiting the production of CFCs and halons and setting a timetable for their eventual phase-out. The Montreal Protocol was modified several times, including by the London Amendments (1990), Copenhagen Amendments (1992), and Montreal Amendments (1997), to expedite the phase-out of

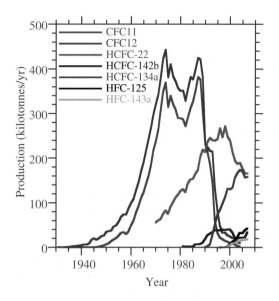

Figure 11.20. Production of selected CFCs, HCFCs, and HFCs by reporting companies only, between 1931 and 2007. Nonreported emissions are estimated to account for 60 percent of CFC production, 10 percent of HCFC-22 production, and 0 percent of HFC-134a production. CFC, chlorofluorocarbon; HCFC, hydrochlorofluorocarbon; HFC, hydrofluorocarbon. From Alternative Fluorocarbons Environmental Acceptability Study (AFEAS) (n.d.).

CFCs. The modified protocol called for the phase-out of CFC-11 and -12 by 1996, as well as a phase-out of other CFCs and halons by 2010.

The major effect of the Montreal Protocol, later amendments, and previous regulations to ban the use of CFCs in spray has been the near elimination of CFC emissions but a corresponding increase in CFC substitutes, namely, HCFCs and HFCs. Figure 11.20 shows the emissions from major CFCs, HCFCs, and HFCs produced by reporting companies between 1931 and 2007. Reporting companies include eleven major manufacturers who are estimated to account for 40 percent of the worldwide production of CFCs, 90 percent of the worldwide production of HCFC-22, and 100 percent of the worldwide production of other HCFCs and all HFCs. Most remaining CFCs and HCFC-22 are produced in Russia and China.

Figure 11.20 indicates that HCFC emission rates increased from 1970 to the early 2000s but have since declined. The reason for the decline is that the Montreal Protocol and its amendments required the phase-out of HCFCs starting in 1996. Although HCFCs break down

Table 11.4. Percentage of 1989 baseline hydrochlorofluorocarbon production and consumption allowed in developed countries under the Montreal Protocol and its amendments by year

Year	Percent reduction in consumption and production in developed countries	Percent reduction in consumption in developing countries
2004	30	
2010	75	
2013		0
2015	90	
2016		10
2020	99.5	35
2025		67.5
2030	100	97.5
2040		100

For developed countries, the baseline consumption equals 2.8 percent of the 1989 chlorofluorocarbon (CFC) consumption plus 100 percent of the 1989 hydrochlorofluorocarbon (HCFC) consumption by the country. For these countries, the baseline production equals the average of (a) the 1989 HCFC production plus 2.8 percent of the 1989 CFC production and (b) the 1989 HCFC consumption plus 2.8 percent of the 1989 CFC consumption. For developing countries, the baseline consumption in 2013 is the average consumption in 2009 and 2010.

more readily in the troposphere than do CFCs, HCFCs still contain chlorine, some of which can reach the stratosphere and damage the ozone layer. Table 11.4 shows the phase-out schedule of HCFCs, as modified by amendments through 2010 to the Montreal Protocol Developed countries are given less time to eliminate consumption than developing countries.

As CFC and HCFC emissions have declined, emissions of HFCs have taken their place (Figure 11.20). HFCs contain no chlorine and do not pose a known threat to the ozone layer. Unfortunately, they are strong absorbers of thermal-IR radiation and contribute to global warming, as indicated by their global warming potentials (Table 11.2). As such, efforts are also now under way to phase out HFCs as a result of their global warming effects.

Because CFC emissions have declined due to the Montreal Protocol, tropospheric mixing ratios of CFC-11 have begun to decrease and those of CFC-12 have begun to level off (Figure 11.21a). Steady decreases in carbon tetrachloride [$CCl_4(g)$], and dramatic decreases in methyl chloroform [$CH_3CCl_3(g)$] have also occurred (Figure 11.21b). HCFC-22 (Figure 11.21b) and -142b (Figure 11.21c) mixing ratios continued to increase through 2010, as did the replacement compound, $SF_6(g)$ (Figure 11.21d).

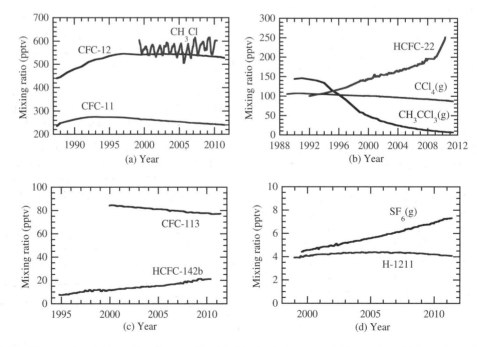

Figure 11.21. Change in mixing ratio of several (a, b) chlorinated gases at Mauna Loa Observatory, Hawaii, 1987–2011, (c) fluorinated gases since 1994 at Mace Head, Ireland, and (d) a halon and sulfur hexafluoride at Mauna Loa since 1999. Data from Mauna Loa Data Center (n.d.).

Because stratospheric mixing ratios of CFCs require a long time to respond to emission changes from the surface, *the stratospheric ozone layer is not expected to recover fully until 2050 or beyond* (World Meteorological Organization, 1998). The annual occurrence of the Antarctic ozone hole is expected to last for several decades as well. Global warming, which is characterized by a warming of the troposphere but cooling of the stratosphere, will likely delay the recovery of the ozone layer further, as discussed in Chapter 12.

11.11. Summary

In this chapter, the stratospheric ozone layer is discussed, with an emphasis on long-term global ozone reduction and seasonal Antarctic ozone depletion. Both problems are caused by enhanced levels of chlorine and bromine in the stratosphere. Damage to the ozone layer is of concern because decreases in ozone permit UV-B radiation to penetrate to the Earth's surface, where it harms humans, animals, microorganisms, and plants. The story of environmental damage to the ozone layer may have a positive ending. CFCs, emitted since the 1930s, slowly diffused into the middle stratosphere with an overall transport time of 50 to 100 years. Most CFCs emitted during the 1930s did not reach the middle stratosphere until the 1970s, and those emitted during the 1990s will not reach the middle stratosphere until 2040 to 2090. In 1974, a relationship between CFCs and ozone loss was hypothesized. From the 1970s until today, measurements indicate a global ozone reduction and a seasonal Antarctic depletion. In 1978, regulations banning the release of CFCs from spray cans were implemented in the United States. Concerned that the ban on CFCs from spray cans was not enough, the international community agreed to regulate all CFC emissions through the Montreal Protocol in 1987. In 2000, reported CFC emissions were less than one-tenth those in 1976. In 2010, they were nearly zero. Global ozone levels may be replenished to their original levels by the year 2050 as CFCs currently in the atmosphere are slowly removed. Recovery, though, will likely be delayed due to global warming, which warms the troposphere and cools the stratosphere. A cooler stratosphere increases the effectiveness of chlorine destruction of ozone, as discussed in Chapter 12.

11.12. Problems

11.1. Explain why an ozone maximum occurs immediately outside the polar vortex during the same season that the ozone hole occurs over the Antarctic.

11.2. Briefly summarize the steps resulting in the Antarctic ozone hole and its recovery. Explain why the ozone dent is never so severe as the ozone hole.

11.3. List the chlorine catalytic cycles responsible for ozone reduction (a) over the Antarctic and (b) in the global stratosphere. Why is the Antarctic mechanism unimportant on a global scale?

11.4. What effect do you think volcanic aerosol particles that penetrate into the stratosphere have on stratospheric ozone during the daytime when chlorine reservoirs are present? What about during the nighttime?

11.5. If the entire global stratospheric ozone layer takes only 1 year to form from scratch chemically in the presence of oxygen, why will today's ozone layer, which has deteriorated a few percent since 1979, take up to 50 years to regenerate to its 1979 level?

11.6. Why does bromine destroy ozone more efficiently than does chlorine?

11.7. Why do CFCs not cause damage to ozone in the troposphere?

11.8. Some people argue that natural chlorine from volcanic eruptions and from the ocean is responsible for global ozone reduction and Antarctic ozone depletion. What argument contradicts this contention?

11.9. Explain why the ratio of UV-B to total solar radiation reaching the top of the atmosphere is greater than is that reaching the ground, as seen in Figure 11.5.

11.10. Explain how UV-B levels at the ground in polluted air can be lower than those in unpolluted air, if both the polluted air and clean air are situated under an ozone layer that is partially depleted by the same amount.

11.11. Why might you be exposed to more UV radiation on a sunny June day at the Equator in Brazil than on the same sunny day and at the same elevation in Mexico, even though radiation at the top of the atmosphere is greater on that day over Mexico? Ignore air pollution effects.

Chapter 12

The Greenhouse Effect and Global Warming

The two major global-scale environmental threats of international concern since the 1970s have been global stratospheric ozone layer loss and global warming. As discussed in Chapter 11, the global ozone layer is expected to recover by the mid–twenty-first century because national and international regulations have required the chemical industry to use alternatives to chloro- and bromocarbons, which are the chemicals primarily responsible for stratospheric ozone loss. Regulations are similarly responsible for improvements in air quality and acid deposition problems in many parts of the world since the 1970s (e.g., the U.S. Clean Air Act Amendments of 1970 motivated U.S. automobile manufacturers to develop the catalytic converter in 1975, which led to improvements in urban air quality; regulations through the 1979 Geneva Convention on Long-Range Transboundary Air Pollution led to the amelioration of some acid deposition problems in the 1980s and 1990s). However, progress toward solving the second major issue of international concern, global warming, has been slow. In this chapter, global warming is described and distinguished from the natural greenhouse effect. In addition, historical and recent temperature trends, both in the lower and upper atmosphere, are addressed, and climate responses to increased pollution are examined. The chapter also discusses potential effects of global warming and international and national efforts to curtail it.

12.1. Temperature on Earth in the Absence of a Greenhouse Effect

The **natural greenhouse** effect is the warming of the Earth's troposphere due to an increase in natural greenhouse gases. **Greenhouse gases** are gases that are largely transparent to the sun's visible radiation but absorb and reemit the Earth's thermal-IR radiation at selective wavelengths. They cause a net warming of the Earth's atmosphere similar to the way in which a glass house causes a net warming of its interior. Because most incoming solar radiation can penetrate a glass house but a portion of outgoing thermal-IR radiation cannot, air inside a glass house warms during the day as long as mass (e.g., plant mass) is present within the house to absorb the solar radiation, to heat up the air, and to reemit thermal-IR radiation. The surface of the Earth, like plants, absorbs solar radiation and reemits thermal-IR radiation. Greenhouse gases, like glass, are transparent to most solar radiation but absorb a portion of the Earth's thermal-IR radiation at selective wavelengths.

Global warming is the increase in the Earth's temperature above that from the natural greenhouse effect due to the addition of anthropogenically emitted greenhouse gases and aerosol particle components that directly absorb solar radiation. Important absorbing aerosol particle components include **black carbon (BC)** and certain organic carbon compounds, collectively referred to as **brown carbon (BrC)**. BC absorbs across the entire solar spectrum. BrC strongly absorbs UV and short visible wavelengths of light, but not solar-IR wavelengths (Chapter 7). Because particles containing BC and BrC contain some nonwarming components but predominantly cause warming, they are referred to as **warming particles**.

Whereas BC and BrC warm the climate, most other aerosol constituents – sulfate, nitrate, ammonium, sodium, potassium, calcium, magnesium, nonabsorbing organic material, and liquid water attached to these ions and molecules – reflect solar radiation and reduce near-surface temperatures. The sources of the cooling components generally differ from the sources of warming components, so particles containing predominantly cooling components are referred to as **cooling particles**. Cooling particles offset part of the warming due to greenhouse gases and warming particles (Section 12.2.3).

In the absence of greenhouse gases or warming particles, the temperature of the Earth can be estimated, to first order, with a simple radiation balance model that considers solar radiation coming into and thermal-IR radiation leaving the Earth. This model can be applied to other planets as well. The difference between the temperature predicted by the energy balance model and the real temperature observed at the surface of the Earth before anthropogenic emissions is caused primarily by the greenhouse effect. The simple energy balance model is described next.

12.1.1. Incoming Solar Radiation

The sun emits radiation with an effective photosphere temperature of about $T_p = 5{,}785$ K (Chapter 2). Thus, the energy flux (joules per second per square meter or watts per square meter) emitted by the sun's photosphere can be calculated from the Stefan-Boltzmann law (Equation 2.2) as

$$F_p = \varepsilon_p \sigma_B T_p^4 \qquad (12.1)$$

where ε_p is the emissivity of the photosphere. The emissivity is near unity because the sun is essentially a blackbody (a perfect absorber and emitter of radiation). Multiplying the energy flux by the spherical surface area of the photosphere, $4\pi R_p^2$, where $R_p = 6.96 \times 10^8$ m (696,000 km) is the effective **radius of the sun** (the distance from the center of the sun to the top of the photosphere), gives the total energy per unit time (J s^{-1} or W) emitted by the photosphere as $4\pi R_p^2 F_p$. Energy emitted from the photosphere propagates through space

on the edge of an ever-expanding concentric sphere originating from the photosphere. Because conservation of energy requires that the total energy per unit time passing through a concentric sphere any distance from the photosphere equals that originally emitted by the spherical photosphere, the total energy per unit time passing through a sphere with a radius corresponding to the **Earth-sun distance** (R_{es}) must be

$$4\pi R_{es}^2 F_s = 4\pi R_p^2 F_p \qquad (12.2)$$

where F_s is the solar energy flux (J s^{-1} m^{-2} or W m^{-2}) on a sphere with a radius corresponding to the Earth-sun distance. Rearranging Equation 12.2 and combining the result with Equation 12.1 gives

$$F_s = \left(\frac{R_p}{R_{es}}\right)^2 F_p = \left(\frac{R_p}{R_{es}}\right)^2 \sigma_B T_p^4 \qquad (12.3)$$

which indicates that the energy flux from the sun decreases proportionally to the square of the distance away from the sun. This can be illustrated by putting your hand over a light bulb. Close to the bulb, you will feel the heat from the bulb; however, as you move your hand away from the bulb, the heat that you feel decreases proportionally to the square of the distance away from the bulb.

The average Earth-sun distance is about 1.49598×10^{11} m (150 million km), giving the average energy flux at the top of the Earth's atmosphere from Equation 12.3 as $F_s = 1{,}365$ W m^{-2}, which is the **solar constant**. Figure 12.1 shows that the Earth-sun distance is 147.1 million km in December (Northern Hemisphere winter) and 152.1 million km in June (Northern Hemisphere summer) due to the fact that the Earth rotates around the sun in an elliptical orbit with the sun at one focus. If these distances are used in Equation 12.3, $F_s = 1{,}411$ W m^{-2} in December and 1,321 W m^{-2} in June. Thus, a difference of 3.4 percent in Earth-sun distance between

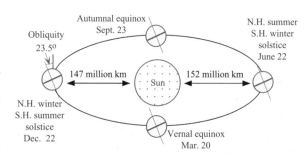

Figure 12.1. Relationship between sun and Earth at the times of solstices and equinoxes. The sun is positioned at one focus of the ellipse.

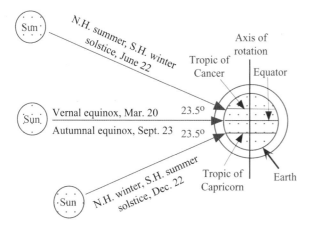

Figure 12.2. Positions of the sun relative to the Earth during solstices and equinoxes. Of the four times shown, the Earth-sun distance is greatest at the Northern Hemisphere summer solstice.

December and June corresponds to a difference of 6.9 percent in solar radiation reaching the Earth between these months. In other words, 6.9 percent more radiation falls on the Earth in December than in June.

Despite the excess radiation reaching the top of the Earth's atmosphere in December, the Northern Hemisphere (NH) winter still starts in December because the Southern Hemisphere (SH) is tilted toward the sun in December, as shown in Figure 12.1. It also shows that the axis of rotation of the Earth is currently tilted all year by ∼23.5 degrees from a line perpendicular to the plane of the Earth's orbit around the sun. This angle is called the **obliquity** of the Earth's axis of rotation. As a result of the Earth's obliquity, the sun's direct rays hit their farthest point south, 23.5°S latitude, on December 22 (**NH winter solstice, SH summer solstice**) and their farthest point north, 23.5°N latitude, on June 22 (**NH summer solstice, SH winter solstice**) (Figures 12.1 and 12.2). The latitude at 23.5°S is called the **Tropic of Capricorn** and that at 23.5°N is the **Tropic of Cancer**. Winter and summer solstices are the shortest and longest days of the year, respectively. On March 20 (**vernal equinox**) and September 23 (**autumnal equinox**), the sun is directly over the Equator and the length of day equals that of night.

For three reasons, temperatures in the NH are warmer in June than in December, even though the Earth is closer to the sun in December (Figure 12.1). First, the sun's rays are directly over the NH in June. In December, they are directly over the SH; thus, they are slanted, making them diffuse and less intense over the NH (Figure 12.2). Second, NH days are longer in

June than in December. Finally, the NH summer is about a week longer than is the NH winter because the Earth takes longer to pass between equinoxes from March to September than from September to March (Figure 12.1).

From the sun's point of view, the Earth appears as a circular disk (rather than a sphere) absorbing the sun's radiation. Thus, the quantity of incoming solar radiation received by the Earth is the solar constant multiplied by the cross-sectional area of the Earth, πR_e^2 (m²), where $R_e = 6.378 \times 10^6$ m (6,378 km) is the **Earth's radius**. Not all incoming solar radiation is absorbed by the Earth. Some is reflected by snow, sea ice, deserts, and other light-colored ground surfaces, as well as clouds. The fraction of incident energy reflected by a surface is the **albedo** or **reflectivity** of the surface. The albedo varies for different-colored surfaces and changes with wavelength. Table 12.1 gives mean albedos in the visible spectrum for several surface types, showing that the albedo of the Earth and atmosphere together (**planetary albedo**) is about 30 percent. More than two-thirds of the Earth's surface is covered with water, which has an albedo of 5 to 20 percent (with a typical value of 8 percent), depending largely on the angle of the sun relative to the surface. Soils and forests also have low albedos. Much of the Earth-atmosphere reflectivity is due to clouds and snow, which have high albedos.

Taking into account the cross-sectional area of the Earth and the Earth's albedo (A_e, fraction), the total

Table 12.1. Solar albedos and thermal-infrared emissivities for several surface types

Type	Albedo (fraction)	Emissivity (fraction)
Earth and atmosphere	0.3	0.90–0.98
Liquid water	0.05–0.2	0.92–0.96
Fresh snow	0.7–0.9	0.82–0.995
Old snow	0.35–0.65	0.82
Thick clouds	0.3–0.9	0.25–1.0
Thin clouds	0.2–0.7	0.1–0.9
Sea ice	0.25–0.65	0.96
Soil	0.05–0.2	0.9–0.98
Grass	0.16–0.26	0.9–0.95
Desert	0.20–0.40	0.84–0.91
Forest	0.10–0.25	0.95–0.97
Concrete	0.1–0.35	0.71–0.9

Sources: Estimates from Sellers (1965); Oke (1978); Liou (1992); and Hartmann (1994).

energy per unit time, or power (W), absorbed by the Earth in the simple energy balance model is

$$P_{in} = F_s \left(1 - A_e\right) \left(\pi R_e^2\right) \tag{12.4}$$

This incoming radiation must be equated with radiation emitted by the Earth to derive an equilibrium temperature.

12.1.2. Outgoing Thermal-Infrared Radiation

The Earth emits radiation at all points on its surface, which can roughly be represented as a sphere. The Stefan-Boltzmann law (Equation 2.2) gives the energy flux emitted by any object, including the Earth. Applying this law and multiplying through by the surface area of the Earth, $4\pi R_e^2$ (m^2), gives the power (W) emitted by the Earth as

$$P_{out} = \varepsilon_e \sigma_B T_e^4 \left(4\pi R_e^2\right) \tag{12.5}$$

where ε_e is the globally averaged thermal-IR emissivity of the Earth (dimensionless) and T_e is the **equilibrium temperature** (K) of the Earth's surface, which is the temperature of the Earth without considering the greenhouse effect. Table 12.1 gives thermal-IR emissivities for different surfaces. The actual globally averaged emissivity of the Earth is about 0.9 to 0.98; however, for the basic calculation presented here, it is assumed to equal 1.0.

12.1.3. Equilibrium Temperature of the Earth

Equating the net incoming solar power from Equation 12.4 with the outgoing thermal-IR power from Equation 12.5 and solving for the **equilibrium temperature** of the Earth's surface in the absence of an atmosphere gives

$$T_e = \left[\frac{F_s \left(1 - A_e\right)}{4\varepsilon_e \sigma_B} \right]^{1/4} \tag{12.6}$$

Example 12.1
Estimate the equilibrium temperature of the Earth given $F_s = 1,365$ W m^{-2}, $A_e = 0.3$, and $\varepsilon_e = 1.0$.

Solution
From Equation 12.6, the equilibrium temperature of the Earth is $T_e = 254.8$ K, which would be the temperature of the Earth in the absence of an atmosphere.

The equilibrium temperature of the Earth, 255 K (Example 12.1), is 18 K below the freezing temperature of water and would not support most life on Earth. Fortunately, the actual globally averaged surface air temperature is about 288 K. The 33 K difference between the predicted equilibrium temperature and the actual temperature of the Earth results from the fact that the Earth has an atmosphere that is transparent to most incoming solar radiation but selectively absorbs many wavelengths of outgoing thermal-IR radiation. Some of the thermal-IR radiation absorbed by the atmosphere is reemitted to the surface, warming it. The resulting 33 K increase in temperature over the equilibrium temperature of the Earth is the **natural greenhouse effect**.

Table 12.2 compares equilibrium temperatures calculated from Equation 12.6 with actual temperatures of several planets. The planet with the largest difference between its actual and equilibrium temperatures is Venus (Figure 12.3a). Venus's surface temperature is more than 470 K warmer than is its equilibrium temperature. Because Venus is closer to the sun than is the Earth (Figure 12.3b), Venus receives more solar radiation than does the Earth, so Venus's temperature, early in its evolution, was higher than was that of the Earth. As a result, liquid water and ice on the surface of Venus, if ever present, evaporated and sublimated, respectively. The resulting water vapor, exposed to intense far-solar UV radiation, photolyzed to atomic hydrogen [H(g)] and the hydroxyl radical [OH(g)]. Over time, atomic hydrogen escaped Venus's gravitational field to space, depleting the atmosphere of its ability to reform water vapor. Because the surface of Venus lost all liquid water to vapor, there was no mechanism to dissolve and convert its atmospheric $CO_2(g)$, which built up due to volcanic outgassing, back to carbonate rock. As its mixing ratio increased, $CO_2(g)$ absorbed more thermal-IR radiation, heating the atmosphere and preventing both condensation of water and additional removal of $CO_2(g)$. The nearly endless positive feedback cycle that occurred on Venus is called a **runaway greenhouse effect**. Today, Venus has a surface air pressure ninety times that of the Earth, and its major atmospheric constituent is $CO_2(g)$.

Table 12.2 shows that Mercury, Mars, Pluto, and the moon all have thin atmospheres and little greenhouse effect. Their atmospheres are thin because light gases have escaped their weak gravitational fields. Although Mars's surface pressure is less than 1 percent of the Earth's, Mars's $CO_2(g)$ partial pressure is about twenty times that of the Earth. Because $CO_2(g)$ is relatively heavy, it has not entirely escaped Mars's atmosphere, and because Mars has no oceans, $CO_2(g)$ cannot be

Table 12.2. Equilibrium and actual temperatures of the surfaces of planets, dwarf planets, and the moon

Object	Distance from sun (10^9 m)	Equatorial radius (10^6 m)	Surface pressure (bars)	Major atmospheric gases (volume mixing ratios)	Albedo	Equilibrium temperature (K)	Actual temperature (K)
Mercury	57.9	2.44	2×10^{-15}	He (0.98), H (0.02)	0.11	436	440
Venus	108	6.05	90	CO_2 (0.96), N_2 (0.034), H_2O (0.001)	0.65	252	730
Earth	150	6.38	1	N_2 (0.77), O_2 (0.21), Ar (0.0093)	0.30	255	288
Moon	150	1.74	2×10^{-14}	Ne (0.4), Ar (0.4), He (0.2)	0.12	270	274
Mars	228	3.39	0.007	CO_2 (0.95), N_2 (0.027), Ar (0.016)	0.15	217	218
Jupiter	778	71.4	>100	H_2 (0.86), He (0.14)	0.52	102	129
Saturn	1,427	60.3	>100	H_2 (0.92), He (0.74)	0.47	77	97
Uranus	2,870	26.2	>100	H_2 (0.89), He (0.11)	0.51	53	58
Neptune	4,497	25.2	>100	H_2 (0.89), He (0.11)	0.41	45	56
Pluto	5,900	1.5	1×10^{-5}	?	0.30	41	50

Sources: Data compiled from Lide (1998), except that equilibrium temperatures were calculated from Equation 12.6.

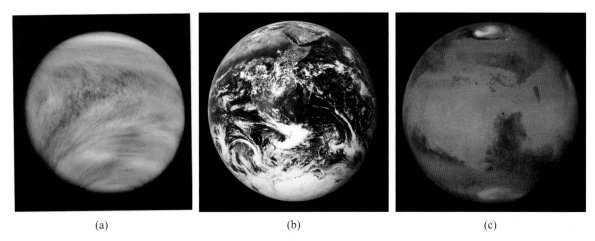

(a) (b) (c)

Figure 12.3. (a) Ultraviolet image of Venus's clouds as seen by the Pioneer Venus Orbiter, February 26, 1979. Available from National Space Science Data Center. (b) Earth as seen from the Apollo 17 spacecraft on December 7, 1972. Available from National Space Science Data Center. (c) Image of Mars taken by Hubble telescope on March 10, 1997. The North Pole dry ice cap is rapidly sublimating during the Martian Northern Hemisphere spring. David Crisp/WFPC2 Science Team, Jet Propulsion Laboratory/California Institute of Technology; available from National Space Science Data Center.

removed by dissolution. Some $CO_2(g)$ deposits seasonally over Mars's poles as **dry ice** (solid carbon dioxide) (Figure 12.3c). Dry ice forms from the gas phase when the temperature decreases to 194.65 K. Despite its abundance of $CO_2(g)$, Mars has only a small greenhouse effect because its surface emits little thermal-IR due to its low surface temperature and its atmosphere contains no water vapor.

The main gases on Jupiter, Saturn, Uranus, and Neptune are molecular hydrogen and helium. These planets are so large that their gravitation prevents the escape of even light gases. Because neither hydrogen nor helium is a strong absorber of thermal-IR radiation, the greenhouse effect on these planets is small. The high surface pressure on these planets compresses hydrogen into oceans of liquid hydrogen. Also, high pressures in the interior of Jupiter result in the formation of solid hydrogen and, possibly, metallic solid hydrogen.

12.1.4. The Goldilocks Hypothesis

Why does the Earth support life while its nearest neighbors, Venus and Mars (Figure 12.3), do not? The **Goldilocks hypothesis** suggests simply that Venus is too hot, Mars is too cold, and the Earth is just right. Venus is too hot because of its proximity to the sun, Mars is too cold because of its distance from the sun, and the Earth is the ideal distance. However, the answer is more complicated than this because the surface temperature of a planet depends not only on incoming solar radiation, but also on the amount of heat-trapping gases in the atmosphere, as illustrated in Table 12.2.

The primary natural greenhouse gas on the Earth is water vapor. However, most water on Earth does not exist as vapor; most is liquid, and some is ice. If all liquid water in the oceans of the Earth evaporated, the Earth's greenhouse effect would create an unlivable planet. Fortunately, the Earth's temperature (288 K) is such that it is far below the boiling point of water (373 K). Conversely, if the Earth did not have water vapor or other greenhouse gases in its atmosphere, its equilibrium temperature (255 K) (Table 12.2) would be below the freezing point of water, too cold to support either liquid water or water vapor and also too cold to support life. However, Earth's atmosphere contains water vapor, initially outgassed from volcanos and now also evaporated from the oceans. Water vapor absorbs thermal-IR radiation, keeping the planet warm naturally. The atmosphere also contains carbon dioxide gas, which in background levels, contributes to the Earth's natural warmth.

Venus is not so lucky. At its surface temperature (730 K), all water is vapor. The absence of liquid water on Venus has prevented carbon dioxide, outgassed by volcanos, from being processed into carbonate rock, as it is in the Earth's oceans. As such, carbon dioxide built up in Venus's atmosphere so much that it comprises 96 percent of its composition (Table 12.2), creating a runaway greenhouse effect (Section 12.1.3).

Mars, too, is unlucky, because its surface temperature (218 K) is so low that all water, if ever present, was in the form of ice. The cold temperatures prevented gases from vaporizing, inhibiting the development of a greenhouse effect.

In sum, whereas Venus is too hot and Mars is too cold, the Earth's temperature is ideal for supporting liquid water. The presence of liquid water and amiable temperatures have allowed life on Earth to flourish.

12.2. The Greenhouse Effect and Global Warming

Greenhouse gases are relatively transparent to incoming solar radiation but opaque to selective wavelengths of IR radiation. The term "relatively" transparent is used because all greenhouse gases absorb far-UV radiation (which is a trivial fraction of incoming solar radiation). In addition, ozone strongly absorbs UV-B and -C radiation and weakly absorbs visible radiation. Water vapor and carbon dioxide absorb solar-IR radiation. However, as shown in Figure 11.5, gas absorption affects only a fraction of total solar radiation incident at the top of the Earth's atmosphere.

12.2.1. Greenhouse Gases and Particles

The **natural greenhouse effect** is the warming of the Earth's atmosphere due to the presence of background greenhouse gases, primarily water vapor, carbon dioxide, methane, ozone, nitrous oxide, and methyl chloride. The natural greenhouse effect is responsible for the 33 K warming of the Earth's near-surface air temperature above its equilibrium temperature of 255 K, giving the Earth's surface an average temperature of 288 K. Without the natural greenhouse effect, Earth's surface would be too cold to support most life. Thus, the presence of natural greenhouse gases is beneficial.

Global warming is the increase in the Earth's temperature above that due to the natural greenhouse effect (288 K) as a result of the atmospheric buildup of anthropogenic greenhouses gases and absorbing aerosol particle constituents, namely, black carbon (BC) and brown carbon (BrC) (light-absorbing organic carbon).

Table 12.3. Estimated percentages of natural greenhouse effect and global warming temperature changes due to greenhouse gases and black carbon since the mid-1800s

Compound Name	Formula	2010 Total tropospheric mixing ratio (ppmv) or loading (Tg)	Percentage of current total mixing ratio or loading that is anthropogenic	Percentage of natural greenhouse effect temperature change (33 K) due to component	Percentage of global warming temperature change due to component
Water vapor	$H_2O(g)$	10,000	<1	88.9	0.5
Carbon dioxide	$CO_2(g)$	390	29.5	7.5	46.4
Black carbon	$C(s)$	0.15 Tg	90	0.2	17
Methane	$CH_4(g)$	1.9	58	0.5	14
Ozone	$O_3(g)$	0.02–0.07	20–50	1.1	11
Nitrous oxide	$N_2O(g)$	0.325	15	1.5	4.2
Methyl chloride	$CH_3Cl(g)$	0.00058	<0.1	0.3	~0
CFC-11	$CFCl_3(g)$	0.00024	100	0	1.8
CFC-12	$CF_2Cl_2(g)$	0.00053	100	0	4.2
HCFC-22	$CF_2ClH(g)$	0.00025	100	0	0.6
Carbon tetrachloride	$CCl_4(g)$	0.00009	100	0	0.3

CFC, chlorofluorocarbon; HCFC, hydrochlorofluorocarbon. 1 Tg = 10^6 metric tonnes = 10^6 g.
Sources: Derived from Jacobson (2010b) and IPCC (2007).

Whereas greenhouse gases transmit solar radiation and absorb thermal-IR radiation, BC strongly absorbs all wavelengths of solar radiation, and BrC strongly absorbs UV and short visible wavelengths of solar radiation. BC and BrC both weakly absorb thermal-IR radiation. Thus, greenhouse gases and BC/BrC both warm the air, but by different mechanisms.

Table 12.3 shows that the most important natural greenhouse gas is **water vapor**, which accounts for approximately 29.4 K (89 percent) of the 33 K temperature increase resulting from natural greenhouse warming. Carbon dioxide is the second most important and abundant natural greenhouse gas, accounting for about 7.5 percent of the natural greenhouse effect. Black carbon, whose major natural source is forest fires, is estimated to be responsible for only 0.2 percent of the Earth's natural warming above its equilibrium temperature.

Figure 12.4 gives the percentage absorption of incident thermal-IR radiation by different greenhouse gases versus wavelength. Water vapor is a strong absorber at several wavelengths between 0.7 μm and 8 μm and above 12 μm. Carbon dioxide absorbs strongly at 2.7 and 4.3 μm and above 13 μm. Little thermal-IR absorption occurs between 8 μm and 12 μm. This wavelength region is called the **atmospheric window**. In the

atmospheric window, gases are relatively transparent to the Earth's outgoing thermal-IR radiation, allowing the radiation to escape to space. Of the natural gases in the Earth's atmosphere, ozone and methyl chloride absorb radiation within the atmospheric window.

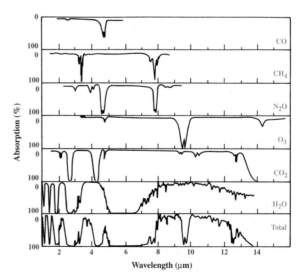

Figure 12.4. Percentage absorption of radiation by greenhouse gases at infrared wavelengths. Adapted from Valley (1965).

Nitrous oxide and methane absorb at the edges of the window. An increase in the mixing ratio of any thermal-IR–absorbing gas, regardless of whether it absorbs in the window, increases atmospheric heating. However, an increase in the mixing ratio of a gas that absorbs in the window heats the air more than does an increase in the mixing ratio of a gas that absorbs outside the window because more radiation currently passes through the window than outside it.

12.2.2. Historical Aspects of Global Warming

The first scientist to consider the Earth as a greenhouse was **Jean Baptiste Fourier** (1768–1830), a French mathematician and physicist known for his studies of heat conductivity and diffusion. In 1827, Fourier suggested that the atmosphere behaved like the glass in a hot house, letting through "light" rays of the sun but retaining "dark rays" from the ground.

The first to recognize that specific gases in the atmosphere selectively absorbed thermal-IR radiation was **John Tyndall** (1820–1893; Figure 12.5), the English experimental physicist who was also known for studying the interactions of light with small particles (Section 7.1.5). Around 1865, Tyndall discovered that water vapor absorbs more thermal-IR radiation than does dry air and postulated that water vapor moderates the Earth's climate.

Figure 12.5. John Tyndall (1820–1893). Edgar Fahs Smith Collection, University of Pennsylvania Library.

Figure 12.6. Svante August Arrhenius (1859–1927). Edgar Fahs Smith Collection, University of Pennsylvania Library.

The first to propose the theory of global warming was Swedish physical chemist **Svante August Arrhenius** (1859–1927; Figure 12.6). In 1896, he suggested that a doubling of $CO_2(g)$ mixing ratios, which could occur due to the rapid increase in coal combustion since the beginning of the Industrial Revolution in the eighteenth century, might lead to temperature increases of 5°C (Arrhenius, 1896). This estimate is higher than recent estimates that a doubling of $CO_2(g)$ will result in a 2°C temperature rise, but the theory is still consistent. Arrhenius also theorized that reductions in $CO_2(g)$ caused glacial periods to occur. This theory was incorrect because changes in the Earth's orbit have been responsible for glacial maxima. The decrease in $CO_2(g)$ mixing ratios during a glacial period is an effect rather than a cause of temperature changes.

12.2.3. Leading Causes of Global Warming

Carbon dioxide is the leading cause of global warming (Table 12.3), which is the temperature change of the lower atmosphere above and beyond that from the natural greenhouse effect. The second leading cause of historic near-surface global warming may be particulate black carbon and its associated brown carbon (Jacobson, 2000, 2001b, 2010b; Chung and Seinfeld,

2005; Hansen et al., 2005; Ramanathan and Carmichael, 2008). Black carbon is emitted during coal, diesel and jet fuel, natural gas, kerosene, biofuel, and biomass burning. Other anthropogenic pollutants that contribute to global warming include methane, nitrous oxide, ozone, CFCs, HCFCs, chlorocarbons, and water vapor.

Although it is the most important natural greenhouse gas, *water vapor's direct anthropogenic emission contribution to global warming is relatively small*. The major anthropogenic emission source of water vapor is evaporation of water used to cool coal, nuclear, natural gas, and biofuel power plants and industrial facilities, which heat up due to combustion or nuclear reaction within them. Evaporation also occurs during the irrigation of crops, but much of this water would evaporate in any case. The second largest anthropogenic emission source of water vapor is fossil fuel, biofuel, and anthropogenic biomass burning combustion. Such combustion releases carbon dioxide, water vapor, and air pollutants. Of the evaporation plus combustion sources of water vapor, which totaled $\sim$64 GT-H_2O/yr in 2005, about two-thirds was evaporation. However, to put this in perspective, the total anthropogenic emission rate of water vapor was only about 1/8,000th the natural emission rate of $\sim$500,000 GT-H_2O/yr. Nevertheless, because natural water vapor causes $\sim$29.4 K of the natural greenhouse effect, a simple scaling suggests that global warming due to anthropogenic water vapor emissions may be 1/8,000th of this, $\sim$0.0038 K, which represents $\sim$0.47 percent of global warming to date (Table 12.3).

Water vapor is emitted not only directly, but also during evaporation from the oceans and soils as a climate response to higher temperatures caused by the emissions of other greenhouse gases and BC/BrC. Such climate feedback is discussed in Section 12.4.2. The additional water vapor causes warming due to some feedbacks and cooling due to others, but the net effect of water vapor is warming. However, *the additional warming due to water vapor resulting from such feedbacks is attributable to the emitted global warming agent triggering the higher temperatures, not to water vapor*. The global warming attributable to water vapor is the warming caused by the direct anthropogenic emissions of water vapor.

Although CO_2(g) is the most abundant and overall most important anthropogenically emitted agent triggering global warming, several other chemicals are more efficient, molecule for molecule, at heating the air. For example, black carbon in fossil fuel soot heats the air more than 1 million times more per unit mass in the atmosphere than does CO_2(g). CH_4(g) heats the air

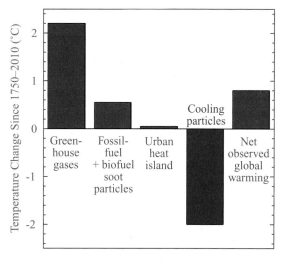

Figure 12.7. Primary contributors to net observed global warming from 1750 to 2010 based on global model calculations (Jacobson, 2010b). Greenhouse gases include CO_2(g), CH_4(g), N_2O(g), O_3(g), chlorofluorocarbons, and hydrochlorofluorocarbons. Cooling particles consist of sulfate, nitrate, chloride, ammonium, sodium, potassium, calcium, magnesium, nonabsorbing organic carbon, and water. Fossil fuel plus biofuel soot particles (warming particles) contain black carbon, brown carbon, nonabsorbing organic carbon, and sulfate. Biofuel soot particles additionally include many of the same components as do cooling particles. The urban heat island effect is described in Section 6.7.3.

about 25 times more per unit mass than does CO_2(g). N_2O(g) and $CFCl_3$(g) cause about 270 and 4,750 times more warming per unit mass, respectively, than does CO_2(g). Although other chemicals are more effective at warming per unit mass than is carbon dioxide, the emission rate and mixing ratio of carbon dioxide are much greater than are those of the other chemicals. As such, *CO_2(g) is the leading cause of global warming*.

As illustrated in Figure 12.7, net observed global warming to date is due primarily to the summed heating of greenhouse gases, absorbing aerosol particles, and urban surfaces (the urban heat island effect) minus the cooling from reflective aerosol particles. Figure 12.7 indicates that cooling particles cause more cooling than warming particles cause warming. In fact, cooling particles are **masking** more than half of actual global warming. In other words, if only emissions of cooling particles were eliminated, the net observed global warming would double. Even if both cooling and warming particles were reduced, temperatures would increase substantially.

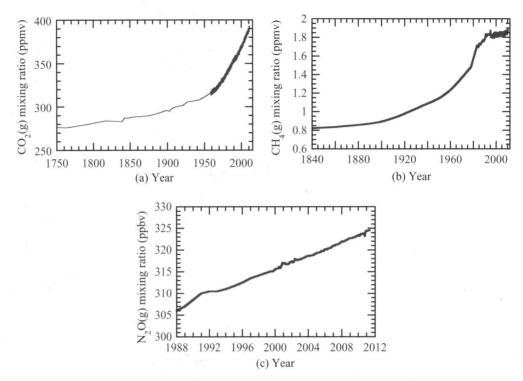

Figure 12.8. Temporal changes in tropospheric mixing ratios of (a) carbon dioxide, (b) methane, and (c) nitrous oxide. Data for carbon dioxide were from Siple Station, Antarctica, ice core for 1744–1953 (Friedli et al., 1986) and from Mauna Loa Data Center (n.d.) for 1958–2009. Data for methane were from Law Dome ice core, Antarctica, for 1841–1978 (Ethridge et al., 1992); from Carbon Dioxide Information Analysis Center (CDIAC; n.d.) at Barrow, Alaska, for 1983–1995; and from Prinn et al. (2000) at Trinidad Head, California, for 1995–2010. Nitrous oxide data were from Mauna Loa Data Center (n.d.).

Because aerosol particles are the leading cause of air pollution mortality, reducing both cooling and warming particles is desirable from a public health perspective. As such, a strategy that involves reducing particle and greenhouse gas emission simultaneously is beneficial for reducing both health and climate problems simultaneously (Chapter 13). Alternatively, because the sources of warming particles, fossil fuel and biofuel soot, differ from sources of cooling particles, a strategy of selectively controlling fossil fuel and biofuel soot together with greenhouse gases would address both climate and air pollution health problems simultaneously.

12.2.4. Trends in Mixing Ratios and Emissions of Gases and Particles

Since the mid-1800s, the tropospheric mixing ratios of $CO_2(g)$, $CH_4(g)$, and $N_2O(g)$ have increased substantially. These gases are relatively well mixed in the lower atmosphere. In addition, black carbon emissions have increased.

Figure 12.8 shows changes in the ambient mixing ratios of $CO_2(g)$, $CH_4(g)$, and $N_2O(g)$ since 1750, 1840, and 1988, respectively. The historical $CO_2(g)$ and $CH_4(g)$ data originate from ice core measurements. The more recent data in all cases originate from ground-based outdoor measurements. In 1958, **Charles David Keeling** began tracking the mixing ratio of $CO_2(g)$ at Mauna Loa Observatory, Hawaii. His record, shown in detail in Figure 3.11 and in less detail in Figure 12.8a, is the longest continuous ground-based record of the gas. Whereas $CO_2(g)$ mixing ratios have increased continuously through 2011, $CH_4(g)$ mixing ratios began to level off in the 1990s and 2000s. However, they increased again from 2007 to 2010 (Frankenberg et al., 2011; Figure 12.8b).

The reason for the increases in tropospheric mixing ratios of $CO_2(g)$, $CH_4(g)$, and $N_2O(g)$ over time is the increase in their anthropogenic emissions coupled with a slow removal rate. Figure 12.9 shows the anthropogenic emission rates of $CO_2(g)$ between 1750 and 2007 (Boden et al., 2011). During this period,

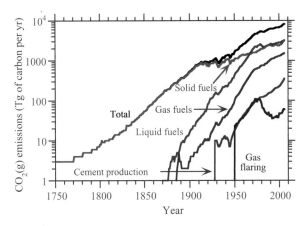

Figure 12.9. Global anthropogenic emissions of carbon dioxide from 1750 to 2007. 1 Tg (teragram) = 10^6 metric tonnes = 10^{12} g. Data from Boden et al. (2011).

337,000 teragrams (1 Tg = 10^6 metric tonnes = 10^6 g) of $CO_2(g)$ were emitted. Half of the total emissions occurred between 1982 and 2007. The 2007 emission rate was 8,400 Tg/yr, an increase of 1.7 percent over the 2006 emission rate and the largest yearly emission rate up to that time. Liquid and solid fuel combustion accounted for 76 percent of anthropogenic $CO_2(g)$ emissions in 2007. Cement production and gas flaring (burning of natural gas emitted from coal mines and oil wells) contributed to about 5 percent of the total $CO_2(g)$ emissions. Rates of gas flaring have decreased as a result of improved technologies enabling the capture of natural gas during coal and oil mining.

A major anthropogenic source of $CO_2(g)$ not accounted for in Figure 12.9 is permanent deforestation. **Deforestation** is the clear-cutting of forests for their wood and the burning of forests to make room for farming and cattle grazing. Permanent deforestation occurs when cut and burned forests are not permitted to regrow, such as during conversion of a forest to grazing land. The loss of trees resulting from deforestation prevents photosynthesis from converting atmospheric $CO_2(g)$ to organic material. Deforestation rates are high in tropical rainforests of South America, Africa (Figure 12.10), and Indonesia. Clear-cutting of forests for their wood in the Pacific Northwest of the United States, in parts of Canada, and in other forested regions of the world is also common. Permanent deforestation results in a global emission rate of 1,500 to 2,700 Tg-C/yr (Houghton, 1994; Lobert et al., 1999; Andreae and Merlet, 2001), or one-fourth of the total annual anthropogenic emission rate of the carbon in carbon dioxide.

Permanent deforestation is one mechanism of **desertification**, the conversion of viable land into desert. Desertification also results from overgrazing and is a problem at the border of the Sahara Desert, which is in continuous expansion. It increases the Earth's albedo and decreases the specific heat of the ground. Desert sand is more reflective than is forest or vegetation; therefore, desertification tends to cool the ground when only albedo change is considered. However, sand has a lower specific heat than do the trees that it replaces; thus, expansion of deserts can also warm the ground when only specific heat changes are considered. The net effect of desertification on temperatures is uncertain.

Figure 12.11 shows the increase in anthropogenic emissions of $CH_4(g)$ from different sources between 1860 and 1994. Total anthropogenic methane emissions increased during that period from 29.3 to 371 Tg/yr. The largest component of anthropogenic methane emissions in 1994 was livestock farming, which overtook rice farming as the leading methane emission source.

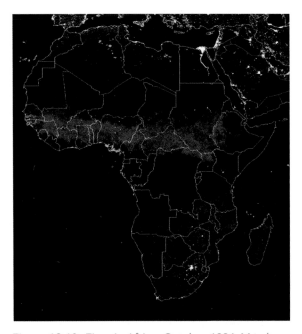

Figure 12.10. Fires in Africa, October 1994–March 1995. Fires are denoted by red; flares from oil and gas exploration/extraction by green. The more intense the color, the greater the frequency of fires. Data collected by U.S. Air Force Weather Agency; image and data processing by NOAA's National Geophysical Data Center.

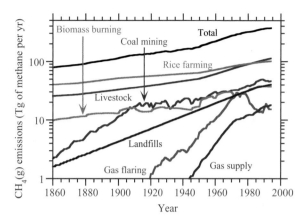

Figure 12.11. Global anthropogenic emissions of methane from different sources from 1860 to 1994. 1 Tg = 10^6 metric tonnes = 10^{12} g. Data from Stern and Kaufman (1998).

Anthropogenic emissions of $N_2O(g)$ also increased in the nineteenth and twentieth centuries. Nitrous oxide is emitted mostly by bacteria in fertilizers and sewage, as a combustion product during biofuel, biomass, and fossil fuel burning, and as a result of nylon production.

In the 1930s and 1940s, CFCs and HCFCs, synthetic long-lived chlorinated compounds, were developed for industrial uses (Chapter 11). These compounds contribute to both stratospheric ozone destruction and global warming because many absorb thermal-IR radiation within the atmospheric window. Whereas stratospheric mixing ratios of CFC-11 and -12 and $CCl_4(g)$ are leveling off or decreasing, those of HCFC-22, other HCFCs, and HFCs are increasing (Figure 11.21). HFCs and other fluorine-containing compounds, such as sulfur hexafluoride [$SF_6(g)$] and perfluoroethane [$C_2F_6(g)$], are strong absorbers in the atmospheric window; thus, their buildup is a cause for concern. Table 11.2 provides the global warming potential of CFCs, HCFCs, and HFCs.

Figures 12.12a and 12.12b show changes in the global emissions of black carbon by country and sector from 1850 to 2005. Black carbon emissions increased worldwide during this period, although its emissions in the United States and Europe first increased and then decreased. Emissions of sulfur dioxide, a precursor to particulate sulfate aerosol, a cooling agent, have remained relatively steady between 1970 and 2005 (Figure 12.12c). The increase in global BC emissions and relatively constant sulfur dioxide emissions imply a net warming due to anthropogenic aerosol particles since 1970.

12.3. Recent and Historical Temperature Trends

To assess the extent and seriousness of global warming today, it is necessary to investigate the temperature record of past climates on the Earth. In the following subsections, recent and historic changes in global temperature are examined.

12.3.1. Recent Temperature Record

Three types of datasets are used to assess global and regional air temperature changes during the past century. All three show evidence of near-surface global warming.

The first type of dataset is one that uses temperature measurements taken between 2 m and 10 m above the surface at land-based meteorological stations and fixed-position weather ships. Two worldwide datasets using such measurements include the U.S. Global Historical Climate Network (GHCN) dataset (Peterson and Vose, 1997) and the UK Meteorological Office dataset (Brohan et al., 2006). Both datasets include measurements since the 1850s. The number of measurement stations included in each dataset changes yearly. The GHCN dataset includes data from fewer than 500 stations prior to 1880, a high of 5,464 stations in 1966, and about 2,000 stations more recently. The Brohan et al. (2006) dataset includes data from fewer than 250 stations prior to 1880, up to 1,800 stations in the 1950s, and fewer than 1,000 stations more recently.

Figure 12.13 shows changes in globally averaged near-surface air temperatures between 1880 and 2010 from the GHCN dataset. Temperatures were relatively stable between 1860 and 1910, but steadily increased from 1910 to 1940. Temperatures were stable or slightly decreased between 1940 and the mid-1970s, but they increased rapidly from the mid-1970s to 2010. Nine of the ten warmest years were in the 2000s, and the years 2010 and 2005 were tied for the warmest years on record. The data indicate that the average global near-surface air temperature has warmed by 0.7°C to 0.9°C since 1880.

A second type of dataset is one that uses temperature measurements at different altitudes taken from balloons (radiosondes) twice daily at about 1,000 locations over land worldwide. Figure 12.14 shows the vertical change in the globally averaged temperature between 1958 and 2007 from the dataset. It indicates that temperatures in the troposphere (pressures greater than 200 hPa) increased and temperatures in the stratosphere (pressures less than 200 hPa) decreased during this

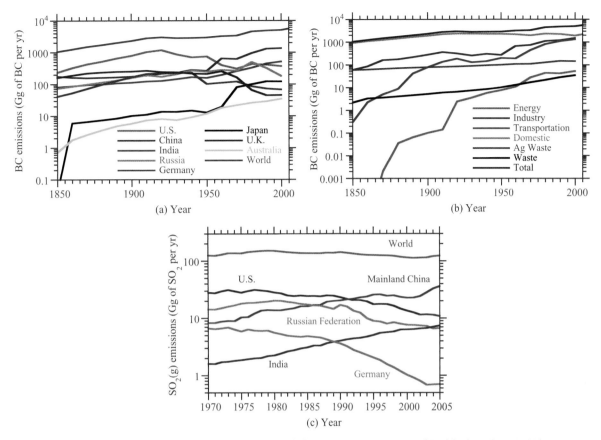

Figure 12.12. Global anthropogenic, nonpermanent deforestation emissions of (a) black carbon (BC) by country from 1850 to 2005, (b) BC by sector worldwide from 1850 to 2005, and (c) sulfur dioxide by country from 1970 to 2005. From Lamarque et al. (2010).

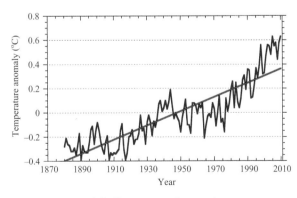

Figure 12.13. Globally averaged annual temperature variations between 1880 and 2010 relative to the 1961–1990 mean temperature. Measurements were from land-based meteorological stations and fixed-position weather ships. The slope of the linear fit through the curve is 0.007°C per year. The slope of the data from 1955 to 2010 is 0.013°C per year. Data from Hansen et al. (2011).

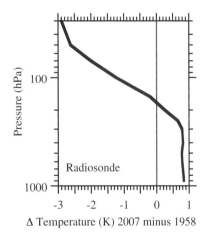

Figure 12.14. Globally and annually averaged, radiosonde-based air temperature difference versus altitude between 2007 and 1958. Data from Randel et al. (2009).

period. Both trends (near-surface warming and stratospheric cooling) are consistent with the theory of global warming.

The reason for the stratospheric cooling relative to the surface warming in Figure 12.14 is that absorption of rising thermal-IR radiation by greenhouse gases, particularly carbon dioxide, in the lower and midtroposphere reduces transfer of that radiation to the upper troposphere and stratosphere, where some of it would otherwise be absorbed by natural and anthropogenic greenhouse gases. In other words, the addition of greenhouse gases to the air lowers the altitude of heating that occurs in the atmosphere, warming the surface and cooling the stratosphere. Surface warming and stratospheric cooling, observed in the radiosonde data, are simulated in models when carbon dioxide and other greenhouse gases are added to the air.

A third type of temperature dataset is the microwave sounding unit (MSU) satellite dataset, compiled since 1979 (Spencer and Christy, 1990). Temperatures for this dataset are derived from a comparison of the brightness or intensity of microwaves (with a wavelength of 5,000 μm) emitted by molecular oxygen in the air. Reported temperatures from this dataset are averages over a horizontal footprint of 110 km in diameter and a vertical thickness at 0 to 8 km (1,000–350 hPa). Thus, the satellite dataset represents temperatures high above the boundary layer (e.g., on the order of 4 km), not at the surface of the Earth. Some errors in the MSU measurements arise because a portion of the radiation measured by the MSU originates from the Earth's surface, not from the atmosphere (Hurrell and Trenberth, 1997). In addition, raindrops and large ice crystals can cause the MSU instrument to slightly underpredict temperatures in rain-forming clouds. Finally, temperatures from 0 to 8 km are not measured directly but estimated from an equation that relies on temperatures estimated by the satellite at different altitudes centered around 7 km. The equation does not account for the variation of air density with altitude. Nevertheless, the MSU dataset shows global temperature increases from 0 to 8 km since 1979 and stratospheric cooling from 14 to 22 km, relatively consistent with the radiosonde data and the theory of global warming.

In sum, three types of global temperature measurements all show global warming. The two types that measure near-surface temperatures (surface and radiosonde) both show strong global warming. The two types that measure stratospheric temperatures (radiosonde and satellite) both show cooling, which is consistent with the theory of global warming. The underlying reason for the surface warming and simultaneous stratospheric cooling is the increase in human-emitted pollutants. This physical phenomenon has been demonstrated repeatedly by computer models since the early 1970s.

12.3.2. Historical Temperature Record

Whereas Figures 12.13 and 12.14 indicate that recent increases in the Earth's near-surface temperature have occurred, it is important to put the temperature changes in perspective by examining past climates of the Earth. Two techniques of estimating historic temperatures are ice core analysis and ocean floor sediment analysis. Other methods include analyses of lake levels, lake bed sediments, tree rings, rocks, pollens in soil deposits, and pollens in deep sea sediments.

12.3.2.1. Ice Core Analysis

When snow accumulates over the Antarctic, it slowly compacts and recrystallizes into ice by **sintering**, which is the chemical bonding of a material by atomic or molecular diffusion. During the ice formation process, air becomes isolated in bubbles or pores within the ice. The trapped air contains roughly the composition of the air at the time the ice was formed. Among the constituents of the air trapped in the ice bubbles are carbon dioxide, methane, and molecular oxygen. When vertical ice cores are extracted, the composition of their air bubbles can be analyzed. Whereas most oxygen atoms in molecular oxygen contain eight protons and eight neutrons in their nucleus, giving them an atomic mass of 16, about 1 in every 1,000 such atoms contains ten neutrons, giving it an atomic mass of 18. Such atoms are called **isotopically enriched atomic oxygen** (^{18}O), which is heavier than standard atomic oxygen (^{16}O). Generally, the warmer the snow or air temperature, the greater the relative ratio of ^{18}O to ^{16}O in molecular oxygen trapped in snow. In fact, a relatively linear relationship has been found between the annual average snow surface temperature and the ^{18}O-to-^{16}O ratio. Thus, from ice core measurements of ^{18}O, past air temperatures can be estimated (Jouzel et al., 1987, 1993).

12.3.2.2. Ocean Floor Sediment Analysis

A related method of obtaining temperature data is to drill deep into sediment of ocean floors and analyze the composition and distribution of calcium carbonate shells. When water, which contains hydrogen and oxygen, evaporates from the ocean surface, the ratio of ^{18}O to ^{16}O in the ocean increases because ^{16}O is lighter than

^{18}O, and water containing ^{16}O is more likely to evaporate than is water containing ^{18}O. Because calcium carbonate in shells contains oxygen from seawater and because higher seawater temperatures correlate with higher ^{18}O-to-^{16}O ratios in seawater, higher ^{18}O-to-^{16}O ratios in shells correlate with higher water temperatures. Because certain shelled organisms live within specified temperature ranges, the distribution of shells in a vertical core sample also gives insight into historical water temperatures.

12.3.2.3. From the Origin of the Earth to 542 Million Years Ago

Figure 12.15 shows the Earth's geologic time scale. Since the formation of the Earth 4.6 b.y.a., near-surface air temperatures have gone through great swings.

Between 4.6 and 4.0 b.y.a., surface temperatures escalated due to the conduction of energy from the Earth's core to its surface, creating magma oceans (Section 2.3.1) that resulted in air temperatures 300°C to 400°C greater than those today (Crowley and North, 1991). During that time, energy released by meteorite impacts also contributed to high surface temperatures.

Between 4.6 and 2.5 b.y.a. (the **Hadean and Archean eons**), the intensity of solar radiation incident upon the Earth was about 20 to 30 percent lower than it is today because the solar output of a young star is low and gradually increases over time. Yet, even after the magma oceans solidified to form the Earth's crust 3.8 to 4 b.y.a., air temperatures were still much higher than today (58°C vs. 15°C today). The low sun intensity coupled with the high air temperatures following crustal formation is referred to as the **Faint Young Sun Paradox** (Sagan and Mullen, 1972; Ulrich, 1975; Newman and Rood, 1977; Gilliland, 1989; Crowley and North, 1991). The reason for the high temperatures may have been an enhanced greenhouse effect due to both carbon dioxide and methane. During solidification of the Earth's crust and mantle, outgassing, particularly of water vapor, carbon dioxide, and methane, occurred. Whereas much of the water vapor condensed to form the oceans, most carbon dioxide and methane remained in the air. The oceans had not formed sufficiently by 3.5 b.y.a. to remove carbon dioxide by chemical weathering. Furthermore, anoxygenic photosynthesis (Section 2.3.3.4), which removes $CO_2(g)$ from the air and converts it to cell material, did not develop until 3.5 b.y.a. Figure 2.11 shows that the Earth's atmosphere prior to the oxygen age (2.3 b.y.a.) may have contained 10 to 80 percent carbon dioxide by volume. In terms of absolute quantities, partial pressures of $CO_2(g)$ 4.6 to 2.5 b.y.a. may have been 0.01 to 0.1 atm, or 30 to 300 times their current values (Garrels and Perry, 1974; Pollack and Yung, 1980). The high partial pressures of $CO_2(g)$ enhanced the greenhouse effect of the young Earth.

Over time, chemical weathering in the newly formed oceans converted $CO_2(g)$ to carbonate rocks. Bacteria also consumed $CO_2(g)$ and fossilized when they died. Carbonate rock formation may have been enhanced by cyanobacteria and other autotrophic bacteria, which use carbon dioxide as an energy source. Dead cyanobacteria pile up to form laminated, bounded structures of trapped carbonaceous material called **stromatolites**, which are usually found in shallow marine waters in warm regions, and some are still being formed (Figure 12.16).

The **Proterozoic eon** (2.5–0.542 b.y.a.) was a period during which the Earth was generally warm and ice free, except for two periods of extended continental glaciation. The first was the **Huronian glaciation** (2.4–2.1 b.y.a), due in part to the **Great Oxygenation Event** (**GOE**) (Section 2.3.4), during which oxygen produced by photosynthesizing bacteria increased slightly in the air because rocks had become saturated with oxygen. The increase in oxygen increased $OH(g)$ and $O(^1D)(g)$, which converted $CH_4(g)$, a strong greenhouse gas, to $CO_2(g)$, a weaker one, cooling the climate. Simultaneously, $CO_2(g)$ decreased due to an increase in carbonate rock formation. The drops in $CH_4(g)$ and $CO_2(g)$, coupled with the low solar intensity at that time, may have triggered the glaciation.

The second period of glaciation involved two separate ice ages that occurred during the **Cryogenian period** (850–635 m.y.a.). These were the **Sturtian glaciation** (750–700 m.y.a.) and the **Marinoan/Varanger glaciation** (660–635 m.y.a.). The two ice ages were the most severe in the Earth's history, causing glaciers to extend on continents down to low latitudes, as evidenced by rock analysis (Williams, 1975). Some scientists speculate that the glaciations resulted in a **snowball Earth**, but there is no evidence of sea ice appearing at low latitudes (Crowley and North, 1991), indicating that the glaciations occurred over land only. The glaciations may have been caused by a drop in $CO_2(g)$ following a surge in weathering of exposed volcanic rock due to the rifting and breakup of the barren and lifeless supercontinent, **Rodinia**, 750 m.y.a. The remnants of Rodinia merged again 600 m.y.a. to form the supercontinent **Pannotia**, which itself split into **Gondwana** (Africa, South America, Antarctica, Australia), **Laurentia** (present-day North America),

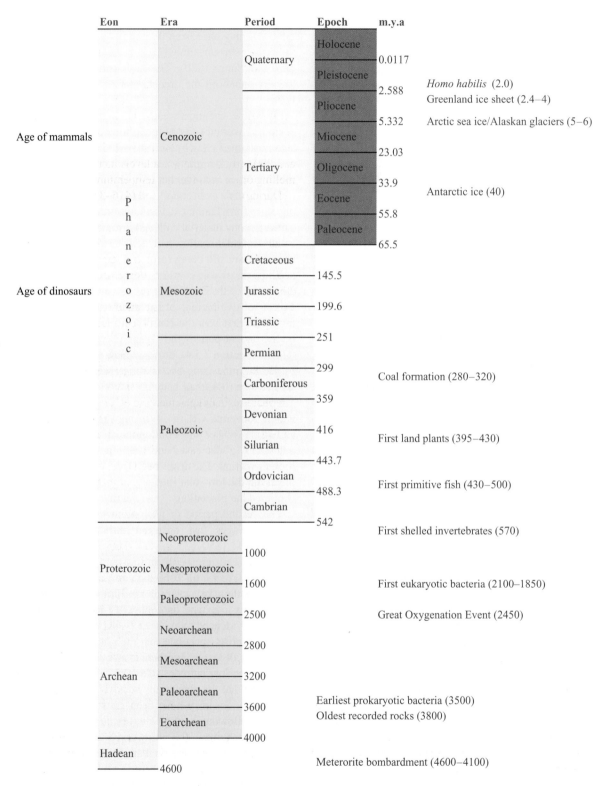

Figure 12.15. Geologic time scale and important events since the formation of the Earth. Time before present is measured in million years ago (m.y.a.).

Figure 12.16. Stromatolites in Hamelin Pool, Shark Bay, Western Australia. © Marek Pilar/Dreamstime.com.

Siberia, and **Baltica** (present-day northern Europe) near 540 m.y.a.

12.3.2.4. From 542 to 100 Million Years Ago

Between 542 and 100 m.y.a., which encompasses the **Paleozoic era** (542–251 m.y.a) and much of the **Mesozoic era** (251–65.5 m.y.a.), the Earth's average temperature was relatively high, but with interruptions by two important glaciations.

From 570 to 460 m.y.a., global $CO_2(g)$ mixing ratios were thirteen to twenty times their current value, sea levels were at an all-time high, and the Northern Hemisphere was covered with water above 30°N latitude due to the configuration of the continents, which were concentrated in the Southern Hemisphere. About 480 m.y.a., the high $CO_2(g)$ resulted in an average air temperature of ~22°C, compared with ~15°C today.

At the beginning of the **Ordovician period** (488.3–443.7 m.y.a.), Gondwana was centered near the Equator, whereas Laurentia, Siberia, and Baltica were separate island continents. During the period, Gondwana moved south, extending between the Equator and the South Pole, causing the warm Earth to cool so that, by 460 m.y.a., temperatures were similar to those today. Continued cooling caused an ice age to set in by 447 to 443 m.y.a. The average temperature during the ice age was ~12°C, resulting in a mass extinction. The ice age lasted only 0.5 to 1.5 million years. The glaciation was precipitated by a drop in $CO_2(g)$ from 7,000 to 4,400 ppmv, still more than ten times its current level. The strong cooling in the presence of high $CO_2(g)$

was due to the configuration of the continents at the time.

During the **Silurian period** (443.7–416 m.y.a.), which followed the glaciation, Gondwana continued to migrate southward, and most of the Northern Hemisphere was covered in ocean. Globally averaged temperatures recovered to a high of ~22°C. Temperatures remained high throughout the Silurian period due to the strong greenhouse conditions, with $CO_2(g)$ persisting at 3,000 to 4,000 ppmv. Sea levels increased due to the melting of ice at the higher temperatures.

During the **Denovian period** (416–359 m.y.a.), **land-plant coverage of the Earth surged**. The resulting green plant photosynthesis caused a sudden reduction in atmospheric $CO_2(g)$ that continued throughout the period, causing an initial drop in temperature. Simultaneously, the spread of plants decreased the continental albedo by 10 to 15 percent (Posey and Clapp, 1964). The lower albedo caused temperatures to recover so that they were similar at the end of the period to those at the beginning, ~22°C.

The **Carboniferous period** (359–299 m.y.a.) is a period before the dinosaurs during which giant plants and insects inhabited the continents and coal beds formed in regions now known as eastern North America and western Europe. Prior to the period, $CO_2(g)$ levels were ~1,500 ppmv and global temperatures were ~20°C, which compares with 393 ppmv and ~15°C, respectively, today. At the beginning of the period, the Earth dropped into the **Karoo Ice Age**, which lasted 100 million years (360–260 m.y.a.), with temperatures ultimately decreasing to ~12°C. This ice age was caused primarily by the restriction in circulation of polar and tropical waters caused by the reorganization of land masses so that they stretched from pole to pole and by the existence of a polar land mass (at the South Pole) that supported the accumulation of a large amount of ice. The ice age was also enhanced by the continued drop in $CO_2(g)$ due to a surge in plant growth during the period.

The Karoo Ice Age persisted into the **Permian period** (299–251 m.y.a.), dissipating around 260 m.y.a. At that time, the continents merged together to form one supercontinent, **Pangaea** (Figure 12.17a), which broke apart around 200 to 170 m.y.a. (Figure 12.17b). The Pangaean climate, on average, was drier and warmer than today (Crowley and North, 1991). The warm temperatures allowed the dinosaurs to evolve starting 232 to 234 m.y.a. and thrive throughout the **Triassic** (251–199.6 m.y.a.) and **Jurassic** (199.6–145.5 m.y.a.; Figure 12.17c) periods and beyond, until 65.5 m.y.a.

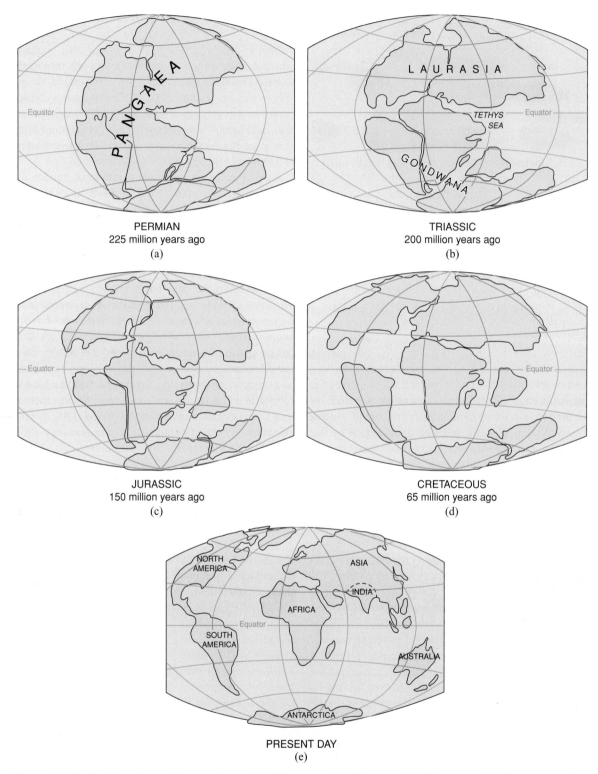

Figure 12.17. Configuration of the continents (a) 225 m.y.a., (b) 200 m.y.a., (c) 150 m.y.a., (d) 65 m.y.a., and (e) today. From Kious and Tilling (1996). http://pubs.usgs.gov/gip/dynamic/historical.html.

12.3.2.5. From 100 to 3 Million Years Ago

Between 120 and 90 m.y.a., in the mid-**Cretaceous period** (145.5–65.5 m.y.a.), temperatures increased above those today. This may have been the last period of an icefree globe. Sea levels were high, covering about 20 percent of continental areas (Barron et al., 1980).

At the end of the Cretaceous period (65.5 m.y.a.) (Figure 12.17d), a mass extinction of 75 percent of all plant and animal species, including the dinosaurs, occurred. This extinction is called the **Cretaceous-Tertiary (K-T) extinction** and may have been due to a 10-km-wide asteroid hitting the Earth (Alvarez et al., 1980). Evidence for the asteroid theory includes a worldwide layer of the element iridium (Ir) in clays at depths below the Earth's surface corresponding to the K-T transition. Although iridium has been measured in a Hawaiian volcanic eruption plume (Zoller et al., 1983), these plumes are not known to penetrate to the stratosphere, which would be necessary for the iridium to be distributed globally (Crowley and North, 1991). Currently, the most accepted explanation for the iridium layer at the K-T boundary is an extraterrestrial source, such as an asteroid. The asteroid impact is believed to have created a large dust cloud that blocked the sun for a period of weeks to months, lowering the surface temperature by tens of degrees (Toon et al., 1982). The reduction in surface temperature may have been responsible for the mass extinction. Because some extinctions occurred before the K-T transition, the asteroid theory is still open to debate.

Between 90 m.y.a. and the **Paleocene epoch** (65.5–55.8 m.y.a.), global temperatures declined from their mid-Cretaceous highs. Temperatures abruptly increased between the late Paleocene to the early **Eocene epoch** (55.8–33.9 m.y.a.). At the onset of the early Eocene warming, the carbon content of deep sea bulk sediments decreased, indicating an increase in atmospheric $CO_2(g)$ (Berner et al., 1983; Shackleton, 1985). Also, the Norwegian-Greenland Sea opened (Talwani and Eldholm, 1977). This and other tectonic activity may have resulted in enhanced volcanism, increasing $CO_2(g)$.

Following the early Eocene temperature maximum, a gradual global cooling occurred from 50 to 3 m.y.a. During this period, continents drifted toward higher latitudes, cooling the land. A sharp cooling event occurred near 40 to 38 m.y.a., in the late Eocene epoch, resulting in one of the largest extinctions during the **Cenozoic era** (65.5 m.y.a.–present).

Cenozoic era ice may have first appeared over the Antarctic ~40 m.y.a., forming the base of the present-day **Antarctic ice sheets**. An **ice sheet** is a broad, thick sheet of glacier ice that covers an extensive land area for a long time. An ice sheet can also form on top of sea ice if the sea ice is adjacent to land. A **glacier** is a large mass of land ice formed by the compaction and recrystallization of snow. Glaciers flow slowly downslope or outward in all directions under their own weight. **Sea ice** is ice formed by the freezing of sea water (Figure 12.18). Today, the Antarctic is covered by two ice sheets, the **East and West ice sheets**, which are separated by the Transantarctic Mountains. The West ice sheet, which lies over the Ross Sea and over land (Figure 12.19), is an order of magnitude smaller than is the East ice sheet, which lies exclusively over land.

(a)

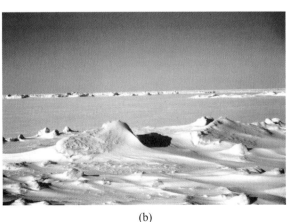

(b)

Figure 12.18. (a) Spring (1950) melt of sea ice in the Beaufort Sea, near Tigvariak Island, Alaska North Slope. (b) Winter (1950) sea ice near the same location. Rear Admiral Harley D. Nygren, NOAA Corps (ret.), available from NOAA Central Library, www.photolib.noaa.gov/.

 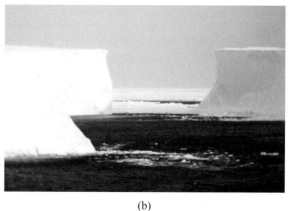

(a)   (b)

Figure 12.19. (a) Edge of Ross Sea ice sheet, December 1996. (b) Icebergs grounded on Pennel Bank, Ross Sea, Antarctica, January 1999. Michael Van Woert, NOAA NESDIS, ORA. available from NOAA Central Library, www.photolib.noaa.gov/.

During the **Oligocene epoch** (33.9–23.03 m.y.a.), temperatures continued the decline that started during the Eocene epoch. The temperature drop triggered the expansion of Antarctic ice cover starting 31 to 29 m.y.a. From 26 to 15 m.y.a., during the early **Miocene epoch** (23.03–5.332 m.y.a.), temperatures temporarily stabilized. From 15 to 10 m.y.a., they dropped more, increasing Antarctic ice coverage and decreasing sea levels. The cause of this temperature decrease may have been a decrease in $CO_2(g)$ due to changes in ocean water vertical mixing (Crowley and North, 1991).

From 5 to 4 m.y.a., during the early **Pliocene epoch** (5.332–2.588 m.y.a.), temperatures rebounded slightly, reducing Antarctic ice sheet areas. Although much of the Northern Hemisphere was warm during the early Pliocene, seasonal **sea ice over the Arctic Ocean** and **glaciers in Alaska** began to form 5 to 6 m.y.a. and persisted thereafter (Clark et al., 1980; Zubakov and Borzenkova, 1988). From 2.6 to 2.4 m.y.a., a sharp decrease in temperature caused trees to disappear in Northern Greenland and tundra to expand in Siberia. The **Greenland ice sheet** may have formed during this cooling period, although evidence for its formation between 3 and 4 m.y.a. has also been suggested (Leg 105 Shipboard Scientific Party, 1986).

12.3.2.6. From 3 Million Years Ago to 20,000 Years Ago

The past 3 million years have been characterized by advances and retreats of Northern Hemisphere ice sheets. The relatively cold temperatures and the continuous presence of at least one ice sheet (over the Antarctic) during the **Quaternary period** (2.588 m.y.a.–present) have resulted in our present overall period being referred to as the **Quaternary Ice Age**. This ice age ranks among the four longest in the Earth's history, along with the Karoo, Sturtian-Marinoan/Varanger, and Huronian ice ages.

Between the late Pliocene epoch and 700,000 years ago (y.a.), during the **Pleistocene epoch** (2.588 m.y.a.–11,700 y.a.), fluctuations in the advances and retreats of ice sheets occurred within a period of about 40,000 years (Shackleton and Opdyke, 1976). Between 700,000 y.a. and the present, such advances and retreats have occurred within periods of about 100,000 years. These fluctuations can be explained as follows.

Figure 12.20 shows temperature changes during the past 450,000 years, a period covering nearly one-fourth of the Pleistocene epoch. It also shows carbon dioxide and methane trends over the past 420,000 years. Data for the table were obtained from the Vostok, Antarctica, ice core. The Vostok core was obtained by drilling through ice down to a depth of more than 3.6 km, a depth representing 450,000 years of atmospheric history.

The Earth has gone through four glacial and interglacial periods every 100,000 years over the past 450,000 years. Within these major periods are minor oscillations. The major and minor temperature oscillations can be explained for the most part by three cycles related to the Earth's orbit, called **Milankovitch cycles**, which are named after Serbian astronomer **Milutin Milankovitch** (1930, 1941). They are caused by gravitational attraction between the planets of the solar system and the Earth. Milankovitch was the first to calculate the effects of these cycles on incident solar radiation reaching the Earth. The cycles are briefly discussed next.

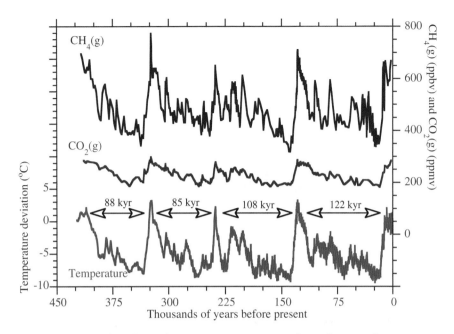

Figure 12.20. Methane, carbon dioxide, and temperature variations from the Vostok ice core over the past 425,000 to 450,000 years. Temperature variations are relative to a modern surface air temperature over the ice of −55°C. From Jouzel et al. (1987, 1993, 1996); Petit et al. (1999).

12.3.2.7. Changes in the Eccentricity of the Earth's Orbit

Figure 12.21 shows that the Earth travels around the sun in an elliptical pattern, with the sun at one focus. In the figure, a and b are the lengths of the major and minor semi-axes, respectively, of the ellipse. The distance between the center of the ellipse (point C) and either focus is c, which is related to a and b by the Pythagorean relation, $a^2 = b^2 + c^2$. The **eccentricity** (e) of the ellipse is

$$e = \frac{c}{a} \tag{12.7}$$

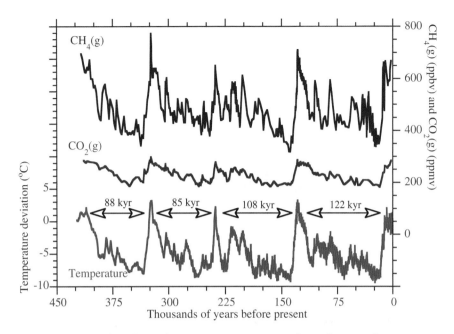

Figure 12.21. Earth's orbit around the sun during periods of high and low orbital eccentricity. The eccentricities are exaggerated in comparison with real eccentricities of the Earth's orbit.

The eccentricity varies between 0 and 1. A circle has an eccentricity of 0. Earth's eccentricity is currently low, 0.017 (Example 12.2), indicating that the Earth's orbit is nearly circular but still noncircular enough to create a 3.4 percent difference in Earth-sun distance between June and December.

> **Example 12.2**
> Calculate the current eccentricity of the Earth.
>
> **Solution**
> Figure 12.1 shows that the Earth-sun distance during the winter solstice (when the Earth is closest to the sun) is currently $a - c = 147.1$ million km and during the summer solstice (when the Earth is furthest from the sun) is $a + c = 152.1$ million km. Solving these two equations gives $c = 2.5$ million km and $a = 149.6$ million km. Substituting these numbers into Equation 12.7 gives the current eccentricity of the Earth's orbit around the sun as $e \approx 0.017$.

The eccentricity of the Earth varies sinusoidally with a period of roughly 100,000 years. The minimum and maximum eccentricities during each 100,000-year period vary as well. The minimum eccentricity is usually greater than 0.01, and the maximum is usually less than 0.05. Whereas today the Earth is in an orbit of relatively low eccentricity (0.017), the eccentricity will decrease during the next 10,000 years before increasing again. In 50,000 years, the Earth will again be in an orbit of high eccentricity. During orbits of high eccentricity, the Earth-sun distance is about 10 percent greater in June than in December, resulting in incident solar radiation being about 21 percent lower in June than in December. Today, the Earth-sun distance is about 3.4 percent greater in June than in December, resulting in incident solar radiation being about 6.9 percent lower in June than in December.

Because the yearly averaged distance between the Earth and the sun is less in a period of low eccentricity than in a period of high eccentricity, yearly averaged temperatures are higher in periods of low eccentricity than in periods of high eccentricity. This can be seen in Figure 12.20, which shows that natural interglacial temperature maxima occurred 122,000, 230,000, 315,000, and 403,000 years ago, all times of low eccentricity.

12.3.2.8. Changes in the Obliquity of the Earth's Axis

The **obliquity** of the Earth's axis of rotation is the angle of the axis relative to a line perpendicular to the plane of the Earth's orbit around the sun. Figure 12.22a shows the range in the Earth's obliquity. Every 41,000 years, the Earth goes through a complete cycle where the obliquity changes from 22 to 24.5 degrees and back to 22 degrees. Currently, the obliquity is 23.5 degrees and decreasing. When the obliquity is low (22 degrees), more sunlight hits the Equator, increasing the south-north temperature contrast. When the obliquity is high (24.5 degrees), more sunlight reaches higher latitudes, reducing the temperature contrast. Thus, the obliquity affects seasons and temperatures at each latitude. Superimposed on the large temperature variations

in Figure 12.20 are smaller variations, some resulting from changes in obliquity.

12.3.2.9. Precession of the Earth's Axis of Rotation

Precession is the angular motion (wobble) of the Earth's axis of rotation about an axis fixed in space (Figure 12.22b). It is caused by the gravitational attraction between the Earth and other bodies in the solar system. Currently, the Northern Hemisphere is farther from the sun in summer than in winter. In 11,000 years, angular motion of the Earth's axis of rotation will cause the Northern Hemisphere to be closer to the sun in summer than in winter. The complete cycle of the precession of the Earth's axis is 22,000 years. Precession of the Earth's axis does not change the yearly or globally averaged incident solar radiation at the top of the Earth's atmosphere. Instead, it changes the quantity of incident radiation at each latitude during a season. In 11,000 years, for example, Northern Hemisphere summers will be shorter but possibly warmer than today because the Northern Hemisphere will be closer to the sun in summer. Southern Hemisphere summers will be longer but possibly cooler than they are today because the Southern Hemisphere will be farther from the sun during its summer. Because seasonal changes in temperature result in yearly averaged changes in temperature at a given latitude, some of the cyclical changes in temperatures seen in the Antarctic data shown in Figure 12.20 are due to changes in precession.

12.3.2.10. Effects of the Milankovitch Cycles

The Milankovitch cycles are responsible for most of the cyclical changes in the Earth's temperature in Figure 12.20, thereby causing advances and retreats of glaciers

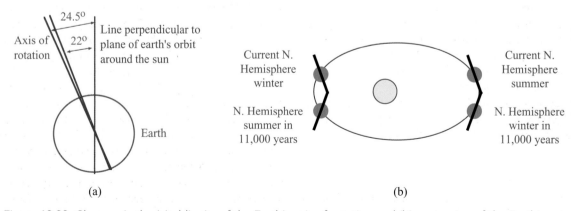

Figure 12.22. Changes in the (a) obliquity of the Earth's axis of rotation and (b) precession of the Earth's axis relative to a point in space.

during the Pleistocene epoch. By changing temperatures, the Milankovitch cycles also caused the corresponding cyclical changes in $CO_2(g)$ and $CH_4(g)$ in Figure 12.20.

For example, higher temperatures decrease the solubility of $CO_2(g)$ in seawater, increasing evaporation of $CO_2(g)$ from seawater to the air. Changes in temperature also change vertical mixing rates of ocean water, nutrient uptake rates by phytoplankton, and rates of erosion of continental shelves (which affect biomass loadings), all of which feed back to change $CO_2(g)$ in the air (e.g., Crowley and North, 1991).

Similarly, higher temperatures increase the microbiological production of methane by methanogenic bacteria (Section 2.3.3.2). Also, higher temperatures melt ice faster in **permafrost soil**, which is soil that remains below the freezing point of water, 0°C, for 2 or more years. The melting of permafrost soil releases methane stored under it to the air. Similarly, warmer ocean temperatures cause **methane hydrates**, which are compounds containing methane trapped in ice that form under high pressure and low temperature deep in the ocean, to melt and release their methane into ocean water. Dissolved methane then bubbles to the surface and escapes to the atmosphere.

In sum, the natural changes in carbon dioxide and methane seen in Figure 12.20 were primarily a response to changes in temperature rather than a cause of changes in temperature. The situation today differs, where increases in carbon dioxide, methane, and other chemicals emitted by humans are causing warming above what nature would cause and at a rate much more rapid than in the historical record, as discussed shortly.

Figure 12.20 shows that a temperature minimum two ice ages ago occurred about 150,000 y.a. Near that time, glaciers extended down to Wisconsin in the United States, and possibly farther south in Europe (Kukla, 1977). About 130,000 y.a., temperatures increased again, causing deglaciation. Over the Antarctic, temperatures rose 2°C to 3°C above what they are today (Figure 12.20). As the eccentricity of the Earth's orbit increased, temperatures decreased again, causing renewed glaciation. During this period (**the last glacial period**), two major stages of glaciation occurred, the first starting 115,000 y.a. and the second 75,000 y.a. The second stage continued until about 12,000 years ago.

12.3.2.11. From 20,000 to 9,000 Years Ago

The **last glacial maximum** occurred from 22,000 to 12,000 y.a. (Figure 12.23). For glaciation over eastern North America, western Europe, and the Alps, this

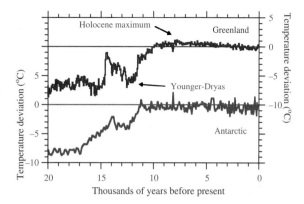

Figure 12.23. Temperature variation in the Northern Hemisphere (top line) and in the Antarctic (bottom line) during the past 20,000 years. The deviations for the ice core data are relative to a modern surface air temperature over the ice of −55°C. The Greenland data are from North Greenland Ice Core Project Members (2004). Antarctic data are from the Vostok ice core (Jouzel et al., 1987, 1993, 1996; Petit et al., 1999).

period is referred to as the **Wisconsin**, **Weichselian**, and **Würm**, respectively. During the last glacial maximum, an ice sheet called the **Laurentide ice sheet** covered North America, and another called the **Fennoscandian ice sheet** covered much of northern Europe. These ice sheets were about 3,500 to 4,000 m thick and drew up enough ocean water to decrease the sea level by about 120 m (CLIMAP Project Members, 1981; Fairbanks, 1989). The decrease in sea level was sufficient to expose land connecting Siberia to Alaska, creating the **Bering land bridge**. This land bridge allowed humans to migrate from Asia to North America and, ultimately, to Central and South America. The Laurentide ice sheet extended from the Rocky Mountains in the west to the Atlantic Ocean in the east, but only as far south as the Missouri and Ohio valleys. The melting of the Laurentide ice sheet resulted in the flooding of significant land off the East Coast of North America, creating shallow ocean water out to a great distance offshore.

Temperatures during the last glacial maximum were 5°C to 8°C lower than they are today over the Northern Hemisphere and 8°C lower than they are today over the Antarctic (Figure 12.23). During the last glacial maximum, Antarctic ice area expanded as did Arctic sea ice area. In the tropics, precipitation decreased, reducing inland lake and river levels. Globally, near-surface winds may have been 20 to 50 percent faster than those today. $CO_2(g)$ mixing ratios were about 200 ppmv (Figure 12.20), almost half their current value. CH_4(g) mixing ratios were about 0.35 ppmv, 20 percent of their current value.

Figure 12.23 shows temperature changes from ice core data in Greenland and the Antarctic during the past 20,000 years. Deglaciation from the last Ice Age caused temperature to rise over Antarctica between 17,000 y.a. and 10,000 y.a., with a hiatus between 13,500 y.a. and 12,000 y.a. Temperatures increased over Greenland started around 15,500 y.a. Northern Hemisphere deglaciation was slow at first, but 14,500 y.a. an abrupt increase in temperature hastened it. Around 12,800 y.a., temperature began to drop into ice age conditions again until 11,500 y.a. This period of strong cooling is called the **Younger Dryas period**, where Dryas is the name of an Arctic wildflower. The Younger Dryas cooling period followed a shorter Older Dryas cooling period from 14,000 to 13,700 y.a.

The Younger Dryas cooling may have been due to changes in atmospheric circulation resulting from large water releases from melting glaciers. During glacial retreat, bursts of meltwater flooded the Columbia Plateau in eastern Washington, the largest flood in known history. Up to forty similar bursts may have occurred during this period (Waitt, 1985). Some of the bursts may have caused freshwater to flow northward over land through Canada into the Arctic Ocean, before flowing to the North Atlantic Ocean.

Normally, warm **Gulf Stream** ocean water heading northeastward from the Gulf of Mexico along the Atlantic Coast sinks when it reaches the North Atlantic Ocean. The reason is that high winds in the North Atlantic evaporate water, increasing the saltiness and density of remaining ocean water, causing it to sink faster. Fast-sinking water in the North Atlantic pulls more Gulf Stream water northeastward. The injection of fresh water into the North Atlantic during the Younger Dryas had the opposite effect, decreasing the density and inhibiting the sinking of ocean water, slowing the Gulf Stream. Because the Gulf Stream current is responsible for transporting enough warm water to northern Europe to keep it warmer than the same latitudes over North America, the weakening of the current caused an average Northern Hemisphere cooling. The ice core record does not show a similar cooling signal in Antarctica, due to the distance of Antarctica from this event.

Following the Younger Dryas period, temperatures increased again. Most of the Northern Hemisphere ice sheets disappeared by 9,000 y.a., although remnants of the Laurentide ice sheet remained until 6,000 y.a. In sum, the bulk of the two major ice sheets in the Northern Hemisphere disappeared over a period of 5,000 years, from 14,000 to 9,000 y.a.

12.3.2.12. The Holocene Epoch

The time between 11,700 y.a. and the present is the **Holocene epoch**. From 10,000 to 5,000 y.a., temperatures were warm. During this warm period, humans developed agriculture and domesticated animals (~8,000 y.a.). The **mid-Holocene maximum** in temperatures occurred 6,000 to 5,000 y.a., during which Northern Hemisphere temperatures were ~1°C warmer than at the end of the ice core record. Corresponding mid-Holocene warming over the Antarctic is less obvious (Figure 12.23).

Following the mid-Holocene, Northern Hemisphere temperatures declined. However, between 950 AD and 1250 AD, they increased again to their greatest peak since the mid-Holocene but lower than those today. During this relatively warm period, called the **medieval climate optimum**, Arctic ice retreated, Iceland and Greenland were colonized by the Vikings, alpine passes between Germany and Italy became ice free, and grapes were harvested for wine production in England (Le Roy Ladurie, 1971; Crowley and North, 1991).

Between 1450 and 1890 AD, temperatures in North America, particularly in Europe, decreased, and mild glaciation reappeared. This period is the **Little Ice Age** (Matthes, 1939). The lowest temperatures during the Little Ice Age were in the mid- to late 1600s, early 1800s, and late 1800s (Crowley and North, 1991). During the Little Ice Age, glaciers advanced in the European Alps; Sierra Nevada Mountains; Rocky Mountains; Himalayas; southern Andes Mountains; and mountains in eastern Africa, New Guinea, and New Zealand. The area of sea ice around Iceland also increased (Bergthorsson, 1969). In addition, from 1607 to 1814, the River Thames froze every winter. In 1658, the Swedish Army marched across the frozen Oresund to invade Copenhagen. In 1780, New York Harbor froze, allowing pedestrians to walk from Manhattan to Staten Island. In 1794, the French Army took the Netherlands, whose fleet of ships was locked in ice.

Although temperatures decreased over much of the world, temperatures over the Antarctic warmed during part of the Little Ice Age, as seen in Figure 12.24. Temperature decreases during the Little Ice Age were smaller in magnitude than those during the Younger Dryas period.

From April 5 to 12, 1815, during the Little Ice Age, the most deadly volcano in human history, **Tambora**, Indonesia, erupted, killing an estimated 92,000

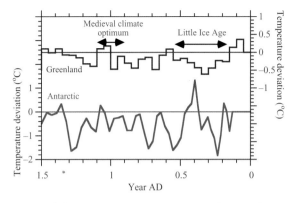

Figure 12.24. Temperature variation in the Northern Hemisphere summer (top line) and in the Antarctic (bottom line) during the past 1,500 years. The deviations for the ice core data are relative to a modern surface air temperature over the ice of −55°C. Greenland data are from North Greenland Ice Core Project Members (2004). Antarctic data are from the Vostok ice core (Jouzel et al., 1987, 1993, 1996; Petit et al., 1999).

islanders directly and through disease and famine. The volcano emitted an estimated 1.7 million tonnes of ash and aerosol particles that traveled globally, cooling North America and Europe over the next 2 years (Stommel and Stommel, 1981; Stothers, 1984). In 1816, frost in the spring and summer killed crops in the United States, Canada, and Europe, sparking famine in some areas of Europe. In the northeast United States and Canada, 1816 was known as the **year without a summer**. The artist Joseph M. W. Turner (1775–1851) captured colorful sunsets caused by the volcanic particles in paintings during this period.

In the dark summer of 1816, George Gordon (Lord) Byron (1788–1824) met and became friends with Percy Bysshe Shelley (1792–1822) in Geneva, Switzerland. Because of the depressing nature of the weather, Lord Byron and Shelley's wife, Mary Wollstonecraft Shelley (1797–1851) entered into a competition to write the most depressing work. Lord Byron wrote the poem *Darkness*, and Mary Shelley started the novel *Frankenstein* (*The Modern Prometheus*), which she finished in 1818.

12.3.2.13. Comparison of Current with Historical Temperatures

The current high temperatures are not unprecedented in Earth's history. In fact, temperatures throughout most of the planet's history have been even higher than today, except during the major and minor ice ages. However, *during previous periods of high temperature, the Earth did not need to support the infrastructure to feed, clothe, and house billions of people*. Previous changes in Earth's temperature were also gradual, not sudden. The *global warming that is now occurring has resulted in a rate of temperature increase that is higher than nearly all known historic rates of temperature increase*.

Table 12.4 shows that the rate of temperature increase from 1880 to 2010 was about 0.7°C per 100 years and that from 1955 to 2010 was 1.3°C per 100 years. For comparison, the rate of temperature increase during the past millennium was about 0.052°C per 100 years, during the deglaciation after the Younger Dryas period was 0.3°C per 100 years, and during the deglaciation leading to the last interglacial period was 0.13°C per 100 years. The rate of temperature increase today is greater than the rate of historic temperature increase due to natural events.

Yet, an examination of the Vostok core data indicates that during deglaciations, temperature increases over

Table 12.4. Rates of temperature change during different historical periods

Period	Years ago	Temperature change (K) per 100 years
1958–2007 (radiosonde records, Figure 12.14)	52–4	+1.3
1955–2010 (land and ship measurements, Figure 12.13)	56–1	+1.3
1880–2010 (land and ship measurements, Figure 12.13)	121–1	+0.7
During the past 1,000 years (Vostok core, Figure 12.24)	1,000–0	+0.052
Deglaciation after Younger Dryas (Vostok core, Figure 12.23)	12,632–11,191	+0.3
Last years of deglaciation after Younger Dryas (Vostok core, Figure 12.23)	11,237–11,191	+2.2
Deglaciation leading to last interglacial period (Vostok core, Figure 12.20)	138,000–128,000	+0.13

short periods (as opposed to averaged over the entire deglaciation period) could be quite rapid. For example, during the last years of deglaciation after the Younger Dryas period, the rate of temperature increase was 2.2°C per 100 years. The difference today is that the Earth is in an interglacial period, not a deglaciation. *The current rapid rate of increase in temperature is abnormally high for an interglacial period.*

12.4. Feedbacks and Other Factors That May Affect Global Temperatures

Although near-surface air temperatures have increased recently at abnormal rates compared with historic rates, and these temperature increases have occurred concurrently with increases in $CO_2(g)$, $BC(s)$, $CH_4(g)$, and $N_2O(g)$ emissions, some have questioned whether any factor aside from emitted air pollutants could cause the global warming seen in the recent record. This topic is discussed next.

12.4.1. Arguments Mistakenly Used to Explain Global Warming as a Natural Phenomenon

An argument mistakenly used to explain global warming as a natural rather than anthropogenic phenomenon is the suggestion that higher temperatures are due to natural Milankovitch cycle variations. One argument is that because the current eccentricity of the Earth's orbit (0.017) is in a declining stage, temperatures should naturally increase over the next thousand years or more, and such increases might explain global warming. However, because the eccentricity has also been declining during the past 1,000 years, and the rates of temperature increase today are much higher than were those during the past 1,000 years (Table 12.4), this argument does not explain global warming.

Another proposed explanation for global warming has been that it is due to the natural variation in solar output. When sunspots appear, the intensity of the sun's output increases. A **sunspot** is a large magnetic storm that consists of a dark, cool central core called an **umbra** and is surrounded by a ring of dark fibrils called a **penumbra**. As a result of the magnetic activity associated with sunspots, regions near the umbra are hot, resulting in more net energy emitted by the sun when sunspots are present than when they are absent. Sunspot number and size peak every 11 years; however, because the sun's magnetic field reverses itself every 11 years, a complete sunspot cycle actually takes 22 years. The

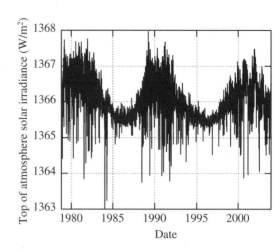

Figure 12.25. Top-of-the-atmosphere measured solar irradiance, November 1, 1978–October 1, 2003. Data from Fröhlich and Lean (2004).

difference in solar intensity at the top of the Earth's atmosphere between times of sunspot maxima and minima is about 1.4 W m^{-2} (e.g., Figure 12.25), or only 0.1 percent of the solar constant.

Although sunspots have been linked to subtle climate changes during the course of a sunspot cycle, the fact that sunspot intensity varies relatively consistently from cycle to cycle indicates that sunspots cannot cause multidecade increases in temperatures as have occurred since the 1950s in particular. Furthermore, as shown in Figure 12.25, peak sunspot intensity declined rather than increased between 1978 and 2003, whereas Figure 12.13 indicates that global near-surface air temperature increased sharply during this period. Thus, sunspots and changes in solar intensity are not correlated with recent sustained global temperature increases.

A third proposed explanation for global warming is that it is due to natural **internal variability** of the ocean atmosphere system due to chaotic feedbacks among meteorological variables that characterize the atmosphere (e.g., temperature, pressure, winds, humidity, clouds). Whereas the random nature of the climate explains part of the warming and cooling cycles between the 1850s and the present (seen in Figure 12.13), changes due to internal variability are on the order of ±0.3°C (Stott et al., 2000) and thus could cause either cooling or warming, with the maximum warming smaller than global observed temperature changes of +0.7°C to +0.9°C since 1880.

A fourth argument is that global warming is due to the urban heat island (UHI) effect. Urban areas covered about 0.128 percent of the Earth in 2005 and are generally warmer than vegetated areas around them because urban surfaces reduce evapotranspiration and have sufficiently different heat capacities, thermal conductivities, albedos, and emissivities than vegetated land to enhance urban warming compared with vegetated land they replace (Section 6.7.3). However, data analysis studies found that the globally averaged UHI effect may contribute ≤0.1 K to global temperature changes since the preindustrial era (Easterling et al., 1997; Hansen et al., 1999; Peterson, 2003; Parker, 2006; Intergovernmental Panel on Climate Change (IPCC), 2007). A computer model simulation that accounted for the effects of urban areas at their actual resolution found a similar result (Jacobson and ten Hoeve, 2012).

12.4.2. Feedback of Gases to Climate

The simple radiative energy balance model introduced at the beginning of this chapter describes the equilibrium temperature of the Earth in the absence of natural greenhouse gases. However, temperature changes in the real atmosphere depend not only on the trapping of heat radiation by greenhouse gases, but also on the feedback, or **climate response**, of the heating to water vapor, clouds, winds, and the oceans.

If an increase in greenhouse gas emissions initially forces a warming of the climate, the climate may respond either positively, enhancing the warming, or negatively, diminishing the warming. A **positive feedback mechanism** is a climate response mechanism that causes temperature to change farther in the same direction as that of the initial temperature perturbation. A **negative feedback mechanism** is a mechanism that causes temperature to change in the opposite direction from that of the initial perturbation. Positive feedback can lead to a runaway greenhouse effect, such as the one on Venus, whereas negative feedback tends to mitigate potential effects of global warming. Next, some positive and negative feedback mechanisms resulting from an initial increase in temperatures by greenhouse gases are itemized.

12.4.2.1. Water Vapor-Temperature Rise Positive Feedback

If air temperatures initially increase due to a greenhouse gas, more water evaporates from the oceans, lakes, and rivers, and sublimates from snow and sea ice, increasing atmospheric water vapor (another greenhouse gas), raising temperatures more.

12.4.2.2. Snow-Albedo Positive Feedback

If air temperatures initially increase, sea ice and glaciers melt, uncovering darker ocean or land surfaces below, decreasing the Earth-atmosphere albedo, increasing solar radiation absorbed by the surface, raising temperatures more.

12.4.2.3. Water Vapor-High Cloud Positive Feedback

If air temperatures initially increase, more water evaporates from the oceans, lakes, and rivers or sublimates from snow or sea ice. Some of this water vapor produces more high clouds, which consist primarily of large ice crystals that are relatively transparent to solar radiation but absorb thermal-IR radiation. An increase in high cloud cover due to the initial rise in temperature increases the absorption and reemission of the Earth's thermal-IR radiation, raising temperatures more.

12.4.2.4. Solubility-Carbon Dioxide Positive Feedback

If air temperatures initially increase, the solubility of carbon dioxide in ocean water decreases, increasing the transfer of carbon dioxide from the ocean to the atmosphere, raising temperatures more.

12.4.2.5. Saturation Vapor Pressure-Water Vapor Positive Feedback

If air temperatures initially increase, the saturation vapor pressure of water increases, reducing the ability of water to condense, increasing the quantity of water vapor in the air, raising temperatures more.

12.4.2.6. Bacteria-Carbon Dioxide Positive Feedback

If air temperatures initially increase, the rate at which soil bacteria decompose dead organic matter into carbon dioxide and methane increases, increasing atmospheric carbon dioxide and methane levels, raising temperatures more.

12.4.2.7. Permafrost-Methane Positive Feedback

If air temperatures initially increase, permafrost over land and methane hydrates deep in ocean water both melt, increasing the release to the atmosphere of methane stored under the permafrost and in the methane hydrates, raising temperatures more.

12.4.2.8. Water Vapor-Low Cloud Negative Feedback

If air temperatures initially increase, water evaporates or sublimates from ocean or ice surfaces, increasing the cover of low clouds (made of small liquid water drops, which reflect solar radiation), increasing the effective Earth-atmosphere albedo, and decreasing downward solar radiation and temperatures.

12.4.2.9. Plant-Carbon Dioxide Negative Feedback

If air temperatures initially increase, plants and trees flourish and photosynthesize more, decreasing the quantity of carbon dioxide in the air, offsetting some of the original temperature increases.

The only practical way to elucidate the relative importance of the different feedback mechanisms is through computer modeling of the climate. To date, climate models accounting for feedbacks have found that increases in the atmospheric greenhouse gases increase global near-surface temperatures and reduce stratospheric temperatures, consistent with theory and observations described in Section 12.3.1.

12.4.3. Effects of Aerosol Particles on Climate

Although all greenhouse gases warm near-surface air, some aerosol particle components warm the air but most cool the air. The main particle components of warming particles are black carbon (BC), brown carbon (BrC), and soil dust components. Constituents in BrC that cause warming include primarily aromatic organic compounds, nitrated organic compounds, and tar balls. Constituents of soil dust that cause warming include iron and aluminum (Section 7.1.3.1). Cooling particles contain primarily sulfate, nitrate, ammonium, and liquid water, among other compounds (Section 12.1).

Warming aerosol particles absorb solar radiation, convert the sunlight to heat, and then reradiate the heat (thermal-IR radiation) to the air around them. Absorbing aerosol particles that become coated by other aerosol material (either nonabsorbing or absorbing) heat the air more than do uncoated absorbing particles. This enhancement is due to the **optical focusing effect**. When an absorbing aerosol particle becomes coated, the overall particle becomes larger, so more sunlight hits and refracts into the particle. The additional photons reflect internally within the particle multiple times, and many are ultimately absorbed by the absorbing core in the particle.

Absorbing aerosol particles can either be emitted with a coating or obtain a coating by internal mixing as they age. An example of an emitted coated particle is a jet fuel particle, which contains black carbon coated by lubricating oil, unburned fuel oil, sulfate, and some metals. **Internal mixing** occurs after emissions due either to the collision of two particles followed by their coalescence (coagulation) or to the diffusional transfer of a gas onto a particle surface followed by a change of state (condensation). Internal mixing of black carbon is important because internally mixed black carbon can cause twice as much global heating as **externally mixed** (uncoated) black carbon (Jacobson, 2000, 2001b).

All aerosol particle components absorb thermal-IR radiation, but thermal-IR absorption is weaker than solar absorption for those components that absorb solar radiation. Thermal-IR absorption by cooling particles is also weaker than is solar scattering by such particles. Thermal-IR absorption is most important when the size of a particle approaches the wavelength of thermal-IR radiation (e.g., 10 μm). Thus, good thermal-IR–absorbing particles are in the coarse aerosol size mode (Section 5.1). They primarily include soil dust, sea spray, and nitrate-containing and volcanic particles.

In sum, warming aerosol particles reduce sunlight to the ground by absorbing and converting it to heat that is released into the air around them. Cooling aerosol particles also reduce sunlight reaching the ground by reflecting incoming solar radiation back to space. Because both absorbing and scattering aerosol particles reduce solar radiation reaching the surface of the Earth, they all cause a **dimming** of the sunlight reaching the Earth.

A major issue is the extent to which warming from warming aerosol particles offsets cooling from cooling particles in the global and regional average. If only instantaneous radiative effects are considered and time-dependent effects are ignored, cooling caused by cooling particles exceeds warming by warming particles. Instantaneous radiative effects, however, tell only part of the story. A true estimate of the effect of aerosol particles on climate requires the consideration of **time-dependent climate responses**, summarized next.

12.4.3.1. Daytime Stability Effect and Aerosol Reduction of Wind

Aerosol particles that absorb solar radiation heat the air by absorbing sunlight and converting the light into thermal-IR (heat) radiation that is emitted back to the air around it. Although the absorbing particles are short lived, the heated air molecules last longer and are advected long distances. *Warming particles, which absorb sunlight, thus heat the air in a manner different from greenhouse gases, which absorb the Earth's thermal-IR radiation and reemit it to the air around them.*

Because warming particles heat the air relative to the ground, they increase the air's stability, reducing the vertical mixing of pollutants, energy, and other atmospheric properties (Section 6.6.1.2). Because cooling particles reflect downward solar radiation to space, they prevent the radiation from reaching the surface, cooling the surface relative to the air and increasing the air's stability (Bergstrom and Viskanta, 1973; Ackerman, 1977; Venkatram and Viskanta, 1977). In sum, *all aerosol particles increase the stability of the air.*

An increase in the stability of the air slows near-surface winds and increases wind speeds above the surface. The reason is that wind speeds generally increase with increasing height, starting with zero wind speed at the surface. The more stable the air, the less that air rises and sinks buoyantly. The rising and sinking of air is one form of **turbulence**. The less the vertical turbulence, the less that fast winds from above the surface mix toward the surface, slowing near-surface winds and increasing winds above the surface. As such, *all aerosol particles reduce surface near-wind speeds* (Jacobson and Kaufman, 2006).

The reduction in near-surface wind speed due to aerosol particles reduces the wind speed–dependent emission rates of sea spray, soil and road dust, pollens, spores, and gas-phase aerosol precursors. The reduction in the concentration of these particles reduces daytime solar reflectivity and day- and nighttime thermal-IR heating near the surface. As such, aerosol particles affect temperatures not only directly, but also by changing the emission rates, and thus concentrations, of other particles and gases. This feedback is referred to as the **daytime stability effect** (Jacobson, 2002a).

12.4.3.2. The Smudge Pot Effect

During the day and night, all aerosol particles trap the Earth's thermal-IR radiation, warming the surface air (Bergstrom and Viskanta, 1973; Zdunkowski et al., 1976). This warming is well known to citrus growers, who, at night, used to burn crude oil in smudge pots to fill the air with smoke and trap thermal-IR radiation, preventing their citrus crops from freezing. The warming of the air relative to a surface below increases the stability of air, slowing near-surface winds and increasing them aloft, reducing the wind speed–dependent

emission rates of other aerosol particles and gases in a manner similar to the daytime stability effect. The effect of thermal-IR absorption by aerosol particles on surface heating and emissions of other particles and gases, and the resulting feedback to climate, is the **smudge pot effect** (Jacobson, 2002a). Unlike the daytime stability effect, the smudge pot effect operates during the day and night but is relatively more important at night because it does not compete with the daytime stability effect.

12.4.3.3. Indirect Effects

All cloud drops form on top of aerosol particles. Aerosol particles that can potentially form cloud drops are **cloud condensation nuclei** (CCN). When pollution particles are emitted, many serve as CCN along with natural aerosol particles. For the same total liquid water content of a cloud, the addition of more CCN produces more small cloud drops and less large cloud drops. A larger number of small drops has a greater cross-sectional area, summed among all drops, than does a smaller number of large drops. The greater cross-sectional area of small drops in comparison with large drops means that adding CCN to the air increases the reflectivity of sunlight by the cloud (**first indirect effect**), cooling the ground during the day (Twomey, 1977).

Example 12.3
For a cloud liquid water content of 1 g m^{-3}, calculate the aggregate cross sectional area among cloud drops if the water were spread over equally sized drops with a concentration of (a) $n = 1,000$ drops cm^{-3} and (b) $n = 500$ drops cm^{-3}.

Solution
Dividing the cloud liquid water content by the number of drops in each case gives the volume of each drop as 10^{-9} cm^3 and 2×10^{-9} cm^3, respectively. The radii of spherical drops of these volumes are $r = 6.20$ and $r = 7.82$ μm, respectively. The summed cross-sectional area among all drops in each case is, therefore, $n\pi r^2 = 121,000$ μm^2 cm^{-3} and 96,000 μm^2 cm^{-3}, respectively. Thus, for the same mass of liquid water, a greater number of small drops has a larger summed cross-sectional area than does a lesser number of larger drops.

An increase in the number of small drops and a decrease in the number of large drops due to the addition of pollution particles also reduces the rate of drizzle in a low cloud, thereby increasing the liquid water content and fractional cloudiness of such a cloud (**second indirect effect**), further cooling the surface during the day (Albrecht, 1989; Gunn and Phillips, 1957). Thus, the indirect effects of anthropogenic aerosol particles are effects of their emission on clouds' reflectivity, lifetime, and precipitation, and, consequently, surface temperature.

Different particle components activate cloud drops to different degrees. For example, newly emitted diesel soot particles, which contain black carbon coated by lubricating oil, unburned fuel oil, and some sulfate and metals, do not serve as good CCN because they are largely hydrophobic when emitted. As they age, though, soot particles gradually become coated with sulfuric acid or another hygroscopic material, increasing their ability to serve as a CCN. Sodium chloride particles and sulfuric acid-ammonia-water particles, however, are good CCN upon emissions or formation.

12.4.3.4. The Semidirect Effect

Absorption of solar radiation by large liquid water or ice particles in a cloud increases the stability of air below the cloud, reducing the vertical mixing of moisture from the surface to the cloud and the relative humidity in the cloud, thinning the cloud (Nicholls, 1984). In fact, any decreases in the relative humidity near a cloud correlates with a decrease in low cloud cover (Bretherton et al., 1995; Klein, 1997). Similarly, absorbing aerosol particles below, within, or above a cloud warm the air relative to the surface, decreasing the near-cloud relative humidity and increasing atmospheric stability, both of which reduce cloud cover (e.g., Hansen et al., 1997; Ackerman et al., 2000). Reduced cloud cover increases sunlight reaching the Earth's surface, warming the surface in a positive feedback process called the **semidirect effect** (Hansen et al., 1997).

12.4.3.5. Cloud Absorption Effects

Aerosol particles enter cloud drops in two major ways. Cloud drops either form on top of aerosol particles (**nucleation scavenging**) or coagulate with aerosol particles interstitially between them (**aerosol-hydrometeor coagulation**). Thus, absorbing aerosol particles within clouds can be present either as **inclusions** within cloud drops or **interstitially** between cloud drops.

Effects from cloud heating due to the direct absorption of solar radiation by aerosol inclusions and interstitial aerosol particles within a cloud are **cloud absorption effects** (Jacobson, 2012). Absorption by drop inclusions has been used to explain in part why

data indicate that thick clouds often have a low albedo (Danielson et al., 1969). It has also been used to suggest through scaling arguments that the upper limit of the annual, global average of BC heating due to cloud absorption may be 1 to 3 W m^{-2}, depending on the position of the BC in cloud drops (Chylek et al., 1996). Global-scale calculations accounting for absorbing inclusions have suggested that it is a strong contributor to the global warming caused by BC (Jacobson, 2006, 2010b).

Figure 12.26 compares heating rates due to absorption by BC inclusions within cloud drops with heating rates due to absorption by the same total volume of BC intestinally between cloud drops and with heating rates due to BC in the clear sky (outside a cloud). It indicates that heating of the air by BC that is interstitial between cloud drops exceeds that due to BC in clear sky. Clear sky BC receives light primarily from direct rays of the sun. Interstitial BC, however, receives light scattered from many solar rays that hit cloud drops and are scattered within the cloud many times, eventually hitting the BC.

Figure 12.26 also shows that cloud heating by BC inclusions in cloud drops exceeds heating by BC interstitially between drops. The reason is that BC inclusions receive not only light scattered by cloud drops, but also light that refracts into the drop in which they reside. Light refracted into a drop internally reflects multiple times, increasing the chance that it will hit and be absorbed by BC in the drop.

Particle inclusions can be represented in different ways within a cloud drop. Figure 12.26b illustrates results from two representations. One is the assumption that BC exists as several randomly distributed inclusions within each drop. This representation is known as the **dynamic effective medium approximation** (DEMA). The second is the assumption that BC exists as a single core surrounded by a shell of water. This representation is referred to as the **core shell approximation (CSA)** The DEMA is arguably a more physical representation than is the CSA for treating inclusions in clouds drops because BC can enter drops by both nucleation scavenging and aerosol-hydrometeor coagulation; thus, multiple BC inclusions usually exist in one drop. The CSA assumes that only one BC inclusion exists in each drop. Figure 12.26b indicates that the multiple inclusion treatment of BC heats the cloud more than does the single-core treatment.

The effect of absorbing aerosol inclusions within and between cloud drops is to darken and thus heat a cloud, causing it to burn off more rapidly. The darkening of

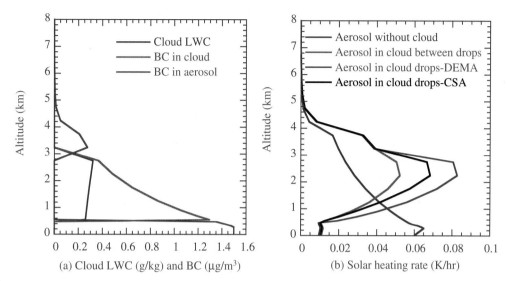

Figure 12.26. (a) Cloud liquid water content (LWC), clear sky black carbon (BC) concentration, and in-cloud BC concentration as a function of height for the calculations in (b). (b) Modeled solar heating rates for four cases: (i) aerosol particles containing BC in the clear sky ("aerosol without cloud") (in this case, the BC profile in (a) is the same as the cloudy sky BC profile); (ii) the same aerosol distribution as in (i) but, in this case, interstitially between cloud drops, above the cloud, and below the cloud ("aerosol in cloud between drops"); (iii) the same total volume of aerosol BC as in the other cases but with the BC treated as inclusions randomly distributed within cloud drops using the dynamic effective medium approximation (DEMA) of Chylek et al. (1984) ("aerosol in cloud drops-DEMA"); and (iv) the same as (iii) but with the aerosol BC treated as inclusions within cloud drops using the core shell approximation (CSA) of Ackerman and Toon (1981) ("aerosol in cloud drops-CSA") From Jacobson (2012).

clouds by biomass burning aerosol particles in Brazil has been observed (e.g., Kaufman and Nakajima, 1993). The effect on global climate of absorbing aerosol inclusions within clouds has also been calculated to be significant (Jacobson, 2006, 2010b).

12.4.3.6. Combined Indirect, Semidirect, and Cloud Absorption Effects

The indirect, semidirect, and cloud absorption effects are tightly coupled and must be considered together when examining the effects of aerosol particles on climate. This is illustrated in Figure 12.27, which shows the change in **cloud optical depth (COD)** with increasing **aerosol optical depth (AOD)**, obtained separately from satellite data and a computer model. COD and AOD are measures of how much cloud and aerosol particles, respectively, redirect or reduce sunlight due to scattering plus absorption of light.

Figure 12.27 shows that, at low AOD, an increase in AOD increases COD. This is because an increase in AOD is generally caused by an increase in the number of CCN, and an increase in the number of CCN triggers the first and second indirect effects, resulting

in more reflective clouds (a greater cross-sectional area summed over all cloud drops) and hence a higher COD. However, as AOD continues to increase, the semidirect

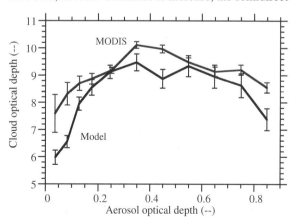

Figure 12.27. Boomerang curves showing the correlation between aerosol optical depth and cloud optical depth over a biomass burning region of Brazil during September, obtained both from MODIS Aqua satellite data and separately from a three-dimensional computer model, GATOR-GCMOM. The hash marks are uncertainty bars. From ten Hoeve et al. (2011).

and cloud absorption effects become more important, burning off clouds. In fact, at high AOD, the semidirect and cloud absorption effects overpower the indirect effects, decreasing COD with increasing AOD. The shapes of the curves in Figure 12.27 are similar to that of a boomerang, so the curves are referred to as **boomerang curves** (Koren et al., 2004; Kaufman and Koren, 2006; ten Hoeve et al., 2011).

12.4.3.7. Effect of Airborne Absorbing Particles on Surface Albedo

Black carbon, brown carbon, and soil dust in the air above snow, sea ice, clouds, and other reflective surfaces not only absorb downward sunlight, but also absorb light reflected upward from these surfaces. The warming of the air due to particle absorption in such cases feeds back to melt snow and ice, uncovering darker surfaces below, increasing temperatures in a positive feedback (Jacobson, 2002a). This enhanced warming is most efficient at high latitudes, where more snow and sea ice exist, rather than at lower latitudes. It is also most efficient where snow and sea ice cover are thin.

12.4.3.8. Snow Darkening Effect

When aerosol particles containing absorbing material deposit onto snow and sea ice, they reduce the albedo or reflectivity of the surface, increasing the solar heating of the surface (e.g., Hansen and Nazarenko, 2004; Jacobson, 2004; Flanner et al., 2007). If snow or sea ice thins or completely melts as a result, the darker surface underneath (land or ocean) is revealed, triggering an even greater warming of the surface in a positive feedback loop, referred to as the **snow darkening effect**.

Aerosol particles reach snow and sea ice surfaces primarily by **wet deposition** (the formation of snow and rain particles on top of aerosol particles, followed by the deposition of snow and rain to the surface, and the scavenging of aerosol particles by snow and rain as they are falling; Section 5.3.3). On a global scale, a small fraction of particles (<10 percent) (Jacobson, 2010b) also fall to the surface by their own weight or by winds driving them to the surface (**dry deposition**; Section 5.3.3).

12.4.3.9. Self-Feedback Effect

When aerosol particles are emitted, they change air temperature and thus the relative humidity. For example, warming aerosol particles decrease the relative humidity. Because the uptake of liquid water by aerosol particles (hydration; Section 5.3.2.3) increases with increasing relative humidity, a decrease in the relative humidity causes most aerosol particles to shrink, reducing their surface area and liquid water content. As such, less gas, such as sulfuric acid, nitric acid, and ammonia, condenses on or dissolves in the aerosol particles. The resulting reduction in the size of particles increases the penetration of solar radiation to the surface in a positive feedback referred to as the **self-feedback effect** (Jacobson, 2002a).

12.4.3.10. Photochemistry Effect

By intercepting UV light in the air, aerosol particles alter photolysis rates of gases, affecting their concentrations and the subsequent chemical production or destruction of other gases. Because many gases absorb UV radiation, changing the concentration of some gases by changing the photolysis rates of other gases affects temperatures. The process by which aerosol particles change photolysis coefficients, thereby affecting temperatures, is the **photochemistry effect** (Jacobson, 2002a).

12.4.3.11. Particle Effect through Large-Scale Meteorology

Aerosol particles affect local temperatures, which thus affect local air pressures, winds, relative humidities, and clouds. Changes in local meteorology slightly shift the locations and magnitudes of semipermanent and thermal pressure systems and jet streams. The effect of local particles on large-scale temperatures is the **particle effect through large-scale meteorology**.

12.4.3.12. Rainout Effect

When aerosol particles reduce precipitation due to the second indirect effect, the semidirect effect, or the cloud absorption effect, they reduce the removal of aerosol particles by wet deposition, increasing the concentration of aerosol particles in the air. The resulting climate effect of this process is referred to as the **rainout effect** (Jacobson, 2002a).

12.5. Consequences of Global Warming

Projections suggest that carbon dioxide mixing ratios may increase from about 393 ppmv in 2011 to 730 to 1,040 ppmv in 2100. During this period, global near-surface temperatures may increase by a mean estimated range of 1.8 K to 4 K, depending on the future emission scenario assumed (IPCC, 2007). The possible consequences of additional global warming are discussed next.

12.5.1. Loss of Ice/Rise in Sea Level

Increases in temperatures affect sea levels in at least two ways. First, higher temperatures enhance the melting of sea ice, ice sheets, and glaciers, adding water to the oceans. Second, because liquid-water density decreases with increasing temperature, higher temperatures cause sea water to expand and sea levels to rise. Historical changes in global temperature correlate with changes in sea levels. When the Earth's temperature peaked during the mid-Cretaceous period 120 to 90 m.y.a., the Earth's polar caps melted, sea levels rose to unprecedented levels, and 20 percent of continental land flooded.

Today, snow, sea ice, and glaciers cover 3.3 percent of the Earth's total surface area. The volume of ice worldwide in all forms is about 25 million km^3, representing about 70 m of sea level, or just less than 2 percent of the average worldwide ocean depth of 3,800 m. Ice is stored in the East and West Antarctic ice sheets, the Greenland ice sheet, sea ice over the Arctic, the large valley and piedmont glaciers of southeast Alaska, and the glaciers of central Asia. About 55 m of sea level is stored in the East Antarctic ice sheet, about 5 m in the west sheet, about 7 m in the Greenland ice sheet, and about 3 m in the remaining ice. If all this ice melts, the *sea level will rise by about 70 m* above its current level. This will result in about *7 percent of the current world's land area that is above sea level being below sea level*.

During the twentieth century, sea levels rose 1.8 mm yr^{-1}; however, from 1993 to 2009, the rate increased to 3.4 ± 0.4 mm yr^{-1}. If this pace continues, sea levels will rise by 31 cm by 2100, putting about 0.015 percent of the current world's land below sea level. If warming accelerates, sea levels will rise more.

The **Mendenhall Glacier**, Alaska (Figure 12.28a), provides an example of the rate of receding glaciers. The glacier, located about 19 km from downtown Juneau, is about 19 km long. The terminus of the glacier receded about 580 m from 1951 to 1958 and another 2.82 km from 1958 to 2011. Many other glaciers, such as the beautiful **Grey Glacier**, Chile (Figure 12.28b), are also at risk. The Greenland ice sheet, which covers 1.71 million km^2, is also dwindling.

Although the melting of sea ice, ice sheets, and glaciers and the corresponding rise in sea level are of concern, a worst-case scenario sea level rise is unlikely to occur soon because the largest source of ice, the East Antarctic ice sheet, based over land, is stable right now. In fact, the East ice sheet may be increasing in size currently because of an increased supply of water vapor to the sheet resulting from higher global temperatures (Bentley, 1984). Extended global warming could reverse this trend and ultimately cause a collapse of the East ice sheet. Such a process was originally likely to take thousands of years (Crowley and North, 1991) but could occur sooner if temperatures rise faster than anticipated (e.g., Figure 12.28c).

The West Antarctic ice sheet, based over water, is less stable than is the East ice sheet (Stuiver et al., 1981). If extreme global warming occurs, the West ice sheet, as a result of its relative instability, is likely to collapse much more quickly than is the East ice sheet.

Of more immediate concern is the Arctic sea ice, which suffered a 32 percent loss in area in August 2010 relative to the 1979 to 2008 mean (Polar Resource Group, n.d.). The reduction in sea ice uncovers the low-albedo Arctic Ocean surface below, accelerating global warming in a positive feedback loop.

The loss of sea ice also reduces the survival rate of polar bears. The **polar bear**, which lives within the Arctic Circle, fishes from the sea, hunting primarily from the edge of sea ice (Figure 12.28d). During the summer, when low-latitude sea ice melts, polar bears migrate farther north to where the ice persists. The accelerated loss of sea ice during summer and winter forces polar bears to swim longer distances to the source of their food, depleting their energy stores, increasing their risk of drowning, reducing their reproduction rates, and reducing their survival rates.

The main effect of sea level rise, even in small quantities, is the flooding of low-lying coastal areas and the elimination of a few flat islands that lie just above sea level. Bangladesh, the most densely populated country in the world, is particularly at risk. A 1-m rise in sea level would displace about 17 million people from their homes there. New Orleans, Louisiana, which already lies below sea level (Figure 12.29a), would similarly face a danger of flooding, as it did following Hurricane Katrina. **Tuvalu** is a chain of nine coral atolls in the South Pacific Ocean, about halfway between Hawaii and Australia. Tuvalu has a total land area of 26 km^2, about 0.1 times the size of Washington, DC, a coastline that stretches for 24 km, and a population of about 10,000. An increase in sea level of 2 m could eliminate the country. Such a sea level rise would similarly drown small islands such as those in the San Blas Archipelago, a Caribbean chain of 400 small islands off the coast of Panama (Figure 12.29b).

(a) (b)

(c) (d)

Figure 12.28. (a) Glacier in Mendenhall National Park, Alaska, June 5, 2010. Alita Bobrov/Dreamstime.com. (b) Turquoise-blue textured ice found in Grey Glacier, Torres Del Paine National Park, in the Southern Patagonian ice field of the Andes Mountains, Chile, February 22, 2009. © Achim Baque/Dreamstime.com. (c) Iceberg with window in Antarctica, February 29, 2008. © Achim Baque/Dreamstime.com. (d) "The Last Polar Bear" on a shrinking ice floe in the Arctic Ocean. © Jan Martin Will/Dreamstime.com.

12.5.2. Changes in Regional Climate, Severe Weather, and Agriculture

Global warming is causing regional and temporal climate variations. The number of extremely hot days is increasing, and the number of extremely cold days is decreasing. By increasing evaporation, global warming is *increasing precipitation particularly over land*. Precipitation changes over the ocean are more modest by comparison. In some cases, higher precipitation is in the form of more snow, resulting from more water vapor at those altitudes in the atmosphere where temperatures are below the freezing point. Global warming is also increasing droughts in some areas and floods in others (IPCC, 2007).

Higher sea-surface temperatures due to global warming *increase hurricane intensity but not necessarily their number*. Data analysis suggests a correlation between higher sea-surface temperatures and increased hurricane intensity in the North Atlantic and western North Pacific oceans (Emanuel, 2005) and an increase in the number of the most intense hurricanes (Categories 4 and 5) since the 1970s (Webster et al., 2005). However, changes in the overall number of hurricanes appear to be due more to a natural cycle than anthropogenic influence.

Changes in regional climate cause a *shift in the location of viable agriculture*. Although some crops now flourish in areas that were once too cold or too dry, others have died in regions that have become too hot or

(a)   (b)

Figure 12.29. (a) New Orleans, Louisiana, a city that lies below sea level. © Olivierl/Dreamstime.com. (b) An island in the San Blas Archipelago, a Caribbean chain of nearly 400 small islands where native Kuna Indians live semiautonomously from Panama, their host country. © Erikgauger/Dreamstime.com.

too wet. Citrus crops, in particular, require a certain number of cooling degree days to survive. Many citrus farms are no longer viable in their original location due to winter temperature increases that have already occurred.

Because plants grow faster when temperatures, carbon dioxide levels, or water vapor levels mildly increase (plant–carbon dioxide negative feedback), some forms of agriculture are expected to flourish in areas where only mild increases in temperature and moisture occur. In areas where extreme temperature increases occur, agriculture will die out. Of particular concern are subtropical desert regions of Africa, where temperatures are already high. In these regions, agriculture is subject to the whims of the climate, and millions of people depend on the local food supply. *Small increases in temperature could trigger famine* in these areas, as has occurred in the past.

12.5.3. Changes in Ocean Acidity and Ecosystems

The increase in atmospheric $CO_2(g)$ increases the amount of dissolved and dissociated carbonic acid in the ocean by Reactions 3.15 and 3.16. The dissociation of carbonic acid increases the H^+ concentration in the ocean, decreasing the ocean pH, resulting in **ocean acidification**. The pH of the ocean today is above 8 because of the high concentrations of cations (Na^+, Ca^{2+}, Mg^{2+}, K^+) in ocean water. Due to $CO_2(g)$ buildup in the atmosphere since the Industrial Revolution, the pH of the ocean declined from about 8.25 to 8.14 between 1751 and 2004 (Jacobson, 2005b). If

atmospheric $CO_2(g)$ continues to increase unabated, the pH is estimated to decline to about 7.85 by 2100. This pH decrease represents a factor of 2.5 increase in the H^+ concentration between 1751 and 2100.

Ocean acidification damages **coral reefs**, which are made of calcium carbonate secreted by corals. **Corals** consist of living polyps, attached at their base to the coral reef, that cluster in groups and secrete calcium carbonate to form a solid exoskeleton. The greater the acidity of water, the more readily calcium carbonate dissolves by the reverse of Reaction 3.17.

Between 1990 and 2008, the coral growth of the Great Barrier Reef off Australia declined by about 13 percent, much greater than during any time in the past 400 years (De'ath et al., 2009). Ocean acidification similarly makes it more difficult for noncoral sea life to form and maintain shells from calcium carbonate. It also directly damages fish that are accustomed to a narrow pH range.

Over land, rapid, continuous increases in temperature will lead to the extinction of some species that are accustomed to narrow climate conditions and are unable to migrate faster than the rate of global warming.

Although enhanced $CO_2(g)$ levels generally invigorate forests, sharp increases in temperature associated with higher $CO_2(g)$ can lead to forest dieback in tropical regions, affecting the rates of $CO_2(g)$ removal by photosynthesis and emission by respiration.

12.5.4. Changes in Heat Stress

Global warming affects human health by increasing heat stress. In warm climates, higher temperatures increase

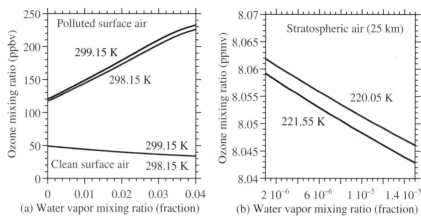

Figure 12.30. Ozone mixing ratio as a function of water vapor mixing ratio under (a) polluted and clean surface conditions and (b) stratospheric conditions at 25 km. For each condition, two temperatures are shown, a base temperature (blue line) and a temperature 1 K higher than the base temperature (red line). For clean air surface conditions, the two lines are nearly on top of each other. Results were obtained considering chemistry alone over (a) a 12-hour daytime period and (b) a 36-hour period. From Jacobson (2008a).

heat stress–related health problems, including mortality, more than they do in milder climates (e.g., Medina-Ramon and Schwartz, 2007). People currently living in cold climates are likely to experience little or no heat stress. Heat-related health problems, such as heat rash, heat stroke, and death, generally affect the elderly and those suffering from illnesses more than they affect the general population.

12.5.5. Changes in Disease

Increases in land precipitation as a result of global warming increase the populations of mosquitoes and several other insects that carry disease. Furthermore, because **malaria** is not transmitted below a certain temperature, rising temperatures will increase the spread of malaria to places previously too cold for it to thrive, including higher latitudes and mountains. Similarly, **influenza** occurs year-round in the tropics because of the warm climate. Higher year-round temperatures at higher latitudes will likely lengthen the flu season. In addition, drought in rural areas may drive populations toward cities, where diseases transmit more readily than in rural areas.

12.5.6. Changes in Air Pollution

Most major greenhouse gases, including carbon dioxide, methane, nitrous oxide, and CFCs, do not cause direct harmful human health problems at normal ambient mixing ratios (Section 3.6). However, other greenhouse gases, such as ozone and carbon monoxide, do.

In addition, soot particles containing black carbon are dangerous to health. However, global warming itself affects the concentrations of health-affecting air pollutants, such as ozone and particulate matter.

Because greenhouse gases are long lived, they eventually become well mixed in the global atmosphere. Near emission sources, such as over cities, though, greenhouse gas mixing ratios are higher than they are in the global atmosphere because emission rates over cities exceed dilution rates to the global environment. In the case of carbon dioxide, the high mixing ratio over a city is referred to as a **carbon dioxide dome**. $CO_2(g)$ domes enhance the formation rate of air pollution (Jacobson, 2010a). Methane can also form a dome over a city and have a similar impact.

Both well-mixed greenhouse gases away from cities and carbon dioxide domes over cities increase not only tropospheric temperatures, but also water vapor due to evaporation caused by the higher temperatures. *If chemistry alone is considered, higher temperatures and higher water vapor both independently increase surface ozone in polluted air but cause little change in surface ozone in clean air* (Jacobson, 2008a), as illustrated in Figure 12.30a.

To illustrate, a 1 K rise in air temperature increases ozone by about 6.7 ppbv at 200 ppbv ozone, but the same temperature rise increases ozone by only about 0.1 ppbv at 40 ppbv ozone when only chemistry's effects on ozone are considered. The reason that high temperatures increase ozone more in polluted air than in clean air is the fast increase in the thermal dissociation of PAN to nitrogen dioxide and an organic radical with increasing temperature. For example, at high temperature, PAN thermally dissociates by the reverse of

$$CH_3C(=O)O_2(g) + NO_2(g) \Leftrightarrow CH_3C(=O)O_2NO_2(g)$$

Peroxyacetyl radical   Nitrogen dioxide   Peroxyacetyl nitrate (PAN)

(12.8)

whereas, at low temperature, the formation of PAN (forward reaction) is favored. When PAN thermally dissociates, it produces an organic radical and nitrogen dioxide, both of which produce ozone. The thermal dissociation at high temperature is more important in polluted air than it is in clean air because, in clean air, PAN mixing ratios are low; thus, less $NO_2(g)$ and organics are available for release upon a temperature increase. The general increase in surface ozone with higher temperatures is supported not only by data analysis (e.g., Olszyna et al., 1997), but also by many computer model studies (e.g., Sillman and Samson, 1995; Mickley et al., 2004; Stevenson et al., 2005; Steiner et al., 2006; Brasseur et al., 2006; Kleeman, 2008; Jacobson, 2008a, 2010a).

Higher temperatures due to increases in carbon dioxide and other global warming agents also increase evaporation of ocean, lake, and soil water to produce water vapor. An increase in water vapor increases surface ozone in polluted air but can slightly reduce ozone in clean air, as illustrated in Figure 12.30a. In both clean and polluted air, an increase in water vapor increases the hydroxyl radical, OH(g), by

$$H_2O(g) + O(^1D)(g) \Leftrightarrow 2OH(g)$$

Water vapor   Excited atomic oxygen   Hydroxyl radical

(12.9)

Much of the OH(g) converts to $HO_2(g)$ by reaction of OH(g) with aldehydes and other organic gases.

In polluted surface air, $NO_x(g)$ levels are high, so the resulting $HO_2(g)$ reacts with NO(g) by

$$NO(g) + HO_2(g) \Leftrightarrow NO_2(g) + OH(g)$$

Nitric oxide   Hydroperoxy radical   Nitrogen dioxide   Hydroxyl radical

(12.10)

This reaction increases the $NO_2(g):NO(g)$ ratio, increasing ozone. As such, increases in water vapor increase ozone in polluted surface air (Figure 12.30a).

However, in clean surface air, $NO_x(g)$ levels are low, so $HO_2(g)$ reacts primarily with itself by

$$HO_2(g) + HO_2(g) + M \Leftrightarrow H_2O_2(g) + O_2(g) + M$$

Hydroperoxy radical   Hydroperoxy radical   Hydrogen peroxide   Molecular oxygen

(12.11)

The rate coefficient of this reaction is proportional to the concentration of water vapor; thus, the more water vapor, the faster the reaction and the faster the loss

of $HO_2(g)$. Because $HO_2(g)$ decreases with increasing $H_2O(g)$ through Reaction 12.11 when $NO_x(g)$ is low, the NO(g) that is available under such conditions produces little $NO_2(g)$ by Reaction 12.10 with increasing $H_2O(g)$. As such, ozone decreases slightly with increasing $H_2O(g)$. In addition, because OH(g) increases with increasing $H_2O(g)$ by Reaction 12.9, OH(g) speeds the conversion of existing $NO_2(g)$ to organic nitrates under low $NO_x(g)$ conditions, decreasing the $NO_2(g):NO(g)$ ratio, thereby decreasing ozone.

Temperature and water vapor affect surface ozone through other processes aside from chemistry. For example, higher temperatures increase the emission rates of biogenic organic gases, such as isoprene and monoterpenes, both of which increase ozone. Higher temperatures also increase electric power demand for air conditioning and, thus, summer daytime emissions from the energy sector. Conversely, warmer nighttime and winter temperatures reduce natural gas and electricity usage for heating, offsetting some of the increased daytime and summer increases. Third, higher temperatures increase carbon monoxide and hydrocarbon emissions from vehicles (e.g., Rubin et al., 2006; Motallebi et al., 2008;). In contrast, warmer winter and nighttime temperatures tend to offset some of the higher temperature increases (U.S. EPA, 2006).

High temperatures can increase not only ozone, but also particulate matter. First, higher temperatures increase the rates of wildfire occurrence due to dryer conditions in many locations (Westerling et al., 2006). Wildfires are a major source of particulate matter (Jaffe et al., 2008).

Second, the increase in biogenic organic gas emissions that occurs when temperatures rise increases the rate of secondary organic matter particle formation.

Third, higher water vapor due to higher temperatures can increase the relative humidity in some locations, whereas higher temperatures can decrease the relative humidity in others. An increase in the relative humidity increases the uptake of liquid water onto aerosol particles, increasing the surface area on which condensation of gases occurs and the volume in which dissolution of other gases occurs. Thus, an increase in the relative humidity due to global warming can increase particulate matter, whereas a decrease can cause the reverse. Similarly, higher temperatures can reduce particulate matter by evaporating some organic and inorganic aerosol constituents to the gas phase (Aw and Kleeman, 2003).

Fourth, global warming is usually characterized by an increase in surface air temperatures and a slightly lesser increase in ground temperatures. This increase in stability decreases vertical dilution of pollution, which increases the near-surface concentration of pollution, including particulate matter.

Human mortality due to enhanced air pollution from global warming can be estimated, but with uncertainty. One estimate, derived from model simulations, is that a temperature rise of 1 K causes ~1,000 (350–1,800) additional deaths per year in the United States (a 1.1 percent increase in the death rate due to all air pollution in the United States), with about 40 percent of the additional deaths due to ozone and the rest due to particulate matter (Jacobson, 2008a). A simple extrapolation from U.S. to world population (from 301.5 to 6,800 million in 2011) gives 22,250 (7,600–40,200) additional deaths per year worldwide per 1 K temperature rise due to $CO_2(g)$ above the baseline air pollution death rate.

12.5.7. Changes in Stratospheric Ozone

Whereas greenhouse gases warm the troposphere, they cool the stratosphere (Figure 12.14). The reason is that greenhouse gases in the troposphere prevent significant thermal-IR radiation emitted by the Earth's surface from reaching the stratosphere, where such radiation would otherwise be absorbed by ozone and background carbon dioxide, warming the stratosphere. Black carbon and brown carbon particles also warm the troposphere by absorbing solar radiation, but tropospheric BC and BrC have relatively little temperature effect in the stratosphere because of their relatively modest absorption of the Earth's thermal-IR radiation (Jacobson, 2002a, 2010b). However, aircraft emissions of black carbon go directly into the stratosphere at high latitudes, particularly over the Arctic. Such emissions affect Arctic stratospheric and tropospheric temperatures.

Tropospheric warming due to greenhouse gases and absorbing aerosol particles increases the evaporation of water from the oceans and soils, and some of this water vapor reaches the stratosphere. Between 1954 and 2000, for example, stratospheric water vapor increased 1 percent per year (0.45 ppmv per decade; Rosenlof et al., 2001). This trend reversed itself slightly in the lower stratosphere from 2001 to 2005, possibly due to changes in atmospheric circulation (Randel et al., 2006). However, stratospheric water vapor appears to have increased again from 2006 to 2010 (Hurst et al., 2011). Stratospheric cooling and water vapor increase affect the ozone layer in at least four ways.

First, in the stratosphere (at 25 km), a decrease in temperature due to tropospheric global warming increases ozone when only the temperature dependence of gas chemistry is considered (Figure 12.30b). As stratospheric temperature decreases, ozone increases, due primarily to the slower loss rate of ozone at higher temperature by the reaction $O(g) + O_3(g)$ (Evans et al., 1998). In addition, although most reactions proceed more slowly when temperature decreases, the reaction $O(g) + O_2(g) + M \rightarrow O_3(g) + M$ occurs more rapidly when temperature decreases. Thus, *when only temperature effects on chemistry are considered, a cooling of the stratosphere slightly increases global stratospheric ozone.*

Second, as water vapor increases in the stratosphere due to global warming, $OH(g)$ increases by Reaction 12.9, accelerating the $HO_x(g)$ catalytic ozone destruction cycle (Reactions 11.16–11.18) and decreasing ozone (Figure 12.30b) (e.g., Dvortsov and Solomon, 2001). Thus, *when only the effect of the increase in stratospheric water vapor on chemistry is considered, global warming slightly destroys stratospheric ozone.* This loss mechanism of ozone with increasing water vapor is important in the stratosphere, but not at the surface.

Third, stratospheric cooling decreases the saturation vapor pressure of water, allowing more water vapor to condense onto stratospheric sulfuric acid-water aerosol particles, causing them to grow larger. The increase in the size of these aerosol particles increases the rates of chemical reaction on their surfaces. Because such reactions convert CFC and HCFC by-products to more active chlorine gases that photolyze to products that destroy ozone, *a decrease in stratospheric temperature reduces global stratospheric ozone when only the effect of cooling on aerosol particle size is considered.*

Fourth, stratospheric cooling increases the occurrence, size, and lifetime of PSCs (Section 11.8.1). Type I PSCs form below 195 K, and Type II PSCs form below 187 K. Stratospheric cooling increases the frequency of temperatures below these critical levels during winter, increasing Types I and II PSC lifetime and size, enhancing polar ozone loss. *Some polar ozone loss to date is due to increased PSC formation caused by stratospheric cooling that accompanies near-surface global warming.*

In sum, stratospheric cooling resulting from near-surface global warming has opposing effects on global stratospheric ozone, but causes a net destruction of

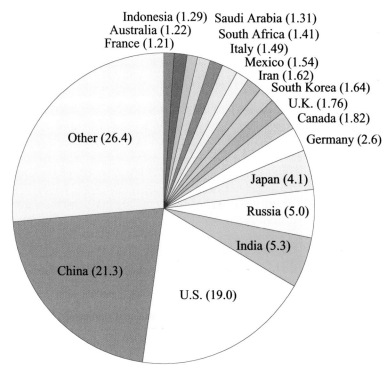

Figure 12.31. Percentage of world carbon emissions by country or continent, 2007. Total emissions were 8,365 million metric tonnes of carbon (not carbon dioxide). Data from Boden et al. (2011).

ozone over the Antarctic and Arctic. Although emissions of CFCs are decreasing and global and Antarctic stratospheric ozone losses are expected to recover during the next 50 years or so, stratospheric cooling due to global tropospheric warming will delay the recovery by one to two decades or more.

12.6. Regulatory Control of Global Warming

Global warming is an international, national, and local problem. All nations emit greenhouse gases and soot particles. Figure 12.31 shows that the top three countries emitting carbon dioxide in 2007 were China, the United States, and India. China's share of the world total increased from 13.9 to 21.3 percent between 1997 and 2007.

Figure 12.32 shows carbon emissions by country and per capita in 2007. Per capita emissions were generally highest in oil-producing countries, particularly Qatar, the United Arab Emirates, and Kuwait (not shown). Yet, the populations of these countries are small, so the total emission from them is also relatively small. The United States stands out as having a high per capita emission rate and a large total emission rate.

12.6.1. Indirect Regulations

Global warming is a scientific issue, but its control has economic ramifications, causing it to be a divisive political issue. Many industries and energy companies currently rely on combustion of fossil fuels for their viability, so they resist regulations that might increase their costs or cause them to go out of business. They resist changes even though total costs to society as a whole, which include air pollution and climate costs, would decrease if fossil fuels were eliminated in favor of cleaner energy sources (Chapter 13).

Furthermore, many newly industrialized nations find that increasing the use of fossil fuels is the easiest method of expanding their economies. However, an expansion of an economy with fossil fuel use comes at the price of higher air pollution, health costs, climate costs, and other environmental costs, with no demonstrable benefit in job creation because jobs would be created in clean industries as well.

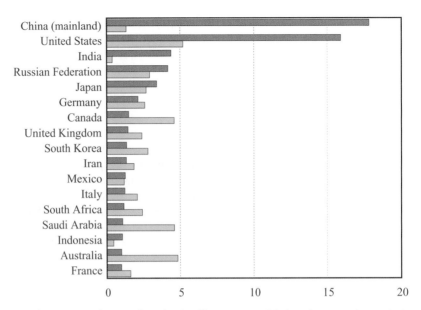

Figure 12.32. National emissions (brown, hundred-million tonnes-C/yr) and per capita emissions (green, tonnes-C/person/yr) of carbon (C) in carbon dioxide in 2007 from the top seventeen countries in terms of total emissions. Data from Boden et al. (2011).

Unlike regulations addressing urban air pollution, acid rain, and global ozone loss, regulations addressing global warming have been relatively weak. Worldwide, vehicle emission standards for CO(g), BC, and ozone precursors (e.g., Table 8.1) based on health grounds have unintentionally but fortuitously reduced some of the pollution causing global warming. Also, **CAFÉ standards** (Section 8.1.8) in the United States have indirectly reduced some carbon dioxide emissions from the transportation sector.

U.S. federal tax code incentives since the late 1970s for the development of renewable energy and improvements in energy efficiency have also indirectly addressed the issue. Other renewable energy incentives, such as the **feed-in-tariff** (Section 13.10) in Germany and approximately thirty other countries starting in 2000, similarly provided a small benefit. Nevertheless, simultaneous tax incentives have existed for the development of fossil fuel energy sources in many countries. In the 1980s, tax incentives for clean energy sources in the United States were severely reduced, whereas those for fossil fuels were enhanced.

Although all countries tax fuels to some extent, Denmark, the Netherlands, Finland, Norway, and Sweden specifically implemented carbon taxes on fuel in the mid-1990s. Denmark taxed all carbon dioxide emission sources, except gasoline, natural gas, and biofuels. The Netherlands taxed all energy sources used as fuel, and Finland taxed all fossil fuels. Norway taxed mineral oil, gasoline, gas burned in marine oil fields, coal, and coke. Sweden taxed oil, kerosene, natural gas, coal, coke, and other sources. Nevertheless, these efforts have done little to stop the rapid growth in greenhouse gas and particle black carbon emissions worldwide.

12.6.2. The Kyoto Protocol

On May 9, 1992, an international agreement addressing global warming, hashed out in Rio de Janeiro, Brazil, was adopted at the United Nations. The agreement, the **United Nations Framework Convention on Climate Change**, called on signatory nations to develop current and projected emission inventories for greenhouse gases, devise policies (to be implemented at a later meeting) for reducing greenhouse gas emissions, and promote technologies for reducing emissions. By 1994, 184 nations had signed the agreement, and most had ratified it. In 1995, the nations involved in the convention met in Berlin, Germany, to discuss details of the proposed policies and target dates for implementing them.

In December 1997, the nations met again for an 11-day conference in Kyoto, Japan, to finalize the policies proposed at the Berlin meeting. The conference resulted in the **Kyoto Protocol**, an international agreement designed to fight global warming by controlling

Table 12.5. Percentage change in emissions allowed for industrialized countries under Kyoto protocol

Country(ies)	Percentage change in emissions
Switzerland, central Europe, European Union	−8
United States	−7
Canada, Hungary, Japan, Poland	−6
Russia, New Zealand, Ukraine	0
Norway	+1
Australia	+8
Iceland	+10

the emission of greenhouse gases. According to the agreement, industrialized countries were required to reduce greenhouse gas emissions for the first commitment period (2008–2012). Such gases included carbon dioxide, methane, nitrous oxide, hydrofluorocarbons (HFCs), perfluorocarbons (PFCs), and sulfur hexafluoride [$SF_6(g)$]. Reductions in $CO_2(g)$, $CH_4(g)$, and $N_2O(g)$ emissions would be relative to 1990 emissions. Reductions in the emission of others gases would be relative to either 1990 or 1995 emissions. Emissions reductions did not apply to emissions from international shipping and aviation.

Table 12.5 lists the allocation of emissions reductions originally mandated for industrialized countries. Some countries were allowed to increase emissions. The net change in emissions, weighted over all industrialized countries, was 5.2 percent. Countries required to reduce emissions were allowed to meet the reductions in one of many ways. One mechanism was to finance emission reduction projects in developing countries, which were not subject to emission limits. This mechanism is referred to as **emission trading**. Tree planting, protecting forests, improving energy efficiency, using renewable energy sources, reforming energy and transportation sectors, creating technologies with zero emissions, and reducing emissions at their sources with existing technologies were other mechanisms.

If a country that ratifies the protocol does not meet its emission reduction commitment on time, the country is penalized by being required to reduce emissions by an additional 30 percent beyond that of its original commitment. The country is also suspended from using an emission trading program to make emission transfers with another country.

By the end of April 2010, 191 nations had signed and ratified the Kyoto Protocol. One country, the United States, had signed but not ratified the protocol. Although the United States had signed the protocol in 1997, it chose not to ratify the protocol in 2001. One rationale for the pullout was that controlling $CO_2(g)$ emissions would damage the U.S. economy. The same argument was made by the automobile industry prior to the passage of the Clean Air Act Amendments of 1970, which required 90 percent reductions of three pollutants from automobiles by 1976: $CO(g)$, $NO_x(g)$, and hydrocarbons. Despite opposition to them, the 1970 amendments motivated U.S. automobile manufacturers to invent the catalytic converter by 1975. The invention not only eventually reduced pollutant emissions as mandated, but also produced profitable patents. Neither the U.S. economy nor the automobile industry suffered as a result of the 1970 regulations. From 1970 to 2000, for example, the number of vehicles in the United States doubled, whereas the population increased by only one-third. The U.S. gross domestic product in fixed dollars also doubled, and the unemployment rate decreased from 4.9 to 4.0 percent during this period. Stringent air pollution regulations under the Clean Air Act Amendments had no overall detrimental effect on the U.S. economy.

Since the Kyoto Protocol, many of the countries that promised to reduce emissions have done so, but others have increased emissions. As indicated in Figure 12.9, global $CO_2(g)$ emissions continue to rise unabated, indicating the insufficiency of the Kyoto Protocol in tackling global warming.

During December 2009, nations of the world met for the fifteenth time under the United Nations Framework Convention on Climate Change, this time in Copenhagen, Denmark. The **Copenhagen Summit** resulted in little substantial progress because both China and the United Stated resisted commitments to mandatory emission limits. They did agree through the **Copenhagen Accord** to recognize the importance of climate change and to take action to keep temperature increases in the global average below 2°C.

A sixteenth meeting occurred in Cancún, Mexico, from November 29 to December 10, 2010. Again, little progress was made because no new emission reduction targets were set. One outcome was an agreement to develop a **Green Climate Fund**, which would be overseen by a board of twenty-four members. Wealthy countries would contribute to the fund, which would be used by poorer countries to pay costs associated with reducing emissions and adapting to climate change. However, no agreement was reached on how to raise money for the fund.

Despite the lack of commitment by the United States to the Kyoto Protocol, several U.S. states legislated **renewable portfolio standards (RPS)** (also called renewable electricity standards) in the 2000s, whereby a certain fraction of electric power generation was required to come from specified clean energy sources by a certain date. The electric power devices allowed to compete in the RPS market were limited to clean energy technologies. However, the RPS market is a private market and thus subject to free competition. By 2011, RPS had been set in more than thirty U.S. states, the UK, Italy, Belgium, and Chile. Worldwide, other policy mechanisms such as the **feed-in-tariff** (Section 13.10) have similarly spurred a conversion to clean energy and thus a reduction in air pollution and climate-relevant emissions.

Additional progress was made in the United States in April 2007, when the U.S. Supreme Court ruled in *Massachusetts v. Environmental Protection Agency* that carbon dioxide and other greenhouse gases were air pollutants covered under the Clean Air Act Amendments of 1970. Thus, the U.S. EPA had the authority to *consider* regulating these gases. This prompted the U.S. EPA ultimately to grant California a **waiver of Clean Air Act preemption** (Section 8.1.12) on June 30, 2009. This allowed the state to set carbon dioxide emission standards for 2009 model-year passenger vehicles, light-duty trucks, and medium-duty passenger vehicles sold in the state. Under the Clean Air Act Amendments of 1970, other states are permitted to set emission regulations as stringent as California's standards. Despite this regulation, the control of global warming is a process still in its infancy. A proposed large-scale solution is provided in Chapter 13.

12.6.4. Fastest Methods of Slowing Global Warming

Policies directed at controlling global warming need to consider not only the total reduction in global temperature from the policy, but also the speed at which the temperature reduction is obtained. For example, Arctic sea ice may disappear within a few decades, so a control measure that results in a rapid reduction in temperature may be desirable in addition to a control measure that results in the slower but larger reduction in temperature.

Because greenhouse gases and aerosol particles have different atmospheric **e-folding lifetimes** (the time required for an initial concentration of a pollutant to decrease to $1/e$ its initial value; Section 1.5), reducing emissions of some pollutants results in a faster climate response than reducing emissions of others. However, even for the shortest-lived pollutants, complete climate responses take longer than the time required to remove the pollutant from the air because temperature changes in the deep ocean take decades to equilibrate with those in the surface ocean, which takes years to equilibrate with those in the air.

The data-constrained overall e-folding lifetime of carbon dioxide in the atmosphere against all loss processes (primarily dissolution into oceans, photosynthesis, and weathering) is 30 to 50 years (Section 3.6.2). That of methane, due primarily to chemical oxidation by the hydroxyl radical [OH(g)], is 8 to 12 years, with a mean of 10 years. That of fossil fuel soot and solid biofuel soot aerosol particles is approximately 4.5 days (Jacobson, 2010b). Fossil fuel soot is emitted during diesel, jet fuel, kerosene, bunker fuel, oil, gasoline, natural gas, and coal combustion. Solid biofuel soot is emitted during the burning of wood, grass, agricultural waste, and dung during home heating and cooking, primarily in developing countries (Section 9.1.14). Soot particles contain black carbon and primary organic carbon (POC), along with other minor constituents.

These e-folding lifetime data alone suggest that *controlling soot aerosol particle emissions may be the fastest method of slowing global warming* relative to controlling either methane or carbon dioxide. This result is also borne out by three-dimensional global model computer simulations that account for emissions and numerous physical processes and feedbacks (Figure 12.33).

Figure 12.33 indicates that controlling continuous anthropogenic emissions of $CO_2(g)$ causes a greater reduction in globally averaged near-surface temperatures after 100 years than does controlling either fossil fuel soot (FS), FS plus biofuel soot and gas (BSG), or anthropogenic $CH_4(g)$ emissions. Thus, although controlling FS or FS + BSG is the fastest method of slowing global warming, FS + BSG is only the second leading cause of global warming after $CO_2(g)$.

In the limit of infinite time, the magnitude of cooling due to controlling the emissions of a chemical depends on both the emission rate of the species and the temperature change of the species per unit emission.

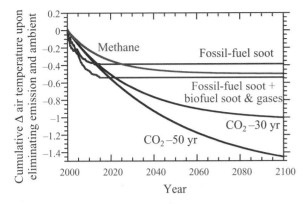

Figure 12.33. Estimated changes in globally averaged near-surface air temperature resulting from the elimination of all continuous anthropogenic emissions of each $CO_2(g)$, methane, fossil fuel soot, and fossil fuel soot plus biofuel soot and gas between 2000 and 2100, from Jacobson (2010b). In the case of $CO_2(g)$, results bounding the data-constrained lifetimes of 30 to 50 years are shown. The lifetime of methane is approximately 10 years. That of fossil fuel soot and solid biofuel soot is about 4.5 days. Because of soot's short lifetime, the elimination of its emission cools climate quickly, but a slow climate response by the oceans to atmospheric temperatures results in its removal affecting temperatures over a longer period. Emission rates for the figure are as follows: carbon dioxide, $\sim$10,200 Tg-C yr^{-1} from fossil fuels and permanent deforestation in 2010; methane, 263 Tg-C yr^{-1} in 2005; solid biofuel soot, 2.5 Tg-C yr^{-1} BC plus 9.9 Tg-C yr^{-1} POC; fossil fuel soot, 3.4 Tg-C yr^{-1} BC plus 2.4 Tg-C yr^{-1} POC (Bond et al., 2004). BC, black carbon; POC, primary organic carbon.

Carbon dioxide has the highest emission rate, followed by methane among the chemicals in Figure 12.33. The **surface temperature response per unit emission (STRE)** of chemical X is the cumulative average near-surface air temperature change after 20 or 100 years per unit continuous emission of the chemical relative to the same for $CO_2(g)$. The STRE is similar to another metric, the **global warming potential (GWP)**, which is the 20- or 100-year integrated instantaneous direct radiative forcing change resulting from a pulse emission of a chemical relative to the same for $CO_2(g)$. An instantaneous **direct radiative forcing** is an instantaneous net downward solar plus thermal-IR radiation change (W/m^2), usually evaluated at the tropopause, when a component is present versus absent. The STRE is a more physical metric than is the GWP because

emissions from, for example, vehicles are continuous rather than pulse emissions. GWP also does not account for feedback from one pulse to another because it does not account for multiple pulses. Furthermore, whereas the time average of instantaneous direct radiative forcing is relatively proportional to temperature changes for greenhouse gases, it is not proportional for aerosol particles; thus, the GWP does not give accurate information about the climate effects of aerosol particles. However, GWP is somewhat similar to STREs for greenhouse gases. For example, the 20- and 100-year GWPs for $CH_4(g)$ from IPCC (2007) are 72 and 25, respectively. This compares with STREs for $CH_4(g)$ of 52 to 92 and 29 to 63, respectively (Table 12.6).

Table 12.6 indicates that the STRE of FS is greater than that of BSG. The reason is that BSG has a much higher ratio of POC to BC than does FS, and POC causes much less heating than does BC. Also, BSG contains other material aside from BC and POC, such as sulfate, nitrate, ammonium, potassium, and magnesium, all of which cause cooling; thus, they offset much of the warming due to the BC and POC. As such, controlling a unit continuous emission of FS has a greater impact on reducing temperatures than does controlling a unit continuous emission of BSG. Because both STREs are much larger than are STREs of $CH_4(g)$ or

Table 12.6. 20- and 100-year surface temperature response per unit emission (STRE) functions for fossil fuel soot (FS), biofuel soot and gas (BSG), black carbon (BC) in FS or BSG, and methane

X	20-year STRE	100-year STRE
FS	2,400–3,800	1,200–1,900
BC in FS	4,500–7,200	2,900–4,600
BSG	380–720	190–360
BC in BSG	2,100–4,000	1,060–2,020
Methane	52–92	29–63

The STRE is defined as the near-surface air temperature change after 20 or 100 years per unit continuous emission of X relative to the same for $CO_2(g)$. For comparison, the 20- and 100-year global warming potentials (GWPs) for methane from IPCC (2007) are 72 and 25, respectively. Multiply the STRE in the table by 12/44 to obtain the STRE relative to $CO_2(g)$-C. The calculations are derived from Figure 12.33 and assume a continuous $CO_2(g)$ emission rate from fossil fuels plus permanent deforestation of 29,700 Tg-$CO_2(g)$ yr^{-1}, 284 Tg-$CH_4(g)$/yr, 5.8 Tg-C yr^{-1} FS (BC + POC), and 12.4 Tg-C yr^{-1} BSG (PC + POC).
Source: From Jacobson (2010b).

$CO_2(g)$ (=1), controlling both FS and BSG causes greater cooling than controlling the same unit continuous emission of $CH_4(g)$ or $CO_2(g)$ over either 20- or 100-year time frames.

Example 12.4

If a diesel vehicle emits 99 g-CO_2(g) km^{-1} and the Euro 5 particulate matter emission standard of 5 mg-FS-C km^{-1} (where FS is fossil fuel soot), estimate the total CO_2(g)-equivalent [CO_2(g)-eq] emissions of the exhaust over 20 and 100 years. CO_2(g)-eq emissions are those of CO_2(g) plus of other components of global warming, each multiplied by the ratio of the global warming potential of the component relative to that of CO_2(g). This parameter allows different sources of pollution to be evaluated relative to each other in terms of their potential contribution to global warming. Do the same calculation for a gasoline vehicle that emits 103 g-CO_2(g) km^{-1} and 1 mg-FS-C km^{-1}. Which vehicle emits more CO_2(g)-eq over each period?

Solution

Multiply the 20- and 100-year STREs for FS from Table 12.6 by the change in FS-C emissions and add the result to the CO_2(g) emissions to obtain the CO_2(g)-eq emissions from the vehicle. For example, the low estimate of the 20-year CO_2(g)-eq emissions for diesel is 99 g-CO_2(g) km^{-1} + 0.005 g-CO_2(g) km^{-1} × 2,400 g-CO_2(g)-eq g^{-1}-FS-C = 111 g-CO_2(g)-eq km^{-1}.

Using this methodology, the overall ranges are

Diesel: 111–118 g-CO_2(g)-eq km^{-1} over 20 years and 105–108.5 g-CO_2(g)-eq km^{-1} over 100 years.

Gasoline: 105.4–106.8 g-CO_2(g)-eq km^{-1} over 20 years and 104.2–104.9 g-CO_2(g)-eq km^{-1} over 100 years.

Thus, this gasoline vehicle emits less CO_2(g)-eq emissions than the diesel vehicle. However, both vehicles emit significant levels of CO_2(g)-eq compared, for example, with an electric vehicle, which emits 0 g-CO_2(g)-eq km^{-1} from the vehicle itself.

Table 12.6 indicates that the 20- and 100-year STREs of black carbon alone in FS and BSG is greater than is that of FS or BSG as a whole, respectively. The reason is that BC is a much stronger warming agent than are the other materials in each FS and BSG, so averaging the warming due to BC with lesser warming or cooling due to other components results in a decrease in the STRE of FS and BSG relative to BC alone in FS and BSG.

In sum, FS causes more warming per unit emission than does BSG, and both cause more warming per unit emission than do methane or carbon dioxide. However, *BSG causes more than eight times the mortality (causing more than 1.5 million premature deaths per year) than FS* because, although both result in particulate matter, the most deadly component of air pollution, BSG is emitted in much more densely populated regions of the world. Both FS and BSG cause hundreds of thousands of times greater mortality per unit mass emission than do CO_2(g) or CH_4(g). However, CO_2(g) causes the greatest overall warming of climate among the chemicals.

Thus, *controlling soot emissions slows global warming and reduces human mortality faster than does any other mechanism considered.* Due to the speed of its climate response, controlling soot may be the only method of preventing the elimination of the Arctic sea ice, which may occur within two to three decades. However, *also controlling CO_2(g) and CH_4(g) is essential for stabilizing global temperatures, particularly because CO_2(g) is the largest component of global warming.*

12.7. Summary

In this chapter, the greenhouse effect; historical temperature trends; and causes, characteristics, effects, and regulatory control of global warming are discussed. Greenhouse gases selectively absorb thermal-IR radiation but are transparent to visible radiation. Without the presence of natural greenhouse gases, particularly water vapor and carbon dioxide, the Earth would be too cold to support higher life forms. Global warming is the increase in the Earth's temperature above that caused by natural greenhouse gases. Greenhouse gases that contribute the most to global warming are carbon dioxide, methane, and nitrous oxide. The second most important component of near-surface global warming, after carbon dioxide and before methane, may be particulate BC, which absorbs solar and thermal-IR radiation. Anthropogenic emissions and ambient levels of greenhouse gases and BC have increased since the mid-1800s. Air temperatures have also increased. Although temperatures throughout Earth's history have frequently been higher than they are today, the current rate of temperature increase is higher than during any time since deglaciation at the end of the last glacial maximum. The consequences of increasing temperatures over the next 100 years are expected to be a rise in sea level, shifts in agriculture, damage to ecosystems, more extreme weather, heat-related health problems, enhanced malaria and influenza, and air pollution mortality. Efforts to date to control emissions of greenhouse gases have been weak internationally, although such emissions continue to grow worldwide. Some recent efforts on both national and international scales, though, provide hope for a breakthrough in emission controls in the near future. Another method of addressing the problem is through expansion of existing clean energy technologies. This method is discussed in Chapter 13.

12.8. Problems

12.1. Calculate the effective temperature of the Earth's surface in the absence of a greenhouse effect assuming $A_e = 0.4$ and $\varepsilon_e = 1.0$. By what factor does the result change if the Earth-sun distance is doubled?

12.2. What is the relative ratio of energy received at the top of Mars's and Venus's atmosphere compared with that at the top of the Earth's atmosphere?

12.3. Explain, in terms of atmospheric components and distance from the sun, why Mars's actual temperatures are colder and Venus's actual temperatures are much warmer than are those of the Earth.

12.4. What would the equilibrium temperature of the Earth be if it were in Mars's orbit? What about if it were in Venus's orbit? Assume no other characteristics of the Earth changed.

12.5. Discuss at least two ways that deforestation can affect global warming.

12.6. Explain how CFCs, HCFCs, and HFCs might contribute to global warming.

12.7. Explain why greenhouse gases may cause an increase in near-surface temperatures but a decrease in stratospheric temperatures.

12.8. Explain why the Earth's temperature during the Archean eon might have been much warmer than today's temperature, although that period's solar output was lower than today's solar output.

12.9. Why are scientists more concerned about the collapse of the West than the East Antarctic ice sheet?

12.10. Explain the common theory as to how the dinosaurs became extinct.

12.11. Explain the fundamental reason for the four glacial periods that have occurred during the past 450,000 years. Did carbon dioxide cause these events, or did its mixing ratio change in response to them? Why?

12.12. What is unusual about the rate of change in the Earth's near-surface temperature since the late 1950s compared with historic rates of temperature change?

12.13. How would a change in the Earth's obliquity to zero (no tilt) affect seasons? How would it affect the relative amount of sunlight over the South Pole during the Southern Hemisphere winter in comparison with today?

12.14. If the Earth's temperature initially decreases and only the snow-albedo feedback is considered, what will happen to the temperature subsequently?

12.15. If the Earth's temperature initially increases and only the plant-carbon dioxide feedback is considered, what will happen to the temperature subsequently?

12.16. How will absorbing aerosol particles in the boundary layer affect air pollution buildup and cloud formation if only the effects of the aerosol particles on atmospheric stability are considered?

12.17. Explain in your own words why absorbing aerosol particles might increase cloudiness when they are in low concentration but decrease cloudiness when they are in high concentration, as observed in data. Identify specific feedback effects in your explanation.

12.18. Identify three benefits of controlling black carbon emissions and discuss whether each benefit is greater than, the same as, or less than controlling carbon dioxide emissions.

12.19. Using STREs from Table 12.6, compare the 20- and 100-year $CO_2(g)$-eq emissions of a diesel vehicle that emits 112 g-$CO_2(g)$/km and 8 mg-FS-C/km (in which FS is fossil-fuel soot) and a gasoline vehicle that emits 10 percent more $CO_2(g)$/km, but one-sixteenth the FS-C/km. Calculate low and high estimates for each case. Which vehicle emits more $CO_2(g)$-eq over 20 years? Which emits more over 100 years? Try to interpret the result. Which vehicle produces more health-affecting particulate matter?

12.9. Essay Questions

12.20. Identify at least six activities that you do or products that you consume that result in the release of one or more greenhouse gases, and identify ways that you can reduce your emissions.

12.21. Identify two arguments against and two arguments for global warming.

12.22. Discuss three possible effects of global warming. Would any of these effects affect your life or lifestyle?

12.23. If you believe that the global warming problem is an important issue, what specific steps should your national and local governments take to address the issue? If you do not believe that it is an important issue, what steps should scientists take to understand the issue better?

Chapter 13

Energy Solutions to Air Pollution and Global Warming

This book addresses local to global atmospheric problems, including outdoor and indoor air pollution, acid deposition, stratospheric ozone loss, and global warming. It also discusses historic regulatory actions aimed at incrementally reducing emissions of gases and particles. In this chapter, the focus shifts to a different type of solution to these problems, namely, the complete and large-scale conversion of the current combustion-based energy infrastructure to one based on electricity and an energy carrier derived from electricity, hydrogen, with the simultaneous implementation of energy efficiency measures. The electricity, in all cases, is produced by clean technologies that take advantage of the natural and renewable resources of **wind, water, and sunlight (WWS)**. An analysis is provided for replacing current worldwide energy for all purposes (electric power, transportation, heating/cooling) with energy from WWS. The analysis includes a presentation of WWS energy characteristics, current and future energy demand, and availability of WWS resources. It also discusses the number of WWS devices needed to power the world, the physical land or ocean footprint and spacing needed for such devices, and materials required to build the devices. It then introduces methods of addressing the variability of WWS energy to ensure that power supply reliably matches demand, the direct and social costs of WWS generation and transmission versus the costs of fossil fuels, and policy measures needed to enhance the viability of a WWS system. The discussion draws primarily from Jacobson and Delucchi

(2011), Delucchi and Jacobson (2011), and references therein.

13.1. Clean, Low-Risk, Sustainable Energy Systems

Because it is not possible to eliminate all gas and particle emissions from combustion of carbon-based fuels, a solution to the problems of global warming, air pollution, and energy insecurity requires a large-scale conversion to clean, perpetual, and reliable energy, together with improvements in energy efficiency. Here, a proposed solution to these problems is discussed. It involves the conversion, by 2030 to 2050, of all sectors of the world's energy infrastructure, including the **electric power, transportation, industrial, and heating/cooling sectors**, to energy derived solely from WWS.

The complete transformation of the energy infrastructure would not be the first large-scale project undertaken in world history. During World War II, the United States transformed motor vehicle production facilities to produce **more than 300,000 aircraft**, and the rest of the world manufactured 486,000 more aircraft. In the United States, production increased from about 2,000 units in 1939 to almost 100,000 units in 1944. In 1956, the United States began work on the **Interstate Highway System**, which now extends for 47,000 miles and is one of the largest public works project in history. The **Apollo program**, widely considered the world's greatest engineering and technological accomplishment, put a human on the moon in less than 10 years. Although these projects differ from the one discussed here, they suggest that the complete transformation of the world's energy system is not an insurmountable barrier.

Because global warming (particularly loss of the Arctic sea ice), air pollution, and energy insecurity are current and growing problems that require rapid action to prevent catastrophic damage, and because several decades are needed for new technologies to become fully adopted, this analysis considers only technologies that exist and that can be ramped up without further major development. To ensure that the energy system remains clean in the long run, even in the presence of population and economic growth, this analysis also considers only technologies that have near-zero emissions of greenhouse gases and air pollutants per unit of energy output over their whole **life cycles**, from manufacturing through operational use to decommissioning. Similarly, it considers only those technologies that have little impact on land footprint and availability, water pollution, and water resources; those that do not have significant waste disposal, air or water contamination, or terrorism risks associated with them; and those based on primary resources that are indefinitely renewable or recyclable. Energy systems running on wind, water, and solar power satisfy these criteria.

A previous study reviewed and ranked several energy systems proposed to address the problems mentioned with respect to their impacts on global warming; air, thermal, and water-chemical pollution; water supply; land use; wildlife; and nuclear weapons proliferation (Jacobson, 2009). The overall rankings for electric power and vehicle options are given in Table 13.1. All recommended electric power options are driven by WWS. The recommended vehicle options included **battery electric vehicles** (BEVs) and **electrolysis-produced (electrolytic) hydrogen fuel cell vehicles** (EHFCVs), also powered by the WWS options. These and other existing technologies for the heating/cooling sectors are presented here. Other clean WWS electric power sources, such as ocean or river current power, could also be deployed, but they are not discussed in this book for simplicity's sake.

Some technologies, such as nuclear power and coal with carbon capture and sequestration (CCS) for electricity, and corn and cellulosic ethanol fuel for transportation, solid biofuels for electricity production, other types of biofuels for transportation, and natural gas were not recommended because they were found to be

Table 13.1. Cleanest solutions to global warming, air pollution, and energy security

Electric power	Vehicles
Recommended	
Wind	Battery electric vehicles (BEVs)
Concentrated solar power (CSP)	Electrolytic hydrogen fuel cell vehicles (EHFCVs)
Geothermal	
Tidal	
Solar photovoltaics (PVs)	
Wave	
Hydroelectric	
Not Recommended	
Nuclear	Corn ethanol fueled vehicles
Coal with carbon capture	Cellulosic ethanol fueled vehicles
Natural gas	Sugarcane ethanol fueled vehicles
Coal without carbon capture	Soy or algae biodiesel fueled vehicles
Solid biofuels	Compressed natural gas fueled vehicles

Source: Jacobson (2009).

moderately or significantly worse than WWS options with respect to air and water pollution, global warming, land use impacts, and/or water supply. More specific reasons why several of these technologies are not recommended are discussed next.

13.1.1. Why Not Nuclear Energy?

For several reasons, the analysis did not recommend **nuclear energy** (conventional fission, breeder reactors, thorium, or fusion) as an energy source that should be expanded in the future. First, the growth of *nuclear energy has historically increased the ability of nations to obtain or enrich uranium for nuclear weapons.* A large-scale worldwide increase in nuclear energy facilities would exacerbate this problem, putting the world at greater risk of a nuclear war or terrorism catastrophe. The historic link between energy facilities and weapons is evidenced by the development or attempted development of weapons capabilities secretly in nuclear energy

Table 13.2. Projected end-use power demand (TW) in 2030, by sector for world and United States if conventional fossil fuel and wood use continues as projected and if 100 percent of conventional fuels are replaced with wind, water, and sunlight (WWS) technologies

Energy sector	Conventional fossil fuels and wood		Replacing fossil fuels and wood with WWS	
	World	United States	World	United States
Residential	2.26	0.43	1.83	0.35
Commercial	1.32	0.38	1.22	0.35
Industrial	8.80	0.92	7.05	0.74
Transportation	4.53	1.10	1.37	0.33
TOTAL	16.92	2.83	11.47	1.78

For comparison, world and U.S. end-use power demands from conventional fossil fuels plus wood in 2008 were 12.5 and 2.5 TW, respectively.
Source: Jacobson and Delucchi (2011).

facilities in Pakistan, India, Iraq (prior to 1981), Iran, and, to some extent, North Korea. If the world were converted to electricity and electrolytic hydrogen by 2030, the 11.5 trillion watts (TW) in resulting end-use power demand (Table 13.2) would require ~15,800 850-MW nuclear reactors, or one installed every day for 43 years. Even if only 5 percent of these were installed, the number of nuclear reactors worldwide would nearly double the number of reactors in 2011 (about 440). Many more countries would possess nuclear facilities, increasing the likelihood that these countries would use the facilities to hide the development of nuclear weapons, as has occurred historically.

Second, *nuclear energy results in nine to twenty-five times more carbon dioxide–equivalent emissions (defined in Example 12.4) per unit energy generated than does wind energy.* This is due in part to emissions from uranium refining and transport and reactor construction (e.g., Lenzen, 2008; Sovacool, 2008) and in part due to the longer time required to site, permit, and construct a nuclear plant compared with a wind farm, resulting in greater emissions from the fossil fuel electricity sector during this period (Jacobson, 2009). Not accounted for in the emissions number is the slight increase in global temperature resulting from the water evaporated during the cooling of nuclear facilities (Section 12.2.3). *Such evaporation also occurs in coal, natural gas, and biofuel energy facilities.*

The longer the time between the planning and operation of an energy facility, the more the emissions from the background electric power grid enhance global warming and air pollution. Although recent nuclear reactor **construction times** worldwide are shorter than the 9-year median construction times in the United States since 1970 (Koomey and Hultman, 2007), they still averaged 6.5 years worldwide in 2007 (Ramana, 2009). Construction time must be added to the site permit time (~3 years in the United States) and **construction permit and issue time** (~3 years). The overall historic and present range of nuclear planning-to-operation times for new nuclear plants has been 11 to 19 years, compared with an average of 2 to 5 years for wind and solar installations.

The *long period of time required between planning and operation of a nuclear power plant* poses a significant risk to the Arctic sea ice. Sea ice records indicate a 32 percent loss in the August 2010 sea ice area relative to the mean from 1979 to 2008 (Polar Research Group, n.d.). Such rapid loss indicates that solutions to global warming must be implemented quickly. Technologies with long lead times will allow the high-albedo Arctic ice to disappear, triggering more rapid positive feedbacks to warmer temperatures by uncovering the low-albedo ocean below.

Third, accidents at nuclear power plants have been either catastrophic (Chernobyl, Russia, in 1986; Fukushima Dai-ichi, Japan, in 2011) or damaging (Three Mile Island, Pennsylvania, in 1979; Saint-Laurent, France, in 1980). The nuclear industry has improved the safety and performance of reactors and proposed new reactor designs that they suggest are safer. However, these designs are generally untested, and there is no guarantee that the reactors will be designed, built, and operated correctly or that a natural disaster or act of terrorism, such as an airplane flown into a reactor, will not cause them to fail, resulting in a major disaster.

Through 2011, about 1.5 percent of all nuclear reactors in history had a partial or significant core meltdown. On March 11, 2011, an earthquake measuring 9.0 on the Richter scale, and the subsequent tsunami that knocked out backup power to a cooling system, caused six nuclear reactors at the **Fukushima 1 Daiichi plant** in northeastern Japan to shut down. Three reactors experienced a significant meltdown of nuclear fuel rods and multiple explosions of hydrogen gas that formed during efforts to cool the rods with seawater. Uranium fuel rods in a fourth reactor also lost their cooling. As a result cesium-137, iodine-131, and other radioactive particles and gases were released into the air. Locally, tens of thousands of people were exposed to the radiation, and 170,000 to 200,000 people were evacuated from their homes. The radiation release created a dead zone around the reactors that may not be safe to inhabit for decades to centuries. The radiation also poisoned the water and food supplies in and around Tokyo. The radiation plume from the plant spread worldwide within a week. Concentrations in Japan within 100 km of the plant were very high, whereas those across the Pacific Ocean were more modest. The lesson from this and similar events is that, even if the risks of catastrophe from nuclear power are small, they are not zero. Catastrophic risks with wind and solar power are zero.

Fourth, **conventional nuclear fission reactors**, which are nuclear reactors in which only about 1 percent of the uranium in the nuclear fuel is used and the rest disposed of, produce **radioactive waste** that must be stored for up to 200,000 years. This gives rise to concerns about how to prevent leakage of the waste for such a vast period and the long-term costs of storage. Due to their inefficient use of uranium, conventional nuclear reactors could exhaust uranium reserves in roughly a century if a large-scale nuclear program were undertaken.

Alternate types of reactors (e.g., breeder reactors), nuclear fuels (e.g., thorium), and processes (e.g., fusion) have been proposed. **Breeder reactors** reuse spent nuclear fuel, thereby consuming a much higher percentage of uranium, producing less waste, and resulting in lower uranium requirements than conventional reactors. However, breeder reactors produce nuclear material that can be reprocessed more readily into nuclear weapons than can material from conventional reactors.

Thorium, like uranium, can be used to produce nuclear fuel in breeder reactors. The advantage of thorium is that it produces less long-lived radioactive waste than does uranium. Its products are also more difficult to convert into nuclear weapons material. However, thorium still produces ^{232}U, which was used in one nuclear bomb core produced during the **Operation Teapot** bomb tests in 1955. Thus, thorium is not free of nuclear weapons proliferation risk.

Nuclear fusion of light atomic nuclei (e.g., protium, deuterium, or tritium; Section 1.1.1) could theoretically supply power indefinitely without long-lived radioactive waste because the products are isotopes of helium. However, there is little prospect for fusion to be commercially available for at least 50 to 100 years, if ever.

13.1.2. Why Not Coal with Carbon Capture?

Carbon capture and sequestration (CCS) is the diversion of $CO_2(g)$ from a point emission source, such as a coal-fired power plant exhaust stack, to an underground geological formation (e.g., saline aquifer, depleted oil and gas field, or unminable coal seam). Geological formations worldwide may theoretically store up to 2,000 Gt-CO_2(g), which compares with a fossil fuel emission rate today of $\sim$30 Gt-CO_2(g) yr^{-1}. To date, CO_2(g) has been diverted underground following its separation from mined natural gas in several operations and from gasified coal in one case. However, no large power plant currently captures CO_2(g). Several options of combining fossil fuel combustion for electricity generation with CCS technologies have been considered. In a standard model, CCS equipment is added to an existing or new coal-fired power plant. CO_2(g) is then separated from other gases and injected underground after coal combustion. The remaining gases are emitted into the air.

Other CCS methods include injection into the deep ocean and production of carbonate minerals. Ocean storage, however, results in ocean acidification. Dissolved CO_2(g) in the deep ocean eventually equilibrates with that in the surface ocean, reducing ocean pH and simultaneously supersaturating the surface ocean with CO_2(g), forcing some of it into the air.

Producing carbonate minerals has a long history. Joseph Black (Section 1.2.2.6) named carbon dioxide **fixed air** because it fixed to quicklime [(CaO(s)] to form $CaCO_3$(s). However, the natural process is slow and requires massive amounts of quicklime for large-scale CO_2(g) reduction. The process can be hastened by increasing temperature and pressure, but this requires additional energy.

The use of CCS may reduce $CO_2(g)$ emissions from the smokestacks of coal-fired power plants by 85 to 90 percent or more. However, the CCS process has no effect on reducing other pollutants from the smokestack, nor does it reduce emissions of any pollutant during the mining and transport of the coal. In fact, because *a coal-CCS plant requires 25 percent more energy to operate than does a conventional coal plant* (Intergovernmental Panel on Climate Change (IPCC), 2005), a coal-CCS system requires 25 percent more coal mining, transport, processing, and non-$CO_2(g)$ air pollution per unit of net energy delivered than does a conventional coal system. Furthermore, because of the significant $CO_2(g)$ emissions due to coal mining and transport and the time lag between planning and operation of a coal plant with CCS equipment relative to that of a wind farm, a coal-CCS plant still emits about fifty times more $CO_2(g)$ than does a wind farm producing the same quantity of electric power (Jacobson, 2009). As such, the use of coal-CCS represents an opportunity cost in terms of air pollution and climate-relevant emissions compared with clean renewable energy options.

13.1.3. Why Not Natural Gas?

Natural gas is a colorless, flammable gas made of about 90 percent methane plus other hydrocarbon gases that is often found near petroleum deposits (Sections 3.6.4 and 5.2.1.5). Natural gas is combusted directly to produce electricity. It is also compressed to less than 1 percent of its gas volume to form **compressed natural gas (CNG)**, which is burned in vehicles for transportation. Natural gas can be compressed and cooled to $-162°C$ to produce **liquefied natural gas (LNG)**. LNG is used primarily to transport natural gas from a natural gas field to a market or pipeline, where it is regassified. Compressing and/or liquefying natural gas require more energy input; thus, both are less efficient than is just burning raw natural gas.

During combustion, natural gas emits $CO_2(g)$, $CH_4(g)$, other hydrocarbon gases, $NO_x(g)$, $SO_x(g)$, $N_2O(g)$, $CO(g)$, and particulate matter, although generally in lower quantities [aside from $CH_4(g)$] than the combustion of most other fossil fuels. Among these pollutants, $CO_2(g)$, $CH_4(g)$, $N_2O(g)$, and absorbing particulate matter contribute to global warming. All pollutants contribute to human health problems.

The use of **natural gas** to generate electricity (without compressing or liquefying it) in a combined cycle power plant normally results in the combined $CO_2(g)$, $CH_4(g)$, and $N_2O(g)$ emissions of about 500 g-$CO_2(g)$-eq/kWh of electricity generated (Spath and Mann, 2000). About 74 percent of the emissions is due to operating the plant, 25 percent is due to mining and distributing the natural gas, and the rest is due to plant construction and ammonia production. Although more than 99 percent of the mass of all global warming–relevant gases emitted by natural gas combustion is $CO_2(g)$ mass, $CO_2(g)$ contributes to only about 88 percent of natural gas's atmospheric warming. Almost 12 percent of the warming over a 100-year time frame is due to the small amount of methane leaked during the mining, transport, and use of natural gas when 1.4 percent of all natural gas mined is leaked (Spath and Mann, 2000). Such a leakage rate is associated with conventional sources of natural gas, such as wells containing both oil and natural gas. The leakage rate from such wells generally ranges from 1 to 4 percent.

However, when natural gas is extracted from shale rock formations instead of wells, a larger percent of natural gas leaks. **Shale** is sedimentary rock composed of a muddy mix of clay mineral flakes and small fragments of quartz and calcite. Large shale formations containing natural gas can be found in eastern North America, close to population centers. The fraction of natural gas produced from shale is significant and increasing, particularly in the United States. Extraction of natural gas from shale requires large volumes of water forced under pressure to fracture and refracture the rock to increase the flow of natural gas. This process is referred to as **hydraulic fracturing** (or **fracking**). As the water returns to the surface over days to weeks, it is accompanied by methane that escapes to the atmosphere.

One study estimates that 3.6 to 7.9 percent of all natural gas mined from shale formations leaks to the atmosphere (Howarth et al., 2011). If correct, these numbers suggest that the 100-year global warming–relevant natural gas emissions from shale are 600 to 700 g-$CO_2(g)$-eq/kWh. This compares with 308 to 570 g-$CO_2(g)$-eq/kWh for coal-CCS (accounting for both life cycle emissions and the opportunity cost emissions from planning-to-operation delays of coal-CCS plants) and with more than 1,000 g-$CO_2(g)$-eq/kWh for coal without CCS (mining, transport, and use; Jacobson, 2009). Because natural gas carbon-equivalent emissions fall between those of coal-CCS and coal without CCS and because natural gas combustion emits air pollution, it is not recommended as a future source of electricity in a clean sustainable world. Because CNG requires even more energy to produce than does natural gas without compression, it too is not recommended.

13.1.4. Why Not Liquid or Solid Biofuels?

Biofuels are solid, liquid, or gaseous fuels derived from organic matter. Most biofuels are derived from dead plants or animal excrement. **Solid biofuels**, such as wood, grass, agricultural waste, and dung, are burned directly for home heating and cooking significantly in developing countries and for electric power generation in developed and developing countries. **Liquid biofuels** are generally used for transportation as a substitute for gasoline or diesel. The most common transportation biofuels are ethanol, used in passenger cars and other light-duty vehicles, and biodiesel, used in many heavy-duty vehicles.

Ethanol [$C_2H_5OH(aq)$] is produced in a factory, generally from corn, sugarcane, wheat, sugar beet, or molasses. The most common among these sources are corn and sugarcane, resulting in the production of **corn ethanol** and **sugarcane ethanol**, respectively. Microorganisms and enzyme ferment sugars or starches in these crops to produce ethanol, as in Reaction 2.3.

Fermentation of cellulose originating from switchgrass, wood waste, wheat, stalks, corn stalks, or *Miscanthus* can also produce ethanol, but the process is more energy intensive because natural enzyme breakdown of cellulose (e.g., as occurs in the digestive tracts of cattle) is slow. Faster breakdown of cellulose requires genetic engineering of enzymes. The ethanol resulting from these sources is referred to as **cellulosic ethanol**.

Ethanol may be used on its own, as it is frequently in Brazil, or blended with gasoline. A blend of 6 percent ethanol/94 percent gasoline is referred to as **E6**. Other typical blends are **E10**, **E15**, **E30**, **E60**, **E70**, **E85**, and **E100**. In many countries, including the United States, E100 is required to contain 5 percent gasoline as a **denaturant**, which is a poisonous or untasteful chemical added to a fuel to prevent people from drinking it. As such, E85, for example, contains about 81 percent ethanol and 19 percent gasoline.

A proposed alternative to ethanol for transportation fuel is **butanol** [$C_4H_9OH(aq)$]. It can be produced by fermentation of the same crops used to produce ethanol but with a different bacterium, *Clostridium acetobutylicum*. Butanol contains more energy per unit volume of fuel than does ethanol. However, unburned butanol also reacts more quickly in the atmosphere with the OH(g) radical than does unburned ethanol, speeding up ground-level ozone formation relative to ethanol. On average, ethanol, itself, produces more ground-level ozone than does gasoline (Section 4.3.8).

Biodiesel is a liquid diesellike fuel derived from vegetable oil or animal fat. Major edible vegetable oil sources of biodiesel include soybean, rapeseed, mustard, false flax, sunflower, palm, peanut, coconut, castor, corn, cottonseed, and hemp oils. Inedible vegetable oil sources include jatropha, algae, and jojoba oils. Animal fat sources include lard, tallow, yellow grease, fish oil, and chicken fat. Soybean oil accounts for about 90 percent of biodiesel production in the United States. Biodiesel derived from soybean oil is referred to as **soy biodiesel**.

Biodiesel consists primarily of long-chain methyl, propyl, or ethyl esters that are produced by the chemical reaction of an oil or fat (both lipids) with an alcohol. It is a standardized fuel designed to replace diesel in standard diesel engines. It can be used as pure biodiesel or blended with regular diesel. Blends range from 2 percent biodiesel/98 percent diesel (**B2**) to 100 percent biodiesel (**B100**). Generally, only blends B20 and lower can be used in a diesel engine without engine modification. The use of vegetable oil or animal fat directly (without conversion to biodiesel) in diesel engines is also possible; however, it results in more incomplete combustion, and thus more air pollutant by-products, as well as a greater buildup of carbon residue in, and damage to, the engine than biodiesel.

A significant effort has been made to produce **algae biodiesel**, which is biodiesel from algae grown from waste material, such as sewage. However, such efforts have been hampered by the fact that algae can be grown efficiently only when they are exposed to the sun. As such, they cannot be grown efficiently in high density, with one on top of the other, and require a significant surface area. Each volume of oil produced from algae also requires about 100 times that volume of water. Both factors have limited the growth of the algae biodiesel industry.

Neither solid nor liquid biofuels of any type are recommended for a clean energy future. Specifically, solid biofuels for home heating and cooking and for electric power generation are not recommended. Liquid ethanol, butanol, and biodiesel from any sources are not recommended. The main reasons are that (1) nearly all biofuels are combusted to generate energy, resulting in air pollution similar to or greater than that from fossil fuels; (2) solid or liquid biofuel does not reduce global warming–relevant emissions nearly to the extent as do WWS resources; (3) several biofuels increase such emissions relative to fossil fuels; (4) many biofuels require rapacious amounts of land; (5) many biofuels require excessive quantities of water; and (6) many biofuels are derived from food sources, increasing food shortages, food prices, and starvation (Delucchi, 2010; Jacobson, 2009). Even biofuels used for energy production that rely on wood waste that would otherwise decay require energy for gathering and transporting the waste and result in air pollution during its combustion.

Because biofuels are not so good as WWS technologies, and often worse than gasoline or diesel in many respects, they represent opportunity costs. *The key to eliminating air pollution and global warming is eliminating combustion and carbon.*

The issues with liquid biofuels are illustrated by comparing the impacts of using ethanol to provide energy for internal combustion engines with those of using WWS resources to provide energy for BEVs and hydrogen fuel cell vehicles (HFCVs). Figure 13.1 compares the net change in (1) carbon dioxide emissions; (2) air pollution deaths; and (3) water required if all light- and heavy- duty, on-road vehicles in the United States were converted from those powered by liquid fossil fuels to those powered by a different technology. The alternate vehicle options include BEVs powered by electricity from wind, concentrated solar power (CSP), photovoltaics (PVs), geothermal, tidal, wave, hydroelectric, nuclear, or coal-CCS; HFCVs powered by hydrogen electrolyzed using wind power electricity; and internal combustion vehicles powered by corn E85 or cellulosic E85.

In the United States, about 26 percent of all $CO_2(g)$ emissions are from vehicle exhaust and 6.7 percent are from the upstream production of fuel. As such, converting to alternate vehicle technologies could reduce U.S. $CO_2(g)$ emissions by, at most, 32.7 percent. Figure 13.1a shows that converting to wind-BEVs reduces U.S. $CO_2(g)$ emissions by 32.4 to 32.6 percent, which represents 99 to 99.7 percent of the 32.7 percent possible reduction.

Using wind electricity to produce hydrogen for HFCVs results in about three times more emissions than using wind electricity to power BEVs directly, but wind-HFCV emissions are still low (97–98.5 percent of the maximum possible $CO_2(g)$ reduction vs. 99–99.7 percent for wind-BEVs). Wind-HFCVs result in greater emissions than do wind-BEVs because an electric vehicle converts 75 to 86 percent of the wind electricity to motion, whereas an HFCV converts about one-third of this percentage due to losses in the **electrolyzer** (used to produce hydrogen from electricity), **compressor** (used to compress hydrogen), and **fuel cell** (used to convert hydrogen to energy and water). Nevertheless, an HFCV is still more efficient than an internal combustion engine, which converts about 17 to 20 percent of the fuel in its tank to mechanical motion.

Figure 13.1a indicates that other WWS sources powering BEVs also result in significant $CO_2(g)$ reductions. Nuclear-BEVs and coal-CCS–BEVs are less efficient at reducing $CO_2(g)$ than are WWS sources. However, corn E85 and cellulosic E85 vehicles either increase $CO_2(g)$ or cause much less $CO_2(g)$ reduction than do the WWS options or nuclear or coal-CCS. Even in the best case for ethanol, using cellulosic E85, carbon emissions are still much larger than they are for WWS-BEVs. Figure 13.1b also shows that the air pollution mortality associated with either corn or cellulosic ethanol exceeds that associated with the WWS-BEV options, and Figure 13.1c indicates that the water requirements for corn ethanol in particular amount to about 10 percent of the U.S. water supply. Because of the significant land required for either corn or cellulosic ethanol (Figure 13.2), it is also impractical to expect E85 fuels to provide energy for any more than 30 percent of the U.S. fleet. Analyses for other types of liquid biofuels result in similar results in most areas.

For these reasons, the *use of nuclear, coal-CCS, natural gas, and biofuel represents an opportunity cost compared with the use of WWS technologies for transportation and electricity.* Thus, this analysis focuses on WWS technologies. WWS is assumed to supply electric power for transportation, cooking, air and water heating, air conditioning, high-temperature industrial processes, and general electricity.

13.1.5. Demand-Side Energy Conservation

Although the analysis focuses on energy supply, **demand-side energy conservation measures** are also important for reducing the requirements and impacts of energy supply. Demand-side energy conservation measures include **improving the energy-out to energy-in efficiency of end uses** (e.g., with more efficient vehicles and lighting, better insulation in homes, and the use of heat-exchange and filtration systems), **using lower-energy modes of transportation** (e.g., using public transit or telecommuting instead of driving), **improving large-scale planning** to reduce energy demand without compromising economic activity or comfort (e.g., designing cities to facilitate greater use of nonmotorized transport and to have better matching of origins and destinations, thereby reducing the need for travel), and **designing buildings to use solar energy directly** (e.g., using more daylighting, solar hot water heating, and passive solar heating/cooling in buildings).

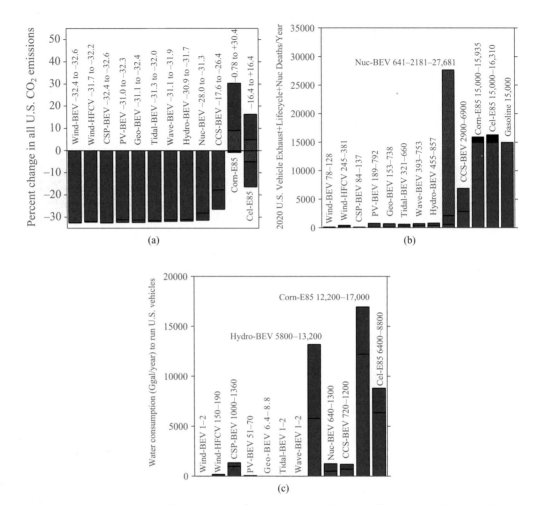

Figure 13.1. (a) Percent changes in U.S. CO_2(g) emissions upon replacing 100 percent of on-road (light- and heavy-duty) vehicles powered by fossil fuels with vehicles powered by different energy technologies. The maximum possible percent reduction in CO_2(g) due to such a conversion is 32.7 percent because 26 percent of U.S. CO_2(g) originates from vehicle exhaust and 6.7 percent from upstream fuel production. Low and high estimates are given. In all cases except the E85 cases, blue represents the low estimate and blue plus red, the high. For corn and cellulosic E85, the full bars represent the range at 100 percent penetration, and the brown bars represent the range at 30 percent penetration. For both ethanol sources, the high estimate occurs when emissions associated with price changes of fuel crops are accounted for. Such emissions occur when the price of corn increases due to its use as a fuel instead of food, and this triggers a conversion of forested or densely vegetated grassland to agricultural land, increasing carbon emissions (e.g., Searchinger et al., 2008).
(b) Estimates of 2020 U.S. premature deaths per year due to emissions from the production and use of energy for vehicles and exhaust from vehicles (where applicable) for the scenario in (a). Low (blue) and high (blue plus red) estimates are given. In the case of nuc-battery electric vehicles (BEVs powered by electricity from nuclear plants), the additional brown bar represents the upper limit of the potential number of deaths, scaled to the U.S. population, due to one potential nuclear weapons catastrophe over 30 years in a megacity caused by the proliferation of nuclear energy facilities worldwide. In the case of corn E85 and cellulosic E85, the red bar is the additional number of deaths due to tailpipe emissions of E85 over gasoline for the United States (Section 4.3.8), and the black bar is the additional number of U.S. deaths per year due to upstream emissions from producing and distributing E85 fuel minus those from producing and distributing gasoline. The estimated number of deaths for gasoline vehicles in 2020 is also shown. (c) Low (blue) and high (blue plus red) estimates of water consumption (gigagallons/year, where 1 gigagallon $= 10^9$ gallons) required to replace all U.S. on-road vehicles with different vehicle technologies. Consumption is net loss of water from water supply. For comparison, the total U.S. water consumption in 2000 was 148,900 gigagallons/year. From Jacobson (2009).

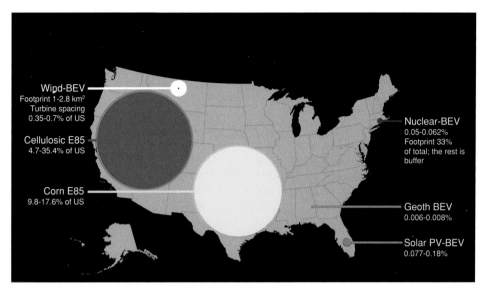

Figure 13.2. Footprint plus spacing area required for a given technology to provide energy for all U.S. vehicles in 2007 as either battery electric vehicles (in the case of wind, solar, geothermal, and nuclear) or E85 vehicles (in the case of corn and cellulosic ethanol). In the case of wind, the white is the spacing area, and the red dot in the center is the footprint. For nuclear, the footprint is one-third of the total. For the rest, the footprint and spacing are effectively the same. The percentages are relative to all fifty U.S. states. From Jacobson (2009).

13.2. Characteristics of Electricity-Generating Wind, Water, and Sunlight Technologies

In this section, the WWS electricity-producing electric power options are briefly described.

13.2.1. Wind

Wind turbines convert the kinetic energy of the wind into electricity. Generally, the slow-turning turbine blade spins a shaft connected to a **gearbox**. Gears of different sizes in the gearbox convert the slow-spinning motion (e.g., 5–10 rotations per minute for modern turbines) to faster-spinning motion (e.g., 1,800 rotations

(a)   (b)

Figure 13.3. (a) Wind farm over land at sunset. © Johannes Gerhardus Swanepoel/Dreamstime.com. (b) Offshore wind farm, Middelgrunden, Denmark. © Rodiks/Dreamstime.com.

per minute) needed to convert mechanical energy to electrical energy in a **generator**. Some modern wind turbines are gearless, with the shaft connected directly to the generator. To compensate for the slow spin rate in the generator in gearless turbines, the radius of rotation is expanded, increasing the speed at which magnets move around a coil in the generator to produce electricity. The efficiency of wind power increases with increasing turbine hub height because wind speeds generally increase with increasing height in the lower atmosphere. As such, larger turbines capture faster winds. Wind farms are often located on flat open land (Figure 13.3a), within mountain passes, on ridges, and offshore (Figure 13.3b). Individual turbines range in size up to 10 MW. High-altitude wind energy capture is also being pursued. Small turbines (e.g., 1–10 kW) are convenient for producing local electricity in the backyards of homes or city street canyons if winds are fast.

13.2.2. Wave

Winds passing over water create surface waves. The faster the wind speed, the longer a wave is sustained, the greater the distance it travels, and the greater its height. **Wave power devices** capture energy from ocean surface waves to produce electricity. One type of device (Figure 13.4), produced by Ocean Power Technologies, Inc., bobs up and down with a wave, creating mechanical energy that is converted to electricity in a generator. The electricity is sent through an underwater transmission cable to shore. Most of the body of this device is submerged under water. Another type of device,

Figure 13.4. Artist rendering of a field of wave converters used to generate electricity, as described in the text. Photo courtesy of Ocean Power Technologies, Inc.

produced by Pelamis Wave Power Ltd., is a floating snakelike device 180 m long by 4 m wide, with cylindrical sections connected by moveable joints. Up-and-down motion along the device increases the pressure on oil within it to drive a hydraulic ram to run a hydraulic motor, whose rotating motion is converted to electricity in a generator.

13.2.3. Geothermal

Geothermal energy is energy extracted from hot water and steam below the Earth's surface. Steam or hot water from the Earth has historically been used to provide heat for buildings, industrial processes, and domestic water. The first use of geothermal energy to generate electricity was by **Prince Piero Conti** of **Larderello**, Tuscany, Italy, in 1904. He lit four light bulbs by using steam from geothermal fields near his palace to drive a steam-driven engine attached to a generator. In 1911, he installed the first geothermal power plant, which had a capacity of 250 kW. This plant grew to 405 MW by 1975. The second electricity-producing plant was built at the Geysers Resort Hotel, California, in 1922. This plant was originally used only to generate electricity for the resort, but it has since been developed to produce electricity for the state.

Today, three major types of geothermal plants are dry steam, flash steam, and binary. **Dry and flash steam geothermal plants** operate where geothermal reservoir temperatures are 180°C to 370°C or higher. In both cases, two boreholes are drilled – one for steam alone (in the case of dry steam) or liquid water plus steam (in the case of flash steam) to flow up, and the second for condensed water to return after it passes through the plant. In the dry steam plant, the pressure of the steam rising up the first borehole powers a turbine, which drives a generator to produce electricity. About 70 percent of the steam recondenses after it passes through a condenser, and the rest is released to the air. Because $CO_2(g)$, $NO(g)$, $SO_2(g)$, and $H_2S(g)$ in the reservoir steam do not recondense along with water vapor, these gases are emitted to the air. Theoretically, they could be captured, but this has not been done to date. In a flash steam plant, the liquid water plus steam from the reservoir enters a flash tank held at low pressure, causing some of the water to vaporize ("flash"). The vapor then drives a turbine. About 70 percent of this vapor is recondensed. The remainder escapes with $CO_2(g)$ and other gases. The liquid water is injected back to the ground.

Binary geothermal plants are developed when the reservoir temperature is 120°C to 180°C. Water rising

(a)   (b)

Figure 13.5. (a) Steam from geothermal hot water in a natural river, Hveragerdi, Iceland, May 1, 2010. © Svobodapavel/Dreamstime.com. (b) Geothermal power station in Iceland, October 6, 2007. © Steve Allen/Dreamstime.com.

up a borehole is kept in an enclosed pipe and heats a low-boiling-point organic fluid, such as isobutene or isopentane, through a heat exchanger. The evaporated organic turns a turbine that powers a generator, producing electricity. Because the water from the reservoir stays in an enclosed pipe when it passes through the power plant and is reinjected to the reservoir, binary systems produce virtually no emissions of $CO_2(g)$, $NO(g)$, $SO_2(g)$, or $H_2S(g)$. About 15 percent of geothermal plants today are binary plants.

13.2.4. Hydroelectric

Water can generate electricity when it drops gravitationally, driving a turbine and generator. Although most **hydroelectric power** is produced by water falling from a reservoir behind a large dam (**large hydroelectricity**; Figure 13.6a), some is produced by water flowing from a river directly through a pipe or tunnel, past a turbine, and back to the river (Figure 13.6b). The latter is referred to as **run-of-the-river hydroelectricity**.

(a)   (b)

Figure 13.6. (a) Three Gorges Dam, China, on July 31, 2010, with air pollution in the background. Construction began in 1994 and was completed in 2009. It is 2.335 km long, 185 m high, 18 m wide at the top, and 130 m wide at the bottom. © Jjspring/Dreamstime.com. (b) Run-of-the-river hydroelectric plant, January 30, 2011. © Paura/Dreamstime.com.

The main advantage of both large hydro and run-of-the river hydro with a modest storage pond behind it is that they can provide electricity within 15 to 30 seconds of a need as long as other water needs are met. Hydroelectricity is thus ideal for providing **peaking power** (i.e., electric power at times of maximum and unexpected demand during the day) and filling in gaps when winds are not blowing or the sun is not shining. It is also used to provide **baseload power** (constant power all day less than or equal to minimum demand during the day).

13.2.5. Tidal

Tides are oscillating currents in the ocean caused by the rise and fall of the ocean surface due to the gravitational attraction among the Earth, moon, and sun. The rising and sinking motion of the ocean surface forces water below the surface to move horizontally as a current. A **tidal turbine** captures the kinetic energy of the ebbing and flowing current just as a wind turbine captures the kinetic energy of the wind. Like a wind turbine, a tidal turbine consists of a rotor, which provides rotating kinetic energy to a generator and converts it to electrical energy that is transmitted to shore. The turbine is generally mounted on the sea floor (Figure 13.7) and may or may not extend to the surface. The tidal turbine

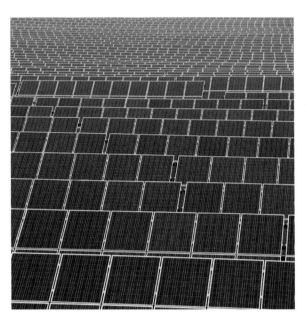

Figure 13.8. Solar photovoltaic farm, March 2011. © View7/Dreamstime.com.

rotor, which lies under water, may be fully exposed to the water or placed within a narrowing duct that directs water toward it. Because tides run about 6 hours in one direction before switching direction for another 6 hours, they are fairly predictable, so tidal turbines may be used to supply baseload energy.

13.2.6. Solar Photovoltaics

Solar photovoltaics (PVs) are arrays of cells (Figure 13.8) containing a material that converts solar radiation into **direct current (DC)** electricity. An inverter is then used to convert DC electricity into **alternating current (AC) electricity**. Photovoltaics can be mounted on roofs or combined in large-scale power plants. In power plants, panels are sometimes mounted on trackers that rotate to follow the sun. The presence of clouds, shadow-casting buildings or trees, dust that accumulates on solar panels, and extremely high temperatures can reduce the efficiency of a solar PV panel. However, when the sun is behind a tree or at an angle relative to the panels, the presence of thin clouds can significantly enhance solar radiation incident upon the panels by scattering light from multiple locations in the cloud more directly onto the panels.

13.2.7. Concentrated Solar Power

Concentrated solar power (CSP) is a technology by which sunlight is focused (concentrated) by mirrors or

Figure 13.7. Illustration of tidal turbine. Courtesy Hammerfest-Strøm. www.hammerfeststrom.com/products/tidal-turbines/.

Figure 13.9. Reflectors focusing solar energy onto a 10-megawatt receiver power tower at Solar One, a concentrated solar power plant in Barstow, California. Photo by Sandia National Laboratory Staff, available from the National Renewable Energy Laboratory, U.S. Department of Energy, www.nrel.gov.

reflective lenses to heat a fluid in a collector at high temperature. The heated fluid (e.g., pressurized steam, synthetic oil, molten salt) flows from the collector to a heat engine where a portion of the heat (up to 30 percent) is converted to electricity. Some types of CSP allow the heat to be stored for many hours so that electricity can be produced at night.

One type of collector is a set of **parabolic trough (long U-shaped) mirror reflectors** that focus light onto a pipe containing oil that flows to a chamber to heat water for a steam generator that produces electricity. A second type is a **central tower receiver** with a field of mirrors surrounding it (Figure 13.9). The focused light heats a circulating thermal storage medium, such as **molten sodium nitrate or potassium nitrate salts**, within an insulated reservoir to more than 500°C. The heat is then used to evaporate water flowing adjacent to the reservoir to produce steam for a steam turbine, which generates electricity. By storing heat in an insulated thermal storage media, the parabolic trough and central tower CSP plants can delay the heating of water, and thus electricity production, until nighttime. In fact, the heat can be stored for up to 15 hours before it is completely used, allowing for 24 hours of electricity production during and following a sunny day, as demonstrated by the Gemasolar CSP plant in Seville, Spain, in July 2011. During cloudy and winter days, electricity is also produced at night, but for fewer hours.

A third type of CSP technology is a **parabolic dish** (e.g., satellite dish) reflector that rotates to track the sun and reflects light onto a receiver, which transfers the energy to hydrogen in a closed loop. The expansion of hydrogen against a piston or turbine produces mechanical power used to run a generator or alternator to produce electricity. The power conversion unit is air cooled; thus, water cooling is not needed. Parabolic dish CSP is not coupled with thermal storage.

CSP plants require either air or water cooling. The use of air cooling, which is desirable in water-constrained locations, reduces overall CSP plant water requirements by 90 percent, with a cost of only about 5 percent less electric power produced (USDOE, 2008).

13.2.8. Use of Wind, Water, and Sunlight Power for Transportation

The cleanest and most efficient transportation technologies proposed for use on a large scale with WWS electricity include **battery electric vehicles (BEVs)** (Figure 13.10), **hydrogen fuel cell vehicles (HFCVs)**, and hybrid BEV-HFCVs. For ships, a WWS option is a hybrid hydrogen fuel cell-battery system, and for aircraft, liquefied hydrogen combustion.

BEVs store electricity in and draw power from batteries to run an electric motor that drives the vehicle. Because BEVs have zero tailpipe emissions, their only possible emissions beyond production and decommissioning of the vehicle are associated with the electricity they use. As long as the electricity they use is from a WWS source, the air pollution and global

Figure 13.10. Battery electric 2010 Tesla Roadster, which uses thousands of lithium ion laptop batteries. The vehicle can travel 244 miles (395 km) on one charge and requires 3.5 hours for a full charge. Photo by Mark Z. Jacobson.

warming–relevant gas and particle emissions from such vehicles over their lifetimes is nearly zero (Figure 13.1).

Another advantage of a BEV is that it can travel up to five times further per unit of input energy than can an internal combustion vehicle (km/kWh-outlet vs. km/kWh-gasoline). As such, BEVs require less energy than do internal combustion vehicles, and a conversion from internal combustion to BEVs will reduce energy demand for transportation fuel by a factor of up to 5. Contemporary BEVs use lithium ion batteries, which do not contain the toxic chemicals associated with lead acid or nickel cadmium batteries.

HFCVs use a fuel cell to convert hydrogen fuel and oxygen from the air into electricity that is used to run an electric motor. A **fuel cell** is a device that converts chemical energy from a fuel into electric energy. HFCVs are truly clean only if the hydrogen used by them is generated by **electrolysis** (passing electricity through water to split water, producing hydrogen), where the electricity is obtained from clean WWS technologies. To reduce storage space in a vehicle, the hydrogen must be compressed. Because energy is required during electrolysis, hydrogen compression, and fuel cell use, HFCVs are less efficient than are BEVs; however, they are more efficient than internal combustion vehicles. HFCVs emit only water vapor during their operation. The only other emissions associated with them aside from those resulting from vehicle and electrolyzer manufacturing and decommissioning are emissions associated with manufacturing and operating (if any) the WWS technology producing the electricity for electrolysis.

Liquid (cryogenic) hydrogen for air transportation is not a novel idea. Such hydrogen has been used in the space shuttle for decades and was tested in demonstration flights for the Soviet Union's commercial aircraft Tupolev Tu-154B, beginning on April 15, 1988. In the latter case, the aircraft was fitted with a thermally insulated fuel tank behind the passenger cabin that contained liquid hydrogen at a temperature of 20 K. The hydrogen powered a third engine on the aircraft.

Liquid hydrogen requires 4.2 times the volume of jet fuel for the same energy. As such, a liquid hydrogen aircraft requires a larger fuel tank than does a jet-fueled aircraft, increasing drag. However, jet fuel weighs 2.9 times more than does liquid hydrogen for the same energy, so a liquid hydrogen aircraft weighs much less than does a jet-fueled aircraft, offsetting much of the efficiency loss due to the greater drag. The advantage of a liquid hydrogen aircraft is the elimination of carbon dioxide, black carbon, carbon monoxide, reactive hydrocarbon, and sulfur dioxide emissions from the exhaust. Because the aircraft still requires combustion, nitrogen oxides are produced, water vapor is emitted, and thus contrails still form.

13.2.9. Use of Wind, Water, and Sunlight Power for Heating and Cooling

For water and air heating in buildings, WWS energy technologies include heat pumps and electric resistance heaters. Heat pumps are efficient devices that extract heat from the air or ground, even at low temperature. The heat is either used to heat water or forced through vents into a building to heat air. Heat pumps can act in reverse as air conditioners. A heat pump that extracts heat from the ground is a **ground-source heat pump**, whereas one that extracts heat from the air is an **air-source heat pump** (Figure 13.11). One that extracts heat from water, such as a swimming pool, is a **water-source heat pump**. The advantage of a ground-source heat pump over an air-source heat pump is that temperatures under the ground are relatively stable and warm, even when the air outside is cold; thus, ground-source heat pumps are more efficient under cold air conditions. The advantage of an air-source heat pump is that it is easier to install and maintain because none of its parts is buried underground.

Electric resistance heaters can also be used for air and water heating, but they are less efficient than are heat pumps. **Rooftop solar hot water heaters** have

Figure 13.11. Air-source heat pump, installed outdoors but used for indoor air heating and cooling. Photo by Mark Z. Jacobson.

been used worldwide for decades. In a WWS world, they would be used to preheat water in a heat pump water heater or electric resistance water heater.

For high-temperature industrial processes, high temperatures can be obtained by combusting electrolytic hydrogen or with electric resistance heating. The electricity used to run a heat pump or resistance heater or to produce hydrogen would be produced by WWS technologies.

13.3. Energy Needed to Power the World

The power required in 2008 to satisfy all end-use power demand worldwide for all purposes was about 12.5 TW. End-use power excludes losses incurred during the production and transmission of power. About 35 percent of primary energy worldwide in 2008 was from oil; 27 percent was from coal; 23 percent was from natural gas; 6 percent was from nuclear power; and the rest was from biofuel, sunlight, wind, and geothermal power. Delivered electricity was about 2.2 TW of the all-purpose, end-use power.

If the world follows the current trajectory of fossil fuel growth, all-purpose, end-use power demand will increase to almost 17 TW by 2030, and U.S. demand will increase to almost 3 TW (Table 13.2). The breakdown in terms of primary energy will be similar to that today, that is, heavily dependent on fossil fuels. What would the world's power demand look like if the energy infrastructure for all end uses was converted to a sustainable WWS infrastructure?

Table 13.2 estimates the global and U.S. end-use power demand, by sector, in a world powered entirely by WWS, with zero fossil fuel or biofuel combustion. It is assumed that all end uses that can feasibly be electrified will use WWS power directly and that the remaining end uses will use WWS power indirectly in the form of electrolytic hydrogen (hydrogen produced by splitting water with WWS power), as described in Section 13.2.

Table 13.2 indicates that a conversion to a *WWS infrastructure reduces worldwide end-use power demand by 30 percent*. The main reason is that the use of electricity for heating and electric motors is considerably more efficient than is fuel combustion in the same applications. Also, the use of WWS electricity to produce hydrogen for fuel cell vehicles, although less efficient than the use of WWS electricity to run BEVs, is more efficient and cleaner than is combusting liquid fossil fuels for transportation. Combusting electrolytic hydrogen is slightly less efficient but cleaner than is combusting fossil fuels for direct heating. However, the lower efficiency of direct hydrogen combustion is accounted for in Table 13.2. Some power demand reductions in a WWS world in Table 13.2 are due to modest energy conservation measures and to the elimination of the energy requirement for petroleum refining.

13.4. Wind, Water, and Sunlight Resources Available to Power the World

How do the power requirements of a WWS world, shown in Table 13.2, compare with the availability of WWS power? Table 13.3 gives the estimated power available worldwide from renewable energy in terms of raw resources, resources available in high-power locations, resources that can feasibly be extracted in the near term considering cost and location, and contemporary resources used. Table 13.3 indicates that only wind and solar resources can provide more power on their own than energy demand worldwide. Wind in likely developable locations can power a WWS world about 3.5 to 7 times over and solar about 20 to 30 times over.

Figure 13.12 shows the modeled world wind resources at 100 m, which is in the range of the hub height of modern wind turbines. Globally, $\sim$1,700 TW of wind power would theoretically be available over the world's land plus ocean surfaces at 100 m if winds at all speeds were used to power wind turbines and if wind speed losses due to turbine energy extraction were limited to turbine wakes (Table 13.3). In reality though, the world maximum, or saturation, wind potential is lower due to array losses (Section 13.5) that grow with increasing wind penetration. The wind power over land and near shore where the wind speed is 7 m s^{-1} or faster (the speed necessary for cost-competitive wind energy; Jacobson and Masters, 2001) is around 72 to 170 TW (Archer and Jacobson, 2005; Lu et al., 2009; Jacobson and Delucchi, 2011). Data analyses indicate that 15 percent of the data stations (and, thus, statistically land area) in the United States (and 17 percent of land and coastal offshore data stations) have wind speeds 7 m s^{-1} or faster. Globally, 13 percent of stations are above that threshold (Archer and Jacobson, 2005). More than half of the land and offshore power in high-energy locations could be practically developed. Large regions of fast winds worldwide include the Great Plains of the United States and Canada; northern Europe; the Gobi and Sahara Deserts; much of the Australian desert areas; and parts of South Africa, and southern South America. In the United States, wind from the Great Plains and offshore along the East Coast (Kempton et al., 2007) could supply all U.S. power needs. Other windy offshore regions include the North Sea, the West Coast of

Table 13.3. Power available in wind, water, and sunlight (WWS) energy resources worldwide if energy were used in conversion devices

Energy technology	(a) Power worldwide (TW)	(b) Power in high-energy locations (TW)	(c) Power in likely developable locations (TW)	(d) Power delivered as electricity 2009 (GW)[a]
Wind	1,700	72–170	40–85	49
Wave	>2.7	2.7	0.5	0.0007
Geothermal	45	2	0.07–0.14	7.6
Hydroelectric	1.9	<1.9	1.6	410
Tidal	3.7	0.8	0.02	0.066
Solar PV	6,500	1,300	340	3.8
CSP	4,600	920	240	0.119
TOTAL				469

The power available is before transmission, distribution, or array (in the case of wind, wave, and tidal) losses are taken into account. For wind in likely developable locations, such losses are about 10–15 percent of the available power (Section 13.5). For wind turbines covering the world, array losses are much larger due to increasing interference among wind farms. Also shown is 2009 delivered electricity from WWS resources. The total delivered electricity (469 GW) by WWS resources represents 21.3 percent of the world end-use electric power consumption in 2008 (2.2 TW). CSP, concentrated solar power; PV, photovoltaic.
[a] Renewable Energy Policy Network for the 21st Century (REN21; 2010).
Sources: From Jacobson and Delucchi (2011) and references therein.

the United States (Dvorak et al., 2010), and the east coast of Asia.

Figure 13.13 shows the distribution of solar energy at the Earth's surface. Globally, 6,500 TW of solar energy are available over the world's land and ocean surfaces, assuming only sunlight was used to power PV devices (Table 13.3); however, the deliverable solar power over land in locations where solar PV could be developed practically is about 340 TW. Alternatively, CSP could provide about 240 TW of the world's power output, less than PV because the land area required for CSP without storage is about one-third greater than is that for PV. With thermal storage, the land area for CSP increases because more solar collectors are needed to provide energy for storage, but energy output does not change and can be used at night.

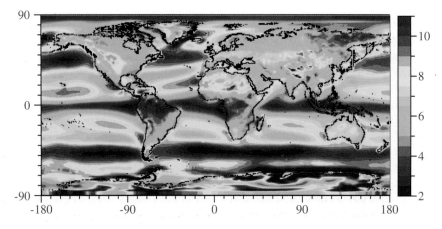

Figure 13.12. Yearly averaged wind speed (m s^{-1}) at 100 m above ground level, modeled by computer at 1.5 × 1.5 degrees horizontal resolution. The globally, land-, and ocean-averaged values are 7.0, 6.1, and 7.3 m s^{-1}, respectively. From Jacobson and Delucchi (2011).

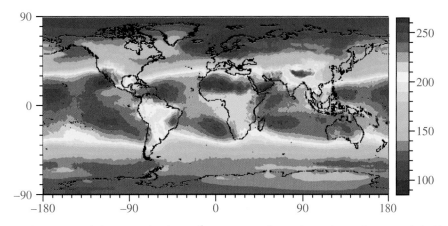

Figure 13.13. Yearly averaged downward solar radiation (W m^{-2}) reaching the surface, modeled by computer at 1.5 × 1.5 degrees horizontal resolution. The globally and land-averaged values are 193 and 183 W m^{-2}, respectively. From Jacobson and Delucchi (2011).

The other WWS technologies have less resource availability in likely developable locations than does wind, CSP, or PV (Table 13.3). Nevertheless, they will still make important contributions to the WWS solution. Wave power can be extracted practically near coastal areas, which limits its worldwide potential. Although the Earth has a very large reservoir of geothermal energy below the surface, most of it is too deep to extract practically. Even though hydroelectric power today exceeds all other sources of WWS power, its future potential is limited because most of the large reservoirs suitable for generating hydropower are already in use.

13.5. Number, Footprint, and Spacing of Plants and Devices Required

How many WWS power plants or devices are needed to power the world, assuming end-use power requirements from Table 13.2? Table 13.4 provides one of several possible future scenario for 2030. In this scenario, wind is assumed to comprise 50 percent of the world's supply, and solar is assumed to comprise 40 percent. These two resources are the only ones that could feasibly power the world independently, based on the data in Table 13.3. Although more solar is available than wind worldwide, wind is currently less expensive and thus likely to play a greater role. However, because a combination of solar and wind is needed to match power supply with demand effectively (Section 13.8), solar must play a role nearly as large as wind.

Solar energy in Table 13.4 is divided into rooftop PV (supplying 6 percent of the world's energy), PV power plants (supplying 14 percent), and CSP plants (supplying 20 percent). The rooftop PV-versus-power plant PV division is based on an analysis of likely available rooftop area. Rooftop PV does not require an electricity transmission and distribution network or new land area. Although 4 percent of the proposed future supply is hydroelectric power, most of this (70 percent) is already in place. Geothermal power is proposed to supply another 4 percent. Tidal and wave power would each supply 1 percent of the world's end-use power.

A derivation of the number of devices for each energy technology, the footprint the devices occupy, and the spacing they require in Table 13.4 is provided here using wind energy as an example. The number of wind turbines is determined by first calculating the annual energy output of a single wind turbine, accounting for efficiency losses, and then comparing this number to the worldwide energy that wind must supply.

The annual energy output (kWh yr^{-1}) from a single wind turbine is

$$E_t = P_r \times CF \times H \times \eta_t \quad (13.1)$$

where P_r is the rated power (kW) of the wind turbine, CF is the capacity factor of the turbine (dimensionless), $H = 8{,}760$ h yr^{-1} is the number of hours in a year, and η_t is the system efficiency.

The **rated power** of a wind turbine is the maximum instantaneous power that the turbine can produce. Wind turbines produce power as a function of wind speed (Figure 13.14). Between zero wind speed and the **cut-in wind speed** (generally, 2–3.5 m s^{-1}), the turbine is not

Table 13.4. Number of wind, water, and sunlight (WWS) power plants or devices needed to provide the world and U.S. total end-use power demand in 2030 (11.5 and 1.8 TW, respectively, from Table 13.2)

Energy technology	Rated power of one plant/device (MW)	Percent of 2030 power demand met by plant/device	Number of plants/devices needed worldwide	Footprint area (percent of global land area)	Spacing area (percent of global land area)	Number of plants/devices needed in United States
Wind turbine	5	50	3.8 million	0.000033	1.17	590,000
Wave device	0.75	1	720,000	0.00026	0.013	110,000
Geothermal plant	100	4	5,350	0.0013	0	830
Hydroelectric plant	1,300	4	900[a]	0.407[a]	0	140[a]
Tidal turbine	1	1	490,000	0.000098	0.0013	7,600
Roof PV system	0.003	6	1.7 billion	0.042[b]	0	265 million
Solar PV plant	300	14	40,000	0.097	0	6,200
CSP plant	300	20	49,000	0.192	0	7,600
TOTAL		100		0.74	1.18	
TOTAL NEW LAND				0.41[c]	0.59[c]	

This table assumes the given fractionation of demand among plants or devices while accounting for transmission, distribution, and array losses. Also shown are the footprint and spacing areas required to power the world as a percentage of the global land area, 1.446×10^8 km². CSP, concentrated solar power; PV, photovoltaic.

[a] About 70 percent of the hydroelectric plants are already in place. See Jacobson (2009) for a discussion of apportioning the hydroelectric footprint area by use of the reservoir.

[b] The footprint area for rooftop solar PV does not represent an increase in land because the rooftops already exist and are not used for other purposes.

[c] Assumes that 50 percent of the wind is over water, and wave and tidal are in water; 70 percent of hydroelectric is already in place; and rooftop solar does not require new land.

Source: Jacobson and Delucchi (2011).

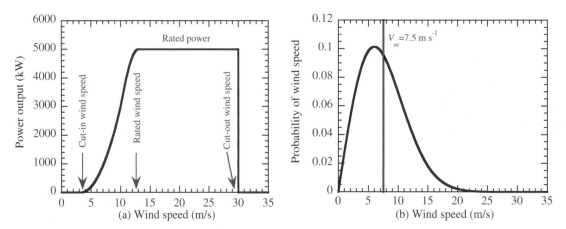

Figure 13.14. (a) Power curve for an RE Power wind turbine rated at 5,000 kW (5 MW). The curve shows the power output of the turbine as a function of instantaneous wind speed. The turbine rotor diameter is 126 m, and the hub height is 100 m above the surface. The cut-in wind speed is 3.5 m s^{-1}, the rated wind speed is 13 m s^{-1}, and the cut-out wind speed is 30 m s^{-1}. From Archer and Jacobson (2007). (b) Rayleigh probability distribution of wind speed when the mean wind speed is 7.5 m s^{-1}.

designed to produce power because the power generated would be low and uneconomical. Above the cut-in wind speed, the instantaneous power output increases roughly proportionally to the cube of the wind speed. At the **rated wind speed**, the power output reaches the rated power (Figure 13.14a). The power output stays at the rated power for all higher wind speeds until the **cut-out wind speed** is reached, at which point the power output drops to zero to prevent damage to the turbine. A turbine can survive wind speeds up to the **destruction wind speed**, which is often 50 to 60 m s^{-1}. For comparison, wind speeds in a Category 4 hurricane are 58.3 to 69 m s^{-1}.

If a wind turbine ran for a full year at its rated power, its energy output would be P_rH. However, because wind speeds are generally lower than the rated wind speed, wind turbines realize only a fraction of their maximum energy output during the year. The fraction of the maximum energy that a wind turbine produces during a year is the **capacity factor** of the turbine. A simple, yet accurate (within 1–3 percent for most and <10 percent for all turbines tested) equation for the capacity factor of a wind turbine is

$$CF = 0.087 \times V_m - P_r/D^2 \qquad (13.2)$$

where V_m is the mean annual wind speed in units of m s^{-1}, P_r is the rated power in kW, and D is the turbine rotor diameter in m (Masters, 2004; Jacobson and Masters, 2001). The units used must be those specified here. Units do not equate because the equation is empirical. The mean annual wind speed in Equation 13.2 differs from the instantaneous wind speed used in Figure 13.14a. The mean wind speed used in Equation 13.2 is the mean of a **Rayleigh probability distribution** of wind speeds, which is a probability distribution of wind speeds that looks similar to a bell curve but skewed toward higher wind speeds (Figure 13.14b). The Rayleigh probability distribution is a specialized case of the **Weibull probability distribution** of wind speeds,

$$f(v) = \frac{k}{c}\left(\frac{v}{c}\right)^{k-1} \exp\left[-\left(\frac{v}{c}\right)^k\right] \qquad (13.3)$$

where $f(v)$ is the fractional occurrence of instantaneous wind speed v (m s^{-1}) among all wind speeds, k is an integer, and $c = 2V_m/\sqrt{\pi}$. For a Rayleigh distribution, $k = 2$. Winds are generally Rayleigh in nature, so measurements of a wind speed frequency distribution over a year often result in a Rayleigh distribution (Archer and Jacobson, 2005).

> **Example 13.1**
> Calculate the range of capacity factors of a 5-MW turbine with a 126-m rotor diameter operating in 7 to 8.5 m s^{-1} annually averaged wind speeds.
>
> **Solution**
> Substituting $P_r = 5{,}000$ kW, $D = 126$ m, and $V_m = 7$–8.5 m s^{-1} into Equation 13.2 gives capacity factors ranging from 0.294 to 0.425. Thus, during the year, this turbine produces 29.4 to 42.5 percent of its maximum possible energy output.

The **system efficiency** of a wind turbine is the ratio of energy delivered to raw energy produced by the turbine; it accounts for array, transmission, and distribution losses. **Array losses** are energy losses resulting from decreases in wind speed that occur at a large wind farm when upstream turbines extract energy from the wind, reducing the wind speed slightly for downstream turbines. Upstream turbines also create vortices or ripples (wakes) that can interfere with downwind turbines. The greater the spacing between wind turbines, the lower the array losses due to loss of energy in the wind, vortices, and wakes. In a wind farm with an array of turbines, the **spacing area** (m^2) occupied by one turbine is typically

$$A_t = 4D \times 7D \quad \text{or} \quad A_t = 3D \times 10D \qquad (13.4)$$

where D is the turbine rotor diameter (m). The first dimension is the distance between turbines in a row, and the second dimension is the distance between rows in the predominant direction of the wind. Spacing between wind turbines is not wasted space. It is often used for agriculture, rangeland, or open space. Over the ocean, it is open water.

With the spacing configurations in Equation 13.4, array losses are generally 5 to 20 percent (Masters, 2004). Frandsen (2007), for example, examined wind speed data from the Norrekaer Enge II wind farm in Denmark, which consists of seven rows of six 300-KW turbines. Spacing within rows was $7D$ to $8D$ and between rows was $6D$. Wind speeds averaged over each successive row were derived from power output. The data indicated that, despite wind speed reductions past turbine rows 1 and 2, wind speeds in subsequent turbine rows stayed constant or increased compared with row 3. The overall reduction in wind speed between rows 1 and 7 was only 7 percent, representing an array efficiency of 93 percent.

Immediately downwind of an individual turbine (in the **turbine's wake**), wind speeds first decrease and then increase, eventually converging into the background wind speed. The reason for the **regeneration of the wind** past a turbine is as follows. As wind speeds first decrease within a wake, the vertical and horizontal gradients in wind speed increase. In other words, outside the wake the wind speed remains fast, whereas within the wake the wind speed is slower, so the gradient in wind speed increases. The larger gradient in wind speed causes faster winds from outside the wake to flow into the wake, both vertically (from aloft, downward) and horizontally. In addition, the deficit in wind speed in the wake increases the horizontal pressure gradient and thus increases the pressure gradient force. The wind speed in the wake must then partly regenerate to ensure geostrophic balance (Section 6.2.1).

Transmission and distribution losses are losses that occur between the power source and end user of electricity. Transmission losses are losses of energy that occur along a transmission line due to resistance. Distribution losses occur due to step-up transformers, which increase voltage from an energy source to a high-voltage transmission line and decrease voltage from the high-voltage transmission line to the local distribution line. The average transmission plus distribution losses in the United States in 2007 were 6.5 percent (Energy Information Administration, 2011b). The overall system efficiency of a wind farm, accounting for array, transmission, and distribution losses, typically ranges from $\eta_t = 0.85$–0.9. Thus, such losses reduce estimated wind energy resources in likely developable locations (e.g., in Table 13.3) by 10 to 15 percent.

Example 13.2
Given a world end-use power demand of 11.5 TW in 2030 for all purposes, calculate the number of wind turbines from Example 13.1 needed to power 50 percent of the world's energy. Assume the system efficiency due to array, transmission, and distribution losses in the low wind speed case is 0.85 and in the high wind speed case is 0.9.

Solution
From Equation 13.1, the annual energy output of the individual turbine in Example 13.1 ranges from $E_t = 11.0$–16.7 million kWh yr^{-1} for the low and high wind speed cases, respectively. Multiplying world end-use power demand of 11.5 TW in 2030 by $H = 8{,}760$ h yr^{-1} gives the world end-use energy demand for all purposes of 1.01×10^{14} kWh yr^{-1}. Multiplying this number by one-half (because wind will supply half the

---

total energy) and then dividing by the energy produced per turbine E_t gives 4.6 million turbines at a wind speed of 7 m s^{-1} and 3.0 million turbines at a wind speed of 8.5 m s^{-1}. The average of these two is 3.8 million wind turbines, the estimate provided in Table 13.4.

The **footprint** on the ground of an energy device is the actual land area of the top surface of soil touched by the device. It differs from the spacing area, which is the area between devices required to reduce interference of one device with the other, as just illustrated for wind. Separation of devices is also needed for wave and tidal farms.

In the case of wind turbines, the footprint consists primarily of the area of the turbine's tubular tower plus that of the base on which it sits (generally 4–5 m in diameter total). Whereas wind turbines have foundations under the ground larger than their bases on the ground, such underground foundation areas are not footprint because the foundations are covered with dirt, allowing vegetation to grow and wildlife to flourish on top of them. The footprint area for wind also does not include temporary or unpaved dirt access roads because most large-scale wind farms will go over areas such as the Great Plains and some desert regions, where photographs of several farms indicate that unpaved access roads blend into the natural environment and are often overgrown by vegetation. Offshore wind farms do not require any roads. In farmland locations, most access roads have dual purposes, serving both agricultural fields and turbines. In cases where paved access roads are needed, 1 km^2 of land provides about 200 km (124 miles) of linear roadway 5 m wide; thus, access roads would not increase the footprint requirements of wind farms more than a small amount.

The footprint area also does not include transmission because the actual footprint area of a transmission tower is smaller than is the footprint area of a wind turbine. This is due to the fact that a transmission tower consists of four narrow metal support rods separated by distance that penetrate the soil to an underground foundation. Many photographs of transmission towers indicate more vegetation growing under the towers than around the towers because areas around the towers are often agricultural or used for other purposes, whereas the area under the tower is vegetated soil. Because the land under transmission towers supports vegetation and wildlife, it is not considered footprint beyond the small area of the support rods. Example 13.3 provides a calculation for the footprint and spacing needed to power the world with 50 percent wind in 2030.

Example 13.3

Calculate the footprint and spacing area required for wind turbines from Example 13.2 to power 50 percent of the world's power for all purposes in 2030, assuming a tubular tower plus base circular diameter of 4 m and a spacing area of $A_t = 4D \times 7D$.

Solution

Under the assumption that each turbine has a circular base, the footprint area occupied by 3.8 million turbines is 48 km², smaller than the area of Manhattan (59.5 km²). The spacing area required is 1.69×10^6 km², or 1.17 percent of the global land area of 1.446×10^8 km².

The footprint area required for rooftop solar PV is zero because rooftops already exist. For nonrooftop solar PV plus CSP, the spacing areas are assumed to be the same as the footprint areas. Under this assumption, Table 13.4 indicates that providing 34 percent of the world's power with nonrooftop solar PV plus CSP requires about one-fourth of the land area for footprint plus spacing as does powering 50 percent of the world with wind, but a much larger footprint area alone than does wind. However, the footprint area estimate required for solar is conservative because solar PV power plant panels, for example, can be elevated above grass with a pedestal, significantly decreasing the footprint area required for PV.

Geothermal power requires a smaller footprint than does solar but a larger footprint than does wind per unit energy generated. The footprint area required for hydroelectric per unit electric power generated is large due to the area required to store water in a reservoir. However, 70 percent of needed hydroelectric power for a WWS system is already in place.

Together, the entire WWS solution would require the equivalent of ~0.74 percent of the global land surface area for footprint and 1.18 percent for spacing (or 1.9 percent for footprint plus spacing). Up to 61 percent of the footprint plus spacing area could be over the ocean if all wind turbines were placed over the ocean, although a more likely scenario is that 30 to 60 percent of wind turbines may ultimately be placed over the ocean given the strong wind speeds there (Figure 13.12).

Considering that 50 percent of wind turbines may be placed over the ocean, that wave and tidal are already 100 percent over the ocean, that 70 percent of hydroelectric power is already in place, and that rooftop solar does not require new land, the additional footprint and spacing areas required for WWS power worldwide are only ~0.41 and ~0.59 percent, respectively, of all land (or 1 percent of all land for footprint plus spacing) worldwide. This compares with ~40 percent of the world's land currently used for agriculture and pasture.

13.6. Material Resources Required

In a global, all-WWS power system, the new technologies produced in the greatest abundance will be wind turbines, solar PVs, CSP systems, BEVs, and HFCVs. This section discusses the availability of materials for these technologies.

13.6.1. Materials for Wind Turbines

The primary materials needed for wind turbines include steel (for towers, nacelles, and rotors), prestressed concrete (for towers), magnetic materials (for gearboxes), aluminum (for nacelles), copper (for generators), wood epoxy (for rotor blades), glass fiber–reinforced plastic (GRP; for rotor blades), and carbon filament–reinforced plastic (CFRP; for rotor blades). In the future, use of composites of GRP, CFRP, and steel will likely increase.

The manufacture of 3.8 million 5-MW or larger wind turbines to power 50 percent of the world's energy in 2030 (Table 13.4) will require large amounts of bulk materials such as **steel and concrete**. However, there do not appear to be significant environmental or economic constraints on expanded production of these materials. The major components of concrete – gravel, sand, and limestone – are abundant, and concrete can be recycled and reused. The world does have somewhat limited reserves of economically recoverable iron ore (on the order of 100 to 200 years at current production rates, but the steel used to make towers, nacelles, and rotors for wind turbines should be 100 percent recyclable).

Copper is used in coils to conduct electricity in wind turbine generators. The production of millions of wind turbines would consume less than 10 percent of the world's low-cost copper reserves. Other conductors could also be used instead of copper.

Table 13.5. Estimated resource by country for $Nd_2O_3(s)$, as of 2009 (Tg)

Country	Resource
United States	2.1
CIS	3.8
India	0.2
Australia	1.0
China	16.0
Other	4.1
WORLD LAND TOTAL	27.3

CIS, Commonwealth of Independent States.
Sources: Jacobson and Delucchi (2011); U.S. Geological Survey (2011).

An element used in wind turbine generator permanent magnets is **neodymium** (Nd). Nd is one of seventeen **rare earth elements (REEs)**. REEs are not actually rare because they comprise a significant fraction of the Earth's crust. For example, Nd comprises $38 \, mg \, kg^{-1}$ of the crust. REEs were named as they were because they were initially found distributed rather than concentrated in economically extractable deposits. However, more deposits have been discovered over time.

Building 3.8 million turbines would require about 3.8 million metric tonnes of Nd, or about 4.4 million metric tonnes of neodymium oxide [$Nd_2O_3(s)$]. Table 13.5 indicates that the estimated world resource of $Nd_2O_3(s)$ is about 27.3 million tonnes. The **resource** of a material is defined as the "concentration of naturally occurring solid, liquid, or gaseous material in or on the Earth's crust in such form and amount that economic extraction of a commodity from the concentration is currently or potentially feasible" (U.S. Geological Survey, 2011). Thus, Nd should not be a limiting factor in worldwide wind power development, although recycling may be needed.

13.6.2. Materials for Solar Photovoltaics

Solar PVs are made of amorphous, polycrystalline, and microcrystalline silicon; cadmium telluride; copper indium selenide/sulfide; and other materials. Because many alternative types of PV systems are available, it is unlikely that the growth of PV will be constrained by one or a few materials. For example, for thin-film PVs, substituting ZnO electrodes for indium thin oxide allows multiterawatt production of PV cells. The limited availability of tellurium (Te) and indium (In) reduces the prospects of cadmium telluride (CdTe) and copper indium gallium selenide (CIGS) thin-film cells.

The power production of silicon PV technologies is limited not by crystalline silicon (because silicon is widely abundant) but by silver, which is used as an electrode. Reducing the use of silver as an electrode would allow virtually limitless production of silicon-based solar cells.

For multijunction concentrator cells, the limiting material is germanium (Ge); however, substitution of gallium (Ga), which is more abundant, would allow terawatt expansion. Overall, it is unlikely that the development of a large global PV system will be limited by the scarcity or cost of raw materials.

13.6.3. Materials for Concentrated Solar Power

CSP plants consist primarily of mirrors, receivers, and thermal storage fluid. Mirrors are generally glass mirrors with a reflective silver layer on the back of the glass. Receivers are stainless steel tubes with a surface that selectively absorbs solar radiation and is surrounded by an antireflective glass tube. The receiver heats the thermal storage fluid circulating through the receiver inner tube. Materials shortages are not expected to limit the large-scale production of CSP, particularly because many alternate mirror types, receivers, and thermal storage fluids are possible.

13.6.4. Materials for Electric Vehicles

For electric vehicles, two types of materials are of most concern: Nd for electric motors and lithium for lithium ion batteries. Some permanent magnet alternating current motors, such as those used in the Toyota Prius hybrid electric vehicle, use Nd. However, the amount required per vehicle is about an order of magnitude less than that required per wind turbine, and Nd is not a limiting factor in wind turbine production (Table 13.5). Furthermore, numerous electric motors do not use Nd.

Table 13.6 shows recent estimates of worldwide **lithium** resources; however, it does not include the recently discovered, potentially large lithium resources in Afghanistan. More than one-fourth of the estimated global lithium resources in 2011 were in Bolivia (Figure 13.15). However, Bolivia to date has produced little lithium, and Chile is currently the world's leading producer.

At 10 kg per vehicle, the current estimated world lithium resource could supply 3.3 billion vehicles, far

Table 13.6. Lithium resources by country, as of 2011 (million metric tonnes)

Country	Resource
United States	4
Argentina	2.6
Afghanistan	?
Australia	0.63
Bolivia	9
Brazil	1
Canada	0.36
Chile	7.5
China	5.4
Congo	1.0
Portugal	0.01
Serbia	1.0
Zimbabwe	0.023
WORLD LAND TOTAL	33
WORLD OCEANS	240

Source: U.S. Geological Survey (2011).

more than the 800 million vehicles that exist today. As such, lithium may not be a limiting factor in electric vehicle production in a 100 percent WWS economy. If it is, additional potential sources of lithium are the oceans, which contain 240 million tonnes, far more than all estimated land resources. However, currently, the cost of extracting ocean lithium is high and energy intensive. Finally, if demand does grow to the size of supply, the price of lithium will rise, spurring the hunt for new sources of lithium and encouraging more recycling.

Figure 13.15. Road through Salar de Uyuni, southwest Bolivia, the largest salt flat in the world. The crust, consisting of halite and gypsum, covers a brine pool containing the largest lithium resource in the world. © Paop/Dreamstime.com.

13.6.5. Materials for Hydrogen Fuel Cell Vehicles

The production of millions of HFCVs would increase demand for **platinum** (Pt) used in fuel cells. However, because HFCVs would replace gasoline and diesel vehicles, many of which use Pt in catalytic converters, the increase in Pt demand for HFCVs would be partly offset by a decrease in Pt demand for combustion vehicles.

Because BEVs would comprise the largest share of vehicles in a WWS world due to their high efficiency, the number of HFCVs required would be limited. If 10 percent of the vehicle fleet were HFCVs, about 100 million 50-kW HFCVs might be needed in 2030. These would require about 1.25 million kg Pt. The estimated resource of Pt group metals in deposits worldwide is more than 100 million kg (U.S. Geological Survey, 2011). **Platinum group metals** (PGMs) are six elements – ruthenium (Ru), rhodium (Rh), palladium (Pd), osmium (Os), iridium (Ir), and Pt – that have similar physical and chemical properties and are often found in the same deposits. The largest known deposit of PGMs is in the Bushveld Igneous Complex of the Transvaal Basin, South Africa. Other large deposits are near Norilsk, Russia, and Sudbury, Canada. Pt comprises at least one-sixth of total PGM deposits; thus, Pt resources should be much larger than the Pt needed for vehicles. In addition, Pt is already recycled, and additional Pt will be available due to the reduced need for Pt in catalytic converters upon conversion to WWS. As such, Pt is not considered a limiting element in a WWS economy.

13.7. Downtime of Wind, Water, and Sunlight versus Conventional Energy Technologies

One goal of any electric power system is to minimize downtime due to system failure or maintenance. Extreme events and unplanned maintenance can shut down plants unexpectedly. For example, unplanned maintenance can shut down a coal plant, extreme heat waves can cause cooling water to warm sufficiently to shut down a nuclear plant, supply disruptions can curtail the availability of natural gas, and droughts can reduce the availability of hydroelectricity.

WWS technologies generally suffer less downtime than do conventional electric power technologies. For example, the average coal plant in the United States from 2000 to 2004 was down 6.5 percent of the year for unscheduled maintenance and 6.0 percent of the year for scheduled maintenance (North American Electric Reliability Corporation (NERC), 2009), but modern wind turbines have a down time of only 0 to 2 percent over land and 0 to 5 percent over the ocean (Dong Energy et al., 2006, p. 133). Similarly, commercial solar projects have downtimes of ~1 percent on average, although some have zero downtime during a year and others up to 10 percent (Banke, 2010).

A difference, though, exists between outages of **centralized power plants** (coal, nuclear, natural gas) and outages of **distributed power plants** (wind, solar, wave). When individual solar panels or wind turbines are down, only a small fraction of electrical production is affected. When a centralized plant is down, a large fraction of the grid is affected. When more than one large, centralized plant is offline at the same time, an entire grid can be affected.

13.8. Reliably Matching Demand with Variable Wind, Water, and Sunlight Resources

Wind, wave, and solar power produce output that varies in time according to the weather. Thus, they are referred to as **variable WWS resources**. Another term commonly used to describe variability is **intermittency**. However, all energy resources are intermittent due to scheduled and unscheduled maintenance (Section 13.7), whereas variable resources are those whose energy outputs vary with the weather in addition to being affected by maintenance.

One concern with the use of variable WWS resources is whether such resources can provide reliable electric power. Any electricity system must respond to changes in demand (also referred to as **load**) over periods of seconds, minutes, hours, seasons, and years, and accommodate unanticipated changes in the availability of generation. It is not possible to control the weather; thus, a sudden change in demand often cannot be met by a variable resource.

The concern about matching demand with supply, though, applies to all energy sources, not just the variable ones. For example, because coal and nuclear provide constant (**baseload**) supplies during the day, they do not match power demand, which varies continuously. As a result, gap-filling resources, such as natural gas and hydroelectric, are used to meet demand peaks. Some WWS technologies, including geothermal and tidal power, can also serve as baseload power sources.

Although concern exists about stability of the current grid, it generally works, increasing the desire of grid operators to maintain the status quo. In Denmark, wind energy supplies about 20 percent of the total electric power, suggesting that a large penetration of WWS power may also work. However, variable renewables have not been used to date to power larger percentages of a grid, and thus concerns persist. Such concerns, though, have not yet been based on real experience or scientific data.

There are at least seven ways to design and operate a WWS energy system so that it will reliably satisfy demand:

1. Interconnect geographically dispersed variable energy sources (e.g., wind, solar, wave).
2. Combine WWS resources as one commodity and use a nonvariable energy source, such as hydroelectric power, to fill in remaining gaps between energy demand and supply.
3. Use demand-response management to shift times of demand to better match the availability of WWS power.
4. Store electric power at the site of generation for later use.
5. Oversize WWS peak generation capacity to minimize the times when available WWS power is less than demand, and provide power to produce hydrogen for transportation and heating when WWS power exceeds demand.
6. Store electric power in electric vehicle batteries.
7. Integrate weather forecasts into system operation.

13.8.1. Interconnecting Geographically Dispersed Generators

Interconnecting geographically dispersed wind, wave, or solar farms to a common transmission grid smoothes out electricity supply significantly (Kahn, 1979). Similarly, the combined energy from colocated wind and wave farms reduces variability of wind and wave power individually (Stoutenburg et al., 2010). For wind alone, interconnection over regions as small as a few hundred kilometers apart can eliminate hours of zero power, accumulated over all wind farms. When nineteen geographically dispersed wind sites in the Midwest, over a region 850 × 850 km, were hypothetically interconnected, about 33 percent of yearly averaged wind power was calculated to be available at the same reliability as a coal-fired power plant (Archer and Jacobson,

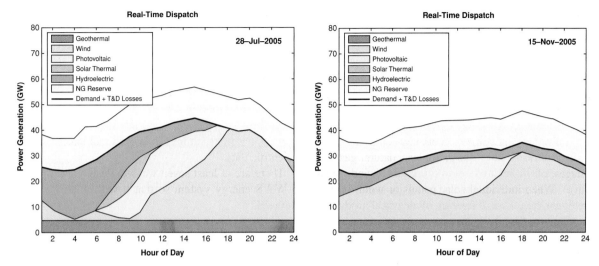

Figure 13.16. Matching California electricity demand plus transmission/distribution losses (black line) with 100 percent renewable supply based on a least-cost optimization calculation for 2 days in 2005. Natural gas was available for backup but was not needed during these days. *Notes:* System capacities: 73.5 GW of wind, 26.4 GW of concentrated solar power (CSP), 28.2 GW of photovoltaics (PV), 4.8 GW of geothermal, 20.8 GW of hydroelectric, and 24.8 GW of natural gas. Transmission and distribution losses were 7 percent of the demand. The least-cost optimization accounted for the day-ahead forecast of hourly resources, carbon emissions, wind curtailment, and thermal storage at CSP facilities, and it allowed for the nighttime production of energy by CSP. The hydroelectric supply was based on historical reservoir discharge data and currently imported generation from the Pacific Northwest. The wind and solar supplies were obtained by aggregating hourly wind and solar power at several sites in California estimated from wind speed and solar irradiance data for those hours applied to a specific turbine power curve, a specific concentrated solar plant configuration (parabolic trough collectors on single-axis trackers), and specific rooftop PV characteristics. The geothermal supply was limited by California's developable resources. From Hart and Jacobson (2011).

2007). The amount of power guaranteed by having the wind farms dispersed over nineteen sites was four times greater than the amount of power guaranteed by having one wind farm.

13.8.2. Using Complementary and Nonvariable Supply to Match Demand

The complementary nature of different renewable energy resources can also be taken advantage of to match minutely through seasonally varying power demand. For example, when the *wind is not blowing, the sun is often shining and vice versa.* This occurs physically because, under a high-pressure system, descending air evaporates clouds, increasing sunlight. At the same time, though, pressure gradients are generally weak, causing winds to be slow, under a high-pressure system. Conversely, under a low-pressure system, rising air increases cloudiness, reducing sunlight penetrating to the surface. Pressure gradients in low-

pressure systems are generally strong, causing winds to be fast (Section 6.6).

Figure 13.16 illustrates how the combined use of wind (variable), solar rooftop PV (variable), CSP (or solar thermal) with storage (variable), geothermal (baseload), and hydroelectric (dispatchable) can be used together to match hourly power demand plus transmission and distribution losses on 2 days in California in 2005. The geothermal power installed was increased over 2005 levels but was limited by California's geothermal resources. The daily hydroelectric generation was determined by estimating the historical generation on those days from reservoir discharge data. Wind and solar capacities were increased substantially over current levels, but they did not exceed maximum levels determined by prior land and resource availability studies.

Figure 13.16 illustrates the potential for matching power demand hour by hour based a computer simulation that accounts for the variable nature of each resource. Although results for only 2 days are shown, results for all hours of all days of both 2005 and 2006 (730 days total) suggest that 99.8 percent of delivered energy during these days could be produced carbon free from WWS technology (Hart and Jacobson, 2011). For these scenarios, natural gas was held as reserve backup and provided energy for the few remaining hours. However, natural gas reserves could be eliminated with the use of demand-response measures, storage beyond CSP, electric vehicle charging and management, and increases in wind and solar capacities beyond those used. In the last case, excess power not needed for the grid would be used to produce hydrogen for industrial processes and transportation. Such a use would reduce wind **curtailment** (shutting down a wind turbine when its electricity is not needed) and thus reduce overall system costs.

13.8.3. Using Demand-Response Management to Adjust Demand to Supply

A third method of addressing the short-term variability of WWS power, **demand-response**, is to use financial incentives to shift times of certain electricity uses, called flexible loads, to times when more energy is available. **Flexible loads** are electricity demands that do not require power in an unchangeable minute-by-minute pattern; instead, they can be supplied in adjustable patterns over several hours. For example, electricity demands for a wastewater treatment plant and for charging BEVs are flexible loads. Electricity demands that cannot be scheduled well, such as electricity use for computers and lighting, are **inflexible loads**. With demand-response, a utility may establish an agreement with a flexible load wastewater treatment plant for the plant to use electricity during only certain hours of the day in exchange for a better electricity rate. In this way, the utility can shift the time of demand to a time when more supply is available. Similarly, the demand for electricity for BEVs is a flexible load because such vehicles are generally charged at night, and it is not critical which hours of the night the electricity is supplied as long as the full power is provided sometime during the night. In this case, a utility can use a smart meter to provide electricity for the BEV when wind availability is high and reduce the power supplied when wind availability is low. Utility customers would sign up their BEVs under a plan by which the utility controlled the nighttime (primarily) or daytime supply of power to the vehicles.

13.8.4. Storing Electric Power at the Site of Generation

A fourth method of dealing with variability is to store excess energy at the site of generation, in batteries, hydrogen gas, pumped hydroelectric power, compressed air (e.g., in underground caverns or turbine nacelles), flywheels, or a thermal storage medium (as is done with CSP). Storage in hydrogen is particularly advantageous because significant hydrogen will be needed in a WWS energy economy for use in fuel cells, aircraft, and high-temperature industrial processes.

13.8.5. Oversizing Wind, Water, and Sunlight Generation to Match Demand Better and Produce $H_2(g)$

Oversizing the peak capacity of wind and solar installations to exceed peak inflexible power demand can reduce the time that available WWS power supply is below demand, thereby reducing the need for other measures to meet demand. The additional energy available when WWS generation exceeds demand can be used to produce hydrogen for heating processes and transportation, which must be produced anyway as part of the WWS solution. Oversizing and using excess energy for hydrogen would also eliminate the current practice of shutting down (curtailing) wind and solar resources when they produce more energy than the grid needs. Curtailment wastes energy; thus, reducing curtailment and using the energy for other purposes should reduce overall system costs.

13.8.6. Storing Electric Power at Points of End Use and in Electric Vehicle Batteries

Another proposed method of better matching power supply with demand is to store electric power in the batteries of BEVs and then to withdraw such power when needed to supply electricity back to the grid. This concept is referred to as **vehicle-to-grid (V2G)** (Kempton and Tomic, 2005a). The utility would enter into a contract with each BEV owner to allow electricity transfers back to the grid any time during a specified period agreed upon by the owner in exchange for a lower electricity price. V2G does have the potential to wear down batteries faster, but one study suggests that only 3.2 percent of U.S. light-duty vehicles, if all converted to BEVs, would need to be under contract for V2G vehicles to smooth out U.S. electricity demand if 50 percent

Table 13.7. Approximate fully annualized generation and short-distance transmission costs for wind, water, and sunlight (WWS) power (2007 U.S. cents/kWh-delivered)

Energy technology	2005–2010	2020+
Wind onshore	4–7	≤4
Wind offshore	10–17	8–13
Wave	≥11	4–11
Geothermal	4–7	4–7
Hydroelectric	4	4
CSP	10–15	7–8
Solar PV	9–13	5–7
Tidal	>11	5–7
New conventional (plus externalities)	7 (+5) = 12	8–9 (+5.5) = 13.5–14.5

Also shown are generation costs and externality costs (from Table 13.8) of new conventional fuels. CSP, concentrated solar power; PV, photovoltaic.
Source: Delucchi and Jacobson (2011), except that solar PV and CSP have been updated based on recent price changes and the upper-end of conventional fuel cost has been updated based on new projections.

of demand were supplied by wind (Kempton and Tomic, 2005b)

13.8.7. Using Weather Forecasts to Plan for and Reduce Backup Requirements

Forecasting the weather (winds, sunlight, waves, tides, and precipitation) gives grid operators more time to plan ahead for a backup energy supply when a variable energy source might produce less than anticipated. Good forecast accuracy can also allow spinning reserves to be shut down more frequently, reducing the overall carbon emissions of the system if natural gas is used as backup (Hart and Jacobson, 2011). Forecasting is done with either a numerical weather prediction model, the best of which can produce minute-by-minute predictions 1 to 4 days in advance with good accuracy, or statistical analyses of local measurements. The use of forecasting reduces uncertainty and makes planning more dependable, thus reducing the impacts of variability.

13.9. Cost of Wind, Water, and Sunlight Electricity Generation and Long-Distance Transmission

An important criterion in the evaluation of WWS systems is to ensure that the full cost of delivered power,

including the capital, land, operating, maintenance, storage, and transmission cost per unit energy delivered, is low. In this section, cost estimates of a reliable WWS generation and transmission system are discussed.

Table 13.7 presents estimates of 2005 to 2010 and future (2020 and beyond) costs of power generation with conventional (but not extra long-distance) transmission for WWS systems. It also shows the average U.S. delivered electricity cost for conventional (mostly fossil) generation, excluding electricity distribution. For fossil fuel generation, the **externality cost**, which includes the hidden cost of air pollution and climate change to society as a whole due to a polluting energy technology, is also shown. The derivation of the externality cost is provided in Table 13.8.

Table 13.7 indicates that onshore wind, hydroelectric, and geothermal plants cost the same or less than typical new conventional technologies (e.g., new coal-fired or natural gas power plants) from 2005 to 2010, without accounting for the additional externality costs of the conventional technologies. The WWS technologies cost significantly less than conventional technologies when the externality costs are accounted for. WWS technologies today cost more than old coal-fired power plants in the United States because many such plants were grandfathered into the Clean Air Act Amendments of 1970. Thus, they are not required to meet the same emission requirements as new plants. As such, the electricity

Table 13.8. Mean (and range) of environmental externality costs of electricity generation in the United States in 2007 (U.S. cents/kWh)

	2005			2030		
	Air pollution	Climate	Total	Air pollution	Climate	Total
Coal	3.2	3.0	6.2 (1.2–22)	1.7	4.8	6.5 (3.3–18)
Natural gas (NG)	0.16	1.5	1.7 (0.5–8.6)	0.13	2.4	2.5 (0.9–8.9)
Coal/NG mix	2.4	2.6	5.0 (1.0–18)	1.3	4.2	5.5 (2.7–15)
Wind, water, sunlight	~0	~0	~0	~0	~0	~0

The coal/NG mix is 73 percent/27 percent in 2005 and 75 percent/25 percent in 2030.
Source: Delucchi and Jacobson (2011).

generation costs of energy from old coal plants can be 2 to 3 U.S. cents/kWh. However, the externality costs of such plants are much greater than those in Table 13.7.

The future costs of onshore wind, geothermal, and hydroelectric power are expected to remain low. Costs of other WWS technologies are expected to decline. If WWS technologies are compared on the basis of direct plus externality costs, WWS options, including solar PVs, should cost less than new conventional fuel generation by 2020 (Table 13.7).

Table 13.7 includes the cost of electricity transmission in a conventionally configured system over distances common today. However, many future wind and solar farms may be far from population centers, requiring extra long-distance transmission. Furthermore, there is an efficiency advantage to increasing interconnection among geographically dispersed wind and solar farms (Section 13.8.1). Namely, the more that dispersed wind and solar generating sites are interconnected, the less the output of the whole interconnected system varies.

A system of interconnected transmission lines among widely dispersed generators and load centers is referred to as a **supergrid**. The configuration and length of transmission lines in a supergrid depends on the balance between the cost of adding more transmission lines and the benefit of reducing system output variability as a result of connecting more dispersed generation sites. A supergrid has been proposed to link Europe and North Africa (e.g., Czisch, 2006). Supergrids are also needed within Australia/Tasmania, North America, South America, Africa, Russia, China, eastern and Southeast Asia, and the Middle East.

For long-distance transmission, **high-voltage direct current** (HVDC) lines are generally used instead of alternating current (AC) lines because of the lower transmission losses per unit distance with HVDC lines over a long distance. The cost of HVDC transmission is a function of the cost of the towers and lines; the distance of transmission; the cost of equipment such as converters, transformers, filters, and switchgear; electricity losses in lines and equipment; the life of the transmission line; maintenance costs; the discount rate; and the efficiency of power generated for the line. The most important and uncertain cost components are the costs of lines and towers. The cost of extra long-distance HVDC transmission on land ranges from 0.3 to 3 U.S. cents/kWh, with a best estimate of about 1 U.S. cent/kWh (Delucchi and Jacobson, 2011). A system with up to 25 percent undersea transmission, which is relatively expensive, would increase the best estimate of the additional long-distance transmission cost by less than 20 percent.

13.10. Policy Mechanisms

Current energy markets, institutions, and policies have been developed to support the production and use of fossil fuels and biofuels. New policies are needed to ensure that clean energy systems develop quickly and broadly and that dirtier energy systems are not promoted. Several economic and noneconomic policies have either been implemented or considered to accomplish these goals.

Renewable portfolio standards (also called renewable electricity standards) are policy mechanisms whereby a certain fraction of electric power generation must come from specified clean energy sources by a certain date (Section 12.6.3).

Feed-in tariffs (FITs) are subsidies to cover the difference between generation cost (ideally including grid connection costs) and wholesale electricity prices.

They have been an effective policy tool for stimulating the market for renewable energy. To encourage innovation and the large-scale implementation of WWS technologies, which will itself lower costs, FITs should be reduced gradually. Otherwise, technology developers have little incentive to improve. One way to reduce FITs gradually is with a **declining clock auction**, in which the right to sell power to the grid goes to the bidders willing to do it at the lowest price, providing continuing incentive for developers and generators to reduce costs. A risk of any auction, however, is that the developer will underbid and be left unable to develop the proposed project at a profit. Regardless of whether the declining clock auction is used, reducing and eventually phasing out the FIT as the cost of producing power from WWS technologies declines (referred to as **tariff regression**) is an important goal.

Output subsidies are payments by government to energy producers per unit of energy produced. For clean renewable energy producers, the justification of such subsidies is to correct the market because fossil fuel and biofuel energy producers are not paying for the pollution they emit, which has health and climate impacts on society as a whole (Table 13.8) through higher taxes, insurance rates, and medical costs. In other words, the subsidies attempt to address the **tragedy of the commons** (Preface), which arises because the air is not privately owned, so it has been polluted without polluters paying the cost of the pollution.

Investment subsidies are direct or indirect payments by government to energy producers for research and development. Historically, such subsidies have been given mostly to conventional fuel producers through legislation and clauses in the tax code that allow for deductions or credits for specified activities. Thus, conventional generators have benefited historically by not paying the externality costs of their pollution (Table 13.8) and by receiving direct subsidies. Investment subsidies are now available to WWS energy sources in many countries.

One type of investment subsidy is a **loan guarantee**, whereby the government provides the guarantee for a loan to a company for construction of a facility. Without such guarantees, many large energy projects could not obtain approval for a construction loan. Such loan guarantees, historically provided to the conventional fuels industry, are beneficial for WWS facilities as well. On a smaller scale, **municipal financing** for residential energy efficiency retrofits or solar installations could help overcome the financial barrier of high upfront cost to individual homeowners. **Purchase incentives and**

rebates can also help stimulate the market for electric vehicles.

A potential policy tool that has not been widely used to date is a **revenue-neutral pollution tax**. This is a tax on polluting energy sources, with the revenue transferred directly to nonpolluting energy sources. In this way, no net tax is collected, so the cost to the public is zero in theory.

A related tool is straight **pollution tax** (e.g., a **carbon tax**). Such a tax is less popular because it is perceived as a cost to the public; that is, polluting companies could simply increase their prices to pay the tax. The revenue-neutral pollution tax avoids this problem because it allows the cost of clean energy to decrease; thus, if polluters increase their prices, electricity buyers can then shift to purchase the cleaner energy.

An important noneconomic policy program is to reduce demand by *improving the efficiency of end-use energy*, or substituting low-energy activities and technologies for high-energy ones. Demand reduction reduces the pressure on energy supply, making it easier for clean energy supply to match demand. Providing incentives for consumers to shift the time of their electricity use (demand-response) (Section 13.8.3) can improve efficiency of an overall electric power infrastructure.

Another noneconomic policy is to **mandate emission limits** for technologies. This is a **command-and-control** policy option implemented widely under the Clean Air Act Amendments and in many countries of the world to reduce vehicle emissions. By tightening emission standards, including for carbon dioxide, policy makers can force the adoption of cleaner vehicles. Such emission limits can and have similarly been set for many other sources of pollution.

Related to mandatory emission limits is the policy mechanism of **cap and trade**. Under this mechanism, mandatory emission limits lower than current emissions are set for an entire industry, and pollution permits are issued corresponding to the total pollution emissions allowed. Polluters in the industry can then buy and sell pollution permits, resulting in an increase in the price of permits if polluters want to continue polluting.

13.11. Summary and Conclusions

A solution to the problems of global warming, outdoor and indoor air pollution, and acid deposition requires a large-scale conversion of the current world's energy infrastructure. In this chapter, a scenario for such a conversion is described. The solution would eliminate these problems, as well as the associated 2.5 to 3 million annual premature deaths and tens of millions of illnesses worldwide. It was based on converting the current fossil fuel and biofuel combustion–based energy infrastructure to one based on clean and safe electricity, on hydrogen derived from electricity, and on energy efficiency. All energy worldwide for electric power, transportation, and heating/cooling would be converted. The ultimate sources of electric power would be WWS technologies.

The scenario included the proposed deployment of 3.8 million 5-MW wind turbines (supplying 50 percent of global end-use power demand in 2030); 49,000 300-MW CSP power plants (supplying 20 percent of demand); 40,000 solar PV power plants (14 percent); 1.7 3-kW rooftop PV systems (6 percent); 5,350 100-MW geothermal power plants (4 percent); 900 1,300-MW hydroelectric power plants (4 percent), of which 70 percent are already in place; 720,000 0.75-MW wave devices (1 percent); and 490,000 1-MW tidal turbines (1 percent). In addition, the scenario assumed an expansion of the transmission infrastructure to accommodate the new power systems of BEVs and HFCVs, ships that run on hydrogen fuel cell and battery combinations, liquefied hydrogen aircraft, air- and ground-source heat pumps, electric resistance heating, and hydrogen production for high-temperature processes.

The equivalent footprint area on the ground for the sum of WWS devices needed to power the world would be ~0.74 percent of global land area. The spacing area – used for agriculture, ranching, and open space – would be ~1.16 percent of global land area. However, if one-half of the wind devices were placed over water, and considering that wave and tidal devices are in water, 70 percent of hydroelectric is already developed, and rooftops for solar PV already exist, the additional footprint and spacing of devices on land required compared with energy today would be only ~0.41 and ~0.59 percent of world land area, respectively.

The development of WWS power systems is not likely to be constrained by the availability of bulk materials such as steel and concrete. In a global WWS system, some materials, such as Nd (in electric motors and generators), lithium (in batteries), and Pt (in fuel cells), although not limited, may need to be recycled.

Several methods exist of matching electric power demand with WWS supply continuously:

1. Interconnect geographically dispersed naturally variable energy sources (e.g., wind, wave, solar).

2. Combine WWS resources as one commodity and use a nonvariable energy source, such as hydroelectric power, to fill in gaps between demand and supply.
3. Use demand-response management to shift times of peak demand to better match times of WWS supply.
4. Store electric power at the site of generation for later use.
5. Oversize WWS peak generation capacity to minimize the times when available WWS power is less than demand and to provide spare power to produce hydrogen for transportation and heat uses.
6. Store electric power in electric vehicle batteries.
7. Forecast the weather to plan for energy supply needs better.

Today, the cost of generating electricity from onshore wind power, geothermal power, and hydroelectric power is less than or equal to that of new conventional fossil fuel generation. By 2030, the cost of generating electricity from remaining WWS power sources, including solar PVs, CSP, wave, and tidal power, is likely to be less than that of conventional fuels when the cost to society is accounted for. This result does not change when the additional cost of a transmission supergrid (about 1 U.S. cent kWh^{-1}) is included. Also, in 2030, the total cost of electric transportation, based either on batteries or hydrogen fuel cells, is likely to be comparable with or less than the direct plus externality cost of transportation based on liquid fossil fuels.

In sum, a concerted international effort can lead to a conversion of the energy infrastructure so that by c. 2030, the world will no longer build new fossil fuel or nuclear electricity generation power plants or new transportation equipment using internal combustion engines. Rather, it will manufacture new wind turbines, solar power plants, and electric and fuel cell vehicles (except for aviation, which will use liquid hydrogen in jet engines). Once this WWS power plant and electric vehicle manufacturing and distribution infrastructure is in place, the remaining existing fossil fuel and nuclear power plants and internal combustion engine vehicles can be gradually retired and replaced with WWS-based systems, so that by 2050, the world is powered by WWS.

The main barriers to a conversion to a worldwide WWS infrastructure are not technical, resource based, or economic. Instead, they are political and social. Because most polluting energy technologies and resources currently receive government subsidies or tax breaks or are not required to eliminate their emissions, they can run inexpensively for a long time while competing economically with WWS technologies. As a result, no mechanism currently exists to encourage the large-scale conversion to WWS. Wisely implemented policy mechanisms can promote such a conversion; however, such mechanisms can be implemented only if policy makers are willing to make changes. Policy makers in democratic countries are elected, whereas those in autocratic countries make decisions dictated by one or a few leaders. Thus, in democratic countries, a large number of people need to be convinced that changes are beneficial; in autocratic countries, only a few need to be convinced. In both cases, those who need to be convinced require a social evolution in their thought. Such an evolution comes from a better understanding of global warming and air pollution science, consequences, and solutions. If the public and policy makers can become confident in understanding the problems and the large-scale solution needed to solve them, such as that outlined in this chapter, they will gravitate toward the solution. When the solution is finally implemented, the air pollution and climate problems outlined in this book will be relegated to the annals of history.

13.12. Problems

13.1. How many 5-MW wind turbines with a rotor diameter of 126 m operating in a mean annual wind speed of 7.5 m s^{-1} are needed to power the U.S. on-road vehicle fleet consisting of BEVs if the end-use energy required to run such a fleet is $E_v = 1.12 \times 10^{12}$ kWh yr^{-1} (2007) and the plug-to-wheel efficiency of an electric vehicle is $\eta_e = 0.85$? Assume the system efficiency of each wind turbine is $\eta_t = 0.9$. *Hint*: First determine the total electrical energy required to run the fleet by dividing the end-use energy required to run vehicles by the plug-to-wheel efficiency.

13.2. If each wind turbine in Problem 13.1 has a tubular tower plus concrete footing base circular diameter of 5 m (footprint diameter), and if each turbine in a wind farm occupies a spacing of $A_t = 4D \times 7D$, where D is the turbine rotor diameter, calculate the footprint and spacing required for all turbines needed to power a U.S. electric vehicle fleet in 2007. What percent of the fifty-state U.S. land area (9.162×10^6 km^2) does this area represent?

13.3. Using the information obtained from Problem 13.1 and the following equations, calculate the number of wind turbines required to power a 100 percent U.S. HFCV fleet, where the hydrogen is produced by wind electrolysis. Account for energy requirements of the electrolyzer, compressor, and fuel cell. Assume the efficiency of the fuel cell converting hydrogen to energy and water is $\eta_f = 0.5$, the lower heating value of hydrogen is $L_h = 33.3$ kWh kg^{-1}-H$_2$(g), the hydrogen leakage rate is $l_h = 0.03$ (fraction), the electrolyzer energy required is $L_z = 53.4$ kWh kg^{-1}-H$_2$(g), and the compressor energy required is $L_c = 5.64$ kWh kg^{-1}-H$_2$(g).

Energy required to move all U.S. vehicles as HFCVs: E_f (kWh yr^{-1}) = E_v/η_f

H$_2$(g) mass to power all vehicles, accounting for leaks: M_h [kg-H$_2$(g) yr^{-1}] = $E_f/[L_h(1 - l_h)]$

Energy needed to power a U.S. HFCV fleet: E_h (kWh yr^{-1}) = $(L_z + L_c) \times M_h$

Number of wind turbines needed to power a U.S. HFCV fleet = E_h/E_t

13.4. Calculate the number of solar PV panels and the fractional area of the United States required to power a 100 percent BEV fleet with PV from PV power plants. Assume that the end-use energy required to run a BEV fleet after the plug-to-wheel efficiency of each vehicle is accounted for is $E_v = 1.12 \times 10^{12}$ kWh yr^{-1} (2007). Also, make the following assumptions: the plug-to-wheel efficiency of an electric vehicle is $\eta_e = 0.85$, a solar panel's rated power is $P_s = 0.232$ kW, the footprint and spacing area of a solar panel are both $A_s = 2$ m^2, the panel's capacity factor is $CF_s = 0.21$, the transmission plus distribution efficiency of energy from a PV power plant is $\eta_s = 0.95$, and the area of the United States is 9.162×10^6 km^2. *Hint*: The single-panel annual energy output (kWh yr^{-1}) is $E_s = P_s \times CF_s \times H/\eta_s$.

13.5. Explain the difference between footprint and spacing for an energy facility. What are some of the uses of the spacing?

13.6. From Figures 13.12 and 13.13, identify three general regions of fast winds (7 m s^{-1} or higher) and three regions of high solar radiation (150 W m^{-2} or higher) worldwide over land or near shore. Among these, identify one area where both wind and solar resources are good.

13.7. Explain why the conversion from an internal combustion engine gasoline vehicle to an electric vehicle should reduce overall power demand?

13.8. Identify four policy measures that could be implemented to encourage the expansion of WWS energy systems.

13.9. Identify four barriers that could slow the large-scale implantation of clean renewable energy.

13.10. Identify four methods of reducing the effect of the variability of wind, solar, and wave power on matching power demand with supply on a continuous basis.

13.13. Group or Individual Project

Develop a "white paper" describing and analyzing the effects on a selected region of three methods of addressing global warming and air pollution simultaneously. In the paper, address the following components:

1. Discuss at least three methods of reducing emission of global warming gases and particles and air pollution gases and particles simultaneously using existing or emerging technologies. Such technologies can include emission control technologies, non- or low-emitting technologies that replace current technologies, energy efficiency measures, or policies encouraging lower emissions. Decide the area over which you will examine the effects (e.g., city, state, province, country, or global).
2. Discuss whether these methods have been implemented to date, and, if so, what the result of their implementation has been.

3. Quantify the estimated emission reductions (total mass per year) in the region of interest of major pollutants during the next 20 years if the methods are implemented.
4. Quantify the estimated reductions in the outdoor mixing ratio or concentration of major primary or secondary pollutants in a city within the region if the methods are implemented. (This requires finding data on current concentrations or mixing ratios and using estimates of species lifetimes and of a mean mixing height.)
5. Estimate the temperature change in the region of interest, as well as on a global scale (use simple scaling with global temperature change information in Chapter 12), due to the changes in emission of the global warming agents.
6. Quantify the health benefits (e.g., reduction in cardiovascular and respiratory disease, asthma, and mortality) of reducing the emissions. Use Equation 5.13.
7. Discuss which industries will and will not benefit as a result of implementing the methods through regulation. Cost estimates may be given here.
8. Briefly summarize the selected methods, results, and overall recommendations in an abstract.

Appendix

Conversions and Constants

A.1.1. Pressure Conversions

1 bar	$= 10^3$ mb	$= 0.986923$ atm	$= 10^5$ N m^{-2}
	$= 10^5$ hPa	$= 10^5$ Pa	$= 10^5$ kg m^{-1} s^{-2}
	$= 10^5$ J m^{-3}	$= 10^6$ g cm^{-1} s^{-2}	$= 750.06$ torr
	$= 750.06$ mm Hg	$= 10^6$ dyn cm^{-2}	
1 atm	$= 1{,}013.25$ hPa	$= 1.01325$ bar	$= 760$ torr
	$= 760$ mm Hg		

A.1.2. Energy Conversions

1 J	$= 1$ N m	$= 10^7$ erg	$= 1$ W s
	$= 10^4$ hPa cm^3	$= 10^7$ dyn cm	$= 0.239$ cal
	$= 1$ kg m^2 s^{-2}	$= 10^7$ g cm^2 s^{-2}	$= 10^{-5}$ bar m^3
	$= 6.25 \times 10^{18}$ eV	$= 1$ C V	

A.1.3. Speed Conversions

1 m s^{-1}	$= 100$ cm s^{-1}	$= 3.6$ km h^{-1}	$= 1.9459$ knots
	$= 2.2378$ mi hr^{-1}		

A.1.4. Constants

A	$=$ Avogadro's number	$= 6.02213 \times 10^{23}$ molec mol^{-1}
c	$=$ speed of light	$= 2.99792 \times 10^8$ m s^{-1}
F_s	$=$ solar constant	$= 1365$ W m^{-2}
k_B	$=$ Boltzmann's constant (R^*/A)	$= 1.3807 \times 10^{-23}$ J K^{-1}
	$= 1.3807 \times 10^{-23}$ kg m^2 s^{-2} K^{-1} molec^{-1}	$= 1.3625 \times 10^{-22}$ cm^3 atm K^{-1}
	$= 1.3807 \times 10^{-16}$ g cm^2 s^{-2} K^{-1} molec^{-1}	$= 3.299 \times 10^{-24}$ cal K^{-1}
	$= 1.3807 \times 10^{-19}$ cm^3 hPa K^{-1} molec^{-1}	$= 1.3625 \times 10^{-25}$ L atm K^{-1} molec^{-1}
	$= 1.3625 \times 10^{-22}$ cm^3 atm K^{-1} molec^{-1}	$= 1.3807 \times 10^{-25}$ m^3 hPa K^{-1} molec^{-1}

g	= gravity	= 9.81 m s^{-2}
m_d	= molecular weight of dry air	= 28.966 g mol^{-1}
$\bar{M}$	= mass of an air molecule (m_d/A)	= 4.8096 × 10^{-26} kg
R^*	= universal gas constant	= 8.3145 J mol^{-1} K^{-1}
	= 8.3145 kg m^2 s^{-2} mol^{-1} K^{-1}	= 0.083145 m^3 hPa mol^{-1} K^{-1}
	= 8.3145 × 10^7 g cm^2 s^{-2} mol^{-1} K^{-1}	= 0.08206 L atm mol^{-1} K^{-1}
	= 8.3145 × 10^4 cm^3 hPa mol^{-1} K^{-1}	= 8.3145 × 10^7 erg mol^{-1} K^{-1}
	= 82.06 cm^3 atm mol^{-1} K^{-1}	= 1.987 cal mol^{-1} K^{-1}
R'	= gas constant for dry air (R^*/m_d)	= 287.04 J kg^{-1} K^{-1}
	= 0.28704 J g^{-1}K^{-1}	= 2.8704 m^3 hPa kg^{-1} K^{-1}
	= 2870.4 cm^3 hPa g^{-1} K^{-1}	= 287.04 m^2 s^{-2} K^{-1}
	= 2.8704 × 10^6 cm^2 s^{-2} K^{-1}	
R_e	= radius of the Earth	= 6.378 × 10^6 m
R_p	= radius of the sun	= 6.96 × 10^8 m
R_{es}	= mean Earth-sun distance	= 1.5 × 10^{11} m
T_p	= temperature of the sun's photosphere	= 5785 K
σ_B	= Stefan-Boltzmann constant	= 5.67051 × 10^{-8} W m^{-2} K^{-4}

References

Ackerman, A. S., O. B. Toon, D. E. Stevens, A. J. Heymsfield, V. Ramanathan, and E. J. Welton, Reduction of tropical cloudiness by soot, *Science*, *288*, 1042–1047, 2000.

Ackerman, T. P., A model of the effect of aerosols on urban climates with particular applications to the Los Angeles Basin, *J. Atmos. Sci.*, *34*, 531–547, 1977.

Ackerman, T. P., and O. B. Toon, Absorption of visible radiation in atmosphere containing mixtures of absorbing and nonabsorbing particles, *Appl. Optics*, *20*, 3661–3667, 1981.

Adachi, K., and P. R. Buseck, Atmospheric tar balls from biomass burning in Mexico, *J. Geophys. Res.*, *116*, D05204, doi:10.1029/2010JD015102, 2011.

Albrecht, B. A., Aerosols, cloud microphysics, and fractional cloudiness, *Science*, *245*, 1227–1230, 1989.

Alexander, D. T. L., P. A. Crozier, and J. R. Anderson, Brown carbon spheres in East Asian outflow and their optical properties, *Science*, *321*, 833–836, 2008.

Alternative Fluorocarbons Environmental Acceptability Study (AFEAS), n.d. www.afeas.org, accessed June 28, 2011.

Alvarez, L. W., W. Alvarez, F. Asaro, and H. V. Michel, Extraterrestrial cause for the Cretaceous-Tertiary extinction, *Science*, *208*, 1095–1108, 1980.

American Lung Association, State of the Air 2010, 2010, www.stateoftheair.org/, accessed February 12, 2011.

Anderson, I., G. R. Lundquist, and L. Molhave, Indoor air pollution due to chipboard used as a construction material, *Atmos. Environ.*, *9*, 1121–1127, 1975.

Andreae, M. O., T. W. Andreae, H. Annegarn, J. Beer, H. Cachier, P. le Canut, W. Elbert, W. Maenhaut, I. Salma, F. G. Wienhold, and T. Zenker, Airborne studies of aerosol emissions from savanna fires in southern Africa: 2. Aerosol chemical composition, *J. Geophys. Res.*, *103*, 32,119–32,128, 1998.

Andreae, M. O., R. J. Charlson, F. Bruynseels, H. Storms, R. Van Grieken, and W. Maenhaut, Internal mixture of sea salt, silicates, and excess sulfate in marine aerosols, *Science*, *232*, 1620–1623, 1986.

Andreae, M. O., and P. Merlet, Emission of trace gases and aerosols from biomass burning, *Global Biogeochem. Cycles*, *15*, 955–966, 2001.

Andrews, E., and S. M. Larson, Effect of surfactant layers on the size changes of aerosol particles as a function of relative humidity, *Environ. Sci. Technol.*, *27*, 857–865, 1993.

Arashidani, K., M. Yoshikawa, T. Kawamoto, K. Matsuno, F. Kayam, and Y. Kodama, Indoor pollution from heating, *Ind. Health*, *34*, 205–215, 1996.

Archer, C. L., and M. Z. Jacobson, Evaluation of global windpower, *J. Geophys. Res.*, *110*, D12110, doi:10.1029/2004JD005462, 2005.

Archer, C. L., and M. Z. Jacobson, Supplying baseload power and reducing transmission requirements by interconnecting wind farms, *J. Appl. Meteorol. Climatol.*, *46*, 1701–1717, 2007.

Ardo, J., N. Lambert, V. Henzlik, and B. N. Rock, Satellite-based estimations of coniferous forest cover changes: Krusne Hory, Czech Republic 1972–1989, *Ambio*, *26*, 158–166, 1997.

Arrhenius, S., On the influence of carbonic acid in the air upon the temperature of the ground, *Philos. Mag.*, *41*, 237, 1896.

Aw, J., and M. J. Kleeman, Evaluating the first-order effect of intraannual temperature variability on urban air pollution, *J. Geophys. Res.*, *108*, D12, doi:10.1029/2002JD002688, 2003.

Ayars, G. H., Biological agents and indoor air pollution, in E. J. Bardana and A. Montanaro, eds., *Indoor Air Pollution and Health*, Marcel Dekker, New York, pp. 11–60, 1997.

Banke, B., Solar electric facility O&M now comes the hard part, part 3, *Renewable Energy World North Am.*, *2* (1), 2010.

Barron, E. J., J. L. Sloan, and C. G. A. Harrison, Potential significance of land-sea distribution and surface albedo variations as a climatic forcing factor: 180 m.y. to the present, *Palaeogeog. Palaeoclim. Palaeoecol.*, *30*, 17–40, 1980.

Bentley, C. R., Some aspects of the cryosphere and its role in climatic change, in *Climate Processes and Climate Sensitivity (Geophysical Monograph)*, *29*, eds. J. E. Hansen and T. Takahashi, American Geophysical Union, Washington, DC, pp. 207–220, 1984.

Bergstrom, R., and R. Viskanta, Modelling of the effects of gaseous and particulate pollutants in the urban atmosphere: part I, thermal structure, *J. Appl. Meteorol.*, *12*, 901–912, 1973.

Bergthorsson, P., An estimate of drift ice and temperature in 1000 years, *Jökull*, *19*, 94–101, 1969.

Berner, A., S. Sidla, Z. Galambos, C. Kruisz, R. Hitzenberger, H. M. ten Brink, and G. P. A. Kos, Modal character of atmospheric black carbon size distributions, *J. Geophys. Res.*, *101*, D14, doi:10.1029/95JD03425, 1996.

Berner, R. A., A. C. Lasaga, and R. M. Garrels, The carbonate-silicate geochemical cycle and its effects on atmospheric carbon dioxide over the last 100 million years, *Am. J. Sci.*, *283* (7), 641–683, 1983.

Bhatti, M. S., A historical look at chlorofluorocarbon refrigerants, *ASHRAE Trans.*, *105*, 1186–1208, 1999.

Blatchford, R., *Dismal England*, Clarion Press, London, 1899.

Boden, T. A., G. Marland, and R. J. Andres, Global, regional, and national CO_2 emissions from fossil-fuel burning, cement manufacture, and gas flaring, in *Trends Online: A Compendium of Data on Global Change*, Carbon Dioxide Information Analysis Center, Oak Ridge National Laboratory, U.S. Department of Energy, Oak Ridge, TN, 2011, http://cdiac.ornl.gov/.

Bond, T. C., D. G. Streets, K. F. Yarber, S. M. Nelson, J.-H. Woo, and Z. Klimont, A technology-based global inventory of black and organic carbon emissions from combustion, *J. Geophys. Res.*, *109*, D14203, doi:10.1029/2003JD003697, 2004.

Bornstein, R., and Q. Lin, Urban heat islands and summertime convective thunderstorms in Atlanta: three case studies, *Atmos. Environ.*, *34*, 507–516, 2000.

Bostedt, G., S. Lofgren, S. Innala, and K. Bishop, Acidification remediation alternatives: exploring the temporal dimension with cost benefit analysis, *Ambio*, *39*, 40–48, 2010.

Bradley, C. E., and A. J. Haagen-Smit, The application of rubber in the quantitative determination of ozone, *Rubber Chem. Technol.*, *24*, 750–755, 1951.

Bretherton, C. S., E. Klinker, A. K. Betts, and J. Coakley, Comparison of ceilometer, satellite, and synoptic measurements of boundary layer cloudiness and the ECMWF diagnostic cloud parameterization scheme during ASTEX, *J. Atmos. Sci.*, *52*, 2736–2751, 1995.

Brewster, D., *Edinburgh Encyclopaedia*, Joseph Parker, Philadelphia, 1832.

Brimblecombe, P., *The Big Smoke*, Methuen, London, 1987.

Brimblecombe, P., Air pollution and health history, in *Air Pollution and Health*, S. T. Holgate, J. M. Samet, H. S. Koren, and R. L. Maynard, eds., Academic Press, San Diego, pp. 5–18, 1999.

Brock, W. H., *The Norton History of Chemistry*, W. W. Norton, New York, 1992.

Brohan, P., J. J. Kennedy, I. Harris, S. F. B. Tett, and P. D. Jones, Uncertainty estimates in regional and global observed temperature changes: a new dataset from 1850, *J. Geophys. Res.*, *111*, D12106, doi:10.1029/2005JD006548, 2006.

Brook, G. A., M. E. Folkoff, and E. O. Box, A world model of soil carbon dioxide, *Earth Surface Proc. Landforms*, *8*, 79–88, 1983.

Brooks, B. O., G. M. Utter, J. A. DeBroy, and R. D. Schimke, Indoor air pollution: an edifice complex, *Clin. Toxicol.*, *29*, 315–374, 1991.

Brown, J. C., *A History of Chemistry*, P. Blakiston's Son & Co., Philadelphia, 1913.

Burr, M. L., A. S. St.-Leger, and J. W. G. Yarnell, Wheezing, dampness, and coal fires, *Community Med.*, *3*, 205–209, 1981.

Caldwell, M. M., L. O. Bjorn, J. F. Bornman, S. D. Flint, G. Kulandaivelu, A. H. Teramura, and M. Tevini, Effects of increased solar ultraviolet radiation on terrestrial ecosystems, *J. Photochem. Photobiol., B*, *46*, 40–52, 1998.

California Air Resources Board (CARB), 2011, Air quality trend summaries, www.arb.ca.gov/adam/trends/trends1.php, accessed February 12, 2011.

Campbell, F. W., and L. Maffel, Contrast and spatial frequency. *Sci. Am.*, *231*, 106–114, 1974.

Carbon Dioxide Information Analysis Center (CDIAC), Data: Atmospheric Trace Gases in Whole-Air Samples, n.d., http://cdiac.ornl.gov/trends/otheratg/blake/data.html, accessed November 23, 2011.

Carter, W. P. L., *Development of Ozone Reactivity Scales for Volatile Organic Compounds*, EPA-600/3-91-050, U.S. Environmental Protection Agency, Research Triangle Park, NC, 1991.

Cass, G. R., On the relationship between sulfate air quality and visibility with examples in Los Angeles. *Atmos. Environ.*, *13*, 1069–1084, 1979.

Catcott, E. J., The effects of air pollution on animals, 1955, http://whqlibdoc.who.int/monograph/WHO_MONO_46_(p221).pdf.

Cattermole, P. and P. Moore, *The Story of the Earth*, Cambridge University Press, New York, 1985.

Centers for Disease Control and Prevention, Annual smoking-attributable mortality, years of potential life lost, and productivity losses – U.S., 1997–2001, *MMWR*, *54*, 625–628, 2005.

Chang, S. G., R. Brodzinsky, L. A. Gundel, and T. Novakov, Chemical and catalytic properties of elemental carbon, in *Particulate Carbon: Atmospheric Life Cycle*, G. T. Wolff and R. L. Klimsch, eds., Plenum Press, New York, pp. 158–181, 1982.

Chapman, S., A theory of upper-atmospheric ozone, *Mem. Royal Met. Soc.*, *3*, 104–125, 1930.

Choi, I. S., Delayed neurological sequelae in carbon monoxide intoxication, *Arch. Neurol.*, *40*, 433–435, 1983.

Chung, S. H., and J. H. Seinfeld, Climate response of direct radiative forcing of anthropogenic black carbon, *J. Geophys. Res.*, *110*, D11102, doi:10.1029/2004JD005441, 2005.

Chylek, P., G. B. Lesins, G. Videen, J. G. D. Wong, R. G. Pinnick, D. Ngo, and J. D. Klett, Black carbon and absorption of solar radiation by clouds, *J. Geophys. Res.*, *101*, 23,365–23,371, 1996.

Chylek, P., V. Ramaswamy, and R. J. Cheng, Effect of graphitic carbon on the albedo of clouds, *J. Atmos. Sci*, *41*, 3076–3084, 1984.

Clapp, B. W., *An Environmental History of Britain*, Longman, London, 1994.

Clark, D. L., R. R. Whitman, K. A. Morgan, and S. D. Mackey, Stratigraphy and glacial-marine sediments of the

Amerasian Basin, central Arctic Ocean, *Geol. Soc. Am. Spec. Pap.*, *181*, 1980.

Clean Air Act Amendments of 1970 (CAAA70, PL 91-604), www.wilderness.net/NWPS/documents/publiclaws/PDF/91-604.pdf, Accessed December 10, 2011.

CLIMAP Project Members, The last interglacial ocean, *Quart. Res.*, *21*, 123–224, 1984.

Cockell, C., R. Corfield, N. Edwards, and N. Harris, *An Introduction to the Earth-Life System*, Cambridge University Press, New York, 2007.

Cohen, B. S., Deposition of charged particles on lung airways, *Health Physics*, *74*, 554–560, 1998.

Crowley, T. J., and G. R. North, *Paleoclimatology*, Oxford University Press, New York, 1991.

Czisch, G., Low Cost but Totally Renewable Electricity Supply for a Huge Supply Area: A European/Trans-European Example, Unpublished manuscript, 2006, www.iset.uni-kassel.de/abt/w3-w/projekte/LowCostEurop-ElSup_revised_for_AKE_2006.pdf, accessed November 26, 2011.

Danielson, R. E., D. R. Moore, and H. C. van de Hulst, The transfer of visible radiation through clouds, *J. Atmos. Sci.*, *26*, 1078–1087, 1969.

Davidson, C. I., W. Tang, S. Finger, V. Etyemezian, M. F. Striegel, and S. I. Sherwood, Soiling patterns on a tall limestone building: changes over 60 years, *Environ. Sci. Technol.*, *34*, 560–565, 2000.

De'ath, G., J. M. Lough, and K. E. Fabricius, Declining coral calcification on the Great Barrier Reef, *Science*, *323*, 116–119, 2009.

Debell, L. J., K. A. Gebhart, W. C. Malm, M. L. Pitchford, B. A. Schichtel, and W. H. White, Spatial and Seasonal Patterns and Temporal Visibility of Haze and Its Constituents in the United States: Report IV, IMPROVE: Interagency Monitoring of Protected Visual Environments, Cooperative Institute for Research in the Atmosphere, Colorado State University, 2006, http://vista.cira.colostate.edu/improve/Publications/Reports/2006/PDF/IMPROVE_Report_IV.pdf, accessed November 26, 2011.

Delucchi, M., A conceptual framework for estimating the climate impacts of land-use change due to energy crop programs, *Biomass Bioenergy*, doi:10.1016/j.biombioe.2010.11.028, 2010.

Delucchi, M. A., and M. Z. Jacobson, Providing all global energy with wind, water, and solar power, part II: reliability, system and transmission costs, and policies, *Energy Policy*, *39* (3), 1170–1190, doi:10.1016/j.enpol.2010.11.045, 2011.

Dentener, F. J., G. R. Carmichael, Y. Zhang, J. Lelieveld, and P. J. Crutzen, Role of mineral aerosol as a reactive surface in the global troposphere, *J. Geophys. Res.*, *101*, D17, doi:10.1029/96JD01818, 1996.

Derwent, R. G., and M. E. Jenkin, Hydrocarbons and the long-range transport of ozone and PAN across Europe, *Atmos. Environ.*, *25A*, 1661–1678, 1991.

Dickerson, R. R., B. G. Doddridge, P. Kelley, and K. P. Rhoads, Large-scale pollution of the atmosphere over the remote Atlantic Ocean: evidence from Bermuda, *J. Geophys. Res.*, *100*, 8945–8952, 1995.

Dickerson, R. R., S. Kondragunta, G. Stenchikov, K. L. Civerolo, B. G. Doddridge, and B. N. Holben, The impact of aerosols on solar UV radiation and photochemical smog, *Science*, *278*, 827–830, 1997.

Didyk, B. M., B. R. T. Simoneit, L. A. Pezoa, M. L. Riveros, and A. A. Flores, Urban aerosol particles of Santiago, Chile: organic content and molecular characterization, *Atmos. Environ.*, *34*, 1167–1179, 2000.

DieselNet, Emission Standards: Summary of Worldwide Diesel Emission Standards, 2011, www.dieselnet.com/standards/#, accessed November 26, 2011.

Dockery, D. W., C. A. Pope III, X. Xu, J. D. Spengler, J. H. Ware, M. E. Fay, B. G. Ferris, Jr., and F. E. Speizer, An association between air pollution and mortality in six U.S. cities, *N. Engl. J. Med.*, *329*, 1753–1759, 1993.

Dong Energy, Vattenfall, Danish Energy Authority, and Danish Forest and Nature Agency, Danish Offshore Wind – Key Environmental Issues, 2006, www.ens.dk/graphics/Publikationer/Havvindmoeller/havvindmoellebog_nov_2006_skrm.pdf, accessed November 26, 2011.

Duce, R. A., On the source of gaseous chlorine in the marine atmosphere, *J. Geophys. Res.*, *70*, 1775–1779, 1969.

Dvorak, M., C. L. Archer, and M. Z. Jacobson, California offshore wind energy potential, *Renewable Energy*, *35*, 1244–1254, doi:10.1016/j.renene.2009.11.022, 2010.

Dvortsov, V. L., and S. Solomon, Response of the stratospheric temperatures and ozone to past and future increases in stratospheric humidity, *J. Geophys. Res.*, *106*, 7505–7514, 2001.

Easterling, D. R., B. Horton, P. D. Jones, T. C. Peterson, T. R. Karl, D. E. Parker, M. J. Salinger, V. Razuvayev, N. Plummer, P. Jamason, and C. K. Folland, Maximum and minimum temperature trends for the globe, *Science*, *277*, 364–367, 1997.

Egloff, G., *Motor Fuel Economy of Europe,* American Petroleum Institute, Washington, DC, 1940.

Emanuel, K., Increasing destructiveness of tropical cyclones over the past 30 years, *Nature*, *436*, 686–688, 2005.

Energy Information Administration, International Energy Outlook 2010, 2011a, www.eia.doe.gov/oiaf/ieo/coal.html, accessed February 22, 2011.

Energy Information Administration, Frequently Asked Questions, How Much Electricity Is Lost in Transmission and Distribution in the United States? 2011b, http://tonto.eia.doe.gov/tools/faqs/faq.cfm?id=105&t=3, accessed November 26, 2011.

Energy Information Administration, *Coal Production by State, Mine Type, and Union Status*, DOE/EIA-0584, 2011c, www.eia.gov/cneaf/coal/page/acr/table7.html, accessed November 26, 2011.

Eriksson, E., The yearly circulation of chloride and sulfur in nature; meteorological, geochemical and pedological implications: part I, *Tellus*, *12*, 63–109, 1960.

Etherington, D. J., D. F. Pheby, and F. I. Bray, An ecological study of cancer incidence and radon levels in south west England, *Eur. J. Cancer*, *32*, 1189–1197, 1996.

Ethridge, D. M., G. I. Pearman, and P. J. and Fraser P. J., Changes in tropospheric methane between 1841 and 1978 from a high accumulation rate Antarctic ice core, *Tellus*, *44B*, 282–94, 1992.

Eurasian Development Bank, http://en.rian.ru/russia/20090213/120116967.html, 2009. accessed February 23, 2011.

European Automobile Manufacturers Association (ACEA), *ACEA Programme on Emissions of Fine Particles from Passenger Cars*, ACEA, Brussels, 1999.

European Commission (EC), Joint Research Centre (JRC)/Netherlands Environmental Assessment Agency (PBL), Emissions Database for Global Atmospheric Research (EDGAR), release version 4.1, 2010, http://edgar.jrc.ec.europa.eu, accessed November 28, 2011.

European Commission (EC), Air Quality Standards, 2011, http://ec.europa.eu/environment/air/quality/standards.htm, accessed November 26, 2011.

European Environment Agency (EEA), Air Pollution (United Kingdom), 2010a, www.eea.europa.eu/soer/countries/uk/soertopic_view?topic=air%20pollution, accessed November 26, 2011.

European Environment Agency (EEA), Air Pollution (France), 2010b, www.eea.europa.eu/soer/countries/fr/soertopic_view?topic=air%20pollution, accessed November 26, 2011.

Evans, S. J., R. Toumi, J. E. Harries, M. P. Chipperfield, and J. M. Russell III, Trends in stratospheric humidity and the sensitivity of ozone to these trends, *J. Geophys. Res.*, *103*, 8715–8725, 1998.

Facchini, M. C., et al., Primary submicron marine aerosol dominated by insoluble organic colloids and aggregates, *Geophys Res. Lett.*, 35, L17814, doi:1029/2008GL034210, 2008.

Fairbanks, R. G., A 17,000-year glacio-eustatic sea level record: influence of glacial melting rates on Younger Dryas event and deep-ocean circulation, *Nature*, *342*, 637–642, 1989.

Fang, M., M. Zheng, F. Wang, K. L. To, A. B. Jaafar, and S. L. Tong, The solvent-extractable organic compounds in the Indonesia biomass burning aerosols – characterization studies, *Atmos. Environ.*, *33*, 783–795, 1999.

Farman, J. C., B. G. Gardiner, and J. D. Shanklin, Large losses of total ozone in Antarctica reveal seasonal ClOx/NOx interaction, *Nature*, *315*, 207–210, 1985.

Federal Trade Commission, Docket No. 2825, Cushing Refining & Gasoline Co., June 19, 1936, Dept. of Justice files, 60-57-107, National Archives, Washington, DC, 1936.

Ferek, R. J., J. S. Reid, P. V. Hobbs, D. R. Blake, and C. Liousse, Emission factors of hydrocarbons, halocarbons, trace gases, and particles from biomass burning in Brazil, *J. Geophys. Res.*, *103*, 32,107–32,118, 1998.

Ferro, A. R., R. J. Kopperud, and L. M. Hildemann, Elevated personal exposure to particulate matter from human activities in a residence, *J. Exposure Analysis Environ. Epidemiol.*, *14*, S34–S40, doi:10.1038/sj.jea.7500356, 2004.

Finlayson-Pitts, B. J., and J. N. Pitts, Jr., *Chemistry of the Upper and Lower Atmosphere*, Academic Press, San Diego, 1999.

Fitzner, C. A., J. C. Schroeder, R. F. Olson, and P. M. Tatreau, Measurement of ozone levels by ship along the eastern shore of lake Michigan, *J. Air Pollut. Control Ass.*, *39*, 727–728, 1989.

Flanner, M. G., C. S. Zender, J. T. Randerson, and P. J. Rasch, Present-day climate forcing and response from black carbon in snow, *J. Geophys. Res.*, *112*, D11202, doi:10.1029/2006JD008003, 2007.

Foster, V. G., Determination of the refractive index dispersion of liquid nitrobenzene in the visible and ultraviolet, *J. Phys. D: Appl. Phys.*, *25*, 525–529, 1992.

Frampton, M. W., P. E. Morrow, C. Cox, F. R. Gibb, D. M. Speers, and M. J. Utell, Effects of nitrogen dioxide exposure on pulmonary function and airway reactivity in normal humans, *Am. Rev. Respir. Dis.*, *143*, 522–527, 1991.

Frandsen, S. T., Turbulence and Turbulence-Generated Structural Loading in Wind Turbine Clusters, Riso-R-1188(EN), Riso National Laboratory, Roskilde, Denmark, 2007.

Frankenberg, C., I. Aben, P. Bergamaschi, E. J. Dlugokencky, R. van Hees, S. Houweling, P. van der Meer, R. Snel, and P. Tol, Global column-averaged methane mixing ratios from 2003 to 2009 as derived from SCIAMACHY: trends and variability, *J. Geophys. Res.*, *116*, D04302, doi:10.1029/2010JD014849, 2011.

Friedli, H., H. Lötscher, H. Oeschger, U. Siegenthaler, and B. Stauffer, Ice core record of 13C/12C ratio of atmospheric CO_2 in the past two centuries. *Nature*, *324*, 237–238, 1986.

Frölich, C., and J. Lean, Solar radiative output and its variability: evidence and mechanisms, *Astron. Astrophys. Rev.*, *12*, 273–320, doi:10.1007/s00159-004-0024-1, 2004.

Garrels, R. M., and E. A. Perry, Cycling of carbon, sulphur, and oxygen through geologic time, in *The Sea, Volume 5: Marine Chemistry*, ed. E. D. Goldberg, Wiley, New York, pp. 303–336, 1974.

Gerritson, S. L., The status of the modeling of ozone formation and geographic movement in the Midwest, in *Cost Effective Control of Urban Smog*, eds. R. F. Kosobud, W. A. Testa, and D. A. Hanson, Federal Reserve Bank of Chicago, Chicago, 1993.

Ghio, A. J., and J. M. Samet, Metals and air pollution particles, in *Air Pollution and Health*, eds. S. T. Holgate, J. M. Samet, H. S. Koren, and R. L. Maynard, Academic Press, San Diego, pp. 635–651, 1999.

Gillette, D. A., E. M. Patterson Jr., J. M. Prospero, and M. L. Jackson, Soil aerosols, in *Aerosol Effects on Climate*, ed. S. G. Jennings, University of Arizona Press, Tucson, pp. 73–109, 1993.

Gilliland, R. L., Solar evolution, *Glob. Plan. Change*, *1*, 35–56, 1989.

Ginnebaugh, D. L., J. Liang, and M. Z. Jacobson, Examining the temperature dependence of ethanol (E85) versus gasoline emissions on air pollution with a largely-explicit chemical mechanism, *Atmos. Environ.*, *44*, 1192–1199, doi:10.1016/j.atmosenv.2009.12.024, 2010.

Glass, N. R., G. E. Glass, and P. J. Rennie, Effects of acid precipitation, *Environ. Sci. Technol.*, *13*, 1350–1355, 1979.

Gold, D. R., Indoor air pollution, *Clin. Chest Med.*, *13*, 215–229, 1992.

Goldstein, I. F., L. R. Andrews, and D. Hartel, Assessment of human exposure to nitrogen dioxide, carbon monoxide, and respirable particles in New York inner-city residences, *Atmos. Environ.*, *22*, 2127–2139, 1988.

Goody, R., *Principles of Atmospheric Physics and Chemistry*, Oxford University Press, New York, 1995.

Graedel, T. E., and C. J. Weschler, Chemistry within aqueous atmospheric aerosols and raindrops, *Rev. Geophys. Space Phys.*, *19*, 505–539, 1981.

Graves, C. K., Rain of troubles, *Science*, *80*, 74–79, 1980.

Green, H., and W. Lane, *Particle Clouds*, Van Nostrand, Princeton, NJ, 1969.

Greenberg, R. R., W. H. Zoller, and G. E. Gordon, Composition and size distributions of articles released in refuse incineration, *Environ. Sci. Technol.*, *12*, 566–573, 1978.

Griffing, G. W., Relations between the prevailing visibility, nephelometer scattering coefficient and sunphotometer turbidity coefficient, *Atmos. Environ.*, *14*, 577–584, 1980.

Groblicki, P. J., G. T. Wolff, and R. J. Countess, Visibility-reducing species in the Denver "brown cloud" – I. Relationships between extinction and chemical composition, *Atmos. Environ.*, *15*, 2473–84, 1981.

Gunn, R., and B. B. Phillips, An experimental investigation of the effect of air pollution on the initiation of rain, *J. Meteorol.*, *14*, 272–280, 1957.

Haagen-Smit, A. J., The air pollution problem in Los Angeles, *Eng. Sci.*, *14*, 7–13, 1950.

Hader, D.-P., H. D. Kumar, R. C. Smith, and R. C. Worrest, Effects on aquatic ecosystems, *J. Photochem. Photobiol., B*, *46*, 53–68, 1998.

Hale, G. M., and M. R. Querry, Optical constants of water in the 200-nm to 200-mm wavelength region, *Appl. Opt.*, *12*, 555–63, 1973.

Hampson, S. E., J. A. Andres, M. E. Lee, L. S. Foster, R. E. Glasgow, and E. Lichtenstein, Lay understanding of synergistic risk: the case of radon and cigarette smoking, *Risk Anal.*, *18*, 343–350, 1998.

Hansen, J., and L. Nazarenko, Soot climate forcing via snow and ice albedos, *Proc. Natl. Acad. Sci.*, *101* (2), 423–428, doi:10.1073/pnas.2237157100, 2004.

Hansen, J., R. Ruedy, J. Glascoe, and M. Sato, GISS analysis of surface temperature change, *J. Geophys. Res.*, *104*, 30,997–31,022, 1999.

Hansen, J. E., R. Ruedy, M. Sato, and K. Lo, NASA GISS Surface Temperature (GISTEMP) Analysis, in *Trends: A Compendium of Data on Global Change.* Carbon Dioxide Information Analysis Center, Oak Ridge National Laboratory, U.S. Department of Energy, Oak Ridge, TN, doi:10.3334/CDIAC/cli.001, http://cdiac.ornl. gov/trends/temp/hansen/hansen.html, 2011.

Hansen, J., M. Sato, and R. Ruedy, Radiative forcing and climate response, *J. Geophys. Res.*, *102*, 6831–6864, 1997.

Hansen, J., et al., Efficacy of climate forcings, *J. Geophys. Res.*, *110*, D18104, doi:10.1029/2005JD005776, 2005.

Hargis, A. M., A review of solar induced lesions in domestic animals, *Compend. Continuing Educ.*, *3*, 287–300, 1981.

Harlap, S., and A. M. Davies, Infant admissions to hospital and maternal smoking, *Lancet*, *1*, 529–532, 1974.

Hart, E. K., and M. Z. Jacobson, A Monte Carlo approach to generator portfolio planning and carbon emissions assessments of systems with large penetrations of variable renewables, *Renewable Energy*, *36*, 2278–2286, doi:10.1016/j.renene.2011.01.015, 2011.

Hartmann, D. L., *Global Physical Climatology.* Academic Press, San Diego, 1994.

Hayami, H., and G. R. Carmichael, Analysis of aerosol composition at Cheju Island, Korea, using a two-bin gas-aerosol equilibrium model, *Atmos. Environ.*, *31*, 3429–3439, 1997.

Heinsohn, R. J., and R. L. Kabel, *Sources and Control of Air Pollution*, Prentice Hall, Upper Saddle River, NJ, 1999.

Henry, W. M., and K. T. Knapp, Compound forms of fossil fuel fly ash emissions, *Environ. Sci. Technol*, *14*, 450–456, 1980.

Henshaw, D. L., J. P. Eatough, and R. B. Richardson, Radon as a causative factor in induction of myeloid leukaemia and other cancers, *Lancet*, *335*, 1008–1012, 1990.

Hering, S. V., and S. K. Friedlander S. K., Origins of aerosol sulfur size distributions in the Los Angeles Basin, *Atmos. Environ.*, *16*, 2647–2656, 1982.

Hill, J. C., Johann Gluber's discovery of sodium sulfate – *Sal Marabile Glauberi*, *J. Chem. Edu.*, *56*, 593–594, 1979.

Hines, A. L., T. K. Ghosh, S. K. Loyalka, and R. C. Warder, eds., *Indoor Air – Quality and Control*, Prentice-Hall, Englewood Cliffs, NJ, 1993.

Hitchcock, D. R., L. L. Spiller, and W. E. Wilson, Sulfuric acid aerosols and HCl release in coastal atmospheres: evidence of rapid formation of sulfuric acid particulates, *Atmos. Environ.*, *14*, 165–182, 1980.

Hitzenberger, R., and H. Puxbaum, Comparisons of the measured and calculated specific absorption coefficients for urban aerosol samples in Vienna, *Aerosol Sci. Technol.*, *18*, 323–345, 1993.

Hong, S., J.-P. Candelone, C. C. Patterson, and C. F. Boutron, Greenland ice evidence of hemispheric lead pollution two millennia ago by Greek and Roman civilizations, *Science*, *265*, 1841–1843, 1994.

Hong, S., J.-P. Candelone, C. C. Patterson, and C. F. Boutron, History of ancient copper smelting pollution during Roman and Medieval times recorded in Greenland ice, *Science*, *272*, 246–248, 1996.

Houghton, R. A., Effects of land-use change, surface temperature and CO_2 concentrations on terrestrial stores of carbon, in *Biotic Feedbacks in the Global Climate System: Will the Warming Speed the Warming?* eds. G. M. Woodwell and F. T. Mackenzie, Oxford University Press, Oxford, UK, 1994.

Howard, L. *The Climate of London*, Vols. I–III, London, 1833.

Howarth, R. W., R. Santoro, and A. Ingraffea, Methane and greenhouse-gas footprint of natural gas from shale formations, *Clim. Change Lett.*, *106* (1), doi:10.1007/s10584-011-0061-5, 2011.

Hughes, J. D., *Pan's Travail: Environmental Problems of the Ancient Greeks and Romans*, The Johns Hopkins University Press, Baltimore, 1994.

Huntrieser, H., et al., Intercontinental air pollution transport from North America to Europe: experimental evidence from airborne measurements and surface observations, *J. Geophys. Res.*, *110*, D01305, doi:10.1029/2004JD005045, 2005.

Hurrell, J. W., and K. E. Trenberth, Spurious trends in satellite MSU temperatures from merging different satellite records, *Nature*, *386*, 164–167, 1997.

Hurst, D. F., S. J. Oltmans, H. Vömel, K. H. Rosenlof, S. M. Davis, E. A. Ray, E. G. Hall, and A. F. Jordan, Stratospheric water vapor trends over Boulder, Colorado: analysis of the 30 year Boulder record, *J. Geophys. Res.*, *116*, D02306, doi:10.1029/2010JD015065, 2011.

Husar, R. B., et al., Asian dust events of April 1998, *J. Geophys. Res.*, *106*, D16, doi:10.1029/2000JD900788, 2001.

IMPROVE (Interagency Monitoring of Protected Visual Environments), IMPROVE and RHR Summary Data, n.d., http://vista.cira.colostate.edu/improve/Data/IMPROVE/summary_data.htm, accessed November 27, 2011.

Intergovernmental Panel on Climate Change (IPCC), Working Group III, *IPCC Special Report on Carbon Dioxide Capture and Storage*, eds. B. Metz, O. Davidson, H. de Coninck, M. Loos, and L. Meyer, Cambridge University Press, Cambridge, UK, 2005.

Intergovernmental Panel on Climate Change (IPCC), Working Group I, *Climate Change 2007: The Physical Science Basis*, eds. M. Manning and D. Qin, Cambridge University Press, New York, 2007.

Islam, M. S., and W. T. Ulmer, Threshold concentrations of SO2 for patients with oversensitivity of the bronchial system, *Wissenchaft Umwelt*, *1*, 41–47, 1979.

Jackman, C. H., E. L. Fleming, S. Chandra, D. B. Considine, and J. E. Rosenfield, Past, present, and future modeled ozone trends with comparisons to observed trends, *J. Geophys. Res.*, *101*, 28,753–28,767, 1996.

Jacob, D. J., Comment on "The photochemistry of a remote stratiform cloud," by W. L. Chameides, *J. Geophys. Res.*, *90*, 5864, 1985.

Jacob, D. J., J. A. Logan, and P. Murti, Effect of rising Asian emissions on surface ozone in the United States, *Geophys. Res. Lett.*, *26*, 22175–2178, 1999.

Jacobson, M. Z., Development and application of a new air pollution modeling system, part II, Aerosol module structure and design, *Atmos. Environ.*, *31*, 131–144, 1997.

Jacobson, M. Z., Studying the effects of aerosols on vertical photolysis rate coefficient and temperature profiles over an urban airshed, *J. Geophys. Res.*, *103*, 10,593–10,604, 1998.

Jacobson, M. Z., Effects of soil moisture on temperatures, winds, and pollutant concentrations in Los Angeles, *J. Appl. Meteorol.*, *38*, 607–616, 1999a.

Jacobson, M. Z., Isolating nitrated and aromatic aerosols and nitrated aromatic gases as sources of ultraviolet light absorption, *J. Geophys. Res.*, *104*, 3527–3542, 1999b.

Jacobson, M. Z., Studying the effects of calcium and magnesium on size-distributed nitrate and ammonium with EQUISOLV II, *Atmos. Environ.*, *33*, 3635–3649, 1999c.

Jacobson, M. Z., A physically-based treatment of elemental carbon optic: implications for global direct forcing of aerosols, *Geophys. Res. Lett.*, *27*, 217–220, 2000.

Jacobson, M. Z., Global direct radiative forcing due to multicomponent anthropogenic and natural aerosols, *J. Geophys. Res.*, *106*, 1551–1568, 2001a.

Jacobson, M. Z., Strong radiative heating due to the mixing state of black carbon in atmospheric aerosols, *Nature*, *409*, 695–697, 2001b.

Jacobson, M. Z., Control of fossil-fuel particulate black carbon plus organic matter, possibly the most effective method of slowing global warming, *J. Geophys. Res.*, *107*, D19, doi:10.1029/2001JD001376, 2002a.

Jacobson, M. Z., Analysis of aerosol interactions with numerical techniques for solving coagulation, nucleation, condensation, dissolution, and reversible chemistry among multiple size distributions, *J. Geophys. Res.*, *107*, D19, doi:10.1029/ 2001JD002044, 2002b.

Jacobson, M. Z., The climate response of fossil-fuel and biofuel soot, accounting for soot's feedback to snow and sea ice albedo and emissivity, *J. Geophys. Res.*, *109*, D21201, doi:10.1029/2004JD004945, 2004.

Jacobson, M. Z., *Fundamentals of Atmospheric Modeling*, 2nd ed. Cambridge University Press, Cambridge, UK, 2005a.

Jacobson, M. Z., Studying ocean acidification with conservative, stable numerical schemes for nonequilibrium air-ocean exchange and ocean equilibrium chemistry, *J. Geophys. Res.*, *110*, D07302, doi:10.1029/2004JD005220, 2005b.

Jacobson, M. Z., Correction to "Control of fossil-fuel particulate black carbon and organic matter," *J. Geophys. Res.*, *110*, D14105, doi:10.1029/2005JD005888, 2005c.

Jacobson, M. Z., Effects of absorption by soot inclusions within clouds and precipitation on global climate, *J. Phys. Chem.*, *110*, 6860–6873, 2006.

Jacobson, M. Z., Effects of ethanol (E85) versus gasoline vehicles on cancer and mortality in the United States, *Environ.*

Sci. Technol., *41* (11), 4150–4157, doi:10.1021/es062085v, 2007.

Jacobson, M. Z, On the causal link between carbon dioxide and air pollution mortality, *Geophys. Res. Lett.*, *35*, L03809, doi:10.1029/2007GL031101, 2008a.

Jacobson, M. Z., Testimony for Hearing on Air Pollution Health Impacts of Carbon Dioxide, U.S. House of Representatives Select Committee on Energy Independence and Global Warming, 2008b, www.stanford.edu/group/efmh/jacobson/040908_testimony.htm, accessed November 27, 2011.

Jacobson, M. Z., Review of solutions to global warming, air pollution, and energy security, *Energy Environ. Sci.*, *2*, 148–173, doi:10.1039/b809990c, 2009.

Jacobson, M. Z., The enhancement of local air pollution by urban CO_2 domes, *Environ. Sci. Technol.*, *44*, 2497–2502, doi:10.1021/es903018m, 2010a.

Jacobson, M. Z., Short-term effects of controlling fossil-fuel soot, biofuel soot and gases, and methane on climate, Arctic ice, and air pollution health, *J. Geophys. Res.*, *115*, D14209, doi:10.1029/2009JD013795, 2010b.

Jacobson, M. Z., Investigating cloud absorption effects: Global absorption properties of black carbon, tar balls, and soil dust in clouds and aerosols. *J. Geophys. Res.*, in press. doi:10.1029/2011JD017218, 2012.

Jacobson, M. Z., and M. A. Delucchi, Providing all global energy with wind, water, and solar power, part I: technologies, energy resources, quantities and areas of infrastructure, and materials, *Energy Policy*, *39*, 1154–1169, doi:10.1016/j.enpol.2010.11.040, 2011.

Jacobson, M. Z., and Y. J. Kaufman, Wind reduction by aerosol particles, *Geophys. Res. Lett.*, *33*, L24814, doi:10.1029/2006GL027838, 2006.

Jacobson, M. Z., and D. L. Ginnebaugh, The global-through-urban nested 3-D simulation of air pollution with a 13,600-reaction photochemical mechanism, *J. Geophys. Res.*, *115*, D14304, doi:10.1029/2009JD013289, 2010.

Jacobson, M. Z., D. B. Kittelson, and W. F. Watts, Enhanced coagulation due to evaporation and its effect on nanoparticle evolution, *Environ. Sci. Technol.*, *39*, 9486–9492, 2005.

Jacobson, M. Z., and G. M. Masters, Exploiting wind versus coal, *Science*, *293*, 1438–1438, 2001.

Jacobson, M. Z., and J. E. ten Hoeve, Effects of urban surfaces and white roofs on global and regional climate, *J. Clim.*, *25*, 1028–1044, doi:10.1175/JCLI-D-11-00032.1, 2012.

Jaffe, D., W. Hafner, D. Chand, A. Westerling, and D. Spracklen, Interannual variations in PM2.5 due to wild-fires in the western United States, *Environ. Sci. Technol.*, *42*, 2812–2818, 2008.

Janerich, D. T., W. D. Thompson, and L. R. Varela, Lung cancer and exposure to tobacco smoke in the household, *N. Engl. J. Med.*, *323*, 632–636, 1990.

Jiang, R. O.-T., K. I.-C. Cheng, V. Acevedo-Bolton, N. E. Klepeis, J. L. Repace, W. R. Ott, and L. M. Hildemann, Measurement of fine particles and smoking activity in a statewide survey of 36 California Indian casinos, *J. Exposure Sci. Environ. Epidemiol.*, *21*, 31–41, doi:10.1038/jes.2009.75, 2011.

John W., S. M. Wall, J. L. Ondo, and W. Winklmayr, Acidic aerosol size distributions during SCAQS. Final Report Prepared for the California Air Resources Board, Contract No. A6-112-32, www.arb.ca.gov/research/apr/past/a6-112-32.pdf, 1989.

John, W., S. M. Wall, J. L. Ondo, and W. Winklmayr, Modes in the size distributions of atmospheric inorganic aerosol, *Atmos. Environ.*, *24A*, 2349–2359, 1990.

Jones, A. P., Indoor air quality and health, *Atmos. Environ.*, *33*, 4535–4564, 1999.

Jorquera, H., W. Palma, and J. Tapia, An intervention analysis of air quality data at Santiago, Chile, *Atmos. Environ.*, *34*, 4073–4084, 2000.

Jouzel, J., N. I. Barkov, J. M. Barnola, M. Bender, J. Chappellaz, C. Genthon, V. M. Kotlyakov, V. Lipenkov, C. Lorius, J. R. Petit, D. Raynaud, G. Raisbeck, C. Ritz, T. Sowers, M. Stievenard, F. Yiou, and P. Yiou, Extending the Vostok ice-core record of palaeoclimate to the penultimate glacial period, *Nature*, *364*, 407–412, 1993.

Jouzel, J., C. Lorius, J. R. Petit, C. Genthon, N. I. Barkov, V. M. Kotlyakov, and V. M. Petrov, Vostok ice core: a continuous isotope temperature record over the last climatic cycle (160,000 years). *Nature*, *329*, 403–408, 1987.

Jouzel, J., C. Waelbroeck, B. Malaize, M. Bender, J. R. Petit, M. Stievenard, N. I. Barkov, J. M. Barnola, T. King, V. M. Kotlyakov, V. Lipenkov, C. Lorius, D. Raynaud, C. Ritz, and T. Sowers, Climatic interpretation of the recently extended Vostok ice records, *Clim. Dyn.*, *12*, 513–521, 1996.

Junge, C. E., Vertical profiles of condensation nuclei in the stratosphere. *J. Meteorol.*, *18*, 501–509, 1961.

Kahn, E., The reliability of distributed wind generators, *Electr. Power Syst.*, *2*, 1–14, 1979.

Kallos, G., V. Kotroni, K. Lagouvardos, and A. Papadopoulos, On the long-range transport of air pollutants from Europe to Africa, *Geophys. Res. Lett.*, *25*, 619–622, 1998.

Kasting, J. F., and J. L. Siefert, Life and the evolution of Earth's atmosphere, *Science*, *296*, 1066–1068, 2002.

Katrlnak, K. A., P. Rez, and P. R. Buseck, Structural variations in individual carbonaceous particles from an urban aerosol, *Environ. Sci. Technol* *26*, 1967–1976, 1992.

Katrlnak, K. A., P. Rez, P. R. Perkes, and P. R. Buseck, Fractal geometry of carbonaceous aggregates from an urban aerosol, *Environ. Sci. Technol.*, *27*, 539–547, 1993.

Kaufman, Y. J., and I. Koren, Smoke and pollution aerosol effect on cloud cover, *Science*, *313*, 655–658, 2006.

Kaufman, Y. J., and T. Nakajima, Effect of Amazon smoke on cloud microphysics and albedo – analysis from satellite imagery, *J. App. Meteorol.*, *32*, 729–744, 1993.

Kempton, W., C. L. Archer, A. Dhanju, R. W. Garvine, and M. Z. Jacobson, Large CO_2 reductions via offshore wind power matched to inherent storage in energy end-uses, *Geophys. Res. Lett.*, *34*, L02817, doi:10.1029/2006GL028016, 2007.

Kempton, W., and J. Tomic, Vehicle-to-grid power fundamentals: calculating capacity and net revenue, *J. Power Sources*, *144*, 268–279, 2005a.

Kempton, W., and J. Tomic, Vehicle-to-grid power implementation: from stabilizing the grid to supporting large-scale renewable energy, *J. Power Sources*, *144*, 280–294, 2005b.

Kious, W. J., and R. I. Tilling, *This Dynamic Earth: The Story of Plate Techtonics*, U.S. Government Printing Office, Washington, DC, 1996.

Kittelson, D. B., Engine and nanoparticles: a review, *J. Aerosol Sci.*, *29*, 575–588, 1998.

Kleeman, M. J., A preliminary assessment of the sensitivity of air quality in California to global change, *Clim. Change*, *87*, S273–S292, 2008.

Klein, S. A., Synoptic variability of low-cloud properties and meteorological parameters in the subtropical trade wind boundary layer, *J. Clim.*, *10*, 2018–2039, 1997.

Koomey, J., and N. E. Hultman, A reactor-level analysis of busbar costs for U.S. nuclear plants,1970–2005, *Energy Policy*, *35*, 5630–5642, 2007.

Koren, I., Y. J. Kaufman, L. A. Remer, and J. V. Martins, Measurement of the effect of Amazon smoke on inhibition of cloud formation, *Science*, *303*, 1342–1345, 2004.

Koschmieder, H., Theorie der horizontalen Sichtweite, *Beitr. Phys. Freien Atm.*, *12*, 33–53, 171–181, 1924.

Kovarik, B., Henry Ford, Charles Kettering and the "Fuel of the Future," 1998, www.radford.edu/wkovarik/papers/fuel.html, accessed November 27, 2011.

Kovarik, B., Charles F. Kettering and the 1921 Discovery of Tetraethyl Lead in the Context of Technological Alternatives, 1999, www.radford.edu/~wkovarik/papers/kettering.html, accessed November 27, 2011.

Krekov, G. M., Models of atmospheric aerosols, in *Aerosol Effects on Climate*, ed. S. G. Jennings, University of Arizona Press, Tucson, pp. 9–72, 1993.

Kukla, G. J., Pleistocene land-sea correlations, I, Europe, *Earth Sci. Rev.*, *13*, 307–374, 1977.

Lagarde, F., G. Pershagen, G. Akerblom, O. Axelson, U. Baverstam, L. Damber, A. Enflo, M. Svartengren, and G. A. Swedjemark, Residential radon and lung cancer in Sweden: risk analysis accounting for random error in the exposure assessment, *Health Phys.*, *72*, 269–276, 1997.

Lalas, D. P., D. N. Asimakopoulos, D. G. Deligiorgi, and C. G. Helmis, Sea-breeze circulation and photochemical pollution in Athens, Greece, *Atmos. Environ.*, *17*, 1621–1632, 1983.

Lamarque, J. F., J. van Aardenne, M. Schultz, S. Smith, and T. Bond, IPCC AR5 Detailed Emission Datasets, ftp://ftp-ipcc.fz-juelich.de/pub/emissions/, 2010.

Larson, S., G. Cass, K. Hussey, and F. Luce, Visibility Model Verification by Image Processing Techniques: Final Report to the California Air Resources Board, Agreement A2-077-32, 1984.

Lary, D. J., Catalytic destruction of stratospheric ozone, *J. Geophys. Res.*, *102*, 21,515–21,526, 1997.

Lazrus, A. L., R. D. Cadle, B. W. Gandrud, J. P. Greenberg, B. J. Huebert, and W. I. Rose, Sulfur and halogen chemistry of the stratosphere and of volcanic eruption plumes, *J. Geophys. Res.*, *84*, 7869, 1979.

Leaderer, B. P., P. Koutrakis, S. L. K. Briggs, and J. Rizzuto, Measurement of toxic and related air pollutants, in *Proceedings of the EPA/Air and Waste Management Association International Symposium (VIP-17)*, Environmental Protection Agency, Washington, DC, p. 567, 1990.

Leaderer, B. P., M. Stowe, R. Li, J. Sullivan, P. Koutrakis, M. Wolfson, and W. Wilson, Residential levels of particle and vapor phase acid associated with combustion sources, in *Proceedings of the Sixth International Conference on Indoor Air Quality and Climate*, Helsinki, Finland, pp. 147–152, 1993.

Leakey, R. E., and R. Lewin, *Origins*, MacDonald and Janes, London, 1977.

Lee, R. J., D. R. Van Orden, M. Corn, and K. S. Crump, Exposure to airborne asbestos in buildings, *Regulatory Toxicology and Pharmacology*, *16*, 93–107, 1992.

Leg 105 Shipboard Scientific Party, High-latitude palaeoceanography, *Nature*, *230*, 17–18, 1986.

Lehmann, C. M. B., V. C. Bowersox, R. S. Larson, and S. M. Larson, Monitoring long-term trends in sulfate and ammonium in U.S. precipitation: results from the National Atmospheric Deposition Program/National Trends Network, *Water Air Soil Pollut. Focus*, *7*, 59–66, 2007.

Lenzen, M., Life cycle energy and greenhouse gas emissions of nuclear energy: a review, *Energy Conserv. Manage.*, *49*, 2178–2199, 2008.

Le Roy Ladurie, E., *Times of Feast, Times of Famine: A History of Climate since the Year 1000*, Doubleday, New York, 1971.

Levy, H., P. Kasibhatla, W. J. Moxim, A. A. Klonecki, A. I. Hirsch, S. J. Oltmans, and W. L. Chameides, The global impact of human activity on tropospheric ozone, *Geophys. Res. Lett.*, *24*, 791–794, 1997.

Lewis, J., Lead poisoning: a historical perspective, *EPA Journal*, 1985, www.epa.gov/aboutepa/history/topics/perspect/lead.html, accessed November 27, 2011.

Li, Y., T. E. Powers, and H. D. Roth, Random effects linear regression meta-analysis models with application to nitrogen dioxide health effects studies. *J. Air Waste Manage. Assoc.*, *44*, 261–270, 1994.

Liang, J., and M. Z. Jacobson, A study of sulfur dioxide oxidation pathways over a range of liquid water contents, pHs, and temperatures, *J. Geophys. Res.*, *104*, 13,749–13,769, 1999.

Lide D. R., ed., *CRC Handbook of Chemistry and Physics*, 79th ed. CRC Press, Boca Raton, FL, 1998.

Likens, G. E., Acid precipitation, *Chem. Eng. News*, *54*, 29–44, 1976.

Liou, K. N., *Radiation and Cloud Processes in the Atmosphere*, Oxford University Press, New York, 1992.

Liu, S. C., M. Trainer, F. C. Fehsenfeld, D. D. Parrish, E. J. Williams, D. W. Fahey, G. Hubler, and P. C. Murphy, Ozone production in the rural troposphere and the implications for regional and global ozone distributions, *J. Geophys. Res.*, *92*, 4191–4207, 1987.

Lobert, J. M., W. C. Keene, J. A. Logan, and R. Yevich, Global chlorine emissions from biomass burning: reactive chlorine emissions inventory, *J. Geophys. Res.*, *104*, 8373–8389, 1999.

Longstreth, J., F. R. de Gruigj, M. L. Kripke, S. Abseck, F. Arnold, H. I. Slaper, G. Velders, Y. Takizawa, and J. C. van der Leun, Health risks, *J. Photochem. Photobiol., B*, *46*, 20–39, 1998.

Lu, R., and R. P. Turco, Air pollution transport in a coastal environment – II. Three-dimensional simulations over Los Angeles Basin, *Atmos. Environ.*, *29*, 1499–1518, 1995.

Lu, X., M. B. McElroy, and J. Kiviluoma, Global potential for wind-generated electricity, *Proc. Natl. Acad. Sci.*, *106*, 10933, doi:10.1073/pnas.0904101106, 2009.

Lurmann, F. W., H. H. Main, K. T. Knapp, L. Stockburger, R. A. Rasmussen, and K. Fung, Analysis of the Ambient VOC Data Collected in the Southern California Air Quality Study: Final Report to the California Air Resources Board, Contract A832-130, 1992.

Lyman, G. H., Radon, in *Indoor Air Pollution and Health*, eds. E. J. Bardana and A. Montanaro, Dekker, New York, 83–103, 1997.

Lyons, W. A., and L. E. Olsson, Detailed mesometeorological studies of air pollution dispersion in the Chicago lake breeze, *Mon. Weath. Rev.*, *101*, 387–403, 1973.

MacNee, W., and K. Donaldson, Particulate air pollution: injurious and protective mechanisms in the lungs, in *Air Pollution and Health*, eds. S. T. Holgate, J. M. Samet, H. S. Koren, and R. L. Maynard, Academic Press, San Diego, pp. 653–672, 1999.

Madronich, S., R. L. McKenzie, L. O. Bjorn, and M. M. Caldwell, Changes in biologically active ultraviolet radiation reaching the Earth's surface, *J. Photochem. Photobiol., B*, *46*, 5–19, 1998.

Manchester, Shefield and Lincolnshire Rail Co. v Wood, 29 Law J. Rep. M.C. 29.

Marcinowski, F., R. M. Lucas, and W. M. Yeager, National and regional distributions of airborne radon concentrations in U.S. homes, *Health Phys.*, *66*, 699–706, 1994.

Maricq, M. M., R. E. Chase, D. H. Podsiadlik, and R. Vogt, Vehicle Exhaust Particle Size Distributions: A Comparison of Tailpipe and Dilution Tunnel Measurements, Technical Paper 1999-01-1461, SAE International, doi:10.4271/1999-01-1461, 1999.

Maroni, M., B. Seifert, and T. Lindvall, Eds., *Indoor Air Quality: A Comprehensive Reference Book*, Elsevier, Amsterdam, 1995.

Marsh, A. R. W., Sulphur and nitrogen contributions to the acidity of rain, *Atmos. Environ.*, *12*, 401–406, 1978.

Martins, J. V., P. Artaxo, C. Liousse, J. S. Reid, P. V. Hobbs, and Y. J. Kaufman, Effects of black carbon content, particle size, and mixing on light absorption by aerosols from biomass burning in Brazil, *J. Geophys. Res.*, *103*, 32,041–32,050, 1998.

Masters, G. M., *Introduction to Environmental Engineering and Science*, 2nd ed, Prentice-Hall, Upper Saddle River, NJ, 1998.

Masters, G. M., *Renewable and Efficient Electric Power Systems*, Wiley-Interscience, Hoboken, NJ, 2004.

Matthes, F. E., Report of committee on glaciers, April 1939, *Trans. Am. Geophys. Union*, *20*, 518–523, 1939.

Mauna Loa Data Center, Data for Atmospheric Trace Gases, n.d., www.esrl.noaa.gov/gmd/hats/insitu/cats/, accessed November 29, 2011.

Mayer, S. J., Stratospheric ozone depletion and animal health, *Veterinary Record*, *131*, 120–122, 1992.

McBride, S. J., A. R. Ferro, W. R. Ott, P. Switzer, and L. M. Hildemann, Investigations of the proximity effect for pollutants in the indoor environment, *J. Exposure Anal. Environ. Epidemiol.*, *9*, 602–621, 1999.

McElroy, J. L., and T. B. Smith, Creation and fate of ozone layers aloft in Southern California, *Atmos. Environ.*, *26*, 1917–1929, 1992.

McElroy, M. B., R. J. Salawitch, S. C. Wofsy, and J. A. Logan, Reduction of Antarctic ozone due to synergistic interactions of chlorine and bromine, *Nature*, *321*, 759–762, 1986.

McGregor, R., 750,000 a year killed by Chinese pollution, *Financial Times*, July 2, 2007, www.ft.com/intl/cms/s/0/8f40e248-28c7-11dc-af78-000b5df10621.html, accessed June 26, 2011.

McKenzie, R., B. Connor, and G. Bodeker, Increased summertime UV radiation in New Zealand in response to ozone loss, *Science*, *285*, 1709–1711, 1999.

McKenzie, R. L., P. J. Aucamp, A. F. Bais, L. O. Bjorn, and M. Ilyas, Changes in biologically-active ultraviolet radiation reaching the Earth's surface, *Photochem. Photobiol. Sci.*, *6*, 218–231, 2007.

McMurry, P. H., and X. Q. Zhang, Size distributions of ambient organic and elemental carbon, *Aerosol Sci. Technol. 10*, 430–437, 1989.

McNeill, J. R., *Something New under the Sun*, W. W. Norton, New York, 2000.

MedicineNet.com, Multiple Chemical Sensitivity, n.d., www.medicinenet.com/sick_building_syndrome/page2.htm, accessed November 27, 2011.

Medina-Ramon, M., and J. Schwartz, Temperature, temperature extremes, and mortality: a study of acclimatization and effect modification in 50 U.S. cities, *Occup. Environ. Med.*, doi:10.1136/oem.2007.033175, 2007.

Mendez, A., J. Perez, E. Ruiz-Villamor, M. P. Martin, and E. Mozos, Clinicopathological study of an outbreak of squamous cell carcinoma in sheep, *Veterinary Record*, *141*, 597–600, 1997.

Mickley, L. J., D. J. Jacob, B. D. Field, and D. Rind, Effects of future climate change on regional air pollution episodes in the United States, *Geophys. Res. Lett.*, *31*, L24103, doi:10.1029/2004GL021216, 2004.

Middleton, W. E. K., *Vision through the Atmosphere*, University of Toronto Press, Toronto, Canada, 1952.

Midgley, T., Jr., Tetraethyl lead poison hazards, *J. Ind. Eng. Chem.*, *17* (8), 1925a, 827.

Midgley, T., Jr., Radium derivative $5,000,000 an ounce/ethyl gasoline defended, *New York Times*, April 7, 1925b, p. 23.

Mie, G., Optics of turbid media, *Ann. Phys.*, *25*, 377–445, 1908.

Milankovitch, M., Mathematische klimalehre und astronomische theorie der Klimaschwankungen, in *Handbuch der Klimatologie*, eds. I. W. Köppen and R. Geiger, Gebruder Borntraeger, Berlin, 1930.

Milankovitch, M., *Canon of Insolation and the Ice-Age Problem* (English translation by the Israel Program for Scientific Translation, published by the U.D. Department of Commerce and the National Science Foundation), Königlich Serbische Akademie, Belgrade, 1941.

Miller, P. D., Maize pollen: Collection and enzymology, in *Maize for Biological Research*, A Special Publication of the Plant Molecular Biology Association, USA, ed. W. F. Sheridan, pp. 279–282, 1985.

Miller, S. L., A production of amino acids under possible primitive Earth conditions, *Science*, *117*, 528–529, 1953.

Miller, S. L., and L. E. Orgel, *The Origins of Life on Earth*, Prentice-Hall, Englewood Cliffs, NJ, 1974.

Molina, L. T., and M. J. Molina, Production of Cl_2O_2 by the self reaction of the ClO radical, *J. Phys. Chem.*, *91*, 433–436, 1986.

Molina, M. J., and F. S. Rowland, Stratospheric sink for chlorofluoromethanes: chlorine atom catalysed destruction of ozone, *Nature*, *249*, 810–2, 1974.

Moody, J. L., J. C. Davenport, J. T. Merrill, S. J. Oltmans, D. D. Parrish, J. S. Holloway, H. Levy II, G. L. Forbes, M. Trainer, and M. Buhr, Meteorological mechanisms for transporting ozone over the western North Atlantic Ocean: a case study for August 24–29, 1993, *J. Geophys. Res.*, *101* (29), 213–229, 1996.

Morales, P., Chilean University Finds Santiago Pollution Has Doubled, 2010, www.santiagotimes.cl/index.php?option= com_content&view=article&id=18110:chilean-university-finds-santiago-pollution-has-doubled&catid=44: environmental&Itemid=40, accessed June 26, 2011.

Mordukhovich, I., E. Wilker, H. Suh, R. Wright, D. Sparrow, P. S. Vokonas, and J. Schwartz, Black carbon exposure, oxidative stress genes, and blood pressure in repeated-measures study, *Environ. Health Perspect.*, *117*, 1767–1772, doi:10.1289/ehp.0900591, 2009.

Motallebi, N., M. Sogutlugil, E. McCauley, and J. Taylor, Climate change impact on California on-road mobile source emissions, *Clim. Change*, *87* (Suppl. 1) S293–S308, doi:10.1007/s10584-007-9354-0, 2008.

Munger, J. W., D. J. Jacob, J. M. Waldman, and M. R. Hoffmann, Fogwater chemistry in an urban atmosphere, *J. Geophys. Res.*, *88*, 5109–5121, 1983.

Murphy, D. M., J. R. Anderson, P. K. Quinn, L. M. McInnes, F. J. Brechtel, S. M. Kreidenweis, A. M. Middlebrook, M. Pósfai, D. S. Thomson, and P. R. Buseck, Influence of sea-salt on aerosol radiative properties in the Southern Ocean marine boundary layer, *Nature*, *395*, 62–65, 1998.

Nagda, N. L., H. E. Rector, and M. D. Koontz, *Guidelines for Monitoring Indoor Air Quality*, Hemisphere, Washington, DC, 1987.

National Atmospheric Deposition Program/National Trends Network, Illinois State Water Survey, 2011, http://nadp. sws.uiuc.edu, accessed June 26, 2011.

National Centers for Environmental Prediction (NCEP), 2.5 degree global final analyses, distributed by the Data Support Section, National Center for Atmospheric Research, 2000.

National Institute for Occupational Safety and Health (NIOSH), Work-Related Lung Disease (WoRLD) Surveillance System, Volume 1: Coal Workers' Pneumoconiosis and Related Exposures, 2008, www.2.cdc.gov/drds/ WorldReportData/SectionDetails.asp?ArchiveID= 1&SectionTitleID=2, accessed November 27, 2011.

National Institute for Occupational Safety and Health (NIOSH), 2010, www.cdc.gov/niosh/, accessed June 26, 2011.

Nazaroff, W. W., and A. V. Nero, Jr., *Radon and Its Decay Products in Indoor Air*, Wiley-Interscience, New York, 1988.

Nero, A. V., M. B. Schwehr, W. W. Nazaroff, and K. L. Revzan, Distribution of airborne radon-222 concentrations in U.S. homes, *Science*, *234*, 992–997, 1986.

Newman, M. J., and R. T. Rood, Implications of solar evolution for the Earth's early atmosphere, *Science*, *194*, 1413–1414, 1977.

Nicholls, S, The dynamics of stratocumulus: aircraft observations and comparisons with a mixed layer model, *Q. J. R. Meteorolog. Soc.*, *110*, 783–820, 1984.

North American Electric Reliability Corporation (NERC), Generating Availability Data System (GADS): Generating Availability Reports, 2011, http://www.nerc.com/page. php?cid=4|43|47, accessed November 27, 2011.

North Greenland Ice Core Project Members, High-resolution record of Northern Hemisphere climate extending into the last interglacial period, *Nature*, *431*, 147–151, 2004.

Nriagu, J. O., A history of global metal pollution, *Science*, *272*, 223–224, 1996.

Oberg, M., M. S. Jaakkola, A. Woodward, A. Perugo, and A. Pruss-Ustun, Worldwide burden of disease from exposure to second-hand smoke: a retrospective analysis of data from 192 countries, *Lancet*, doi:10.1016/S0140-6736(10)61388-8, 2010.

Offenberg, J. H., and J. E. Baker, Aerosol size distributions of elemental and organic carbon in urban and over-water atmospheres, *Atmos. Environ.*, *34*, 1509–1517, 2000.

Oke, T. R., *Boundary Layer Climates*. Methuen, London, 1978.

Oke, T. R., The urban energy balance, *Prog. Phys. Geog.*, *12*, 471–508, 1988.

Olszyna, K. J., M. Luria, and J. F. Meagher, The correlation of temperature and rural ozone levels in southeastern U.S.A., *Atmos. Environ.*, *31*, 3011–3022, 1997.

Oparin, A. I., *The Origin of Life*, Macmillan, New York, 1938.

Ostro, B. D., Tran, H., and Levy, J. I., The health benefits of reduced tropospheric ozone in California. *J. Air Waste Manage. Assoc.*, *56*, 1007–1021, 2006.

Özkatnak, H., and G. D. Thurston, Association between 1980 U.S. mortality rates and alternative measures of airborne particle concentrations. *Risk Anal.*, *7*, 449–461, 1987.

Pandis S. N., Harley R. A., Cass G. R., and Seinfeld J. H., Secondary organic aerosol formation and transport. *Atmos. Environ.*, *26A*, 2269–2282, 1992.

Park, E.-K., K. Takahashi, T. Hoshuyama, T.-J. Cheng, V. Delgermaa, G. V. Le, and T. Sorahan, Global magnitude of reported and unreported mesothelioma, *Environ. Health Perspect.*, *119*, 514–518, 2011.

Parker, D. E., A demonstration that large-scale warming is not urban. *J. Clim.*, *19*, 2882–2895, 2006.

Pasken, R., and J. A. Pietrowicz, Using dispersion and mesoscale meteorological models to forecast pollen concentrations, *Atmos. Environ.*, *39*, 7689–7701, 2005.

Paulson, S. E., and J. H. Seinfeld, Development and evaluation of a photooxidation mechanism for isoprene, *J. Geophys. Res.*, *97* (20), 703–715, 1992.

People's Daily Online, Black Lung Disease Claims 140,000 Lives in China, March 18, 2005, http://english.peopledaily.com.cn/200503/18/eng20050318_177365.html, accessed November 27, 2011.

Peterson, T. C., Assessment of urban versus rural in situ surface temperatures in the contiguous United States: no difference found. *J. Clim.*, *16*, 2941–2959, 2003.

Peterson, T. C., and R. S. Vose, An overview of the Global Historical Climatology Network temperature data base, *Bull. Am. Meteorol. Soc.*, *78*, 2837–2849, 1997.

Petit, J. R., J. Jouzel, D. Raynaud, N. I. Barkov, J.-M. Barnola, I. Basile, M. Bender, J. Chappellaz, M. Davis, G. Delayque, M. Delmotte, V. M. Kotlyakov, M. Legrand, V. Y. Lipenkov, C. Lorius, L. Pepin, C. Ritz, E. Saltzman, and M. Stievenard, Climate and atmospheric history of the past 420,000 years from the Vostok ice core, Antarctica, *Nature*, *399*, 429–439, 1999.

Pilotto, L. S., R. M. Douglas, R. G. Attewell, and S. R. Wilson, Respiratory effects associated with indoor nitrogen dioxide exposure in children, *Int. J. Epidemiol.*, *26*, 788–796, 1997.

Pinto, J. P., R. P. Turco, and O. B. Toon, Self-limiting physical and chemical effects in volcanic eruption clouds, *J. Geophys. Res.*, *94*, 11,165, 1989.

Platts-Mills, T. A. E., and M. C. Carter, Asthma and indoor exposure to allergens, *N. Engl. J. Med.*, *336*, 1382–1384, 1997.

Polar Research Group, Department of Atmospheric Sciences, University of Illinois at Urbana-Champaign, Current Northern Hemisphere Sea Ice Area, *The Cryosphere Today* Web site, n.d., http://arctic.atmos.uiuc.edu/cryosphere/IMAGES/seaice.recent.arctic.png, accessed November 26, 2011.

Pollack, J. B., and Y. L. Yung, Origin and evolution of planetary atmospheres, *Ann. Rev. Earth Planet. Sci.*, *8*, 425–488, 1980.

Polpong, P., and S. Bovornkitti, Indoor radon, *J. Med. Assoc. Thailand*, *81*, 47–57, 1998.

Pooley, F. D., and M. Mille, Composition of air pollution particles, in *Air Pollution and Health*, eds. S. T. Holgate, J. M. Samet, H. S. Koren, and R. L. Maynard, Academic Press, San Diego, pp. 619–634, 1999.

Pope, C. A., Review: epidemiological basis for particulate air pollution health standards, *Aerosol. Sci. Technol.*, *32*, 4–14, 2000.

Pope, C. A., D. V. Bates, and M. E. Raizenne, Health-effects of particulate air-pollution – time for reassessment, *Environ. Health Perspect.*, *103*, 472–480, 1995.

Pope, C. A. III, R. T. Burnett, M. J. Thun, E. E. Calle, D. Krewski, K. Ito, and G. D. Thurston, Lung cancer, cardiopulmonary mortality, and long-term exposure to fine particulate air pollution. *JAMA*, *287*, 1132–1141, 2002.

Pope, C. A. III, and D. W. Dockery, Epidemiology of particle effects, in *Air Pollution and Health*, eds. S. T. Holgate, J. M. Samet, H. S. Koren, and R. L. Maynard, Academic Press, San Diego, pp. 673–705, 1999.

Pope, C. A. III, M. Ezzati, and D. W. Dockery, Fine-particulate air pollution and life expectancy in the United States, *N. Engl. J. Med.*, *360*, 376–386, 2009.

Posey, J. W., and P. F. Clapp, Global distribution of normal surface albedo, *Geofis. Int.*, *4*, 33–48, 1964.

Pósfai, M., J. R. Anderson, P. R. Buseck, and H. Sievering, Soot and sulfate aerosol particles in the remote marine troposphere, *J. Geophys. Res.*, *104*, 21,685–21,693, 1999.

Pósfai, M., A. Gelencsér, R. Simonics, K. Arató, J. Li, P. V. Hobbs, and P. R. Buseck, Atmospheric tar balls: particles from biomass and biofuel burning, *J. Geophys. Res.*, *109*, D06213, doi:10.1029/2003JD004169, 2004.

Poulos, G. S., and R. A. Pielke, Numerical analysis of Los Angeles Basin pollution transport to the Grand Canyon under stably stratified southwest flow conditions, *Atmos. Environ.*, *28*, 3329–3357, 1994.

Prados, A. I., R. R. Dickerson, B. G. Doddridge, P. A. Milne, J. L. Moody, and J. T. Merrill, Transport of ozone and pollutants from North America to the North Atlantic Ocean during the 1996 Atmosphere/Ocean Chemistry Experiment (AEROCE) intensive, *J. Geophys. Res.*, *104*, 26,219–26,233, 1999.

Press Trust of India, Acid Rain Makes Life Hard in 258 Chinese Cities, January 14, 2011, www.ndtv.com/article/world/acid-rains-make-life-hard-in-258-chinese-cities-79213, accessed November 27, 2011.

Prinn, R. G., R. F. Weiss, P. J. Fraser, P. G. Simmonds, D. M. Cunnold, F. N. Alyea, S. O'Doherty, P. Salameh, B. R. Miller, J. Huang, R. H. J. Wang, D. E. Hartley, C. Harth, L. P. Steele, G. Sturrock, P. M. Midgley, and A. McCulloch, A history of chemically and radiatively important gases in air deduced from ALE/GAGE/AGAGE, *J. Geophys. Res.*, *105*, 17,751–17,792, 2000.

Prospero, J. M., and Savoie, D. L., Effect of continental sources on nitrate concentrations over the Pacific Ocean, *Nature*, *339*, 687–689, 1989.

Pruppacher, H. R., and J. D. Klett, *Microphysics of Clouds and Precipitation*, 2nd revised and enlarged ed. Kluwer Academic, Dordrecht, The Netherlands, 1997.

Qin, Y. H., X. M. Zhang, H. Z. Jin, Y. Q. Liu, D. L. Fan, and Z. J. Fan, Effects of indoor air pollution on respiratory illness of school children, *Proceedings of the Sixth International Conference on Indoor Air Quality and Climate*, Helsinki, Finland, 477–482, 1993.

Ramana, M. V., Nuclear power: economic, safety, health, and environmental issues of near-term technologies, *Annu. Rev. Environ. Resour.*, *34*, 127–152, 2009.

Ramanathan, V., and G. Carmichael, Global and regional climate changes due to black carbon, *Nature Geosci.*, *1*, 221–227, 2008.

Ramanathan, V., et al., Indian ocean experiment: an integrated analysis of the climate forcing and effects of the great Indo-Asian haze, *J. Geophys. Res.*, *106*, D22, doi:10.1029/2001JD900133, 2001.

Randel, W. J., F. Wu, H. Vomel, G. E. Nedoluha, and P. Forster, Decreases in stratospheric water vapor after 2001: links to changes in the tropical tropopause and the Brewer-Dobson circulation, *J. Geophys. Res.*, *111*, D12312, doi:10.1029/2005JD006744, 2006.

Randel, W. J., et al., An update of observed stratospheric temperature trends, *J. Geophys. Res.*, *114*, D02107, doi:10.1029/2008JD010421, 2009.

Rando, R. J., P. Simlote, J. E. Salvaggio, and S. B. Lehrer, Environmental tobacco smoke measurement and health effects of involuntary smoking, in *Indoor Air Pollution and Health*, eds. E. J. Bardana and A. Montanaro, Marcel Dekker, New York, pp. 61–82, 1997.

Rayleigh, Lord Baron, On the light from the sky, its polarization and colour, *Phil. Mag.*, *41*, 107–120, 1871.

Reid, J. S., and P. V. Hobbs, Physical and optical properties of young smoke from individual biomass fires in Brazil, *J. Geophys. Res.*, *103*, 32,013–32,030, 1998.

Reid, J. S., P. V. Hobbs, R. J. Ferek, D. R. Blake, J. V. Martins, M. R. Dunlap, and C. Liousse, Physical, chemical, and optical properties of regional hazes dominated by smoke in Brazil, *J. Geophys. Res.*, *103*, 32,059–32,080, 1998.

Renewable Energy Policy Network for the 21st Century (REN21), Renewables 2010: Global Status Report, 2010, www.ren21.net/Portals/97/documents/GSR/REN21_GSR _2010_full_revised%20Sept2010.pdf, accessed November 27, 2011.

Robinson, J. P., and J. Thomas, *Time Spent in Activities, Locations, and Microenvironments: A California-National Comparison*, EPA/600/4-91/006, Environmental Monitoring Systems Laboratory, Las Vegas, NV, 1991.

Rodes, C. E., R. M. Kamens, and R. W. Wiener, The significance and characteristics of the personal activity cloud on exposure assessment measurements for indoor contaminants, *Indoor Air*, *2*, 123–145, 1991.

Rodwell, G. F., Lavoisier, Priestley, and the discovery of oxygen, *Nature*, *27*, 100–101, doi:10.1038/027100c0, 1882.

Rosenberg, N., and L. E. Birdzell, Jr., *How the West Grew Rich*, Basic Books, New York, 1986.

Rosenlof, K. H., et al., Stratospheric water vapor increases over the past half century. *Geophys. Res. Lett.*, *28*, 1195–1198, 2001.

Rubin, J. I., A. J. Kean, R. A. Harley, D. B. Millet, and A. H. Goldstein, Temperature dependence of volatile organic compound evaporative emissions from motor vehicles, *J. Geophys. Res.*, *111*, D03305, doi:10.1029/2005JD006458, 2006.

Rushton, L., and K. Cameron, Selected organic chemicals, in *Air Pollution and Health*, eds. S. T. Holgate, J. M. Samet, H. S. Koren, and R. L. Maynard, Academic Press, San Diego, pp. 813–838, 1999.

Russell, A., J. Milford, M. S. Bergin, S. McBride, L. McNair, Y. Yang, W. R. Stockwell, and B. Croes, Urban ozone control and atmospheric reactivity of organic gases, *Science*, *269*, 491–495, 1995.

Ryan, P. B., M. W. Lee, B. North, and A. J. McMichael, Risk factors for tumours of the brain and meninges: results from the Adelaide adult brain tumour study, *Int. J. Cancer*, *51*, 20–27, 1992.

Sagan, C., and G. Mullen, Earth and Mars: evolution of atmospheres and surface temperatures, *Science*, *177*, 52–56, 1972.

Sakurai, H., H. J. Tobias, K. Park, D. Zarling, K. Docherty, D. B. Kittelson, P. H. McMurry, and P. J. Ziemann, On-line measurements of diesel nanoparticle composition and volatility, *Atmos. Environ.*, *37*, 1199–1210, 2003.

Samet, J. M., M. C. Marbury, and J. D. Spengler, Health effects and sources of indoor air pollution: part 1, *Am. Rev. Respir. Disord.*, *136*, 1486–1508, 1987.

Sanchez, J. A., and F. J. G. Ayala, Recent trend in ozone levels in the metropolitan zone of Mexico City, *J. Mex. Chem. Soc.*, *52*, 256–262, 2008.

Saxena, P., P. K. Mueller, and L. M. Hildemann, Sources and chemistry of chloride in the troposphere: a review, in *Managing Hazardous Air Pollutants: State of the Art*, eds. W. Chow and K. K Connor, Lewis, Boca Raton, FL, pp. 173–190, 1993.

Schichtel, B. A., R. B. Husar, S. R. Falke, and W. E. Wilson, Haze trends over the United States, 1980–1995, *Atmos. Environ.*, *35*, 5205–5210, 2001.

Schick, S., and S. Glantz, Philip Morris toxicological experiments with fresh sidestream smoke: more toxic than mainstream smoke, *Tob. Control*, *14*, 396–404, 2005.

Schlitt, H., and H. Knöppel, Carbonyl compounds in mainstream and sidestream tobacco smoke, in *Present and Future of Indoor Air Quality*, eds. C. J. Bieva, Y. Courtois, and M. Govaerts, Excerpta Medica, Amsterdam, pp. 197–206, 1989.

Schnaiter M., H. Horvath, O. Mohler, K.-H. Naumann H. Saathoff, and OW. Schock, UV-VIS-NIR spectral optical properties of soot and soot-containing aerosols. *J. Aerosol Sci.*, *34*, 1421–1444, 2003.

Schroeder, W. H., M. Dobson, D. M. Kane, and N. D. Johnson, Toxic trace elements associated with airborne particulate matter: a review, *J. Air Pollut. Control Assoc.*, *37*, 1267–1285, 1987.

Schwarzberg, M. N., Carbon dioxide level as migraine threshold factor: hypothesis and possible solutions, *Medical Hypotheses*, *41*, 35–36, 1993.

Searchinger, T., R. Heimlich, R. A. Houghton, F. Dong, A. Elobeid, J. Fabiosa, S. Tokgoz, D. Hayes, and T.-H. Yu, Use of U.S. cropland for biofuels increases greenhouse gases through emissions from land-use change, *Science*, *319*, 1238–1240, 2008.

Seinfeld, J. H., and S. N Pandis, *Atmospheric Chemistry and Physics: From Air Pollution to Climate*, Wiley, Hoboken, NJ, 2006.

Sellers W. D., *Physical Climatology*, The University of Chicago Press, Chicago, 1965.

Shackleton, N. J., Oceanic carbon isotope constraints on oxygen and carbon dioxide in the Cenozoic atmosphere, in *The Carbon Cycle and Atmospheric CO_2: Natural Variations Archean to Present (Geophysical Monograph)*, *32*, eds. E. T. Sundquist and W. S. Broecker, American Geophysical Union, Washington, DC, pp. 412–418, 1985.

Shackleton, N. J., and Opdyke, N. D. Oxygen-isotope and paleomagnetic stratigraphy of Pacific core V28-239: Late Pliocene to latest Pleistocene, in *Investigation of Late Quaternary Paleooceanography and Paleoclimatology*, eds. R. M. Cline and J. D. Hays, Memoir 145, Geological Society of America, Boulder, CO, pp. 449–464, 1976.

Sheridan P. J., C. A. Brock, and J. C. Wilson, Aerosol particles in the upper troposphere and lower stratosphere: elemental composition and morphology of individual particles in northern midlatitudes. *Geophys. Res. Lett.*, *21*, 2587–2590, 1994.

Sillman, S., and P. J. Samson, Impact of temperature on oxidant photochemistry in urban, polluted rural, and remote environments, *J. Geophys. Res.*, *100*, 11,497–11,508, 1995.

Sinitsyna, T., Civilization Versus Nature, 2007, http://russiaprofile.org/politics/a1196417420.html, accessed November 26, 2011.

Smith, K. R., Fuel combustion, air pollution exposure, and health: the situation in the developing countries, *Annu. Rev. Energy Environ.*, *18*, 529–566, 1993.

Sokolik, I., A. Andronova, and C. and Johnson C., Complex refractive index of atmospheric dust aerosols, *Atmos. Environ.*, *27A*, 2495–2502, 1993.

Sollinger, S., K. Levsen, and G. Wünsch, Indoor pollution by organic emissions from textile floor coverings: climate test chamber studies under static conditions, *Atmos. Environ.*, *28*, 2369–2378, 1994.

Solomon, P. A., and R. H. Thuillier, SJVAQS/SUSPEX/SARMAP 1990 Air Quality Field Measurement Project Volume II: Field Measurement Characterization, PG&E Cost Reduction Projects Report 009.2-94.*1*, 1995.

Solomon, S., R. R. Garcia, F. S. Rowland, and D. J. Wuebbles, On the depletion of Antarctic ozone, *Nature*, *321*, 755–757, 1986.

Somerville, S. M., R. J. Rona, and S. Chinn, Passive smoking and respiratory conditions in primary school children, *J. Epidemiol. Community Health*, *42*, 105–110, 1988.

Song, C. H., and G. R. Carmichael, The aging process of naturally emitted aerosol (sea-salt and mineral aerosol) during long range transport, *Atmos. Environ.*, *33*, 2203–2218, 1999.

South Coast Air Quality Management District (SCAQMD), Historic Ozone Air Quality Trends, 2011, www.aqmd.gov/smog/o3trend.html and general information, www.aqmd.gov, accessed November 3, 2011.

Sovacool, B. K. Valuing the greenhouse gas emissions from nuclear power: a critical survey, *Energy Policy*, *36*, 2940–2953, 2008.

Spath, P. L., and M. K. Mann, *Life Cycle Assessment of a Natural Gas Combined-Cycle Power Generation System*, NREL/TP-570-27715, National Renewable Energy Laboratory, Golden, CO, www.nrel.gov/docs/fy00osti/27715.pdf, 2000.

Spencer, R. W., and J. R. Christy, Precise monitoring of global temperature trends from satellites, *Science*, *247*, 1558–1562, 1990.

Spengler, J. D., Nitrogen dioxide and respiratory illnesses in infants. *Am. Rev. Respir. Disord.*, *148*, 1258–1265, 1993.

Spengler, J. D., D. W. Dockery, W. A. Turner, J. M. Wolfson, and B. G. Ferris, Long-term measurements of respirable particles, sulphates, and particulates inside and outside homes, *Atmos. Environ.*, *15*, 23–30, 1981.

Spengler, J. D., and K. Sexton, Indoor air pollution: a public health perspective, *Science*, *221*, 9–17, 1983.

Steiner, D., H. Burtchnew, and H. Grass, Structure and disposition of particles from a spark ignition engine, *Atmos. Environ.*, *26*, 997–1003, 1992.

Steiner, A. L., S. Tonse, R. C. Cohen, A. H. Goldsten, and R. A. Harley, Influence of future climate and emissions on regional air quality in California, *J. Geophys. Res.*, *111*, D18303, doi:10.1029/2005JD006935, 2006.

Stephens, E. R., W. E. Scott, P. L. Hanst, and R. C. Doerr, Recent developments in the study of the organic chemistry of the atmosphere, *J. Air Pollut. Contr. Assoc.*, *6*, 159–165, 1956.

Stern, D. I., and R. K. Kaufmann, Annual Estimates of Global Anthropogenic Methane Emissions: 1860–1994, Trends Online: A Compendium of Data on Global Change, Carbon Dioxide Information Analysis Center, Oak Ridge National Laboratory, U.S. Department of Energy, Oak Ridge, TN, doi:10.3334/CDIAC/tge.001, 1998.

Stevenson, D., R. Doherty, M. Sanderson, C. Johnson, B. Collins, and D. Derwent, Impacts of climate change and variability on tropospheric ozone and its precursors. *Faraday Disc.*, *130*, 1–17, 2005.

Stjern, C. W., A. Stohl, and J. E. Kristjánsson, Have aerosols affected trends in visibility and precipitation in Europe?, *J. Geophys. Res.*, *116*, D02212, doi:10.1029/2010JD014603, 2011.

Stockholm Environment Institute, Regional Air Pollution in Africa, www.sei.se, 1998.

Stommel, H., and E. Stommel, The year without a summer, *Sci. Am.*, *240*, 176–186, 1981.

Stothers, R. B., The great Tambora eruption in 1815 and its aftermath, *Science*, *224*, 1191–1198, 1984.

Stott, P. A., S. F. B. Tett, G. S. Jones, M. R. Allen, J. F. B. Mitchell, and G. J. Jenkins, External control of 20th century temperature by natural and anthropogenic forcings, *Science*, *290*, 2133–2136, 2000.

Stoutenburg, E. D., N. Jenkins, and M. Z. Jacobson, Power output variations of co-located offshore wind turbines and wave energy converters in California, *Renewable Energy*, *35*, 2781–2791, 2010.

Stradling, D., *Smokestacks and Progressives*, The Johns Hopkins University Press, Baltimore, 1999.

Strawa, A. W., K. Drdla, G. V. Ferry, S. Verma, R. F. Pueschel, M. Yasuda, R. J. Salawitch, R. S. Gao, S. D. Howard, P. T. Bui, M. Loewenstein, J. W. Elkins, K. K. Perkins, and R. Cohen, Carbonaceous aerosol (soot) measured in the lower stratosphere during POLARIS and its role in stratospheric photochemistry, *J. Geophys. Res.*, *104*, 26,753–26,766, 1999.

Stuiver, M., G. H. Denton, T. J. Hughes, and J. L. Fastook, History of marine ice sheet in West Antarctica during the last glaciation: a working hypothesis, in *The Last Great Ice Sheets*, eds. G. H. Denton and T. J. Hughes, Wiley-Interscience, New York, pp. 319–436, 1981.

Stull, R. B., *An Introduction to Boundary Layer Meteorology*, Kluwer Academic, Dordrecht, The Netherlands, 1988.

Tabazadeh, A., M. Z. Jacobson, H. B. Singh, O. B. Toon, J. S. Lin, B. Chatfield, A. N. Thakur, R. W. Talbot, and J. E. Dibb, Nitric acid scavenging by mineral and biomass burning aerosols, *Geophys. Res. Lett.*, *25*, 4185–4188, 1998.

Tabazadeh, A., and R. P. Turco, Stratospheric chlorine injection by volcanic eruptions: HCl scavenging and implications for ozone, *Science*, *260*, 1082–1086, 1993.

Tabazadeh, A., R. P. Turco, K. Drdla, and M. Z. Jacobson, A study of Type I polar stratospheric cloud formation, *Geophys. Res. Lett.*, *21*, 1619–1622, 1994.

Talwani, M., and O. Eldholm, Evolution of the Norwegian-Greenland sea, *Geol. Soc. Am. Bull.*, *88*, 969–999, 1977.

Tang, I. N., W. T. Wong, and H. R. Munkelwitz, The relative importance of atmospheric sulfates and nitrates in visibility reduction, *Atmos. Environ.*, *15*, 2463–2471, 1981.

Tang, X., S. Madronich, T. Wallington, and D. Calamari, Changes in tropospheric composition and air quality, *J. Photochem. Photobiol., B*, *46*, 83–95, 1998.

Tegen, I., A. A. Lacis, and I. Fung, The influence on climate forcing of mineral aerosols from disturbed soils, *Nature*, *380*, 419–422, 1996.

Teifke, J. P., and C. V. Lohr, Immunohistochemical detection of p53 over-expression in paraffin wax-embedded squamous cell carcinomas of cattle, horses, cats, and dogs, *J. Compar. Pathol.*, *114*, 205–210, 1996.

ten Hoeve, J. E., L. A. Remer, and M. Z. Jacobson, Microphysical and radiative effects of aerosols on warm clouds during the Amazon biomass burning season as observed by MODIS: impacts of water vapor and land cover, *Atmos. Chem. Phys.*, *11*, 3021–3036, 2011.

Tomlinson, G. G., II, Air pollutants and forest decline, *Environ. Sci. Technol*, *17*, 246–256, 1983.

Toon, O. B., J. Pinto, P. Hamill, and R. P. Turco, Condensation of HNO_3 and HCl in the winter polar stratospheres, *Geophys. Res. Lett. (Suppl.)*, *13*, 1284–1287, doi:10.1029/GL013i012p01284, 1986.

Toon, O. B., J. B. Pollack, T. P. Ackerman, R. P. Turco, C. P. McKay, and M. S. Liu, Evolution of an impact-generated dust cloud and its effects on the atmosphere, *Geol. Soc. Am. Spec. Pap.*, *190*, 187–200, 1982.

Turco, R. P., *Earth under Siege*, Oxford University Press, New York, 1997.

Turco, R. P., O. B. Toon, and P. Hamill, Heterogeneous physiochemistry of the polar ozone hole, *J. Geophys. Res.*, *94*, 16,493–16,510, 1989.

Twomey, S. A., The effect of cloud scattering on the absorption of solar radiation by atmospheric dust, *J. Atmos. Sci.*, *29*, 1156–1159, 1977.

Ulrich, R. K., Solar neutrinos and variations in the solar luminosity, *Science*, *190*, 619–624, 1975.

United Nations Environmental Program, Environmental effects of ozone depletion: 1998 assessment, *J. Photochem. Photobiol., B*, *46*, 1–4, 1998.

U.S. Department of Energy (DOE), Concentrating Solar Power Commercial Application Study: Reducing Water Consumption of Concentrating Solar Power Electricity Generation, Report to Congress, 2008, http://www1.eere .energy.gov/solar/pdfs/csp_water_study.pdf, accessed November 27, 2011.

U.S. Department of Health and Human Services, The Health Consequences of Involuntary Exposure to Tobacco Smoke: A Report of the Surgeon General, U.S. Department of Health and Human Services, Centers for Disease Control and Prevention, Coordinating Center for Health Promotion, National Center for Chronic Disease Prevention and

Health Promotion, Office on Smoking and Health, Atlanta, GA, 2006, www.surgeongeneral.gov/library/secondhandsmoke/report/fullreport.pdf, accessed November 27, 2011.

U.S. Environmental Protection Agency (U.S. EPA), *Air Quality Criteria for Ozone and Other Photochemical Oxidants*, Report No. EPA-600/8-78-004, 1978.

U.S. Environmental Protection Agency (U.S. EPA), Fact Sheet: Respiratory Health Effects of Passive Smoking, EPA Document Number 43-F-93-003, 1993, www.epa.gov/smokefree/pubs/etsfs.html, accessed November 27, 2011.

U.S. Environmental Protection Agency (U.S. EPA), *Air Quality Criteria for Particulate Matter*, EPA/600/P-95/001aF-cF.3v, EPA, Research Triangle Park, NC, 1996.

U.S. Environmental Protection Agency (U.S. EPA), *Latest Findings on National Air Quality: 1997 Status and Trends*, EPA-454/F-98-009, EPA, Office of Air Quality Planning and Standards, Research Triangle Park, NC, 1998.

U.S. Environmental Protection Agency (U.S. EPA), Cold Temperature Effects on Vehicle HC Emissions, EPA420-D-06-001, EPA, Office of Transportation and Air Quality, 2006, www.epa.gov/otaq/regs/toxics/420d06001.pdf, accessed November 27, 2011.

U.S. Environmental Protection Agency (U.S. EPA), Summary of Current and Historical Light-Duty Vehicle Emission Standards, EPA, Office of Transportation and Air Quality, 2007, www.epa.gov/greenvehicles/detailedchart.pdf, accessed November 27, 2011.

U.S. Environmental Protection Agency (U.S. EPA), Endangerment and cause or contribute findings for greenhouse gases under Section 202(a) of the Clean Air Act, *Federal Register*, *74* (239), EPA-HQ-OAR-2009-0171, 2009a, http://epa.gov/climatechange/endangerment.html, accessed November 27, 2011.

U.S. Environmental Protection Agency (U.S. EPA), Integrated Science Assessment for Particulate Matter (Final Report), EPA, Washington, DC, EPA/600/R-08/139F, 2009b, http://cfpub.epa.gov/ncea/cfm/recordisplay.cfm?deid=216546, accessed November 27, 2011.

U.S. Environmental Protection Agency (U.S. EPA), Radon: Health Risks, 2009c, www.epa.gov/radon/healthrisks.html, accessed November 27, 2011.

U.S. Environmental Protection Agency (U.S. EPA), The Benefits and Costs of the Clean Air Act 1990 to 2010, EPA Report to Congress, EPA-410-R-99-001, 2010, www.epa.gov/air/sect812/1990–2010/, accessed November 27, 2011.

U.S. Environmental Protection Agency (U.S. EPA), National Emissions Inventory (NEI) Air Pollutant Emissions Trends Data, 2011a, www.epa.gov/ttn/chief/trends/, accessed November 27, 2011.

U.S. Environmental Protection Agency (U.S. EPA), Technology Transfer Network Clearinghouse for Inventories and Emissions Factors, 2011b, www.epa.gov/ttn/chief/eiinformation.html, accessed November 27, 2011.

U.S. Environmental Protection Agency (U.S. EPA), Air Trends, 2011c, http://www.epa.gov/air/airtrends/, accessed Feb. 12, 2011.

U.S. Geological Survey, *Mineral Commodity Summaries 2011*, U.S. Government Printing Office, Washington, DC, 2011, http://minerals.usgs.gov/minerals/pubs/mcs/2011/mcs2011.pdf, accessed November 27, 2011.

U.S. Public Health Service, The Use of Tetraethyl Lead Gasoline in Its Relation to Public Health, Public Health Bulletin No. 163, Treasury Department, Washington, DC, 1925.

U.S. Public Health Service, Public Health Aspects of Increasing Tetraethyl Lead Content in Motor Fuel, Report on the Advisory Committee on Tetraethyl Lead to the Surgeon General, PHS Publication No. 712, Washington, DC, 1959.

Valley, S. L., Ed., *Handbook of Geophysics and Space Environments*, U.S. Air Force Cambridge Research Laboratories, McGraw-Hill, New York, 1965.

Venkataraman, C., and S. K. Friedlander, Size distributions of polycyclic aromatic hydrocarbons and elemental carbon. 2. Ambient measurements and effects of atmospheric processes, *Environ. Sci. Technol.*, *28*, 563–572, 1994.

Venkataraman, C., J. M. Lyons, and S. K. Friedlander, Size distributions of polycyclic aromatic hydrocarbons and elemental carbon. 1. Sampling, measurement methods, and source characterization, *Environ. Sci. Technol*, *28*, 555–562, 1994.

Venkatram, A., and R. Viskanta, Radiative effects of elevated pollutant layers, *J. Appl. Meteorol.*, *16*, 1256–1272, 1977.

Waggoner, A. P., R. E. Weiss, N. C. Ahlquist, D. S. Covert, S. Will, and R. J. Charlson, Optical characteristics of atmospheric aerosols. *Atmos. Environ.*, *15*, 1891–1909, 1981.

Waitt, R. B., Case for periodic, colossal jökulhlaups from Pleistocene glacial Lake Missoula, *Geol. Soc. Am. Bull.*, *96*, 1271–1286, 1985.

Wakamatsu, S., Y. Ogawa, K. Murano, K. Goi, and Y. Aburamoto, Aircraft survey of the secondary photochemical pollutants covering the Tokyo metropolitan area, *Atmos. Environ.*, *17*, 827–836, 1983.

Wakimoto, R. M., and J. L. McElroy, Lidar observation of elevated pollution layers over Los Angeles, *J. Clim. Appl. Met.*, *25*, 1583–1599, 1986.

Waldman, J. M., J. W. Munger, D. J. Jacob, R. C. Flagan, J. J. Morgan, and M. R. Hoffmann, Chemical composition of acid fog, *Science*, *218*, 677–680, 1982.

Wallace L., Correlations of personal exposure to particles with outdoor air measurements: a review of recent studies, *Aerosol Sci. Technol.*, *32*, 15–25, 2000.

Wallace, L. A., Personal exposure to 25 volatile organic compounds: EPA's 1987 team study in Los Angeles, California, *Toxicol. Industrial Health*, *7*, 203–208, 1991.

Wanner, H. U., Sources of pollutants in indoor air, *IARC Scientific Publications*, *109*, 19–30, 1993.

Webster, P. J., G. J. Holland, J. A. Curry, and H.-R. Chang, Changes in tropical cyclone number, duration, and intensity in a warming environment, *Science*, *309*, 1844–1846, 2005.

Weiss, S. T., Passive smoking and lung cancer: what is the risk? *Am. Rev. Respir. Disord.*, *133*, 1–3, 1986.

Wentzel, M., H. Gorzawski, K.-H. Naumann, H. Saathoff, and S. Weinbruch, Transmission electron microscopial and aerosol dynamical characterization of soot aerosols, *J. Aerosol Sci.*, *34*, 1347–1370, 2003.

Westerholm, R., P. Ahlvik, and H. L. Karlsson, An exhaust characterisation study based on regulated and unregulated tailpipe and evaporative emissions from bi-fuel and flex-fuel light-duty passenger cars fuelled by petrol (E5), bioethanol (E70, E85) and biogas tested at ambient temperatures of + 22C and −7C, Final Report to the Swedish Road Administration, 2008.

Westerling, A. L., H. G. Hidalgo, D. R. Cayan, and T. W. Swetnam, Warming and earlier spring increase western U.S. forest wildfire activity, *Science*, *313*, 940–943, 2006.

Whitney, K., and T. Fernandez, *Characterization of Cold Temperature VOC and PM Emissions from Flex Fuel Vehicles Operating on Ethanol Blends*, 17th CRC On-Road Vehicle Emissions Workshop, San Diego, 2007.

Whitt, D. B., M. Z. Jacobson, J. T. Wilkerson, A. D. Naiman, and S. K. Lele, Vertical mixing of commercial aviation emissions from cruise altitude to the surface, *J. Geophys. Res.*, *116*, D14109, doi:1029/2010JD015532, 2011.

Williams, G. E., Late precambrian glacial climate and the Earth's obliquity, *Geol. Mag.*, *112*, 441–465, 1975.

Woodcock, A. H., Salt nuclei in marine air as a function of altitude and wind force, *J. Meteorol.*, *10*, 362–371, 1953.

World Energy Council, 2010 Survey of Energy Resources, World Energy Council, London, 2010, www.worldenergy.org/documents/ser_2010_report_1.pdf, accessed November 27, 2011.

World Health Organization (WHO), The World Health Report 2002, Annex Table 9, 2002, www.who.int/whr/2002/en/whr2002_annex9_10.pdf, accessed November 27, 2011.

World Health Organization (WHO), WHO Report on the Global Tobacco Epidemic, 2009: Implementing Smoke-Free Environments, 2009, www.who.int/tobacco/mpower/2009/en/, accessed November 27, 2011.

World Health Organization (WHO), Air Quality and Health, Fact Sheet No. 313, 2011a, www.who.int/mediacentre/factsheets/fs313/en/index.html, accessed November 27, 2011.

World Health Organization (WHO), Indoor Air Pollution and Health, Fact Sheet No. 292, 2011b, www.who.int/mediacentre/factsheets/fs292/en/index.html, accessed November 27, 2011.

World Health Organization (WHO), WHO Framework Convention on Tobacco Control, n.d., www.who.int/fctc/en, accessed November 27, 2011.

World Meteorological Organization, *Scientific Assessment of Ozone Depletion: 1994*, WMO Global Ozone Research and Monitoring Project – Report No. 37, Geneva, 1995.

World Meteorological Organization, *Scientific Assessment of Ozone Depletion: 1998*, WMO Global Ozone Research and Monitoring Project – Report No. 44, Geneva, 1998.

Worldwatch Institute, *State of the World 2006: Special Focus: China and India*, Worldwatch Institute, Washington, DC, 2006.

Worsnop, D. R., L. E. Fox, M. S. Zahniser, and S. C. Wofsy, Vapor pressures of solid hydrates of nitric acid: implications for polar stratospheric clouds, *Science*, *259*, 71–74, 1993.

Wright, C. M., et al., Lung asbestos content in lungs resected for primary lung cancer, *J. Thorac. Oncol.*, *3*, 569–576, 2008.

Xu, Z., and G. Han, Chemical and strontium isotope characterization of rainwater in Beijing, China, *Atmos. Environ.*, *43*, 1954–1961, 2009.

Zdunkowski, W. G., R. M. Welch, and J. Paegle, One dimensional numerical simulation of the effects of air pollution on the planetary boundary layer, *J. Atmos. Sci.*, *33*, 2399–2414, 1976.

Zepp, R. G., T. V. Callaghan, and D. J. Erickson, Effects of enhanced solar ultraviolet radiation on biogeochemical cycles, *J. Photochem. Photobiol., B*, *46*, 69–82, 1998.

Zhang, J., S. M. Liu, X. Lu, and W. W. Huang, Characterizing Asian wind-dust transport to the northwest Pacific Ocean: direct measurements of the dust flux for two years, *Tellus*, *45B*, 335–345, 1993.

Zoller, W. H., J. R. Parrington, and J. M. Phelan Kotra, Iridium enrichment in airborne particles from Kilauea Volcano: January 1983, *Science*, *222*, 1118–1121, 1983.

Zubakov, V. A., and I. I. Borzenkova, Pliocene palaeoclimates: past climates as possible analogues of mid–twenty-first century climate, *Palaeogeog. Palaeoclim. Palaeoecol.*, *65*, 35–49, 1988.

Index